AF443393

# SYSTEMATIC ORGANISATION OF INFORMATION IN FUZZY SYSTEMS

**NATO Science Series**

A series presenting the results of scientific meetings supported under the NATO Science Programme.

The series is published by IOS Press and Kluwer Academic Publishers in conjunction with the NATO Scientific Affairs Division.

*Sub-Series*

| | | |
|---|---|---|
| I. | Life and Behavioural Sciences | IOS Press |
| II. | Mathematics, Physics and Chemistry | Kluwer Academic Publishers |
| III. | Computer and Systems Sciences | IOS Press |
| IV. | Earth and Environmental Sciences | Kluwer Academic Publishers |
| V. | Science and Technology Policy | IOS Press |

The NATO Science Series continues the series of books published formerly as the NATO ASI Series.

The NATO Science Programme offers support for collaboration in civil science between scientists of countries of the Euro-Atlantic Partnership Council. The types of scientific meeting generally supported are "Advanced Study Institutes" and "Advanced Research Workshops", although other types of meeting are supported from time to time. The NATO Science Series collects together the results of these meetings. The meetings are co-organized by scientists from NATO countries and scientists from NATO's Partner countries – countries of the CIS and Central and Eastern Europe.

**Advanced Study Institutes** are high-level tutorial courses offering in-depth study of latest advances in a field.
**Advanced Research Workshops** are expert meetings aimed at critical assessment of a field, and identification of directions for future action.

As a consequence of the restructuring of the NATO Science Programme in 1999, the NATO Science Series has been re-organized and there are currently five sub-series as noted above. Please consult the following web sites for information on previous volumes published in the series, as well as details of earlier sub-series:

http://www.nato.int/science
http://www.wkap.nl
http://www.iospress.nl
http://www.wtv-books.de/nato_pco.htm

Series III: Computer and Systems Sciences - Vol. 184                    ISSN: 1387-6694

# Systematic Organisation of Information in Fuzzy Systems

Edited by

## Pedro Melo-Pinto

*Universidade de Trás-os-Montes e Alto Douro/CETAV,
Vila Real, Portugal*

## Horia-Nicolai Teodorescu

*Romanian Academy, Bucharest, Romania
and
Technical University of Iasi, Iasi, Romania*

and

## Toshio Fukuda

*Center for Cooperative Research in Advance Science and Technology,
Nagoya University, Nagoya, Japan*

IOS
*Press*

Ohmsha

Amsterdam • Berlin • Oxford • Tokyo • Washington, DC

Published in cooperation with NATO Scientific Affairs Division

Proceedings of the NATO Advanced Research Workshop on
Systematic Organisation of Information in Fuzzy Systems
24–26 October 2001
Vila Real, Portugal



ISBN 1 58603 295 X (IOS Press)
ISBN 4 274 90546 2 C3055 (Ohmsha)
Library of Congress Control Number: 2002112582


*Publisher*
IOS Press
Nieuwe Hemweg 6B
1013 BG  Amsterdam
Netherlands
fax: +31 20 620 3419
e-mail: order@iospress.nl


*Distributor in the UK and Ireland*
IOS Press/Lavis Marketing
73 Lime Walk
Headington
Oxford OX3 7AD
England
fax: +44 1865 75 0079

*Distributor in the USA and Canada*
IOS Press, Inc.
5795-G Burke Centre Parkway
Burke, VA 22015
USA
fax: +1 703 323 3668
e-mail: iosbooks@iospress.com


*Distributor in Germany, Austria and Switzerland*
IOS Press/LSL.de
Gerichtsweg 28
D-04103 Leipzig
Germany
fax: +49 341 995 4255

*Distributor in Japan*
Ohmsha, Ltd.
3-1 Kanda Nishiki-cho
Chiyoda-ku, Tokyo 101-8460
Japan
fax: +81 3 3233 2426


LEGAL NOTICE
The publisher is not responsible for the use which might be made of the following information.

PRINTED IN THE NETHERLANDS

# Preface

Several developments in recent years require essential progresses in the field of information processing, especially in information organization and aggregation. The Artificial Intelligence (A.I.) domain relates to the way humans process information and knowledge and is aiming to create machines that can perform tasks similar to humans in information processing and knowledge discovery. A.I. specifically needs new methods for information organization and aggregation. On the other hand, the information systems in general, and the Internet-based systems specifically, have reached a bottleneck due to the lack of the right tools for selecting the right information, aggregating information, and making use of the results of these operations. In a parallel development, cognitive sciences, behavioral science, economy, sociology and other human-related sciences need new theoretical tools to deal with and better explain how humans select, organize and aggregate information in relation to other information processing tasks, to goals and to knowledge the individuals have. Moreover, methods and tools are needed to determine how the information organization and aggregation processes contribute building patterns of behavior of individuals, groups, companies and society. Understanding such behaviors will much help in developing robotics, in clarifying the relationships humans and robots may or should develop, and in determining how robotic communities can aggregate in the near future.

Several methods and tools are currently in use to perform information aggregation and organization, including neural networks, fuzzy and neuro-fuzzy systems, genetic algorithms and evolutionary programming. At least two new domains are present today in the field of information processing: analysis of self-organization and information generation and aggregation in dynamical systems, including large dynamical systems, and data mining and knowledge discovery.

This volume should be placed in this context and in relation to the development of fuzzy systems theory, specifically for the development of systems for information processing and knowledge discovery.

The contributors to this volume review the state of the art and present new evolutions and progresses in the domain of information processing and organization *in* and *by* fuzzy systems and other types of systems using uncertain information. Moreover, information aggregation and organization by means of tools offered by fuzzy logic are dealt with.

The volume includes four parts. In the first, introductory part of the volume, in three chapters, general issues are addressed from a wider perspective. The second part of the volume is devoted to several fundamental aspects of fuzzy information and its organization, and includes chapters on the semantics of the information, on information quality and relevance, and on mathematical models and computer science approaches to the information representation and aggregation processes. The chapters in the third part are emphasizing methods and tools to perform information organization, while the chapters in the fourth part have the primary objective to present applications in various fields, from robotics to medicine. Beyond purely fuzzy logic based approaches, the use of neuro-fuzzy systems in information processing and organization is reflected in several chapters in the volume.

The volume addresses in the first place the graduate students, doctoral students and researchers in computer science and information science. Researchers and doctoral students in other fields, like cognitive sciences, robotics, nonlinear dynamics, control theory and economy may be interested in several chapters in this volume.

## Acknowledgments

Thanks are due to the sponsors of the workshop, in the first place NATO Scientific Affairs Division, as well as to the local sponsors, including the University of Trás-os-Montes e Alto Douro/CETAV in Vila Real, Vila Real community, and the Ministry of Science, Portugal. The University of Trás-os-Montes e Alto Douro/CETAV (UTAD) in Vila Real, Portugal, and the Technical University of Iasi, Faculty of Electronics and Telecommunications, Iasi, Romania, have been co-organizers of the NATO workshop. A special mention for the gracious cooperation of the Rector of the University of Trás-os-Montes e Alto Douro/CETAV, Prof. Torres Pereira, for the Vice-Rector of the same university, Prof. Mascarenhas Ferreira, and for the very kind and active support of Prof. Bulas-Cruz, Head of the Engineering Department at UTAD.

We gratefully acknowledge the contribution of the Engineering Department of UTAD and the Group of Laboratories for Intelligent Systems and Bio-Medial Engineering, Technical University of Iasi, Romania. Specifically, we acknowledge the work of several colleagues in the Engineering Department of UTAD and in the Group of Laboratories for Intelligent Systems and Bio-Medial Engineering of the University of Iasi in the organization of this NATO workshop. Special thanks to Mr. J. Paulo Moura, who edited the Proceedings volume of the Workshop, and to Mr. Radu Corban, system engineer, who helped establishing and maintaining the web page of the Workshop. Also, we acknowledge the work of several colleagues in the Group of Laboratories for Intelligent Systems and Bio-Medial Engineering of University of Iasi in reviewing and frequently retyping parts of the manuscript of this book. Thanks are due to the group of younger colleagues helping with the internal reviewing process, namely Mr. Bogdan Branzila, Mrs. Oana Geman, and Mr. Irinel Pletea. The help of Mr. Branzila, who, as a young researcher at the Institute of Theoretical Informatics, has had the task to help correct the final form of the manuscript, has been instrumental in the last months of preparation of the manuscript.

Last but not least, thanks are due to all those in the IOS Press editorial and production departments who helped during the publication process; we had a very fruitful and pleasant cooperation with them and they have always been very supportive.

June 2002

Pedro Melo-Pinto<br>
Horia-Nicolai Teodorescu<br>
Toshio Fukuda

My deepest thanks to Prof. Horia-Nicolai Teodorescu, a friend, for all the help and support along these months, and for all the wonderful work he (and his team) did with this volume.

A special thank you to Prof. Toshio Fukuda for cherishing, from the very first time, this project.

June 2002

Pedro Melo-Pinto

# Contents

# Part I

# Introduction

# Toward a Perception-Based
# Theory of Probabilistic Reasoning

Lotfi A. ZADEH

*Computer Science Division and the Electronics Research Laboratory,*
*Department of EECs, University of California,*
*Berkeley, CA 94720-1776, USA, e-mail: zadeh@cs.berkeley.edu*

The past two decades have witnessed a dramatic growth in the use of probability-based methods in a wide variety of applications centering on automation of decision-making in an environment of uncertainty and incompleteness of information.

Successes of probability theory have high visibility. But what is not widely recognized is that successes of probability theory mask a fundamental limitation – the inability to operate on what may be called perception-based information. Such information is exemplified by the following. Assume that I look at a box containing balls of various sizes and form the perceptions: (a) there are about twenty balls; (b) most are large; and (c) a few are small.

The question is: What is the probability that a ball drawn at random is neither large nor small? Probability theory cannot answer this question because there is no mechanism within the theory to represent the meaning of perceptions in a form that lends itself to computation. The same problem arises in the examples:

Usually Robert returns from work at about 6:00 p.m. What is the probability that Robert is home at 6:30 p.m.?

I do not know Michelle's age but my perceptions are: (a) it is very unlikely that Michelle is old; and (b) it is likely that Michelle is not young. What is the probability that Michelle is neither young nor old?

$X$ is a normally distributed random variable with small mean and small variance. What is the probability that $X$ is large?

Given the data in an insurance company database, what is the probability that my car may be stolen? In this case, the answer depends on perception-based information that is not in an insurance company database.

In these simple examples – examples drawn from everyday experiences – the general problem is that of estimation of probabilities of imprecisely defined events, given a mixture of measurement-based and perception-based information. The crux of the difficulty is that perception-based information is usually described in a natural language--a language that probability theory cannot understand and hence is not equipped to handle.

To endow probability theory with a capability to operate on perception-based information, it is necessary to generalize it in three ways. To this end, let PT denote standard probability theory of the kind taught in university-level courses.

The three modes of generalization are labeled:

(a) f-generalization;
(b) f.g.-generalization, and
(c) nl-generalization.

More specifically:

(a) f-generalization involves fuzzification, that is, progression from crisp sets to fuzzy sets, leading to a generalization of PT that is denoted as PT+. In PT+, probabilities, functions, relations, measures, and everything else are allowed to have fuzzy denotations, that is, be a matter of degree. In particular, probabilities described as low, high, not very high, etc. are interpreted as labels of fuzzy subsets of the unit interval or, equivalently, as possibility distributions of their numerical values.

(b) f.g.-generalization involves fuzzy granulation of variables, functions, relations, etc., leading to a generalization of PT that is denoted as PT++. By fuzzy granulation of a variable, $X$, what is meant is a partition of the range of $X$ into fuzzy granules, with a granule being a clump of values of $X$ that are drawn together by indistinguishability, similarity, proximity, or functionality. For example, fuzzy granulation of the variable *age* partitions its vales into fuzzy granules labeled very young, young, middle-aged, old, very old, etc. Membership functions of such granules are usually assumed to be triangular or trapezoidal. Basically, granulation reflects the bounded ability of the human mind to resolve detail and store information; and

(c) nl-generalization involves an addition to PT++ of a capability to represent the meaning of propositions expressed in a natural language, with the understanding that such propositions serve as descriptors of perceptions. nl-generalization of PT leads to perception-based probability theory denoted as PTp.

An assumption that plays a key role in PTp is that the meaning of a proposition, $p$, drawn from a natural language may be represented as what is called a generalized constraint on a variable. More specifically, a generalized constraint is represented as $X$ isr $R$, where $X$ is the constrained variable; $R$ is the constraining relation; and isr, pronounced ezar, is a copula in which $r$ is an indexing variable whose value defines the way in which $R$ constrains $X$. The principal types of constraints are: equality constraint, in which case *isr* is abbreviated to =; possibilistic constraint, with $r$ abbreviated to blank; veristic constraint, with $r = v$; probabilistic constraint, in which case $r = p$, $X$ is a random variable and $R$ is its probability distribution; random-set constraint, $r = rs$, in which case $X$ is set-valued random variable and $R$ is its probability distribution; fuzzy-graph constraint, $r = fg$, in which case $X$ is a function or a relation and $R$ is its fuzzy graph; and usuality constraint, $r = u$, in which case $X$ is a random variable and $R$ is its usual – rather than expected – value.

The principal constraints are allowed to be modified, qualified, and combined, leading to composite generalized constraints. An example is: usually ($X$ is small) and ($X$ is large) is unlikely. Another example is: if ($X$ is very small) then ($Y$ is not very large) or if ($X$ is large) then ($Y$ is small).

The collection of composite generalized constraint forms what is referred to as the Generalized Constraint Language (GCL). Thus, in PTp, the Generalized Constraint Language serves to represent the meaning of perception-based information. Translation of descriptors of perceptions into GCL is accomplished through the use of what is called the constraint-centered semantics of natural languages (CSNL). Translating descriptors of perceptions into GCL is the first stage of perception-based probabilistic reasoning.

The second stage involves goal-directed propagation of generalized constraints from premises to conclusions. The rules governing generalized constraint propagation coincide with the rules of inference in fuzzy logic. The principal rule of inference is the generalized extension principle. In general, use of this principle reduces computation of desired probabilities to the solution of constrained problems in variational calculus or mathematical programming.

It should be noted that constraint-centered semantics of natural languages serves to translate propositions expressed in a natural language into GCL. What may be called the constraint-centered semantics of GCL, written as CSGCL, serves to represent the meaning of a composite constraint in GCL as a singular constraint $X$ is $R$. The reduction of a composite constraint to a singular constraint is accomplished through the use of rules that govern generalized constraint propagation.

Another point of importance is that the Generalized Constraint Language is maximally expressive, since it incorporates all conceivable constraints. A proposition in a natural language, NL, which is translatable into GCL, is said to be admissible. The richness of GCL justifies the default assumption that any given proposition in NL is admissible. The subset of admissible propositions in NL constitutes what is referred to as a precisiated natural language, PNL. The concept of PNL opens the door to a significant enlargement of the role of natural languages in information processing, decision, and control.

Perception-based theory of probabilistic reasoning suggests new problems and new directions in the development of probability theory. It is inevitable that in coming years there will be a progression from PT to PTp, since PTp enhances the ability of probability theory to deal with realistic problems in which decision-relevant information is a mixture of measurements and perceptions.

Lotfi A. Zadeh is Professor in the Graduate School and director, Berkeley initiative in Soft Computing (BISC), Computer Science Division and the Electronics Research Laboratory, Department of EECs, University of California, Berkeley, CA 94720-1776; Telephone: 510-642-4959; Fax: 510-642-1712;E-Mail: zadeh@cs.berkeley.edu. Research supported in part by ONR Contract N00014-99-C-0298, NASA Contract NCC2-1006, NASA Grant NAC2-117, ONR Grant N00014-96-1-0556, ONR Grant FDN0014991035, ARO Grant DAAH 04-961-0341 and the BISC Program of UC Berkeley.

# Information, Data, and Information Aggregation in relation to the User Model

Horia-Nicolai TEODORESCU
*Romanian Academy, Calea Victoriei 125, Bucharest, Romania*
*and*
*Technical University of Iasi, Fac. Electronics and Tc., Carol 1, 6600 Iasi, Romania*
*e-mail hteodor@etc.tuiasi.ro*

**Abstract**. In this introductory chapter, we review and briefly discuss several basic concepts related to information, data and information aggregation, in the framework of semiotics and semantics.

Computer science and engineering are rather conservative domains, compared to linguistics or some other human-related sciences. However, in the 1970s and 1980s, fuzzy logic and fuzzy systems theory has been the subject of vivid enthusiasm and adulation, or vehement controversy, denial and refutation [1]. This strange page in the history of computer science and engineering domains is largely due to the way fuzzy logic proposed to deal with the very basic foundation of science, namely the logic, moreover to the way it proposed to manipulate information in engineering. Indeed, Lotfi Zadeh has introduced a revolutionary way to represent human thinking and information processing. Today, we are in calmer waters, and fuzzy logic has been included in the curricula for master and doctoral degrees in many universities. Fuzzy logic has reached maturity; now, it is largely considered worth of respect, and respected as a classical discipline. Fuzzy logic and fuzzy systems theory is a branch of the Artificial Intelligence (A.I.) domain and significantly contributes to the processing of information, and to establishing knowledge-based systems and intelligent systems. Instead of competing with other branches, like neural network theory or genetic algorithms, fuzzy logic and fuzzy systems have combined with these other domains to produce more powerful tools.

Information theory has had a smoother development. Its advent has been produced by the development of communication applications. However, because of this incentive and because of the domination of the concept of communication, the field had seen an evolution that may be judged biased and incomplete. In fact, the concept of information is not covering all the aspects it has in common-language and it does not reflect several aspects the semiologists and semanticists would expect. Moreover, it does not entirely reflect the needs of computer science and information science.

The boundary between *data, information,* and *knowledge* has never been more fluid than in computer science and A.I. The lack of understanding of the subject and the disagreement between the experts may lead to disbelief in the field and in the related tools. We should first clarify the meaning of these terms, moreover the meaning of *aggregation.* Although a comprehensive definition of the terms may be a too difficult and unpractical task in a fast evolving field that is constantly enlarging its frontiers, at least "working definitions" should be provided.

While information science is an old discipline, its dynamics has accelerated during recent years, revealing a number of new topics to be addressed. Issues still largely in debate include the relationship and specific differences between data and information, the representation and processing of the meaning contained in the data and information, relationship between meaning and processing method, the quantification of the relevance, significance and utility of information, and relationship between information and knowledge. Questions like "What is data?," "What is information?," "What is meaning?," "What makes the difference between information and data?" "Can information be irrelevant?" have been and still are asked inside both the philosophical and engineering communities.

According to MW [2], *data* means "1: factual information (as measurements or statistics) used as a basis for reasoning, discussion, or calculation ..." and "2: information output by a sensing device or organ that includes both useful and irrelevant or redundant information and must be processed to be meaningful", moreover, 3: information in numerical form that can be digitally transmitted or processed" (selected fragments quoted.). On the other hand, *information* means "1: the communication or reception of knowledge or intelligence" or "2 ...c(1): a signal or character (as in a communication system or computer) representing data" (selected fragments quoted.)

So, data may mean "information in numerical form...", while information may represent "a character representing data". This is just an example of total confusion existing in our field.

Information is based on data and upwards relates to knowledge. Information results by adding meaning to data. Information has to have significance. We need to address semiology, the science of signs, to obtain the definition of information, in contrast with data, which is just not-necessarily informant (meaningful). Meaning and utility play an important role in distinguishing information and data.

We suggest that the difference between data and information consists at least in the following aspects:

- Data can be partly or totally useless (redundant, irrelevant);
- Data has no meaning attached and no direct relation to a subject; data has no meaning until apprehended and processed by a subject.

In contrast, information has the following features:

- assumes existing data;
- communicated: it is (generally) the result of communication (in any form, including communicated through DNA);
- subject-related: assumes a receiving subject that interprets the information;
- knowledge-related: increases the knowledge of the subject;
- usefulness: the use by the recipient of the data or knowledge.

Neither in communication theory, nor in computer science the receiving subject model is included. Although it is essential, the relationship between the information and the subject receiving the information is dealt with empirically or circumstantially, at most. However, if we wish to discuss aggregation of information, the subject model should play an essential part, because the subject's features and criteria finally produce the aggregation. Whenever information and knowledge are present, the model of the user should not miss.

The model of the user cannot and should not be unique. Indeed, at least various different *utilitarian* models can be established, depending on the use the information is given. Also, various *typological, behavioral* models can be established. For example, various ways of reacting to uncertainty may simply reflect the manner the receiving subject behaves. Media know that the public almost always should be considered as formed of different groups and respond in different ways to information. Many of Zadeh's papers actually include suggestions that the user is an active player in the information process, providing meaning to data. Zadeh's approach of including perception as a method to acquire information, as presented in the next chapter, is one more approach to incorporate the user model as *part* of the information.

Also notice that while in communication theory the "communicated" features plays an essential part in defining transmitting data as information, in computer science this feature plays no role.

We suggest that the current understanding of the relationship between data and information is represented by the equation:

$$Information - Data = Meaning + Organization$$

To *aggregate* (from the Latin *aggregare* = to add to, from *ad* = to and *grex* = flock [2]) means [2] "to collect or gather into a mass or whole", "to amount in the aggregate to" [2], while the adjective *aggregate* means "formed by the collection of units or particles into a body, clustered in a dense mass" [2] etc. The broadness and vagueness of these definitions transfer to the concept of "information aggregation". This concept has many meanings, depending on the particular field it is used, moreover depending on the point of view of the researchers using it. For instance, in finance, security, banking, and other related fields, information aggregation means putting together information, specifically information coming from various autonomous information sources (see, for example, [3].) From the sociologic point of view, information aggregation may mean opinion or belief aggregation, the way that opinions, beliefs, rumors etc. sum-up according to specified behavioral rules. From the point of view of a company, data aggregation may mean [4] "any process in which information is gathered and expressed in a summary form, for purposes such as statistical analysis."

Yet another meaning of information aggregation is in linguistics: "Aggregation is the process by which more complex syntactic structures are built from simpler ones. It can also be defined as removing redundancies in text, which makes the text more coherent" [5]. In multi-sensor measurements, "data gathering" and "information aggregation" are related to "data fusion" (i.e., determining synthetic attributes for a set of measurement results performed with several sensors) and finding correlations in the data – a preliminary form of meaning extraction.

Of a different interpretation is *aggregation* in the domain of fuzzy logic. Here, aggregation means, broadly speaking, some form of operation with fuzzy quantities. Several specific meanings are reflected in this volume, and the chapters by George Klir and by Ronald Yager provide an extensive reference list the reader can use to further investigate the subject.

Summarizing the above remarks, the meaning of *information aggregation* (respectively data aggregation) may be:

Information aggregation = gathering various scattered pieces of information (data), whenever the collected and aggregated pieces of information (data) exhibit some kind of relationship, and to generate a supposedly coherent and brief, summary-type result. However, other meanings may emerge and should not be refuted.

Aggregation is not an indiscriminate operation. It must be based on specificity, content, utility or other criteria, which are not yet well stated. Notice that data aggregation may lead to the generation of information, while aggregating information may lead to (new) knowledge.

An example of potentially confusing term in computer science is the phrase "data base". Actually, a database is already an aggregate of data structured according to some meaningful principles, for example relations between the objects in the database. However, users regard the database as not necessarily organized in accordance with what *they* look for as a meaning. Hence, the double-face of databases: structured (with respect to a criterion), yet not organized, unstructured data – with respect to some other criterion.

The information organization topic also covers the clustering of the data and information into hierarchical structures, self-organization of information into dynamical structures, like communication systems, and establishing relationships between various types of information.

Notice that "aggregating information" may also mean that information is structured. Actually, the concept "structure" naturally relates to "aggregation". Quoting again the Merriam-Webster dictionary [1], structure means, among others, "the aggregate of elements of an entity in their relationships to each other". In a structure, constituting elements may play a similar role, and no hierarchy exists. Alternatively, hierarchical structures include elements that may dominate the behavior of the others. Hence, the relationship with hierarchies in structuring information – a topic frequently dealt with in recent years, and also present in this volume.

A number of questions have yet to be answered for we are able to discriminate between data, information and knowledge. This progress can not be achieved in the engineering field only. "Data" may belong to engineers and computer scientists, but "information" and "knowledge" is shared together with semanticists, linguists, psychologists, sociologists, media, and other categories. Capturing various features of the *information* and *knowledge* concept, moreover incorporating them in a sound, comprehensive body of theory will eventually lead to a better understanding of the relationship between the information, the knowledge and the subject (user) concepts [6]. The expected benefit is better tools in information science and significant progresses in A.I.

**Acknowledgments**. This research has been partly supported by the Romanian Academy, and partly by Techniques and Technologies Ltd., Iasi, Romania.

### References

[1].  C.V. Negoita, Fuzzy Sets. New Falcon Publications. Tempe, Arizona, USA, 2000

[2].  Merriam-Webster's Collegiate Dictionary. http://www.m-w.com/cgi-bin/dictionary

[3].  Bhagwan Chowdhry, Mark Grinblatt, and David Levine: Information Aggregation, Security Design And Currency Swaps. Working Paper 8746. http://www.nber.org/papers/w8746. National Bureau of Economic Research. 1050 Massachusetts Avenue. Cambridge, MA 02138. January 2002

[4].  Database.com http://searchdatabase.techtarget.com/sDefinition/0,,sid13_gci532310,00.html

[5].  Anna Ljungberg, An Information State Approach to Aggregation in Multilingual Generation. M.Sc. Thesis, Artificial Intelligence. Natural Language Processing. Division of Informatics. University of Edinburgh, U.K. 2001

[6].  H.-N. L. Teodorescu, Interrelationships, Communication, Semiotics, and Artificial Consciousness. In Tadashi Kitamura (Ed.), What Should Be Computed To Understand and Model Brain Function? From Robotics, Soft Computing, Biology and Neuroscience to Cognitive Philosophy. (Fuzzy Logic Systems Institute (FLSI) Soft Computing Series, Vol. 3. World Scientific Publishers), Singapore, 2001

# Uncertainty and Unsharpness of Information

Walenty OSTASIEWICZ
*Wroclaw University of Economics, Komandorska 118/120, 53-345*
*Wroclaw, Poland*

**Abstract.** Problems addressed in this paper are based on the assumption of the so-called *reistic* point of view on the world about which we know something with certainty and we conjecture about the things when we are not certain. The result of observation and thinking, conceived as an information, about the world must be cast into linguistic form in order to be accessible for analysis as well as to be useful for people in their activity. Uncertainty and vagueness (or by other words, unsharpeness and impreciseness are empirical phenomena, their corresponding representational systems are provided by two theories: probability theory, and fuzzy sets theory. It is argued that **logic** offers the tool for systematic representation of certain information, **stochastic** is the only formal tool to tame uncertainty, and **fuzzy sets** theory is considered as a suitable formal tool (language) for expressing the meaning of unsharpen notions.

## 1. Certainty

The aim of science is on the one hand to make statements that inform us about the world, and on the other hand, to help us how to live happily.

Information that can be proved or derived by means of valid logical arguments is called certain information or knowledge. Apart that, the knowledge, in opposite to information, is sometimes required to posses an ability to be created within a system. Logic provides tools for developing such systems, particularly in the form of a formal or formalized theories.

Let us start with a formal theory.

Suppose that there is a prori some information about a fragment of reality.

Let express this information in the form of two assertions (in the language of the first order logical calculus):

$$A1.\ \forall x:\ \Pi(x, x),$$

$$A2.\ \forall x\ \forall y\ \forall z.\ \Pi(x, z) \wedge \Pi(y, z) \rightarrow \Pi(x, z).$$

These two assertions are considered as specific axioms of the constructed theory. One can see that this theory contains only one primitive notion represented by predicate symbol $\Pi$.

Let us supplement these two expressions, by the system of logical axioms (see [1][2]):

$$L1.\ \alpha \Rightarrow (\beta \Rightarrow \alpha)$$

$$L2.\ (\alpha \Rightarrow (\beta \Rightarrow \gamma)) \Rightarrow ((\alpha \Rightarrow \beta) \Rightarrow (\alpha \Rightarrow \beta))$$

$$L3.\ (\neg\beta \Rightarrow \neg\alpha) \Rightarrow ((\neg\beta \Rightarrow \alpha) \Rightarrow \beta)$$

$$L4.\ \forall x,\ \alpha(x) \Rightarrow \alpha(x\mid t)$$

$$L5.\ \forall x,\ (\alpha \Rightarrow \beta) \Rightarrow (\alpha \Rightarrow \forall x,\ \beta)$$

These five axioms jointly with two basic inference rules (substitution rule and modus ponens rule) form an engine or *machine for creating* (producing) new pieces of information (this means additional information to those given by A1 and A2).

For instance one can easily prove the following assertions (called theorems):

$$T1.\ \forall y\ \forall z:\ \Pi(y, z) \Rightarrow \Pi(z, y),$$

$$T2.\ \forall x\ \forall y\ \forall z:\ \Pi(x, y) \wedge \Pi(y, z) \Rightarrow \Pi(x, z).$$

One can however put the question: what is this theory (set of theorems) about?

The shortest answer is: about nothing.

Any formal theory conveys some information about a fragment of reality only after the interpretation.

The formal theory given above by means of seven axioms (A1, A2, L1, ..., L5) can be interpreted in various domains, conceived as fragments of reality. Interpreted theorems inform us about this reality. As an illustration let us consider a simple example. Suppose that the fragment of reality consists of three things, which are denoted here by the following three signs: $\square$, $\Delta$, O.

Between these three entities there is the following symmetric binary relation:

$$s\,(O, \square) = \mathit{false},\ s\,(\square, \Delta) = \mathit{true},\ s\,(\square, O) = \mathit{false}.$$

which can be read for example as "is similar".

Suppose that predicate symbol $\Pi$ is interpreted as the relation s defined above, then one can easily check that the both axioms A1 and A2 are the true assertions about the world under consideration. This means that all theorems, which can be proved within this theory are surely true statements about our world of three things connected by relation s.

The other approach to construction theory consists in taking a concrete domain, and next to tries to formalize the knowledge about it.

Suppose for example, that the problem consists in ordering cups of coffee according to their sweetness. For some pairs of cups we can definitely decide which of them is sweeter. For some other pairs, we cannot distinguish whether one is sweeter than the other is.

There is probably a tolerance within we allow a cup of coffee to move before we notice any difference. The relation of indifference one can define in terms of ternary relation of betweenness as follows:

$$s(x, y) \Leftrightarrow B(x, y, x).$$

Axioms of the relation B are following (see [3]):

A1. $B(x, y, z) \Rightarrow B(z, y, x)$

A2. $B(x, y, z) \vee B(x, z, y) \vee B(y, x, z)$

A3. $(B(x, y, u) \wedge B(y, z, u) \wedge \neg B(x, y, z)) \Rightarrow s(u, y) \wedge s(u, z)$

A4. $\neg s(u, v) \Rightarrow (B(x, u, v) \wedge B(u, v, y) \Rightarrow B(x, u, y))$

A5. $B(x, y, z) \wedge B(y, x, z) \Rightarrow (s(x, y) \vee (s(z, x) \wedge s(z, y)))$

A6. $s(x, y) \Rightarrow B(x, y, z)$

These axioms are sufficient and necessary for the existence of a function f defined on the set of all cups of coffee such that for some $\varepsilon > 0$ holds following:

$$s(x, y) = \begin{cases} true, & if |f(x) - f(y)| < å \\ false, & otherwise \end{cases}$$

This means that, for some threshold $\varepsilon$, two cups $x$ and $y$ are indistinguishable if the absolute difference $|f(x) - f(x)|$, say between sweetness, is less than $\varepsilon$. One should note that indistinguishability in this case is defined as a usual, crisp binary relation in the terms yes-no.

It seems natural to have the desire to define indistinguishability as a graded relation i. e. as a function taking on values from the unit interval. It turns however up (see [4]) that in this case it is impossible to create a formal theory in a purely syntactic form. Admitting the graduality in our understanding of the reality we must use fuzzy sets concepts as a formal tool to formulate theories in a semantic form.

## 2. Uncertainty

Already Plato in his Republic distinguished between certain information, i. e. knowledge, and uncertain information, called opinion or belief. Certain knowledge is acquired by the tools provided by logic. The ability to obtain this kind of information is also called the **art of thinking**.

Patterned after this name, J. Bernoulli had written the book under similar title, and namely under the title **the art of conjecturing**, or stochastics, intending to provide tools to make belief also an exact science. The art of conjecturing takes over where the art of thinking left off. An exact science has been made by attaching numbers to all our uncertainties.

One distinguishes between the kind of uncertainty that characterizes our general knowledge of the world, and the kind of uncertainty that we discover in gambling. As a consequence, one distinguishes between two kinds of probabilities: **epistemic** probability, and **aleatory** probability. The former is dedicated to assessing degree of belief in propositions, and the later is concerning itself with stochastic laws of chance processes.

Chance regularities or probability laws are usually expressed by means of the so-called cumulative distribution functions (cdf).

## 3. Unsharpeness

The results of thinking processes as well as conjecturing processes only after their casting into linguistic form became available for analysis and for communication. One of the three basic functions of language is to convey information about world.

One of the modes of conveying information is to give appropriate definitions. Most definitions in natural languages are made by examples. As a consequence of this, almost all words are vague.

According to M. Black and N. Rescher a word is vague when its (denotational) meaning is not fixed by sharp boundaries but spread over a range of possibilities, so that its **applicability** in a particular case may be dubious.

Nonsharp or vague words are characterized therefore by the existence of a "gray area" where the applicability of the word is in doubt.

L. Zadeh prefers however to call such words as fuzzy words. The meaning of **fuzzy words** can be precisely defined by means of **fuzzy sets,** which were invented by L. Zadeh (see [5]). By fuzzy set, or more precisely fuzzy subset of a given set $U$, one understands a mapping

$$\mu_X\colon U \to [0, 1],$$

where $X$ stands for a fuzzy word.

The value $\mu_X(x)$ is interpreted as a **grade of applicability** of word $X$ to a particular object $x \in X$. Alternatively $\mu_X(x)$ can be conceived as a perceived psychological distance between an object x and the ideal prototype of a word $X$.

It is worth to notice the essential difference between apparently similar phrases: "fuzzy word" and "fuzzy set". A fuzzy word, in another terminology, is a vague word, so that some words may be fuzzy, while the others are not fuzzy. In the opposite, fuzzy set is a sharp, proper name of some precisely defined mathematical object, so that the term "fuzzy set" is not fuzzy.

## 4. Uncertainty versus vagueness

For methodological convenience, it is useful to make a distinction between an observed world and its representational system, whose typical example is language.

The observed world is as it is, neither certain nor uncertain. Uncertainty pertains an observer because of his (her) ignorance and in their ability understand and to foresee events occurring in world. Those things that the science in its current level of development cannot predict are called contingent or random. Probability, or more generally stochastics, provides tools to tame the all kinds of chance regularities as well as to describe all kinds of uncertainty. After Laplace, one can rightly say that a perfect intelligence would have no need of probability, it is however indispensable for mortal men.

On the other hand, vagueness is a property of signs of representational systems. It should not be confused with uncertainty. Apparently these two empirical phenomena have something in common. In both situations, an observer proclaims: "I do not know. " But these two are very different kinds of not knowing.

A simple example will make this assertion quite clear. Before rolling a die *I do not know* which **number** of spots will result. This kind of uncertainty is called aleatory uncertainty. On the other hand, before rolling a die, or even after the rolling I do knot know weather or not is the die **fair**? This is epistemic uncertainty.

Suppose now that the die is cast, looking at it and seeing the spots *I do not know*, for example, whether or not resulted a **small** number of spots. I have my doubts as to apply the word "small" to the number four, for example. Fuzzy sets theory offers formal tools to quantify the applicability of words to particular objects.

From the above discussion it should be clear enough that probability theory (broadly considered) and fuzzy sets theory are quite different formalisms invented to convey quite different information. By other words one can also say that these are different tools invented to cope with different (incomparable) problems.

For brevity, some distinct features of uncertainty and unsharpeness are summarized in the following table:

| Uncertainty | Unsharpeness |
|---|---|
| Exists because of a lack of biunivocal correspondence between causes and consequences | exists because of a lack of sharp definitions |
| There are limits for certainty | there are no limits for sharpening definitions |
| Pertains the WORLD | pertains WORDS about world |
| Referrers to reasoning and prediction | refers to classification and discrimination |
| It is my defect because of my ignorance | it is my doubt in applicability of words because of our (or your) carelessness in naming things |
| It is quantified by grades of certainty called probability; <br> Probability is warranted by evidence | it is quantified by grades of applicability called membership grade; <br> applicability is warranted by convention |

## 5. Conditional information

As a matter of fact, all information is inherently conditional, because all information has context. Context is nothing else as just another word for condition.

Conditional information is expressed by conditional statements of the following type:

*if A, then B.*

Within the classical logic statements of that type of certain conditional information are formalized by implication, $A \Rightarrow B$, where $A$ and $B$ are binary-valued assertions.

The truth-value of this (material) implication is defined as follows:

$$t(A \Rightarrow B) = \begin{cases} false, & if\ t(A) = true,\ t(B) = false \\ true, & otherwise \end{cases}$$

In classical logic, the material implication $A \Rightarrow B$ can be expressed equivalently in several other ways:

$$\neg A \vee B,\ A \wedge B = A,\ A \vee B = B,\ \neg B \wedge A = 0,$$

where 0 represents the truth value "false".

In case of uncertain conditional information, it would seem natural to expect some formal hints from the side of probability theory. Unfortunately, in probability theory there are only a few proposals, and still debatable, for the definition of conditional information of the type:

*"if event A, then event B".*

Within the traditional probability theory, it is offered the **conditional probability**, but not a probability of conditional event. The conditional probability of event $A$, given event $B$, is defined as follows:

$$P\left(A/B\right) = \frac{P\left(A \cap B\right)}{P\left(B\right)}$$

provided that $P(B) > 0$.

One needs however probability (not conditional!) of the conditional event $(A \mid B)$; that is, one needs to define $P((A \mid B))$.

As it is known, probability measure $P$ is defined on the Boolean space of events, so that in order to enable the calculation of $P((A \mid B))$, one needs to define conditional event $(A \mid B)$ as a Boolean element, i. e. by means of Boolean operations $\wedge$, $\vee$, and $\neg$. It turns however out, that it is impossible (see [6][7]). For that reason, various extensions of the ordinal Boolean operations are proposed (see [6]). The problem with conditional information become more complicated, even almost insuperable, if not only uncertainty, but also vagueness is introduced into conditional statements of the type: "if —, then —".

Suppose A and B are two vague terms, which are modeled by two fuzzy sets:

$$\mu_A: U \to [0, 1] \quad \text{and} \quad \mu_B: U \to [0, 1].$$

Conditional statement

**if** *a is A,* **then** *b is B*

within the fuzzy sets theory is defined as a fuzzy relation

$$\mu_R: U \times U \to [0, 1]$$

defined as:

$$\mu_R\left(x, y\right) = I\left(\mu_A\left(x\right), \mu_B\left(y\right)\right),$$

where $I$ is an operator of the so-called fuzzy implication.

For example, this operator can be defined us follows:

$$I(x, y) = \min\{1, 1 - x + y\}.$$

To a certain extent, the above solution is an analogue to Gordian knot. Alexander the Great of Macedonia, when coming to Phrygia, he cut the Gordian knot with his sward. Similarly, Lotfi Zadeh, the Great of Fuzzy World, when proposing fuzzy sets, he cut the Gordian knot of conditionals by his definition

$$\mu_{\text{if a is A, then b is B}}(x, y) = I(\mu_A(x), \mu_B(y))$$

saying it is thus fuzzy **if – then** rule.

### References

[1] N. Rescher, Introduction to Logic, St. Martin's Press, N. Y., 1964
[2] A. Tarski, Introduction to logic, Oxford University Press, N. Y., 1954
[3] W. Ostasiewicz, Some philosophical aspects of fuzzy sets, *Fuzzy Economic Review*, **1**, 1996, 3-33
[4] J. A. Goguen, The logic of inexact concepts, *Syntheses*, **19**, 1968, 325-373
[5] L. A. Zadeh, Fuzzy sets, *Information and Control*, **8**, 1965, 338-353
[6] P. Calabrese, An algebraic synthesis of the foundation of logic and probability, *Information Sci.* **42**, 1987, 187-237
[7] D. Dubois and H. Prade, Conditional object as nonmonotonic consequence relationships, *IEEE Trans. on Systems, Man and Cybernetics*, **24**, 1994, 1724-1740

# Part II

# Fundamentals

# Uncertainty-Based Information

George J. KLIR
*Center for Intelligent Systems and*
*Dept. of Systems Science & Industrial Eng.*
*Binghamton University - SUNY*
*Binghamton, New York 13902-6000*
*E-mail: gklir@binghamton. edu*

**Abstract.** Uncertainty-based information is defined in terms of reduction of relevant uncertainty. It is shown how the emergence of fuzzy set theory and the theory of monotone measures considerably expanded the framework for formalizing uncertainty and the associated uncertainty-based information. A classification of uncertainty theories that emerge from this expanded framework is examined. It is argued that each of these theories needs to be developed at four distinct levels: (i) formalization of the conceived type of uncertainty; (ii) calculus by which this type of uncertainty can be properly manipulated; (iii) measuring, in a justifiable way, the amount of relevant uncertainty (predictive, prescriptive, etc.) in any situation formalizable in the theory; and (iv) various uncertainty principles and other methodological aspects. Only some uncertainty theories emerging from the expanded framework have been thoroughly developed thus far. They may be viewed as theories of imprecise probabilities of various types. Results regarding these theories at the four mentioned levels (representation, calculus, measurement, methodology) are surveyed.

## 1. Introduction

The recognition that scientific knowledge is organized, by and large, in terms of systems of various types (or categories in the sense of mathematical theory of categories [1]), is an important outcome of systems science [2]. In general, *systems* are viewed as relations among states of some variables. Employing the constructivist epistemological view [3], to which I subscribe, it is recognized that systems are constructed from our experiential domain for various purposes, such as prediction, retrodiction, prescription, planning, control, diagnosis, etc. [2, 4]. In each system, its relation is utilized, in a given purposeful way, for determining unknown states of some variables on the basis of known states of some other variables. Systems in which the unknown states are determined uniquely are called *deterministic*; all other systems are called *nondeterministic*. By definition, each nondeterministic system involves uncertainty of some type. This uncertainty pertains to the purpose for which the system was constructed. It is thus natural to distinguish predictive uncertainty, retrodictive uncertainty, diagnostic uncertainty, etc. In each nondeterministic system, the relevant uncertainty must be properly incorporated into the description of the systems in some formalized language.

Prior to the $20^{th}$ century, science and engineering had shown virtually no interest in nondeterministic systems. This attitude changed with the emergence of statistical mechanics at the beginning of the $20^{th}$ century [5]. Nondeterministic systems were for the first time recognized as useful. However, these were very special nondeterministic systems, in which uncertainty was expressed in terms of probability theory. For studying physical processes at the molecular level, which was a problem area from which statistical mechanics emerged, the use of probability was the right choice. It was justifiable, and hence successful, to capture macroscopic manifestations of enormously large collections of microscopic phenomena (random movements of molecules) via their statistical averages. Methods employed successfully in statistical mechanics were later applied to similar problems in other areas, such as actuarial profession or engineering design of large telephone exchanges, where the use of probability theory was again justifiable by the involvement of large number of random phenomena.

Although nondeterministic systems have been accepted in science and engineering since their utility was demonstrated early in the 20$^{th}$ century, it was tacitly assumed for long time that probability theory is the only framework within which uncertainty in nondeterministic systems can be properly formalized and dealt with. This presumed equality between uncertainty and probability became challenged shortly after World War II, when the emerging computer technology opened new methodological possibilities. It was increasingly realized, as most eloquently described by Weaver [6], that the established methods of science were not applicable to a broad class of important problems for which Weaver coined the suggestive term "problems of *organized complexity*". These are problems that involve considerable numbers of entities that are interrelated in complex ways. They are typical in life, cognitive, social, and environmental sciences, as well as applied fields such as modern technology, medicine, or management. They almost invariably involve uncertainties of various types, but rarely uncertainties resulting from randomness, which can yield meaningful statistical averages.

Uncertainty liberated from its probabilistic confines is a phenomenon of the second half of the 20$^{th}$ century. It is closely connected with two important generalizations in mathematics. One of them is the generalization of classical measure theory [7] to the theory of monotone measures, which was first suggested by Gustave Choquet in 1953 in his *theory of capacities* [8]. The second one is the generalization of classical set theory to *fuzzy set theory*, introduced by Lotfi Zadeh in 1965 [9]. These generalizations enlarged substantially the framework for formalizing uncertainty. As a consequence, they made it possible to conceive of new theories of uncertainty.

To develop a fully operational theory of uncertainty of some conceived type requires that we address relevant issues at each of the following four levels:

- LEVEL 1 — an appropriate mathematical *formalization* of the conceived type of uncertainty must be determined.
- LEVEL 2 — a *calculus* by which the formalized uncertainty is properly manipulated must be developed.
- LEVEL 3 — a justifiable way of *measuring* the amount of uncertainty in any situation formalizable in the theory must be determined.
- LEVEL 4 — *methods* for applying the theory to relevant problems must be developed.

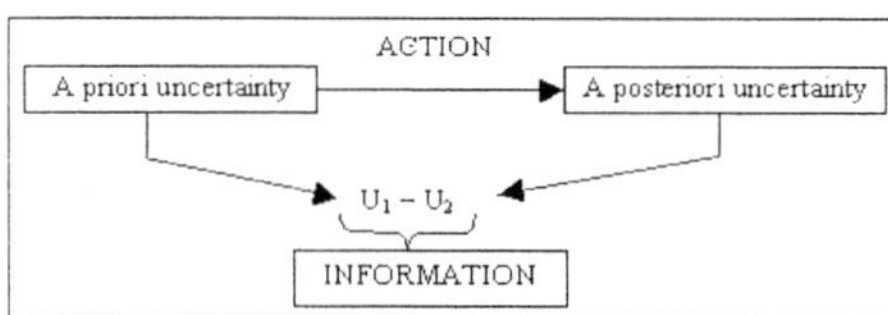


Figure 1. The meaning of uncertainty-based information.

In general, uncertainty is an expression of some information deficiency. This suggests that information could be measured in terms of uncertainty reduction. To reduce relevant uncertainty (predictive, retrodictive, prescriptive, diagnostic, etc.) in a situation formalized within a mathematical theory requires that some relevant action be taken by a cognitive agent, such as performing a relevant experiment, searching for a relevant fact, or accepting and interpreting a relevant message. If results of the action taken (an experimental outcome, a discovered fact, etc.) reduce uncertainty involved in the situation, then the amount of information obtained by the action is measured by the amount of uncertainty reduced — the difference between *a priori* and *a posteriori* uncertainty (Fig. 1).

Measuring information in this way is clearly contingent upon our capability to measure uncertainty within the various mathematical frameworks. Information measured solely by uncertainty reduction is an important, even though restricted, notion of information. To distinguish it from the various other conceptions of information, it is common to refer to it as *uncertainty-based information* [10].

Uncertainty-based information does not capture the rich notion of information in human communication and cognition, but it is very useful in dealing with nondeterministic systems. Given a particular nondeterministic system, it is useful, for example, to measure the amount of information contained in the answer given by the system to a relevant question (concerning various predictions, retrodictions, etc.). This can be done by taking the difference between the amount of uncertainty in the requested answer obtained within the experimental frame of the system [2, 4] in the face of total ignorance and the amount of uncertainty in the answer obtained by the system. This can be written concisely as:

Information $(A_S | S, Q) =$ Uncertainty $( A_{EF_S} | EF_S, Q ) -$ Uncertainty $(A_S | S, Q)$,

where:

- $S$ denotes a given system
- $EF_S$ denoted the experimental frame of system S
- $Q$ denotes a given question
- $A_{EF_S}$ denotes the answer to question Q obtained solely within the experimental     frame $EF_S$
- $A_S$ denotes the answer to question $Q$ obtained by system $S$.

The principal purpose of this chapter is to present a comprehensive overview of *generalized information theory* — a research program whose objective is to develop a broader treatment of uncertainty-based information, not restricted to the classical notions of uncertainty. Although the term "generalized information theory" was coined in 1991 [11], research in the area has been pursued since the early 1980s.

The chapter is structured as follows: After a brief overview of classical information theories in Sec. 2, a general framework for formalizing uncertainty is introduced in Sec. 3. This is followed by a description of the most developed nonclassical theories of uncertainty in Sec. 4, and the measurement of uncertainty and uncertainty-based information in these theories in Sec. 5. Methodological issues regarding the various uncertainty theories, focusing primarily on three general principles of uncertainty, are discussed in Sec. 6. Finally, a summary of main results and open problems in the area of generalized information theory is presented in Sec. 7.

## 2. Classical Uncertainty Theories

Two classical uncertainty theories are recognized. They emerged in the first half of the $20^{\text{th}}$ century and are formalized in terms of classical set theory. The older one, which is also simpler and more fundamental, is based on the notion of *possibility*. The newer one, which has been considerably more visible, is based on the formalized notion of *probability*. For the sake of completeness, they are briefly reviewed in this section.

### *2. 1. Classical Possibility-Based Uncertainty Theory*

To describe this rather simple theory, let $X$ denote a finite set of mutually exclusive alternatives that are of our concern (diagnoses, predictions, etc.). This means that in any given situation only one of the alternatives is true. To identify the true alternative, we need to obtain relevant information (e. g., by conducting relevant diagnostic tests). The most elementary and, at the same time, the most fundamental kind of information is a demonstration (based, for example, on outcomes of the conducted diagnostic tests) that some of the alternatives in $X$ are not possible. After excluding these alternatives from $X$, we obtain a subset $E$ of $X$. This subset contains only alternatives that, according to the obtained information are *possible*. We may say that alternatives in $E$ are supported by evidence.

Let the characteristic function of the set of all possible alternatives, $E$, be called in this context a *possibility distribution function* and be denoted by $r_E$. Then,

$$r_E(x) = \begin{cases} 1 & \text{when } x \in E \\ 0 & \text{when } x \notin E. \end{cases}$$

Using common sense, a possibility function, $Pos_E$, defined on the power set, $\mathscr{P}(X)$, is given by the formula

$$Pos_E(A) = \max_{x \in A} r_E(x),$$

for all $A \in \mathscr{P}(X)$. It is indeed correct to say that it is possible that the true alternative is in $A$ when $A$ contains at least one alternative that is also contained in $E$.

Given a possibility function $Pos_E$ on the power set of $X$, it is useful to define another function, $Nec_E$, to describe for each $A \in \mathscr{P}(X)$ the *necessity* that the true alternative is in $A$. Clearly, the true alternative is necessarily in $A$ if and only if it is not possible that it is in $\overline{A}$, the complement of $A$. Hence,

$$Nec_E(A) = 1 - Pos_E(\overline{A})$$

for all $A \in \mathscr{P}(X)$.

The question of how to measure the amount of uncertainty associated with a finite set $E$ of possible alternatives was addressed by Hartley in 1928 [12]. He showed that the only meaningful way to measure this amount is to use a functional of the form.

$$c \log_b \sum_{x \in X} r_E(x)$$

or, alternatively,

$$c \log_b |E|,$$

where $|E|$ denotes the cardinality of $E$, and $b$ and $c$ are positive constants. Each choice of values $b$ and $c$ determines the unit in which the uncertainty is measured. Requiring, for example, that

$$c \log_b 2 = 1,$$

which is the most common choice, uncertainty would be measured in *bits*. One bit of uncertainty is equivalent to uncertainty regarding the truth or falsity of one elementary proposition. Choosing conveniently $b = 2$ and $c = 1$ to satisfy the above equation, we obtain a unique functional, $H$, defined for any possibility function, $Pos_E$, by the formula

$$H(Pos_E) = \log_2 |E|.$$

This functional is usually called a *Hartley measure* of uncertainty. Its uniqueness was later proven on axiomatic grounds (see Sec. 5. 1) by Rènyi [13]. Observe that

$$0 \leq H(Pos_E) \leq \log_2 |X|$$

for any $E \in \mathcal{P}(X)$ and that the amount of information, $I(Pos_E)$, in evidence expressed by function $Pos_E$ is given by the formula

$$I(Pos_E) = \log_2 |X| - \log_2 |E|.$$

It follows from the Hartley measure that uncertainty associated with sets of possible alternatives results from the lack of specificity. Large sets result in less specific predictions, diagnoses, etc., than their smaller counterparts. Full specificity is obtained when only one alternative is possible. This type of uncertainty is thus well characterized by the term *nonspecificity*.

Consider now two universal sets, $X$ and $Y$, and assume that a relation $R \subseteq X \times Y$ describes a set of possible alternatives in some situation of interest. Assume further that the domain and range of $R$ are sets $R_X \subseteq X$ and $R_Y \subseteq Y$, respectively. Then three distinct Hartley functionals are applicable, defined on the power sets of $X$, $Y$, and $X \times Y$. To identify clearly which universal set is involved in each case, it is useful (and a common practice) to write $H(X)$, $H(Y)$, $H(X, Y)$ instead of $H(Pos_{R_X})$, $H(Pos_{R_Y})$, $H(Pos_R)$, respectively. Functionals

$$H(X) = \log_2 |R_X|,$$
$$H(Y) = \log_2 |R_Y|$$

are called *simple uncertainties*, while function

$$H(X, Y) = \log_2 |R|$$

is called a joint uncertainty.

Two additional Hartley functionals are defined,

$$H(X|Y) = log_2 \frac{|R|}{|R_Y|},$$  (1)

$$H(Y|X) = log_2 \frac{|R|}{|R_X|},$$  (1')

which are called *conditional uncertainties*. Observe that the ratio $|R_X| / |R_Y|$ in $H(X| Y)$ represents the average number of elements of $X$ that are possible alternatives under the condition that an element of $Y$ has been selected. This means that $H(X | Y)$ measures the average nonspecificity regarding alternative choices from $X$ for all particular choices form $Y$. Function $H(X | Y)$ has clearly a similar meaning with the roles of sets $X$ and $Y$ exchanged. Observe also that the conditional uncertainties can be expressed in terms of the joint uncertainty and the two simple uncertainties:

$$H(X| Y) = H(X, Y) - H(Y),$$  (2)

$$H(Y| X) = H(X, Y) - H(X).$$  (2')

If possible alternatives from $X$ do not depend on selections form $Y$, and visa versa, then $R = X \times Y$ and the sets $X$ and $Y$ are called *noninteractive*. Then, clearly,

$$H(X| Y) = H(X),$$  (3)

$$H(Y| X) = H(Y),$$  (3')

$$H(X, Y) = H(X) + H(Y),$$  (4)

In all other cases, when sets $X$ and $Y$ are *interactive*, these equations become the inequalities

$$H(X| Y) < H(X),$$  (5)

$$H(Y| X) < H(Y),$$  (5')

$$H(X, Y) < H(X) + H(Y).$$  (6)

In addition, the functional

$$T_H(X, Y) = H(X) + H(Y) - H(X, Y),$$  (7)

which is usually called an *information transmission*, is a useful indicator of the strength of constraint between sets $R_X$ and $R_Y$. The Hartley measure is applicable only to finite sets. Its counterpart for subsets of the $n$-dimensional Euclidean space $\mathcal{R}^n$ ($n \geq 1$) was not available for long time. It was eventually suggested in 1995 by Klir and Yuan [14] in terms of the functional:

$$HL(Pos_E) = \min_{c \in C} \ln\left[\prod_{i=1}^{n}(1+\mu(E_{i_c})) + \mu(E) - \prod_{i=1}^{n}\mu(E_{i_c})\right], \tag{8}$$

where $E$, $C$, $E_{i_c}$, and $\mu$ denote, respectively, a subset of $\mathcal{R}^n$, the set of all isometric transformations form one orthogonal coordinate system to another, the $i$-th projection of $E$ in coordinate system $c$, and the Lebesque measure. This functional, which is usually referred to as *Hartley-like measure*, was proven to satisfy all mathematical properties that such a measure is expected to satisfy (see Sec. 5. 1) [14, 15], but its uniqueness remains an open problem.

### 2. 2. Classical Probability-Based Uncertainty Theory

The second classical uncertainty theory is based on the notion of classical probability measure [16]. Assuming again a finite set $X$ of mutually exclusive alternatives that are of our concern, evidence is expressed by a *probability distribution function*, $p$, on $X$ such that $p(x) \in [0, 1]$ for each $x \in X$ and

$$\sum_{x \in X} p(x) = 1.$$

Probability measure, *Pro*, is then obtained for all $A \in \mathcal{P}(X)$ via the formula

$$Pro(A) = \sum_{x \in A} p(x).$$

When $p$ is defined on a Cartesian product $X \times Y$, it is called a *joint probability distribution function*. The associated marginal *probability distribution functions* on $X$ and $Y$ are determined, respectively, by the formulas

$$p_X(x) = \sum_{y \in Y} p(x,y)$$

for each $x \in X$ and

$$p_Y(y) = \sum_{x \in X} p(x,y)$$

for each $y \in Y$. The *noninteraction* of $p_X$ and $p_Y$ is defined by the condition

$$p(x, y) = p_X(x) \cdot p_Y(y)$$

for all $x \in X$ and all $y \in Y$. Conditional probabilities, $p(x \mid y)$ and $p(y \mid x)$, are defined in the usual way:

$$p(x \mid y) = \frac{p(x,y)}{p_Y(y)}$$

for all $x \in X$, and

$$p(y|x) = \frac{p(x,y)}{p_X(x)}$$

for all $y \in Y$.

As is well known, the amount of uncertainty in evidence expressed by a probability distribution function p on a finite set $X$ is measured (in bits) by the functional

$$S(p(x) \mid x \in X) = -\sum_{x \in X} p(x)log_2\, p(x). \tag{9}$$

This functional, usually referred to as *Shannon entropy*, was introduced by Shannon in 1948 [17]. It was proven in numerous ways, from several well-justified axiomatic characterizations, that the Shannon entropy is the only functional by which the amount of probability-based uncertainty can be measured (assuming that it is measured in bits) [10].

For probabilities on $X \times Y$, three types of Shannon entropies are recognized: joint, marginal, and conditional. A simplified notation to distinguish them is commonly used in the literature: $S(X)$ instead of $S(p(x)|x \in X)$, $S(X, Y)$ instead of $S(p(x, y)|x \in X, y \in Y)$, etc. Conditional Shannon entropies are defined in terms of weighted averages of local conditional entropies as

$$S(X|Y) = -\sum_{y \in Y} p_Y(y) \sum_{x \in X} p(x|y) \log_2 p(x|y), \tag{10}$$

$$S(Y|X) = -\sum_{x \in X} p_X(x) \sum_{y \in Y} p(y|x) \log_2 p(y|x). \tag{10'}$$

As is well known [10], equations and inequalities (2) - (6) for the Hartley measure have their exact counterparts for he Shannon entropy. For example, the counterparts of (2), (2') are the equations

$$S(X \mid Y) = S(X, Y) - S(Y), \tag{11}$$

$$S(Y \mid X) = S(X, Y) - S(X). \tag{11'}$$

Moreover,

$$T_S(X, Y) = S(X) + S(Y) - S(X, Y) \tag{12}$$

is the *probabilistic information transmission* (the probabilistic counterpart of (7)).

It is obvious that the Shannon entropy is applicable only to finite sets of alternatives. At first sight, it seems suggestive to extend it to probability density functions, $q$, on $R$ (or, more generally, on $R^n$, $n \geq 1$), by replacing in Eq. (9) $p$ with $q$ and the summation with integration. However, there are several reasons why the resulting functional does not qualify as a measure of uncertainty: (i) it may be negative; (ii) it may be infinitely large; (iii) it depends on the chosen coordinate system;

and most importantly, (iv) the limit of the sequence of its increasingly more refined discrete approximations diverges [10]. These problems can be overcome by the modified functional

$$S'[q(x), q'(x) \mid x \in \mathbb{R}] = \int q(x) \log_2 \frac{q(x)}{q'(x)} dx, \tag{13}$$

which involves two probability density functions, $q$ and $q'$. Uncertainty is measured by $S'$ in relative rather than absolute terms.

When $q$ in (13) is a joint probability density function on $\mathbb{R}^2$ and $q'$ is the product of the two marginals of $q$, we obtain the information transmission

$$T_s[q(x,y), q_X(x), q_Y(y) \mid x \in \mathbb{R}, y \in \mathbb{R}] =$$
$$= \int \int q(x,y) \log_2 \frac{q(x,y)}{q_X(x) \cdot q_Y(y)} dx dy. \tag{14}$$

This means that (14) is a direct counterpart of (12).

## 3. General Framework for Formalizing Uncertainty

As already mentioned in Sec. 1, the framework for formalizing uncertainty was considerably enlarged by the generalization of classical (additive) measures to monotone measures, as well as by the generalization of classical sets to fuzzy sets. To describe the enlarged framework, these two generalizations must be explained first.

### 3. 1. Monotone Measures

Given a universal set $X$ and a non-empty family $C$ of subsets of $X$ (usually with an appropriate algebraic structure), a *monotone measure*, $g$, on $\langle X, C \rangle$ is a function

$$g: C \to [0, \infty]$$

that satisfies the following requirements:
    (g1) $g(\varnothing) = 0$ (vanishing at the empty set);
    (g2) for all $A, B \in C$, if $A \subseteq B$, then $g(A) \leq g(B)$ (monotonicity);
    (g3) for any increasing sequence $A_1 \subset A_2 \subseteq \ldots$ of sets in $C$,

$$\text{if } \bigcup_{i=1}^{\infty} A_i \in C, \text{ then } \lim_{i \to \infty} g(A_i) = g\left( \bigcup_{i=1}^{\infty} A_i \right) \text{ (continuity from below);}$$

    (g4) for any decreasing sequence $A_1 \supseteq A_2 \supseteq \ldots$ of sets in $C$,

$$\text{if } \bigcap_{i=1}^{\infty} A_i \in C, \text{ then } \lim_{i \to \infty} g(A_i) = g\left( \bigcap_{i=1}^{\infty} A_i \right) \text{ (continuity from above).}$$

Functions that satisfy requirements (g1), (g2), and either (g3) or (g4) are equally important in the theory of monotone measures. These functions are called *semicontinuous* from below or above, respectively. When the universal set $X$ is finite, requirements (g3) and (g4) are trivially satisfied and may thus be disregarded. If $X \in C$ and $g(X) = 1$, $g$ is called a *regular* monotone measure (or *regular* semicontinuous monotone measure). Observe also that requirement (g2) defines measures that are actually *monotone increasing*. By changing the inequality $g(A) \le g(B)$ in (g2) to $g(A) \ge g(B)$, we can define measures that are *monotone decreasing*. Both types of monotone measures are useful, even though monotone increasing measures are more common in dealing with uncertainty.

For any pair $A, B \in C$ such that $A \cap B = \varnothing$, a monotone measure $g$ is capable of capturing any of the following situations:

(a) $g(A \cup B) > g(A) + g(B)$, called *superadditivity*, which expresses a cooperative action  or synergy between $A$ and $B$ in terms of the measured property;

(b) $g(A \cup B) = g(A) + g(B)$, called *additivity*, which expresses the fact that $A$ and $B$ are noninteractive with respect to the measured property;

(c) $g(A \cup B) < g(A) + g(B)$, called *subadditivity*, which expresses some sort of inhibitory effect or incompatibility between $A$ and $B$ as far as the measured property is concerned.

Observe that probability theory, which is based on classical measure theory [7] is capable of capturing only situation (b). This demonstrates that the theory of monotone measures provides us with a considerably broader framework than probability theory for formalizing uncertainty. As a consequence, it allows us to capture types of uncertainty that are beyond the scope of probability theory.

For some historical reasons of little significance, monotone measures are often referred to in the literature as *fuzzy measures* [18, 19]. This name is somewhat confusing since no fuzzy sets are involved in the definition of monotone measures. To avoid this confusion, the term "fuzzy measures" should be reserved to measures (additive or non-additive) that are defined on families of fuzzy sets.

### 3. 2. Fuzzy Sets

Fuzzy sets are defined on any given universal set of concern by functions analogous to characteristic functions of classical sets. These functions are called *membership functions*. Each membership function defines a fuzzy set on a given universal set by assigning to each element of the universal set its membership grade in the fuzzy set. The set of all membership grades must be at least partially ordered, but it is usually required to form a complete lattice. The most common fuzzy sets, usually referred to as *standard fuzzy sets*, are defined by membership grades in the unit interval [0, 1]. Those for which the maximum (or supremum) is 1 are called *normal*. Fuzzy sets that are not normal are called *subnormal*.

Two distinct notations are most commonly employed in the literature to denote membership functions. In one of them, the membership function of a fuzzy set $A$ is denoted by $\mu_A$ and, assuming that $A$ is a standard fuzzy set, its form is

$$\mu_A : X \to [0, 1],$$

where $X$ denotes the universal set of concern. In the second notation, the membership function is denoted by $A$ and has, of course, the same form

$$A : X \to [0, 1].$$

According to the first notation, the symbol of the fuzzy set involved is distinguished from the symbol of its membership function. According to the second notation, this distinction is not made, but no ambiguity results from this double use of the same symbol since each fuzzy set is completely and uniquely defined by one particular membership function. The second notation is adopted in this paper; it is simpler and, by and large, more popular in the current literature on fuzzy sets.

Contrary to the symbolic role of numbers 0 and 1 in characteristic functions of classical sets, numbers assigned to relevant objects by membership functions of fuzzy sets have a numerical significance. This significance is preserved when classical sets are viewed (from the standpoint of fuzzy set theory) as special fuzzy sets, often referred to as *crisp sets*.

An important property of fuzzy sets is their capability to express gradual transitions from membership to nonmembership. This expressive capability has great utility. For example, it allows us to capture, at least in a crude way, the meaning of expressions in natural language, most of which are inherently vague. Crisp sets are hopelessly inadequate for this purpose. However, it is important to realize that meanings of expressions in natural language are strongly dependent on the context within which they are used.

Among the most important concepts associated with standard fuzzy sets is the concept of an *α-cut*. Given a fuzzy set $A$ defined on $X$ and a number $\alpha$ in the unit interval $[0, 1]$, the α-cut of $A$, denoted by ${}^{\alpha}A$, is the crisp set that consists of all elements of $A$ whose membership degrees in $A$ are greater than or equal to $\alpha$; that is,

$$^{\alpha}A = \{x \in X \mid A(x) \geq \alpha\}.$$

It is obvious that α-cuts are classical (crisp) sets, which for any given fuzzy set $A$ form a nested family of sets in the sense that

$$^{\alpha}A \subseteq {}^{\beta}A, \text{ when } \alpha > \beta.$$

Each standard fuzzy set is uniquely represented by the family of its α-cuts [20]. Any property that is generalized from classical set theory into the domain of fuzzy set theory by requiring that it holds in all α-cuts in the classical sense is called a *cutworthy property*.

In addition to standard fuzzy sets, various nonstandard fuzzy sets have been introduced in the literature. Among the most important and useful in some applications are the following:

- *Interval-valued fuzzy sets*, in which $A(x)$ is a closed interval of real numbers in $[0, 1]$ for each $x \in X$. These sets may also be formulated as pairs $A = (\underline{A}, \overline{A})$ of standard fuzzy sets $\underline{A}, \overline{A}$ such that $\underline{A}(x) \leq \overline{A}(x)$ for all $x \in X$.
- *Fuzzy sets of type 2*, in which $A(x)$ is a fuzzy interval defined on $[0, 1]$ for each $x \in X$ [21]. More general are fuzzy sets of type $k$, in which $A(x)$ is a fuzzy interval of type $k - 1$ $(k \geq 3)$.
- *Fuzzy sets of level 2*, whose domain is a family of fuzzy sets defined on X. More general are fuzzy sets of level $k$, whose domain is a family of fuzzy sets of level $k - 1$ $(k \geq 3)$.
- *Intuitionistic fuzzy sets*, which are defined as pairs $A = (AM, AN)$ of standard fuzzy sets on $X$ such that $0 \leq AM(x) + AN(x) \leq 1$ for all $x \in X$. The values $AM(x)$ and $AN(x)$ are interpreted, respectively, as the degree of membership and the degree of nonmembership of $x$ in $A$.
- *Rough fuzzy sets*, which are defined as rough approximations, $A_R = (\underline{A_R}, \overline{A_R})$, of fuzzy sets $A$ in terms of equivalence classes on $X$ induced by an equivalence relation $R$. For each $\alpha \in [0, 1]$, the α-cuts of $\underline{A_R}$ and $\overline{A_R}$ are defined by the formulas

$$^{\alpha}\underline{A}_R = \cup\left\{[x]_R \mid [x]_R \subseteq {}^{\alpha}A, x \in X\right\}$$

$$^{\alpha}\overline{A}_R = \cup\left\{[x]_R \mid [x]_R \cap {}^{\alpha}A \neq \varnothing, x \in X\right\}$$

where $[x]_R$ denotes the equivalence class that contains $x$. This combination of fuzzy sets with rough sets must be distinguished from another combination, in which a fuzzy equivalence relation is employed in the definition of a rough set. It is appropriate to refer to the sets that are based on the latter combination as *fuzzy rough sets* [22].

- *L-fuzzy sets*, in which $A(x) \in L$, where $L$ denotes a recognized set of membership grades. It is required that $L$ be at least a partially ordered set, but it is usually assumed that it is a complete lattice. This important type of fuzzy sets was introduced very early in the history of fuzzy set theory by Goguen [23].

Observe that the introduced types of fuzzy sets are interrelated in numerous ways. For example, fuzzy set of any type that employs the unit interval [0, 1] can be generalized by replacing [0, 1] with a complete lattice $L$; some of the types (e.g., standard, interval-valued, or type 2 fuzzy sets) can be viewed as special cases of $L$-fuzzy sets; or rough sets can be viewed as special interval-valued sets.

The overall fuzzy set theory is thus a broad class of formalized languages based upon an appreciable inventory of interrelated types of fuzzy sets. The overall fuzzy set theory is now fairly well developed. Many researchers have contributed to it, but perhaps the most important role in its development, not only in its founding, was played by Zadeh. Fortunately, this role is now well documented by two volumes of his selected papers published in the period 1965-95 [24, 25].

### 3. 3. Theories of Uncertainty

The emergence of fuzzy set theory and the theory of monotone measures considerably expanded the framework for formalizing uncertainty. This expansion is two-dimensional. In one dimension, the formalized language of classical set theory is expanded to the more expressive language of fuzzy set theory, where further distinctions are based on special types of fuzzy sets. In the other dimension, the classical (additive) measure theory is expanded to the less restrictive theory of monotone measures, within which various branches can be distinguished by monotone measures with different special properties.

The two-dimensional expansion of possible uncertainty theories is illustrated by the matrix in Fig. 2, where the rows represent various types of monotone measures while the columns represent various types of formalized languages. Under the entry of nonadditive measures in Fig. 2, only a few representative types are listed. Some of them are presented as pairs of dual measures employed jointly in some uncertainty theories. (Sec. 4). Under formalized languages are listed in Fig. 2 not only theories of classical sets and standard fuzzy sets, but also theories based on some of the nonstandard fuzzy sets. An uncertainty theory of a particular type is formed by choosing a formalized language of a particular type and expressing relevant uncertainty (predictive, prescriptive, diagnostic, etc.) involved in situations described in this language in terms of a measure (or a pair of measures) of a certain type. This means that each entry in the matrix in Fig. 2 represents an uncertainty theory of a particular type.

## 4. Nonclassical Uncertainty Theories

The numbered entries in Fig. 2 indicate uncertainty theories that are currently fairly well developed. They include the two classical uncertainty theories outlined in Sec. 2 (identified by numbers 1 and 3 in the bold-line box), as well as the well-known area of *probability measures of fuzzy events* [26] (identified by number 2). An overview of the other, nonclassical uncertainty theories is presented in this section. It is facilitated by Fig. 3, which shows how the theories are ordered by their levels of generality.

### *4. 1. Theory Based on Graded Possibilities*

Classical possibility theory introduced in Sec. 2. 1 can be generalized by allowing graded possibilities [27]. In the generalized form, *possibility distribution functions r* have the form: $r : X \rightarrow [0,1]$ and are required to satisfy the condition $\sup_{x \in X} r(x) = 1$, which is called a *possibilistic normalization*. Any given possibility distribution function r characterizes a unique *possibility measure, Pos*, via the formula:

$$Pos(A) = \sup_{x \in A} r(x)$$

for all nonempty sets $A \in \mathscr{P}(X)$. The associated necessity measure, *Nec*, is then defined for each $A \in \mathscr{P}(X)$ by the equation

$$Nec_E(A) = 1 - Pos_E(\overline{A}),$$

where $\overline{A}$ denotes the complement of $A$. In a more general formulation [28] which is not followed here, *Pos* and *Nec* may be viewed as functions from an ample field on $X$ (a family of subsets of $X$ that is closed under arbitrary unions and intersections, and under complementation in $X$) to a given complete lattice.

When $r$ is defined on $X \times Y$, the associated marginal distribution functions, $r_X$ and $r_Y$, are determined by the formulas

$$r_X(x) = \sup_{y \in Y} r(x,y), \text{ for all } x \in X \text{ and}$$

$$r_Y(y) = \sup_{x \in X} r(x,y), \text{ for all } y \in Y.$$

| UNCERTAINTY THEORIES | | FORMALIZED LANGUAGES | | | | | |
| | | CLASSICAL SETS | NONCLASSICAL SETS | | | | |
| | | | STANDARD FUZZY SETS | NONSTANDARD FUZZY SETS | | | |
| | | | | INTERVAL VALUED | TYPE 2 | LEVEL 3 | ••• |
| M O N O T O N E | A D D I T I V E | CLASSICAL NUMERICAL PROBABILITY | 1 | 2 | | | |
| | N O N A D D I T I V E | POSSIBILITY/ NECESSITY | 3 | 4 | | | |
| M E A S U R E S | | BELIEF/ PLAUSIBILITY | 5 | 6 | | | |
| | | SUGENO λ-MEASURES | 7 | 8 | | | |
| | | INTERVAL-VALUED PROBABILITY DISTRIBUTIONS | 9 | 10 | | | |
| | | CAPACITIES OF VARIOUS ORDERS | 11 | | | | |
| | | ⋮ | | | | | |

Figure 2. Framework for uncertainty theories.

They are *noninteractive* when

$$r(x, y) = \min[r_X(x), r_Y(y)]$$

for all $x \in X$ and $y \in Y$.

It was first recognized by Zadeh [29] that possibility theory is a natural tool for representing and manipulating evidence expressed in terms of fuzzy sets. In this interpretation of possibility theory (entry 4 in Fig. 2), the classical (crisp) possibility and necessity measures (Sec. 2. 1 and entry 3 in Fig. 2) are extended to their fuzzy counterparts via the α-cut representation [20]. When evidence is expressed by a standard fuzzy set $E$.

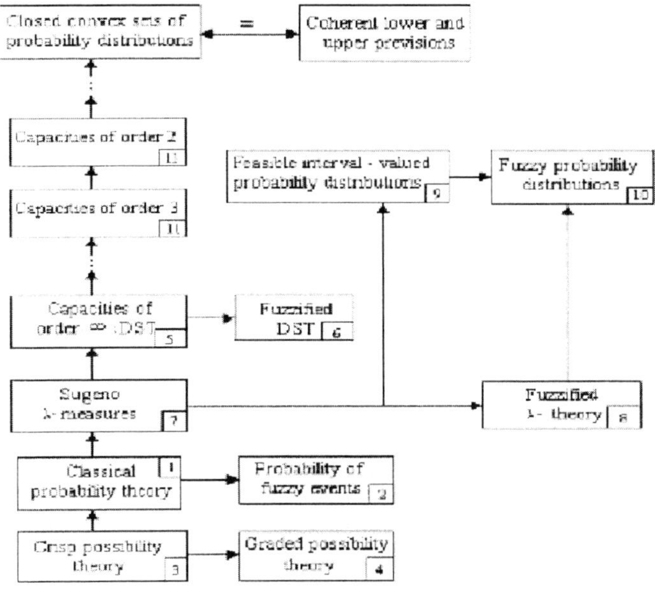

Figure 3. Ordering of theories of imprecise probabilities by levels of their generality.

of possible alternatives, the corresponding possibility distribution function, $r_E$, is defined (as explained in [30]) by the formula

$$r_E(x) = E(x) + 1 - h_E, \tag{15}$$

where

$$h_E = \sup_{x \in X} E(x).$$

### 4. 2. Theories of Imprecise Probabilities

The theory of monotone measures has been instrumental in formalizing the notion of imprecise probabilities, whose importance has been increasingly recognized during the second half of the 20th century. The following are some of the many convincing arguments for imprecise probabilities:

- Imprecision of probabilities is needed to reflect the *amount of information* on which they are based. The imprecision should decrease with the amount of statistical information.
- *Total ignorance* can be properly modeled by vacuous probabilities, which are maximally imprecise, but not by any precise probabilities.
- Imprecise probabilities are *easier to assess and elicit* than precise ones.
- We may be unable to assess probabilities precisely in practice, even if that is possible in principle, because we *lack the time or computational ability*.

- A precise probability model that is defined on some class of events determines only imprecise probabilities for *events outside the class.*
- When *several sources of information* (sensors, individuals of a group in a group decision) are combined, the extent to which they are inconsistent can be expressed by the imprecision of the combined model.

Several theories of imprecise probabilities are now fairly well developed. Among them, two theories, one developed by Walley [31-33] and the other one pursued by Kyburg [34], are currently the most general theories of imprecise probabilities. The former theory is formalized in terms of *lower and upper previsions,* the latter one is based on *closed convex sets of probability distributions.* Since there is a one-to-one correspondence between coherent lower previsions and nonempty closed convex sets of probability distributions, as established by Walley, the two theories are equally general. Some arguments have already been made that these theories should be further generalized. Thus, for example, Kyburg and Pittarelli [35] argue for using sets of probability distributions that are not necessarily convex, and Walley [33] argues for two generalizations of his theory. These various prospective generalizations of existing theories are not well developed as yet and, therefore, they are not covered in this chapter.

All theories of imprecise probabilities that are based on classical set theory share some common characteristics. One of them is that evidence within each theory is fully described by a *lower probability function,* $\underline{g}$, or, alternatively, by an *upper probability function,* $\overline{g}$. These functions are always regular monotone measures that are superadditive and subadditive, respectively, and

$$\sum_{x\in X}\underline{g}(\{x\})\le 1, \quad \sum_{x\in X}\overline{g}(\{x\})\ge 1. \tag{16}$$

In the various special theories of uncertainty, they possess additional special properties.

When evidence is expressed (at the most general level) in terms of an arbitrary closed and convex set $\mathcal{D}$ of probability distribution functions $p$ on a finite set $X$, functions $\underline{g}_{\mathcal{D}}$ and $\overline{g}_{\mathcal{D}}$ associated with $\mathcal{D}$ are determined for each $A \in \mathcal{P}(X)$ by the formulas:

$$\underline{g}_{\mathcal{D}}(A) = \inf_{p\in\mathcal{D}} \sum_{x\in A} p(x), \quad \overline{g}_{\mathcal{D}}(A) = \sup_{p\in\mathcal{D}} \sum_{x\in A} p(x).$$

Since

$$\sum_{x\in A} p(x) + \sum_{x\notin A} p(x) = 1,$$

for each $p \in \mathcal{D}$ and each $A \in \mathcal{P}(X)$, it follows that

$$\overline{g}_{\mathcal{D}}(A) = 1 - \underline{g}_{\mathcal{D}}(\overline{A}). \tag{17}$$

Due to this property, functions $\underline{g}_{\mathcal{D}}$ and $\overline{g}_{\mathcal{D}}$ are called *dual* (or *conjugate*). One of them is sufficient for capturing given evidence; the other one is uniquely determined by (17).

It is common to use the lower probability function $\underline{g}_\mathcal{D}$ to capture the evidence.

As is well known [36, 37] any given lower probability function $\underline{g}_\mathcal{D}$ is uniquely represented by a set-valued function $m_\mathcal{D}$ for which $m_\mathcal{D}(\varnothing) = 0$ and

$$\sum_{A \in \mathcal{P}(X)} m(A) = 1. \tag{18}$$

Any set $A \in \mathcal{P}(X)$ for which $m_\mathcal{D}(A) \neq 0$ is often called a *focal element*, and the set of all focal elements with the values assigned to them by function $m_\mathcal{D}$ is called a *body of evidence*. Function $m_\mathcal{D}$ is called a *Möbius representation* of $\underline{g}_\mathcal{D}$ when it is obtained for all $A \in \mathcal{P}(X)$ via the *Möbius transform*

$$m_\mathcal{D}(A) = \sum_{B \mid B \subseteq A} (-1)^{|A-B|} \underline{g}_\mathcal{D}(B). \tag{19}$$

The inverse transform is defined for all $A \in \mathcal{P}(X)$ by the formula

$$\underline{g}_\mathcal{D}(A) = \sum_{B \mid B \subseteq A} m_\mathcal{D}(B). \tag{20}$$

It follows directly from (17) that

$$\overline{g}_\mathcal{D}(A) = \sum_{B \mid B \not\subseteq \overline{A}} m_\mathcal{D}(B). \tag{21}$$

for all $A \in \mathcal{P}(X)$, and it is easy to check that properties (20) and (21) are consistent with property (18).

Assume now that evidence is expressed in terms of a given lower probability function $\underline{g}$. Then, the set of probability distribution functions that are consistent with $\underline{g}$, $\mathcal{D}(\underline{g})$, which is always closed and convex, is defined as follows:

$$\mathcal{D}(\underline{g}) = \{p(x) \mid x \in X, p(x) \in [0,1], \sum_{x \in X} p(x) = 1, \ \underline{g}(A) \leq \sum_{x \in A} p(x), \text{ for all } A \in \mathcal{P}(X)\}. \tag{22}$$

That is, each given function $\underline{g}$ is associated with a unique set $\mathcal{D}$ and vice-versa. Functions $\underline{g}$ with some special properties, which are the subject of special theories of imprecise probabilities, are surveyed in Secs. 4.3 - 4.6.

### 4. 3. Sugeno λ-measures

Sugeno λ-measures, $g_\lambda$, introduced by Sugeno [18], are special regular monotone measures that satisfy the requirement

$$g_\lambda(A \cup B) = g_\lambda(A) + g_\lambda(B) + \lambda\, g_\lambda(A)\cdot g_\lambda(B) \tag{23}$$

for any given pair of disjoint sets $A$, $B \in \mathcal{P}(X)$, where $\lambda \in (-1, \infty)$ is a parameter by which individual measures in this class are distinguished. It is well known [19] that each λ-measure is uniquely determined by values $g_\lambda(\{x\}) \in [0, 1]$ for all $x \in X$, subject to the condition that at least two of these values are not zero. Given values $g_\lambda(\{x\})$ that conform to this condition, the parameter $\lambda$ is uniquely determined via the equation

$$1 + \lambda = \prod_{x \in X}[1 + \lambda \cdot g_\lambda(\{x\})]. \tag{24}$$

More specifically [18, 19] the following three cases must be distinguished:

1. If $\sum_{x \in X} g_\lambda(\{x\}) < 1$, $g_\lambda$ is a lower probability and, hence, a superadditive measure; $\lambda$ is determined by the root of (24) in the interval $(0, \infty)$, which is unique [19].
2. If $\sum_{x \in X} g_\lambda(\{x\}) = 1$, $g_\lambda$ is a probability measure; $\lambda = 0$, which is the only root of (24).
3. If $\sum_{x \in X} g_\lambda(\{x\}) > 1$, $g_\lambda$ is an upper probability and, hence, a subadditive measure; $\lambda$ is determined by the root of (24) in the interval $(-1, 0)$, that is unique [19].

Given values $g_\lambda(\{x\})$ for all $x \in X$ and the associated value $\lambda$, values $g_\lambda(A)$ are then determined for all $A \in \mathcal{P}(X)$ via (23).

Let $\underline{\lambda} \in (0, \infty)$ and let $\overline{\lambda} = -\underline{\lambda}/(\underline{\lambda}+1)$. Then, $g_{\underline{\lambda}}$ is a lower probability and $g_{\overline{\lambda}}$ is it's dual upper probability [18]. Moreover, the Möbius representation $m$ of $g_{\underline{\lambda}}$ is obtained for all $A \in \mathcal{P}(X)$ by the formula

$$m(A) = \begin{cases} \underline{\lambda}^{|A|-1} \prod_{x \in A} g_{\underline{\lambda}}(\{x\}) & \text{when } A \neq \varnothing \\ 0 & \text{when } A = \varnothing \end{cases},$$

as is shown in [19].

### 4. 4. Feasible Interval-Valued Probability Distributions

In this theory, lower and upper probabilities $g(A)$ and $\overline{g}(A)$ are determined for all sets $A \subset \mathcal{P}(X)$ by intervals $[g(\{x\}), \overline{g}(\{x\})]$ of probabilities on singletons ($x \in X$). Clearly, $g(\{x\}) \in [0,1]$ and $\overline{g}(\{x\}) \in [0,1]$, and inequalities (16) must be satisfied. Each given set of probability intervals $G = \{[g(\{x\}), \overline{g}(\{x\})] \mid x \in X\}$ is associated with a closed convex set, $\mathcal{D}(G)$, of probability distribution functions, $p$, defined as follows:

$$\mathcal{D}(G) = \left\{ p(x) \mid x \in X,\, p(x) \in \left[\underline{g}(\{x\}),\, \overline{g}(\{x\})\right],\, \sum_{x \in X} p(x) = 1 \right\}$$

Sets defined in this way are clearly special cases of sets defined by (22). Their special feature is that they always form an $(n-1)$-dimensional polyhedron, where $n = |X|$. In general, the polyhedron may have $c$ vertices (corners), where $n \le c \le n(n-1)$, and each probability distribution function contained in the set can be expressed as a linear combination of these vertices [38].

A given set $G$ of probability intervals may be such that some combinations of values taken from the intervals do not correspond to any probability distribution function. This indicates that the intervals are unnecessarily broad. To avoid this deficiency, the following concept was introduced in the theory [38]. A given set $G$ is called *feasible* if and only if for each $x \in X$ and every value $v(x) \in [\underline{g}(\{x\}), \overline{g}(\{x\})]$ there exists a probability distribution function $p$ for which $p(x) = v(x)$. The feasibility of any given set $G$ can be easily checked: the set is feasible if and only if it passes the following tests:

$$\sum_{x \in X} \underline{g}(\{x\}) + [\overline{g}(\{y\}) - \underline{g}(\{y\})] \le 1 \text{ for all } y \in X;$$

$$\sum_{x \in X} \overline{g}(\{x\}) - [\overline{g}(\{y\}) - \underline{g}(\{y\})] \ge 1 \text{ for all } y \in X.$$

If $G$ is not feasible, it can be easily converted to the set $G' = \{[\underline{g}(\{x\}), \overline{g}'(\{x\})] \mid x \in X\}$ of feasible intervals by the formulas

$$\underline{g}'(\{x\}) = \max\left\{ \underline{g}(\{x\}),\ 1 - \sum_{y \ne x} \overline{g}(\{y\}) \right\},$$

$$\overline{g}'(\{x\}) = \min\left\{ \overline{g}(\{x\}), 1 - \sum_{y \ne x} \underline{g}(\{y\}) \right\}$$

for all $x \in X$.

Given a feasible set $G$ of probability intervals, the lower and upper probabilities are determined for each $A \in \mathcal{P}(X)$ by the formulas

$$\underline{g}(A) = \max\left\{ \sum_{x \in A} \underline{g}(\{x\}), 1 - \sum_{x \notin A} \overline{g}(\{x\}), \right.$$

$$\overline{g}(A) = \min\left\{ \sum_{x \in A} \overline{g}(\{x\}), 1 - \sum_{x \notin A} \underline{g}(\{x\}). \right.$$

This special theory of imprecise probabilities is covered well in [38]. Its applicability to generalizing Bayesian methodology is the subject of [39].

### *4. 5. Choquet Capacities of Various Orders*

A well-defined category of theories of imprecise probabilities is based on Choquet capacities of various orders [8]. All these theories are generalizations of the theory based on Sugeno $\lambda$-measures [18, 19], but they are not comparable (as shown in [40]) with the theory based on feasible interval-valued probability distributions (Fig. 3). The most general theory in this category is the theory based on *capacities of order 2* [8]. Here, the lower and upper probabilities, $\underline{g}$ and $\overline{g}$, are monotone measures for which

$$\underline{g}(A \cup B) \geq \underline{g}(A) + \underline{g}(B) - \underline{g}(A \cap B),$$
$$\overline{g}(A \cap B) \leq \overline{g}(A) + \overline{g}(B) - \overline{g}(A \cup B),$$

for all $A, B \in \mathcal{P}(X)$. Less general uncertainty theories are then based on *capacities of order k*. For each $k > 2$, the lower and upper probabilities, $\underline{g}$ and $\overline{g}$, satisfy the inequalities

$$\underline{g}(\bigcup_{j=1}^{k} A_j) \geq \sum_{\substack{K \subseteq N_k \\ K \neq \varnothing}} (-1)^{|K|+1} \underline{g}(\bigcap_{j \in K} A_j),$$

$$\overline{g}(\bigcap_{j=1}^{k} A_j) \leq \sum_{\substack{K \subseteq N_k \\ K \neq \varnothing}} (-1)^{|K|+1} \overline{g}(\bigcup_{j \in K} A_j)$$

for all families of $k$ subsets of $X$, where $N_k = \{1, 2, \ldots, k\}$.

Clearly, if $k' > k$. then the theory based on capacities of order $k'$ is less general than the one based on capacities of order $k$. The least general of all these theories is the one in which the inequalities are required to hold for all $k \geq 2$ (the underlying capacity is said to be of order $\infty$). This theory, which was extensively developed by Shafer [41], is usually referred to as *evidence theory* or *Dempster-Shafer theory*. In this theory, lower and upper probabilities are called *belief* and *plausibility* measures. An important feature of this theory is that the Möbius representation of evidence in this theory (usually called a *basic probability assignment function*) is a positive function ($m(A) \geq 0$ for all $A \in \mathcal{P}(X)$), and, hence, it can be viewed as a probability distribution on the power set. This feature makes it possible to develop the theory via "non-traditional and sophisticated application of probability theory" as shown convincingly by Kramosil [42]. Another approach to evidence theory has been pursued by Smets [43].

### *4. 6. Other Types of Imprecise Probabilities*

Theories of imprecise probabilities based on the special monotone measures surveyed in Secs. 4.3 - 4.5 are currently the most visible and the best developed ones. Other types of monotone measures have been introduced in the literature, but the theories of imprecise probabilities based upon them have not been adequately developed as yet. Nevertheless, two broad classes of monotone measures seem worthy of being at least mentioned in this chapter: decomposable measures and k-additive measures.

For any pair of disjoint subsets $A$, $B$ of $X$, a monotone measure, $g_\otimes$, that is *decomposable* with respect to a $t$-conorm $\otimes$ [20] is required to satisfy for every pair $A$, $B \in \mathcal{P}(X)$ the following requirement:

$$g_\otimes (A \cup B) = g_\otimes(A) \otimes g_\otimes(B).$$

The class of Sugeno $\lambda$-measures is a particular class of decomposable monotone measures, in which t-conorms $\otimes_\lambda$ have the form

$$a \otimes_\lambda b = \min[a + b + \lambda ab, 1].$$

For classical probabilities, the $t$-conorm is the arithmetic sum. It seems that decomposable measures were introduced by Dubois and Prade in [44] even though they called them conorm-based measures. In [44], they also introduced dual measures based on t-norms.

The class of *k-additive measures*, which was introduced and investigated by Grabisch [37], is defined as follows: A monotone measure is said to be *k*-additive ($k \geq 1$) if its Möbius representation satisfies $m(A) = 0$ for any $A$ such that $|A| > k$, and there exists at least one subset $B$ of $X$ such that $|B| = k$ and $m(B) \neq 0$. The advantage of k-additive measures, especially for small values of $k$, is that they are characterized via the Möbius representation by a limited number of parameters (focal elements). They are thus suitable for approximating other monotone measures, as investigated by Grabisch [37]. Clearly, the 1-additive measures are classical probability measures.

### 4. 7. Fuzzification of Imprecise Probabilities

Efforts to fuzzify the various theories of imprecise probabilities have been rather limited so far, even though some results have already been obtained for monotone measures defined on fuzzy sets (see, e. g., Ref. [45], which is also reprinted in [19]).

Perhaps the most definitive results have been obtained in fuzzifying the theory based on feasible interval-valued probabilities, which is outlined in Sec. 4. 4. In particular, Bayesian methodology developed for interval-valued probability distributions in [39] was generalized to fuzzy probability distributions in [46]. This generalization is based on extending relevant intervals of real numbers to fuzzy intervals via the $\alpha$-cut representation and on using constrained fuzzy arithmetic [47] to perform required computations.

Numerous approaches to fuzzifying the Dempster-Shafer theory have been proposed in the literature (see, e. g., Ref. [48], which is also reprinted in [19]), but none of them has been adequately developed as yet. Neither any comparative study regarding these various approaches has been conducted so far. Fuzzification of special classes of monotone measures and, in particular, the various theories of imprecise probabilities is thus a little developed area at this time. Moreover, the few above mentioned developments are all restricted to standard fuzzy sets. Fuzzification involving nonstandard fuzzy sets remains in this context by and large unexplored.

## 5. Measures of Uncertainty and Uncertainty-Based Information

### 5. 1. Mathematical Requirements

In each uncertainty theory, uncertainty is represented by a function that assigns to each set of relevant alternatives (predictions, prescriptions, etc.) a number in the unit interval [0, 1], which expresses the *degree of evidence* (likelihood, belief, plausibility, etc.) that the true alternative is in the set. Let this function be called an *uncertainty function*. Examples of uncertainty functions are probability measures, possibility measures, belief measures, or more generally, lower or upper probabilities. Uncertainty functions of each uncertainty theory satisfy certain special requirements, which distinguish them from uncertainty functions of other uncertainty theories.

A *measure of uncertainty* of some conceived type in a given uncertainty theory is a *functional* that assigns to each uncertainty function in the theory a nonnegative real number. This number is supposed to measure, in an intuitively meaningful way, the amount of uncertainty of the considered type that is embedded in the uncertainty function. Examples of measures of uncertainty are the Hartley measure (Sec. 2. 1) and the Shannon entropy (Sec. 2. 2). Uncertainty functions that are directly involved in these measures are, respectively, possibility distribution functions and probability distribution functions. To be acceptable as a measure of the amount of uncertainty of a given type in a particular uncertainty theory, a proposed functional must satisfy several intuitively essential axiomatic requirements. Specific mathematical formulation of each of the requirements depends on the uncertainty theory involved. However, the requirements can be described informally in a generic form, independent of the various uncertainty calculi.

The following axiomatic requirements, each expressed in a generic form, must be satisfied whenever applicable:

1. *Subadditivity* — the amount of uncertainty in a joint representation of evidence (defined on a Cartesian product) cannot be greater than the sum of the amounts of uncertainty in the associated marginal representations of evidence.
2. *Additivity* — the two amounts of uncertainty considered under subadditivity become equal if and only if the marginal representations of evidence are noninteractive according to the rules of the uncertainty calculus involved.
3. *Range* — the range of uncertainty is [0, $M$], where 0 must be assigned to the unique uncertainty function that describes full certainty and $M$ depends on the cardinality of the universal set involved and on the chosen unit of measurement.
4. *Continuity* — any measure of uncertainty must be a continuous functional.
5. *Expansibility* — expanding the universal set by alternatives that are not supported by evidence must not affect the amount of uncertainty.
6. *Branching/Consistency* — when uncertainty can be computed in more ways, all intuitively acceptable, the results must be the same (consistent).
7. *Monotonocity* — when evidence can be ordered in the uncertainty theory employed (as in possibility theory), the relevant uncertainty measure must preserve this ordering.
8. *Coordinate invariance* — when evidence is described within the $n$-dimensional Euclidean space ($n \geq 1$), the relevant uncertainty measure must not change under isometric transformations of coordinates.

When distinct types of uncertainty coexist in a given uncertainty theory, it is not necessary that these requirements be satisfied by each uncertainty type. However, they must be satisfied by an overall uncertainty measure, which appropriately aggregates measures of the individual uncertainty types.

The strongest justification of a functional as a meaningful measure of the amount of uncertainty of a considered type in a given uncertainty theory is obtained when we can prove that it is the only functional that satisfies the relevant requirements and measures the amount of uncertainty in some specific measurement units. A suitable measurement unit is uniquely defined by specifying what the amount of uncertainty should be for a particular (and usually very simple) uncertainty function.

### 5. 2. Generalized Hartley Measures

Nonspecificity, which in classical possibility theory is quantified by the Hartley measure (Sec. 2.1) is a fundamental type of uncertainty. It exists in all uncertainty theories except classical probability theory, as explained later in this section. In each nonclassical theory of uncertainty, the Hartley measure must be appropriately generalized.

A natural generalization of the Hartley measure of nonspecificity to the fuzzy-set interpretation of possibility theory (introduced in Sec. 4.1) was developed by Higashi and Klir [49] under the name *U-uncertainty*. For any possibility distribution function $r_E$ based on evidence expressed in terms of a *normal* fuzzy set $E$ of possible alternatives via Eq. (15), the $U$-uncertainty, $U$, is a functional defined by the formula

$$U(r_E) = \int_0^1 \log_2 \left| {}^\alpha E \right| d\alpha \tag{25}$$

The uniqueness of this functional for measuring nonspecificity in this case was proven on axiomatic grounds by Klir and Mariano [50].

To cover normal as well as subnormal fuzzy sets, Eq. (25) must be replaced with the equation

$$U(r_E) = \int_0^{h_E} \log_2 \left| {}^\alpha E \right| d\alpha + (1 - h_E) \log_2 |X|, \tag{26}$$

where $X$ denotes the universal set of concern. This follows from the general fuzzy-set interpretation of possibility theory derived in [30] and mentioned in Sec. 4. 1. When $E$ is a fuzzy subset of the $n$-dimensional Euclidean space for some $n \geq 1$, the counterpart of (26) is

$$U(r_E) = \int_0^{h_E} HL({}^\alpha E) d\alpha + (1 - h_E) HL(X), \tag{27}$$

where HL denotes the Hartley-like measure defined by (8).

Shortly after the $U$-uncertainty was discovered, Dubois and Prade [51] showed how to further generalize it to measure nonspecificity in evidence theory. The proposed generalized Hartley measure is a functional, $GH$, defined by the formula

$$GH(m) = \sum_{A \in \mathcal{P}(X)} m(A) \log_2 |A|. \tag{28}$$

This functional makes good intuitive sense: it is a weighted average of the Hartley measure of focal elements. Its uniqueness was proven by Ramer [52].

In a recent paper, Abellan and Moral [53] generalized the measure of nonspecificity in evidence theory, expressed by (28), to closed convex sets of probability distributions and, hence, to the various special theories of imprecise probabilities introduced in Secs. 4.3 - 4.6 as well. This generalized functional, $GH$, is defined by the formula

$$GH(m_{\mathcal{D}}) = \sum_{A \in \mathcal{P}(X)} m_{\mathcal{D}}(A) \log_2 |A|, \tag{29}$$

where $m_{\mathcal{D}}$ is the Möbius representation of the lower probability associated with a given closed and convex set $\mathcal{D}$ of probability distributions (Sec. 4.2). Abellan and Moral [53] showed that the functional $GH$ defined by (29) possesses all the essential mathematical properties required for measures of uncertainty. In particular, they proved that the measure has the following properties:

It has the *proper range* $[0, \log_2 |X|]$ when measured in bits; 0 is obtained when $\mathcal{D}$ contains a single probability distribution; $\log_2 |X|$ is obtained when $\mathcal{D}$ contains all probability distributions on $X$ and thus represents total ignorance.

It is *subadditive*: $N(\mathcal{D}) \le N(\mathcal{D}_X) + N(\mathcal{D}_Y)$, where

$$\mathcal{D}_X = \{p_X | p_X(x) = \sum_{y \in Y} p(x, y) \text{ for some } p \in \mathcal{D}\},$$

$$\mathcal{D}_Y = \{p_Y | p_Y(y) = \sum_{x \in X} p(x, y) \text{ for some } p \in \mathcal{D}\}.$$

It is *additive*: $N(\mathcal{D}) = N(\mathcal{D}_X) + N(\mathcal{D}_Y)$ if and only if $\mathcal{D}_X$ and $\mathcal{D}_Y$ are not interactive, which means that for all $A \in \mathcal{P}(X)$ and all $B \in \mathcal{P}(Y)$,

$$m_{\mathcal{D}}(A \times B) = m_{\mathcal{D}_X}(A) \cdot m_{\mathcal{D}_Y}(B)$$

and

$$m_{\mathcal{D}}(R) = 0 \text{ for all } R \ne A \times B.$$

It is *monotonic*: if $\mathcal{D}$ and $\mathcal{D}'$ are closed convex sets of probability distributions on $X$ such that $\mathcal{D} \subseteq \mathcal{D}'$, then $N(\mathcal{D}) \le N(\mathcal{D}')$.

The nonspecificity measure defined by (29) is thus mathematically sound. It is applicable to all the theories of imprecise probabilities that are subsumed under the theory based on closed convex sets of probability distributions (Sec. 4.2).

### 5. 3. Generalized Shannon Measures

In order to get insight into the meaning of the uncertainty that is measured by the Shannon entropy, Eq. (9) may be rewritten as

$$S(p(x) | x \in X) = -\sum_{x \in X} p(x) \log_2 [1 - \sum_{y \ne x} p(y)]. \tag{30}$$

The term $\sum_{y \neq x} p(y)$ in this equation expresses the total evidential claim pertaining to all alternatives that are different from alternative $x$. This evidential claim fully conflicts with the evidential claim $p(x)$. When using

$$-\log_2[1 - \sum_{y \neq x} p(y)]$$

instead of the term alone, as in Eq. (30), the expression of conflict between $p(x)$ and $\sum_{y \neq x} p(y)$ is rescaled from $[0, 1]$ to $[0, \infty]$. It follows from these observations that the Shannon entropy measures the mean (expected) value of the conflict among evidential claims expressed by a probability distribution function on a finite set of mutually exclusive alternatives. The type of uncertainty whose amount is measured by the Shannon entropy is thus *conflict*.

Many studious efforts were made, primarily in the 1980s, by numerous researchers to determine a generalized counterpart of the Shannon entropy in evidence theory. Although many intuitively promising functionals have been proposed for this purpose, each of them was found upon closer scrutiny to violate some of the essential requirements (Sec. 5.1). In most cases, it was the requirement of subadditivity that was violated. A historical summary of these unsuccessful efforts is covered in [10].

### 5. 4. Aggregate Measures

The long, unsuccessful, and often frustrating search for the Shannon-like measure of uncertainty in evidence theory was replaced in the early 1990s with the search for a justifiable aggregate measure, capturing both nonspecificity and conflict. The first attempts were to add the well-justified measure of nonspecificity with one of the proposed candidates for measuring conflict.Unfortunately, all the resulting functionals were again found to violate some of the essential requirements, predominantly subadditivity.

A measure of total uncertainty in evidence theory that possesses all the required mathematical properties was eventually found (independently by several authors in 1993-94), but not as a composite of measures of uncertainty of the two types [10]. This aggregate uncertainty measure, $AU$, is defined for each function Bel defined on $\mathcal{P}(X)$ by the formula

$$AU(Bel) = \max_{P_{Bel}} \left[ -\sum_{x \in X} p(x) \log_2 p(x) \right], \tag{31}$$

where the maximum is taken over the set $P_{Bel}$ of all probability distributions $p$ on $X$ that are consistent with the given belief measure $Bel$, which means that they satisfy the constraints

$$Bel(A) \leq \sum_{x \in A} p(x) \quad \text{for all } A \in \mathcal{P}(X),$$

in addition to the usual axiomatic constraints of probability theory.

Since the aggregate measure $AU$ is defined in terms of the solution to a nonlinear optimization problem, its practical utility was initially questioned. Fortunately, a relatively simple and fully general algorithm for computing the measure was developed and its correctness proven [54].

Observe that $AU$ defined by (31) can be readily generalized to the theory based on closed convex sets of probability distributions. The generalized formula is

$$AU(\mathcal{D}) = \max_{p \in \mathcal{D}} \left[ \sum_{x \in X} - p(x) \log_2 p(x) \right], \tag{32}$$

where $\mathcal{D}$ denotes a given closed convex set of probability distributions on $X$.

Although the functional $AU$ is acceptable on mathematical grounds as an aggregate measure of uncertainty in the theory based on closed convex sets of probability distributions (and thus in the various special theories), it has a severe shortcoming: it is highly insensitive to changes in evidence.

To illustrate this undesirable feature, let us examine a very simple example. Let $X = \{x_1, x_2\}$, $m(\{x_1\}) = a$, $m(\{x_2\}) = b$, and $m(X) = 1 - a - b$, where $a + b \leq 1$. Then, $AU(m) = 1$ for all $a \in [0, .5]$ and $b \in [0, .5]$.

Moreover, when $a > .5$, $AU$ is independent of $b$ and, similarly, when $b > .5$, $AU$ is independent of $a$.

On the basis of this critical appraisal of the aggregate measure $AU$, Smith [55] proposed three complementary measures of total uncertainty in evidence theory. To describe them, let $\overline{S}$ and $\underline{S}$ denote, respectively, the maximum and minimum Shannon measure within all probability distributions that are consistent with the given body of evidence. Observe that in this convenient notation $\overline{S} = AU$.

The first proposed measure of total uncertainty, $TU_1$, is defined as a linear combination of $\overline{S}$ and $GH$,

$$TU_1 = \delta \overline{S} + (1 - \delta)GH \tag{33}$$

where $\delta \in (0, 1)$. This measure has two favorable features. First, it satisfies all essential axiomatic requirements (since both $\overline{S}$ and $GH$ satisfy them). Second, it overcomes the insensitivity of measure $AU (= \overline{S})$. In particular, it distinguishes total ignorance (when $m(X) = 1$) from the uniform probability distribution by values $\log_2 |X|$ and $\delta \log_2 |X|$, respectively. The choice of the value of $\delta$ remains an important open problem regarding this measure. It seems that the proper value of $\delta$ should be determined in the context of each interpretation of evidence theory. This can be easily done, for example, by specifying the desired value of $TU_1$ for the uniform probability distribution. However, it is also conceivable that it may be derived on mathematical grounds.

The second proposed measure, $TU_2$, is defined as the pair consisting of functionals $GH$ and $\overline{S} - GH$:

$$TU_2 = \langle GH, \overline{S} - GH \rangle \tag{34}$$

It is assumed here that the axiomatic requirements are defined in terms of the sum of the two functionals involved, which is always the well-justified aggregate measure $\overline{S}$ ($= AU$). In this sense, the measure satisfies trivially all the requirements. Its advantage is that measures of both types of uncertainty that coexist in evidence theory (nonspecificity and conflict) are expressed explicitly. However, an open problem is to justify that the difference $\overline{S} - GH$ is a genuine generalization of the Shannon measure in evidence theory, and not only a superficial artifact.

The third proposed measure, $TU_3$, is defined as the pair

$$TU_3 = \left\langle GH, \left[\underline{S}, \overline{S}\right]\right\rangle \tag{35}$$

where the second component expresses the whole range of values of the Shannon measure that are obtained for all probability measures that are consistent with the given body of evidence. This measure is more expressive than the other two in the sense that it captures all possible values of the Shannon measure that are associated with the given body of evidence.

However, though the measure makes good sense conceptually, it is not yet clear how to define a meaningful ordering for it, one that would satisfy the requirement of subadditivity. Moreover, no efficient algorithm for computing $\underline{S}$ has been developed as yet.

Due to the mathematically sound generalization of the measure of nonspecificity by Abellan and Moral [53] and due to the fact that the measures $\underline{S}$ and $\overline{S}$ can be defined for any closed convex set of probability distributions, these three measures of total uncertainty are applicable not only to evidence theory, but also to a considerably more general theory based on arbitrary closed convex sets of probability distributions.

This means, in turn, that they are applicable to all theories of imprecise probabilities that are subsumed under this highly general theory (as depicted in Fig. 3).

### 5. 5. *Measures of Fuzziness*

Nonspecifity and conflict are types of uncertainty that coexist in appropriate forms in all nonclassical uncertainty theories. When the theories are fuzzified, they include one additional type of uncertainty: *fuzziness*. However, this type of uncertainty is fundamentally different from the other two. While nonspecificity and conflict result from information deficiency, fuzziness results from the lack of linguistic precision.

When, for example, a given measurement of temperature belongs to the fuzzy set capturing (in a given context) the meaning of the linguistic term "high temperature" with the membership degree 0.7, this degree does not express any lack of information (the actual value of the temperature was measured as is thus known), but rather the compatibility of the known value with the imprecise (vague) linguistic term.

This means that fuzziness does not properly belong to this chapter, whose subject is the relationship between uncertainty and information. It is mentioned here only to clarify the important distinction between information-based uncertainty and linguistic uncertainty.

Various approaches to measuring fuzziness of fuzzy sets have been suggested in the literature. One of them is based on expressing fuzziness of any given fuzzy set in terms of the lack of distinction between the set and its complement. This approach is developed in detail in Ref. [56].

## 6. Principles of Uncertainty

Although the utility of relevant uncertainty measures is as broad as the utility of any relevant measuring instrument, their role is particularly significant in three fundamental principles for managing uncertainty. These principles are a principle of minimum uncertainty, a principle of maximum uncertainty, and a principle of uncertainty invariance [10].

*The principle of minimum uncertainty* is basically an arbitration principle. It facilitates the selection of meaningful alternatives from solution sets obtained by solving problems in which some amount of the initial information is inevitably lost. According to this principle, we should accept only those solutions for which the loss of the information is as small as possible. This means, in turn, that we should accept only solutions with minimum uncertainty.

Examples of problems for which the principle of minimum uncertainty is applicable are simplification problems and conflict resolution problems of various types. Consider, for example, that we want to simplify a finite-state nondeterministic system by coarsening state sets of its variables. This requires that each state set be partitioned in a meaningful way (e. g., preserving a given order of states) into a given number of subsets. This can usually be done in many different ways. The minimum uncertainty principle allows us to compare the various competing partitions by their amounts of relevant uncertainty (predictive, diagnostic, etc.) and consider only those with minimum uncertainty.

As another example, let us consider a set of systems, $S_1$, $S_2$, ..., $S_n$, that share some variables. Assume that these systems are locally inconsistent in the sense that projections from individual systems into variables they share (e.g., marginal probabilities, marginal bodies of evidence, etc.) are not the same. To resolve the local inconsistencies, we need to replace each system $S_i$ with another system, $C_i$, such that systems $C_1$, $C_2$, ..., $C_n$ be locally consistent. This, of course, can be done in many different ways, but the proper way to do that is to minimize the loss of information caused by these replacements. Denoting the relevant uncertainty measure by *UNC*, the principle of minimum uncertainty is applicable to deal with this problem by formulating the following optimization problem:

$$\text{Minimize} \sum_{i=1}^{n} [UNC(C_i) - UNC(S_i)]$$

subject to the following three types of constraints:
- axioms of the theory in which systems $S_i$ and $C_i$ ($i = 1, 2, ..., n$) are formalized;
- equations by which all conditions of local consistency among systems $C_1$, $C_2$, ..., $C_n$ are defined;
- $UNC(C_i) \geq UNC(S_i)$ for all $i = 1, 2, ..., n$, to avoid introducing bias.

The second principle, *the principle of maximum uncertainty*, is essential for any problem that involves *ampliative reasoning*. This is reasoning in which conclusions are not entailed in the given premises. Using common sense, the principle may be expressed as follows: in any ampliative inference, use all information supported by available evidence but make sure that no additional information (unsupported by given evidence) is unwittingly added. Employing the connection between information and uncertainty, this definition can be reformulated in terms of uncertainty: any conclusion resulting from ampliative inference should maximize the relevant uncertainty within constraints representing given premises. This principle guarantees that we fully recognize our ignorance when we attempt to make inferences that are beyond the information domain defined by the given premises and, at the same time, that we utilize all information contained in the premises. In other words, the principle guarantees that our inferences are maximally non-committal with respect to information that is not contained in the premises.

Let $f$ and $UNC(f)$ denote, respectively, a relevant uncertainty function (probability or possibility distribution function, basic probability assignment function, etc.) and the associated measure of uncertainty. Then, the principle of maximum uncertainty is operationally formulated in terms of the following generic optimization problem:

Determine $f$ for which $UNC(f)$ reaches its maximum under the following constraints:

- axioms upon which uncertainty functions are based;
- constraints $E_1$, $E_2$, ..., which represent partial information about $f$ (e.g., marginal distributions, lower or upper bounds, values of $f$ for some sets, etc.).

Ampliative reasoning is indispensable to science and engineering in many ways and, hence, the underlying principle of maximum uncertainty has great utility. For example, whenever we make predictions based on a given scientific model, we employ ampliative reasoning. Similarly, estimating microstates from the knowledge of relevant macrostates and partial knowledge of the microstates (as in image processing and many other problem areas) requires ampliative reasoning. The problem of the identification of an overall system from some of its subsystems is another example that involves ampliative reasoning and, hence, the principle of maximum uncertainty.

The principles of minimum and maximum uncertainty are well developed within classical information theory, where they are referred to as the *principles of minimum and maximum entropy*. The great utility of these principles, particularly in developing predictive models, is perhaps best demonstrated by the work of Christensen [57]. The principle of maximum entropy, which was presumably founded by Jaynes in the mid $20^{th}$ century [58], is now covered extensively in the literature. The book by Kapur [59] is an excellent overview of the astonishing range of applications of the principle.

Optimization problems that emerge from the minimum and maximum uncertainty principles outside classical information theory have yet to be properly investigated and tested in praxis. Each of the three measures of total uncertainty in *DST*, which are introduced in Sec. 5.4, open special challenges in this area of research. In my view, however, it is sufficient to use nonspecificity alone in most applications of the principle.

By minimizing or maximizing nonspecificity, we fully control imprecisions in probabilities in line with given evidence. Using only nonspecificity has two advantages. First, measures of nonspecificity are well justified in all theories of imprecise probabilities. Second, measures of nonspecificity are linear functions and, hence, no methods of nonlinear optimization are needed, contrary to minimum and maximum entropy principles.

The third uncertainty principle, the *principle of uncertainty invariance* (also called the *principle of information preservation*), is of relatively recent origin [60]. It was introduced to facilitate meaningful transformations between the various uncertainty theories. According to this principle, the amount of uncertainty (and the associated uncertainty-based information) should be preserved in each transformation of uncertainty from one mathematical framework to another. Examples of applications of this principle are probability-possibility transformations and approximations of imprecise probabilities formalized in a particular theory by their counterparts in a less general theory.

Thus far, significant results have been obtained for uncertainty-invariant probability-possibility transformations. It was determined by a thorough mathematical analysis [61] that these transformations do exists and are unique only under log-interval scales. They are also meaningful, but not unique, under ordinal scales [62]. In this case, additional, context-dependent requirements may be used to make the transformations unique.

## 7. Conclusions

The principal aim of generalized information theory have been threefold: (i) to liberate the notions of uncertainty and uncertainty-based information from the narrow confines of classical set theory and classical measure theory; (ii) to conceptualize a broad framework within which any conceivable type of uncertainty can be characterized; and (iii) to develop theories for the various types of uncertainty that emerge from the framework at each of the four levels: formalization, calculus, measurement, and methodology. Undoubtedly, these are long-term aims and it is questionable if they can ever be fully achieved. Nevertheless, they serve well as a blueprint for a challenging, large-scale research program.

The basic tenet of generalized information theory, that uncertainty is a broader concept than the concept of classical probability theory, has been debated in the literature since the late 1980s (for an overview, see Ref. [63]). As a result of this ongoing debate as well convincing advances in generalized information theory, limitations of classical probability theory in dealing with uncertainty and uncertainty-based information are increasingly recognized within academic community.

The two-dimensional framework for conceptualizing uncertainty, as illustrated in Fig. 2, is quite instrumental in guiding future research in generalized information theory. However, the framework still needs to be refined by refining each of its dimensions and, perhaps, needs to be extended by introducing additional useful dimensions.

In addition to demonstrating limitations of classical uncertainty theories and conceptualizing a fairly comprehensive framework for studying the full scope of uncertainty, a respectable number of nonclassical uncertainty theories have already been developed within generalized information theory. Notwithstanding the significance of these developments, they represent only a tiny fraction of the whole area, as can be seen from Fig. 2. Most prospective theories of uncertainty and uncertainty-base information are still undeveloped.

The role of information in human affairs has become so predominant that it is now quite common to refer to our society as information society. It is thus increasingly important for us to develop a good understanding of the broad concept of information. In the generalized information theory, the concept of uncertainty is conceived in the broadest possible terms, and uncertainty-based information is viewed as a commodity whose value is its *potential* to reduce uncertainty pertaining to relevant situations. The theory does not deal with the issues of how much uncertainty of relevant users (cognitive agents) is *actually* reduced in the context of each given situation, and how valuable this uncertainty reduction is to them. However, the theory, when adequately developed, will be an adequate base for developing a conceptual structure to capture semantic and pragmatic aspects relevant to information users under various situations of information flow. Only when this is adequately accomplished, a genuine science of information will be created.

## References

[1] G. J. Klir and I. Rozehnal, Epistemological categories of systems: An overview, *Intern. J. of General Systems*, **24**(1-2), pp. 207-224, 1996.

[2] G. J. Klir, Facets of Systems Science (Second Edition). Plenum Press, New York, 2001.

[3] E. von Glasersfeld, Radical Constructivism: A Way of Knowing and Learning. The Farmer Press, London, 1995.

[4] G. J. Klir and D. Elias, Architecture of Systems Problem Solving (Second Edition). Kluwer/Plenum, New York, 2003.

[5] J. W. Gibbs, Elementary Principles in Statistical Mechanics. Yale University Press, New Haven, 1902 (reprinted by Ox Bow Press , Woodbridge, Connecticut in 1981).

[6] W. Weaver, Science and complexity. *American Scientist*, **36**, pp. 536-544, 1948.

[7] P. R. Halmos, Measure Theory. D. Van Nostrand, Princeton, NJ, 1950.

[8] G. Choquet, Theory of capacities. *Annales de L'Institut Fourier*, **5**, pp. 131-295, 1953-54.

[9] L. A. Zadeh, Fuzzy Sets. *Information and Control*, **8**(3), pp. 338-353, 1965.

[10] G. J. Klir and M. J. Wierman, Uncertainty-Based Information: Elements of Generalized Information Theory (Second Edition). Physica-Verlag/Springer-Verlag, Heidelberg and New York, 1999.

[11] G. J. Klir, Generalized information theory. *Fuzzy Sets and Systems*, **40**(1), pp. 127-142, 1991.

[12] R. V. L. Hartley, Transmission of information. *The Bell System Technical J.*, **7**(3), pp. 535-563, 1928.

[13] A. Rènyi, Probability Theory. North-Holland, Amsterdam (Chapter IX, Introduction to Information Theory, pp. 540-616), 1970.

[14] G. J. Klir and B. Yuan, On nonspecificity of fuzzy sets with continuous membership functions. *Proc. 1995 IEEE Intern. Conf. on Systems, Man, and Cybernetics*, Vancouver, 1995.

[15] A. Ramer, Nonspecificity in $\mathbb{R}^n$. *Intern. J. of General Systems*, **30**(4), 2001.

[16] A. N. Kolmogorov, Foundations of the Theory of Probability. Chelsea, New York, 1950. (First published in German in 1933.)

[17] C. E. Shannon, The mathematical theory of communication. *The Bell System Technical J.*, **27**, pp. 379-423, 623-656, 1948.

[18] M. Sugeno, Fuzzy measures and fuzzy integrals: A survey. In: Gupta, M. M., Saridis, G. N. and Gaines, B. R. (eds.), Fuzzy Automata and Decision Processes. North-Holland, Amsterdam and New York, pp. 89-102, 1977.

[19] Z. Wang and G. J. Klir, Fuzzy Measure Theory. Plenum Press, New York, 1992.

[20] G. J. Klir and B. Yuan, Fuzzy Sets and Fuzzy Logic: Theory and Applications. Prentice Hall, PTR, Upper Saddle River, NJ, 1995.

[21] J. M. Mendel, Uncertain Rule-Based Fuzzy Logic Systems. Prentice Hall PTR, Upper Saddle River, NJ, 2001.

[22] D. Dubois and H. Prade, Rough fuzzy sets and fuzzy rough sets. *Intern. J. of General Systems*, **17**(2-3), pp. 191-209, 1990.

[23] J. A. Goguen, L-fuzzy sets. *J. of Math. Analysis and Applications*, **18**(1), pp. 145-174, 1967.

[24] R. R. Yager, S. Ovchinnikov, R. M. Tong and H. T. Nguyen (eds.), Fuzzy Sets and Applications: Selected Papers by L. A. Zadeh. John Wiley, New York, 1987.

[25] G. J. Klir and B. Yuan (eds.), Fuzzy Sets, Fuzzy Logic, and Fuzzy Systems: Selected Papers by Lotfi A. Zadeh. World Scientific, Singapore, 1996.

[26] L. A. Zadeh, Probability measures of fuzzy events. *J. of Math. Analysis and Applications*, **23**, pp. 421-427, 1968.

[27] D. Dubois and H. Prade, Possibility Theory. Plenum Press, New York, 1988.

[28] G. De Cooman, Possibility theory - I, II, III. *Intern. J. of General Systems*, **25**(4), pp. 291-371, 1997.

[29] L. A Zadeh, Fuzzy sets as a basis for a theory of possibility. *Fuzzy Sets and Systems*, **1**(1), pp. 3-28, 1978.

[30] G. J. Klir, On fuzzy-set interpretation of possibility theory. *Fuzzy Sets and Systems*, **108**(3), pp. 263-273, 1999.

[31] P. Walley, Statistical Reasoning With Imprecise Probabilities. Chapman and Hall, London, 1991.

[32] P. Walley, Measures of uncertainty in expert systems. *Artificial Intelligence*, **83**(1), pp. 1-58, 1996.

[33] P. Walley, Towards a unified theory of imprecise probability. *Intern. J. of Approximate Reasoning*, **24**(2-3), pp. 125-148, 2000.

[34] H. E. Kyburg, Bayesian and non-Bayesian evidential updating. *Artificial Intelligence*, **31**, pp. 271-293, 1987.

[35] H. E. Kyburg and M. Pittarelli, Set-based Bayesianism. *IEEE Trans. on Systems, Man, and Cybernetics, Part A*, **26**(3), pp. 324-339, 1996.

[36] A. Chateauneuf and J. Y. Jaffray, Some characterizations of lower probabilities and other monotone capacities through the use of Möbius inversion. *Mathematical Social Sciences*, **17**, pp. 263-283, 1989.

[37] M. Grabisch, The integration and Möbius representations of fuzzy measures on finite spaces, k-additive measures: A survey, In Grabish, M. et al. (eds.), Fuzzy Measures and Integrals: Theory and Application, Springer-Verlag, New York, pp. 70-93, 2000.

[38] K. Weichselberger and S. Pöhlmann, A Methodology for Uncertainty in Knowledge-Based Systems. Springer-Verlag, New York, 1990.

[39] Y. Pan and G. J. Klir, Bayesian inference based on interval probabilities. *J. of Intelligent and Fuzzy Systems*, **5**(3), pp. 193-203, 1997.

[40] A. Sgaro, Bodies of evidence versus simple interval probabilities. *Inter. J. of Uncertainty, Fuzziness and Knowledge-Based Systems*, **5**(2), pp. 199-209, 1997.

[41] G. Shafer, A Mathematical Theory of Evidence. Princeton Univ. Press, Princeton, NJ, 1976.

[42] L. Kramosil, Probabilistic Analysis of Belief Functions. Kluwer Academic/Plenum Publishers, New York, 2001.

[43] P. Smets, Belief functions. In: Smets, P. et al. (eds.), Non-standard Logics for Automated Reasoning, Academic Press, San Diego, pp. 253-286, 1988.

[44] D. Dubois and H. Prade, A class of fuzzy measures based on triangular norms. *Intern. J. of General Systems*, **8**(1), pp. 43-61, 1982.

[45] Z. Qiao, On fuzzy measure and fuzzy integral on fuzzy sets. *Fuzzy Sets and Systems*, **37**(1), pp. 77-92, 1990.

[46] Y. Pan and B. Yuan, Bayesian inference of fuzzy probabilities. *Intern. J. of General Systems*, **26**(1-2), pp. 73-90, 1997.

[47] G. J. Klir and Y. Pan, Constrained fuzzy arithmetic: Basic questions and some answers. *Soft Computing*, **2**(2), pp. 100-108, 1998.

[48] J. Yen, Generalizing the Dempster-Shafer theory to fuzzy sets. *IEEE Transactions on Systems, Man, and Cybernetics.*, **20**(3), pp. 559-570, 1990.

[49] M. Higashi and G. J. Klir, Measures of uncertainty and information based on possibility distributions. *Intern. J. of General Systems*, **9**(1), pp. 43-58, 1983.

[50] G. J. Klir and M. Mariano, On the uniqueness of possibilistic measure of uncertainty and information. *Fuzzy Sets and Systems*, **24**(2), pp. 197-219, 1987.

[51] D. Dubois and H. Prade, A note on measures of specificity for fuzzy sets. *Intern. J. of General Systems*, **10**(4), pp. 279-283, 1985.

[52] A. Ramer, Uniqueness of information measure in the theory of evidence. *Fuzzy Sets and Systems*, **24**(2), pp. 183-196, 1987.

[53] J. Abellan and S. Moral, A non-specificity measure for convex sets of probability distributions. *Intern. J. of Uncertainty, Fuzziness and Knowledge-Based Systems*, **8**(3), pp. 357-367, 2000.

[54] D. Harmanec, G. Resconi, G. J. Klir and Y. Pan, On the computation of uncertainty measure in Dempster-Shafer theory. *Intern. J. of General Systems*, **25**(2), pp. 153-163, 1996.

[55] R. M. Smith, Generalized Information Theory: Resolving Some Old Questions and Opening Some New Ones. Ph. D. Dissertation, Binghamton University - SUNY, Binghamton, 2000.

[56] M. Higashi and G. J. Klir, On measures of fuzziness and fuzzy complements. *Intern. J. of General Systems*, **8**(3), pp. 169-180, 1982.

[57] R. Christensen, Entropy Minimax Sourcebook (4 volumes). Entropy Limited, Lincoln, Mass., 1980-1981.

[58] E. T. Jaynes, Papers on Probability, Statistics and Statistical Physics (edited by R. D. Resenkrantz). Reidel, Dordrecht, 1983.

[59] J. N. Kapur, Maximum Entropy Models in Science and Engineering. John Wiley, New York, 1989.

[60] G. J. Klir, A principle of uncertainty invariance. *Intern. J. of General Systems*, **17**(2-3), pp. 249-275, 1990.

[61] J. F. Geer and G. J. Klir, A mathematical analysis of information preserving transformations between probabilistic and possibilistic formulations of uncertainty. *Intern. J. of General Systems*, **20**(2), pp. 143-176, 1992.

[62] G. J. Klir and B. Parviz, Probability-possibility transformations: A comparison. *Intern. J. of General Systems*, **21**(3), pp. 291-310, 1992.

[63] G. J. Klir, Foundations of fuzzy set theory and fuzzy logic: A Historical Overview. *Intern. J. of General Systems*, **30**(2), pp. 91-132, 2001.

# Organizing Information Using a
# Hierarchical Fuzzy Model

Ronald R. YAGER
*Machine Intelligence Institute, Iona College, New Rochelle, NY 10801*
*ryager@iona.edu*

**Abstract:** We review the basic ideas of fuzzy systems modeling. We introduce a framework for fuzzy systems model called a Hierarchical Prioritized Structure (HPS). We review its structure and operation. We carefully look at the Hierarchical Updation (HEU) operator that is used to combine information from different levels of the HPS. Some alternative forms for this operator are described. We next turn to the issue of constructing the HPS. First we consider the DELTA method which provides an algorithm that can dynamically adapt the model based on observations. We next consider the construction of the HPS where rules are provided by an expert. Here our focus is on obtaining the appropriate ordering the rules.

## 1. Introduction

Fuzzy systems modeling [1] provides a framework for the representation of information about complex relationships between variables. An important feature of this approach is the use of granularity and gradualarity [2]. The basic element used in this approach is a fuzzy if -then rule [3]. These rules are essentially fuzzy relationships or what Kosko calls patches [4]. With the aid of an inference mechanism called fuzzy reasoning one is able to manipulate the information in a fuzzy rule base to generate new knowledge that is only implicitly contained in the model. In [5][6][7] we introduced an extension of this fuzzy modeling technology in which we allowed a hierarchical representation of the rules. This framework is called the **Hierarchical Prioritized Structure** (HPS). In order to use this hierarchical framework to make inferences, generate an output for a given input, we had to provide a new aggregation operator to allow the passing of information between different levels of the hierarchy. This new aggregation operator is called the Hierarchical Updation (HEU) operator.

Essentially the HPS provides a framework using a hierarchical representation of knowledge in terms of fuzzy rules which is equipped with an appropriate machinery for making inferences, generating a system output given an input. An important feature of the inference machinery of the HPS is related to the implicit prioritization of the rules, the higher the rule is in the HPS the higher its priority. The effect of this is that we look for solutions in an ordered way starting at the top. Once an appropriate solution is found we have no need to look at the lower levels. This type of structure very naturally allows for the inclusion of default rules, which can reside at the lowest levels of the structure. It also has an inherent mechanism for forgetting by adding levels above the information we want to forget.

An important issue related to the use of the HPS structure is the building of the model itself. This involves determination of the rules residing in the model as well the determination of the level at which a rule shall appear. This is essentially a kind of learning problem. As in all knowledge based systems learning can occur in many different ways. One extreme is that of being told the knowledge by some, usually human, expert. Early examples of expert systems were generally of this type. At the other extreme is the situation in which we are only provided with input-output observations and we must use these to generate the rules. Many cases lie between these extremes. The type of forgetting mechanism described above provides the HPS with an infrastructure that can allow the implementation of dynamic adaptive learning techniques that continuously modify the model.

Here we shall discuss two instruments useful with respect to the construction/learning of HPS model. In the first situation we initialize the HPS with expert provided default rules and then use input–output observations to modify and adapt this initialization. In the second situation we begin with the total collection of rules that constitute the HPS and are interested in arranging these rules in the appropriate hierarchical configuration.

This new framework was called the **Hierarchical Prioritized Structure (HPS)**. With the aid of this structure one is able to introduce exceptions to more general rules by giving them a priority, introducing them at a higher level in the hierarchy. These exceptions can be themselves rules or specific points. In this work we continue the development of these HPS models by considering some issues related to their construction.

We note another interesting approach to the learning of the HPS, which we shall not discuss here, developed by Rozich, Ioerger and Yager [8]. This methodology called FURL uses a batch learning algorithm. It begins by using the data to construct the simplest rules and then refines these rules and places these refinements higher up in the HPS.

## 2. Introduction to Fuzzy Systems Modeling

Fuzzy systems modeling is a technique for modeling complex nonlinear relationships using a rule based methodology. Central to this approach is a partitioning of the input/output space. In this section we provide a brief introduction to fuzzy systems modeling, more details can be found in [1].

Consider a system or relationship $U = f(V, W)$, $U$ is the output (or consequent) variable and $V$ and $W$ are the input (or antecedent) variables. In fuzzy systems modeling we represent this relationship by a collection, **R**, of fuzzy if then rules of the form

*If V is $A_i$ and W is $B_i$ then U is $D_i$.*

The $A_i's$, $B_i's$ and $D_i's$ are normal fuzzy subsets over the spaces $X$, $Y$ and $Z$. In using fuzzy systems modeling we are essentially partitioning the input space $X \times Y$ into fuzzy regions $A_i \times B_i$ in which we know the output value, $D_i$.

Given values for the input variables $V = x^*$ and $W = y^*$, we calculate the value of $U$ as a fuzzy subset $E$ by using a process called fuzzy inference:

1. For each rule we find the firing level $\lambda_i = A_i(x^*) \wedge B_i(y^*)$.

2. We calculate the effective output of each rule $E_i$.

3. Combine individual effective rule outputs to get overall system output $E$.

Denoting input $V = x^*$ and $W = y^*$, as *INPUT* we denote this process as $E = R \bullet Input$.

Two different paradigms have been typically used for implementing steps two and three in the above procedure. The first paradigm, suggested by Mamdani and his associates [9][10], called the Min-Max inference procedure, uses $E_i(z) = \lambda_i \wedge D_j(z)$ for the effective rule outputs.

It then uses a union of these outputs to get the overall output $E = \bigcup_{i=1}^{n} E_i$ hence $E(z) = Max_i[E_i(z)]$.

The second paradigm uses arithmetic operations instead of the Min-Max operation. In this approach $E_i(z) = \lambda_i * D_i(z)$ and $E(z) = \dfrac{1}{T} \sum_{i=1}^{n} E_i(z)$ where $T = \sum_{i=1}^{n} \lambda_j$. As a simplified expression of this we have $E(z) = \sum_{i=1}^{n} E_j(z)$ where $E_i(z) = w_i \, D_i(z)$ with $w_i = \dfrac{\lambda_i}{T}$. We shall call this the arithmetic inference procedure.

When we desire a crisp output value $z^*$ rather than a fuzzy one we use a defuzzification step [11] such as the center of area (COA) method where we calculate $z^* = \dfrac{\sum_z z E(z)}{\sum_z E(z)}$.

At a meta level this inference process is one in which we start out with the empty set as our possible solutions and then add solutions provided by each rule depending on its firing level. In particular we can look at the fuzzy inference process as an iterative procedure. In the Max-Min paradigm we can express this as

$$H_i(z) = H_{i-1}(z) \vee E_i(z), \quad i = 1, \ldots, n$$

with $H_O(z) = \varnothing$ and with the overall output $E$ equal to $H_n$.

In the arithmetic paradigm we can express this iterative procedure as

$$H_i(z) = H_{i-1}(z) + E_i(z)$$

where we use $E_i(z) = W_i * D_i(z)$. Again in this case we use $H_O(z) = \varnothing$ and overall output $E$ is equal to $H_n$.

A number of modifications of the basic procedure have been suggested. A most important one was suggested by Sugeno [12][13]. Sugeno suggested replacing the consequent fuzzy subset by a crisp value, $D_i = \left\{ \dfrac{1}{d_i} \right\}$. He further suggested the possibility of using linear functional relations in the consequent. Filev [14] suggested a generalization of this idea in which we can use any functional form in the consequent. It is also possible to associate weights or importances with each of the rules. Thus we can associate with each rule a value $\alpha_i$, which indicates the weight of the $i^{th}$ rule. When including these importances we first modify each of the $D_i's$ into $\hat{D}_i's$ before applying the inference procedure. If we are using the max-min model we perform this modification as follows $\hat{D}_i(z) = D_i(z) \wedge \alpha_i$. If we use the arithmetic approach we use $\hat{D}_i(z) = D_i(z) * \alpha_i$.

## 3. HPS Model

In [5][6][7] Yager suggested an extension of the basic fuzzy systems modeling framework. The purpose of this extension was to allow for a prioritization of the rules by using a hierarchical representation of the rules. This formulation is called the **Hierarchical Prioritized Structure** (HPS). In the following we shall briefly describe this structure and its associated reasoning procedure. Figure 1 will be useful in this discussion.

Assume we have a system we are modeling with inputs $V$ and $W$ and output $U$. At each level of the HPS we have a collection of fuzzy *if - then* rules. Thus for level $j$ we have a collection of $n_j$ rules

$$\textit{If } V \textit{ is } A_{ji} \textit{ and } W \textit{ is } B_{ji} \textit{ then } U \textit{ is } D_{ji}, \; i = 1, \ldots, n_j.$$

We shall denote the collection of rules at the *jth* level as $R_j$. We denote the application of the basic inference process with a given input, $V = x^*$ and $W = y^*$, to this sub–rule base as $F_j = R_j$ • *Input*.

In the HPS we denote the output of the *jth* level as $G_j$. $G_j$ is obtained by combining the output the previous level, $G_{j-1}$, with $F_j$ using the Hierarchical Updation (HEU) aggregation operator subsequently to be defined. The output of the last level, $G_n$, is the overall model output $E$. We initialize the process by assigning $G_0 = \emptyset$.

The **HEU** aggregation operator is defined as

$$G_j(z) = G_{j-1}(z) + (1 - \alpha_{j-1})F_j(z).$$

Here $\alpha_{j-1} = Max_z[G_{j-1}(z)]$, the largest membership grade in $G_{j-1}$.

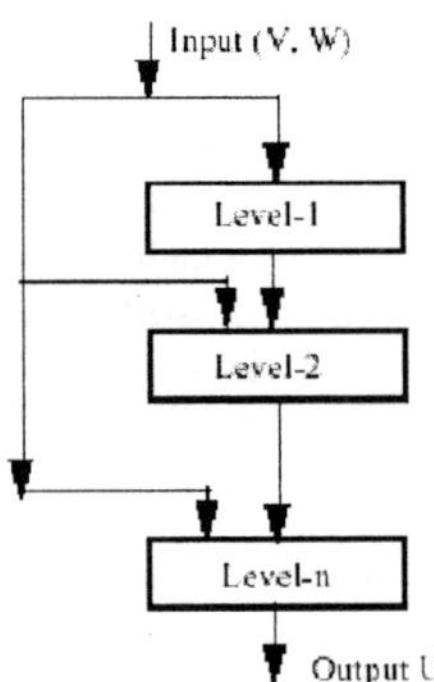


Figure 1. Hierarchical Prioritized Structure

Let us look at the functioning of this operator. First we see that it is not pointwise in that the value of $G_j(z)$ depends, through the function $\alpha_{j-1}$, on the membership grade of elements other than $z$. We also note that if $\alpha_{j-1} = 1$ no change occurs. More generally the larger $\alpha_{j-1}$ the less the effect of the current level. Thus we see that $\alpha_{j-1}$ acts as a kind of choking function. In particular, if for some level $j$ we obtain a situation in which $G_j$ is normal, has as element with membership grade one, the process of aggregation stops. It is also clear that $G_{j-1}$ and $F_j$ are not treated symmetrically. We see that as we get closer to having some elements in $G_{j-1}$ with membership grade one then the process of adding information slows. The form of the HEU essentially implements a prioritization of the rules. The rules at highest level of the hierarchical are explored first if they find a good solution we look no further at the rules.

Figure 2 provides an alternative view of the HPS structure.

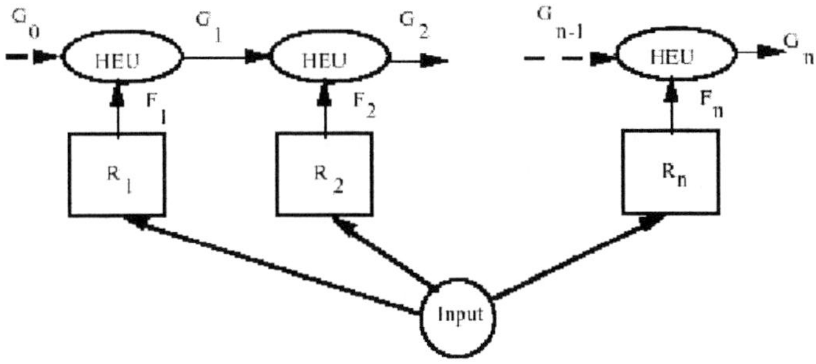

Figure 2. Alternative View of HPS

We shall illustrate the application of this structure with the following example

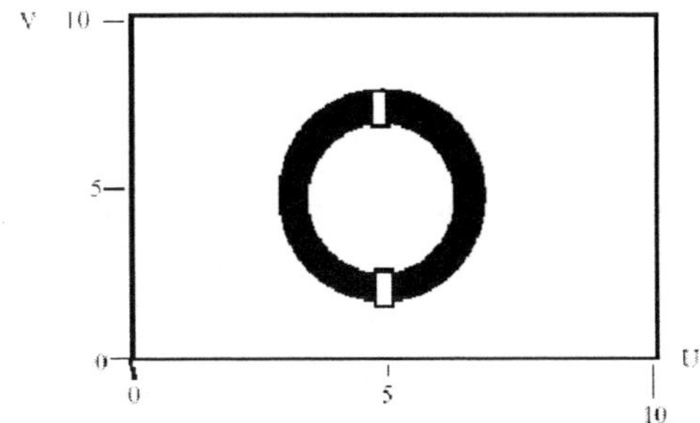

Figure 3. Structure of $F(U, V)$

**Example**: Consider a function $W = F(U, V)$ defined on $U = [0, 10]$ and $V = [0, 10]$. Refer to figure 3 for the following discussion. We shall assume that in the white areas the value of the function is small and in the black area the value of the function is large. The figure could for example be representative of a geospatial mapping in which W is the altitude and the black areas correspond to a mountain range.

We can describe this function relationship by the following three level HPS structure.

**Level-1:** If $U$ is close to five then $W$ is small　　　　(Rule-1)

**Level-2:** If $\left((U-5)^2 + (V-5)^2\right)^{0.5}$ is about two then $W$ is large　　　　(Rule-2)

**Level-3:** If $U$ and $V$ are anything else then $W$ is small　　　　(Rule-3)

For our purposes we define the underlined fuzzy subsets as follows:

$$Small = \left\{\frac{0.3}{5}, \frac{0.6}{6}, \frac{1}{7}, \frac{0.6}{8}, \frac{0.3}{9}\right\}, \quad Large = \left\{\frac{0.3}{21}, \frac{0.6}{22}, \frac{1}{23}, \frac{0.6}{24}, \frac{0.3}{25}\right\}$$

$$close \ to \ five(U) = e^{\frac{-(U-5)^2}{0.25}}, \quad about \ two(r) = e^{-(r-2)^2}$$

Let us look at three special cases. **1.** $U = 5$ and $V = 6$. Here rule one fires to degree 1.

Hence the output of the first level is $G_1 = \left\{\frac{0.3}{5}, \frac{0.6}{6}, \frac{1}{7}, \frac{0.6}{8}, \frac{0.3}{9}\right\}$. Since this has maximal

membership grade equal to one the output of the system is $G_1$.

**2.** $U = 6$ and $V = 6$. Here the firing level of rule one is 0.02 and the output of the first level is

$G_1 = \left\{\frac{0.2}{5}, \frac{0.2}{6}, \frac{0.2}{7}, \frac{0.2}{8}, \frac{0.2}{9}\right\}$ and has maximal firing level 0.2. Applying the input to rule-2 we

get a firing level of 1. Thus $F_2 = \left\{\frac{0.3}{21}, \frac{0.6}{22}, \frac{1}{23}, \frac{0.6}{24}, \frac{0.3}{25}\right\}$.

Thus $G_2 = \left\{\frac{0.2}{5}, \frac{0.2}{6}, \frac{0.2}{7}, \frac{0.2}{8}, \frac{0.2}{9}, \frac{0.24}{21}, \frac{0.46}{22}, \frac{0.8}{23}, \frac{0.46}{24}, \frac{0.24}{25}\right\}$ and therefore $\alpha_2 = 0.8$.

Applying the input to rule three we get firing level 1.

Thus $F_2 = \left\{\frac{0.3}{21}, \frac{0.6}{22}, \frac{1}{23}, \frac{0.6}{24}, \frac{0.3}{25}\right\}$. Since $G_3(z) = G_2(z) + (1 - 0.8) F_3(z)$ we get

$G_3 = \left\{\frac{0.26}{5}, \frac{0.312}{6}, \frac{0.4}{7}, \frac{0.312}{8}, \frac{0.26}{9}, \frac{0.24}{21}, \frac{0.46}{22}, \frac{0.8}{23}, \frac{0.46}{24}, \frac{0.24}{25}\right\}$.

Defuzzifying this value we get $W = 16.3$.

**3.** $U = 9$ and $V = 8$. In this case the firing level of rule one is 0, thus $G_1 = \varnothing$. Similarly the firing level of rule 2 is also 0, and hence $G_1 = \varnothing$. The firing level of rule 3 is one and hence the overall output is *small*.

### 4. Hierarchical Updation Operator

Let us look at some of the properties of this Hierarchical Updation (HEU) operator which we denote as $\gamma$. If $A$ and $B$ are two fuzzy sets then we have $\gamma(A, B) = D$ where $D(z) = A(z) + (1 - \alpha)B(z)$ with $\alpha = Max_{x \in X}(A(x))$. This operator is not pointwise as $\alpha$ depends on $A(x)$ for all $x \in X$. This operator is a kind of union operator, we see that $\gamma(A, \varnothing) = A$ and $\gamma(\varnothing, B) = B$. This operator is not commutative $\gamma(A, B) \neq \gamma(B, A)$. The operator is also not monotonic. Consider $D = \gamma(A, B)$ and $D' = \gamma(A', B)$ where $A(z) \leq A'(z)$ for all $z$, $A \subset A'$. In this case $\alpha \leq \alpha'$. Here $D'(z) = A'(z) + (1 - \alpha') B(z)$ and $D(z) = A(z) + (1 - \alpha) B(z)$. Here $D'(z) - D(z) = A'(z) - A(z) + B(z)(\alpha - \alpha')$. Thus while $A'(z) - A(z) \geq 0$ we have $(\alpha - \alpha') \leq 0$ and therefore there is no guarantee that $D'(z) \geq D(z)$. Also we note that while $\gamma(X, B) = B$ we have $\gamma(A, X) = D$ where $D(z) = A(z) + (1 - \alpha)$.

We can suggest a general class of operators that can serve as hierarchical aggregation operators. Let $T$ be any *t-norm* and $S$ be any *t-conorm*. [15] A general class of hierarchical updation operators can be represented as $D = HEU(A, B)$ where

$$D(z) = S(A(z), T(1 - \overset{\alpha}{\;}, B(z)))),$$

with $\alpha = Max_z(A(z))$.

First let us show that our original operator is a member of this class. Assume $S$ is the bounded sum, $S(a, b) = Min[1, a + b]$ and $T$ is the product, $S(a, b) = a \cdot b$. In this case $D(z) = Min[1, A(z) + \overline{\alpha}\,B(z)]$. Consider the term $A(z) + \overline{\alpha}\,B(z)$. Since $\alpha = Max_z[A(z)]$ then $\alpha < A(z)$ and therefore $A(z) + \overline{\alpha}\,B(z) \leq A(z) + (1 - A(z))B(z) \leq 1$. Thus $D(z) = A(z) + (1 - \alpha)\,B(z)$ which was our original suggestion. We can now obtain other forms for this HEU operator by selecting different instantiations of $S$ and $T$. If $S = Max(\vee)$ and $T = Min(\wedge)$ we get

$$D(z) = A(z) \vee (\overline{\alpha} \wedge B(z)).$$

If $S$ is the algebraic sum, $S(a, b) = a + b - ab$ and $T$ is the product then

$$D(z) = A(z) + \overline{\alpha}B(z) - \overline{\alpha}A(z)B(z) = A(z) + \overline{\alpha}\overline{A}(z)B(z)$$

If we use $S$ as the bounded sum and $T$ as the *Min* we get

$$D(z) = Min\left[1, A(z) + \overline{\alpha} \wedge B(z)\right]$$

Since $\alpha < A(z)$ then $A(z) + \overline{\alpha} \wedge B(z) \leq A(z) + (1 - A(z)) \wedge B(z) \leq A(z) + (1 - A(z)) \leq 1$ hence we get

$$D(z) = A(z) + \overline{\alpha} \wedge B(z).$$

More generally if $S$ is the bounded sum and $T$ is any *t*-norm then

$$D(z) = Min\left[1, A(z) + T(\overline{\alpha} \wedge B(z))\right].$$

Since $T(\overline{\alpha} \wedge B(z)) \leq \overline{\alpha} \leq 1 - A(z)$ then

$$D(z) = A(z) + T(\overline{\alpha}, B(z)).$$

## 5. The DELTA Method for Learning an HPS from Observations

In the preceding we have described the inference mechanism associated with a given hierarchical prioritized structure. We have said nothing about how we obtained the rules in the model. The issue of the construction of the HPS model is an important one. The format of HPS model will allow many different methods for obtaining the model.

In this section we shall outline a dynamic learning approach for the construction of an HPS which allows continuous learning. We shall call this the DELTA (**D**efault **E**xception **L**earning

That's Adaptive) method for HPS. In this approach we initialize the HPS by providing a default representation of the relationship we are trying to model. With this default relationship we allow the system builder to provide an initial model of the system which will be augmented as we get more data about the performance of the actual system. This default model can be as simple or as complex as the designers knowledge of the system can support. In this approach the augmentation of the model will be one in which we add specific observations and rules to the HPS. The addition of knowledge to the structure will be driven by observations that are exceptions to what we already believe the situation to be. The exceptions will be captured and stored at the top level of the hierarchy. Groups of exceptions shall be aggregated to form new rules that will be stored at the next level of the hierarchy.

We shall use a three level HPS model as shown in figure 4. For ease of explanation we shall assume a model having a single input. The extension to multiple inputs is straight forward.

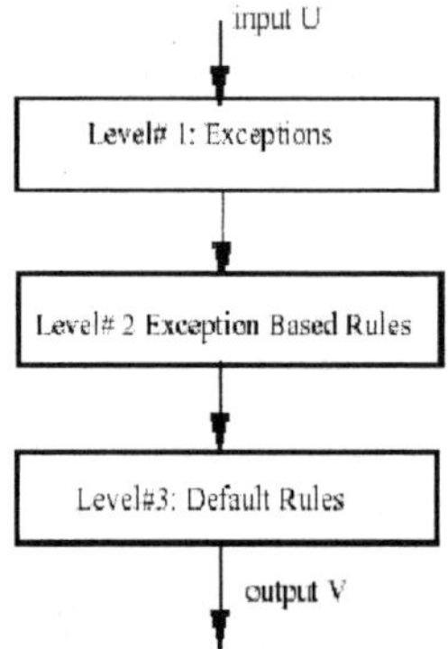


Figure 4. Exception Based Hierarchy

The construction of the structure is initialized with the first and second levels being empty. The third level is initialized with our default information about the structure of the relationship between the input and output variables $V$ and $U$. In particular the third level contains default rules of the form

*If V is A then U is f(V).*

In the above $f(V)$ is some prescribed function relationship and $A$ is a fuzzy subset indicating the range of that default rule. The knowledge in the default can be any manifestation of the prior expectation of the system modeler. It could be a simple rule that says $U = b$ for all values of $V$, a linear relationship that says $U = k_1 + k_2 V$ for all values of $V$ or a collection of more complex rule base on some partitioning of the input space.

The HPS model will be learned from observations presented to it, especially observations which are exceptions to what we already believe. In particular the information in level one and two will be obtained from the observations presented to the model. As we shall see level one will contain facts about individual observations that are exceptions to what we already believe. Level two shall contain rules that aggregate these exceptions. The aggregation process used here is very much in the spirit of the mountain clustering method [14][16][17] introduced by Yager and Filev.

In figure 5 we provide a flow diagram of the basic learning mechanism used in this approach. In the following we describe the basic mechanism for the construction of this type of HPS. An observation $(x, y)$ is presented to the HPS model. We calculate the output for the input $x$, denote this $y^*$. We then compare this calculated output with the desired output. If $y$ and $y^*$ are close to each other we can disregard this data and assume it doesn't provide any learning. If $y$ and $y^*$ are not close we use this data to modify the HPS.

More specifically for the pair $(y, y^*)$ we calculate the value $Close(y, y^*)$, $[0, 1]$ indicating the degree of closeness of the observed value and the calculated value. If $Close(y, y^*) \geq \alpha$, a threshold level, we disregard the data. If $Close(y, y^*) < \alpha$ we use this data to update the model. We denote for this observation $P = 1 - Close(y, y^*)$ as a measure of this observations ability to count as an exception, its strength of exception.

We add to the top level of the current HPS model this observation in the form of a point rule,

*if V is x then U is y.*

For simplicity we shall denote this rule as the **point**$(x, y)$. We further associate with this rule a value $M$ that we initialize as $P$, its strength of exception. As we shall see this $M$ value will be used in the same way as the mountain function is used in the mountain method to help in the aggregation of point exceptions to form exception rules.

We next update the $M$ value for all the other exception rules in the top level of the HPS. Specifically for any point rule, if $V$ is $x_i$ then $U$ is $y_i$, in the top level we update $M_i$ as

$$M_i^{'} = M_i + Pe^{-Distance((x,y)-(x_i,y_i))}$$

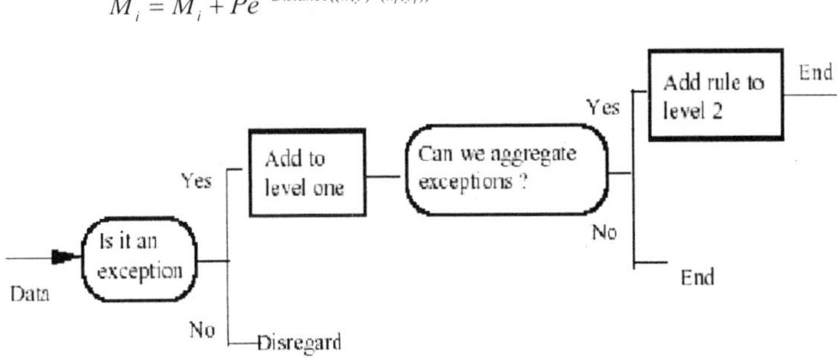


Figure 5. Schematic of Learning Process

Thus we see that as a result of experiencing an observation that is considered an exception we add this observation to a current model and modify the $M$ value of all other exceptions by adding to them a value proportional to the strength of the current exception modulated by its distance to the current exception.

We next check to see if the addition of this new exception has caused a accumulation of the exceptions which can be gathered to form an exception rule, here we use the $M$ values.

Specifically we find the data point in the top level that now has the highest $M$ value. Let us denote this value as $\hat{M}$ and assume it occurs for the point $(\hat{x}, \hat{y})$. If $\hat{M} \geq \beta$, $\beta$ being a threshold value for exception rule formation, we create a new rule of the form

*If V is about $\hat{x}$ then U is about $\hat{y}$.*

Where, as we have noted above, $\hat{x}$ and $\hat{y}$ are the coordinates of the data point with the largest M value. This new rule is added to the second level of HPS. Thus we see that a collection of exceptions close to each other, focused at $(\hat{x}, \hat{y})$, form an exception rule at the second level. We emphasize that it is with the aid of the M function that we measure the power of a exception point in the first level to be the nucleus of an exception rule in the second level.

The final step is the cleansing and reduction of the top level by eliminating the individual exception rules that are now accounted for by the formulation of this new rule at the second level. We first modify our function $M$ at each point $(x, y)$ in the top level to form $M'$ where

$$M'(x, y) = M(x, y) - \hat{M}e^{-Distance((x,y)-(\hat{x},\hat{y}))}$$

We next eliminate all point rules for which $M'(x, y) \le 1 - \alpha$.

Furthermore we let $\hat{A}$ and $\hat{B}$ be the fuzzy subsets about $\hat{x}$ and $\hat{y}$. For each exception point in the top level we calculate $\hat{A}(x_i)$ and $\hat{B}(y_i)$ and let $t_i = Min\ (A(x_i), B(y_i))$. We then eliminate all exceptions for which $t_i \ge \gamma$, a threshold for exception cleansing.

It should be noted that the above procedure has a number of parameters effecting our actions. In particular we introduced $\alpha$, $\beta$ and $\gamma$. It is with the aid of these parameters that we are able to control the uniqueness of the learning process. For example, the smaller we make $\alpha$ the more rigorous our requirements are for indicating a observation as an exception, it is related to our sensitivity to exceptions. The parameter $\beta$ determines are openness to the formulation of new rules. The choice of these parameters is very much in the same spirit as choice of the learning rate used in the classical gradient learning techniques such as back propagation. Experience with the use of this exception base machinery will of course sharpen our knowledge of the effect of parameter selection. At a deeper level the selection of these parameters should be based upon how we desire the learning to perform and gives us a degree of freedom in the design of our learning mechanism. This results, just as in the case of human learning, in highly individualized learning.

It is important to emphasize some salient features of the DELTA mechanism for constructing HPS models. We see this has an adaptive type learning mechanism. We initialize the system with current user knowledge and then modify this initializing knowledge based upon our observations. In particular, like a human being, this has the ***capability*** for continuous learning. That is even while its being used to provide outputs it can learn from its mistakes. Also we see that information enters the systems as observations and moves its way down the system in rules very much in the humans process information in the face of experience.

## 6. Using Priority Relations to Obtain the HPS

One can envision another mode for constructing the HPS. Here we start with a collection of already provided rules and are interested in inserting these rules into the different levels of the HPS. Here we assume we have a set of rules, $H = \{h_1, h_2, \ldots, h_n\}$. In order to construct the HPS we need a ordering of these rules regarding their priorities. The issue of assigning priorities to rules is very complex and highly domain dependent. First we shall briefly touch upon general some considerations useful for distinguishing priorities among rules. In suggesting these guidelines it must be kept in mind that the effect of assigning a higher priority to one rule over the other is that if both rules fire for some input then the higher priority rule can block the lower priority from effecting the solution. We essentially look to the higher priority rule for the answer.

A first consideration is that any default or unqualified rule should have a strictly lower priority than a qualified rule. A second consideration is the certainty associated with a rule, the higher the certainty the more priority you give to the rule. Another consideration is the specificity of the antecedent of the rule, the more specific the antecedent higher the priority. Thus a rule that says "dogs are non-aggressive" should have a lower priority than a rule that says "rapid dogs are aggressive. " Implicit in this guideline is that rules corresponding to a particular object or point should have the highest priority. Thus a rule that says "if $x = 3$ then $U = 15$" should have the highest priority. In this spirit if $R$ is a rule and $\overline{R}$ is a rule indicating an exception to this rule then $\overline{R}$ should have a higher priority.

Even using these guidelines the construction a total ordering over all of the rules is hard. In cases in which there are a large number of rules it may be difficult to directly comprehend the totality of the ordering. A second reason is that their may be an incomparability of different rules regarding their relative priorities, our information about the priorities may not complete. One way to avoid the difficulty of requiring an expert to provide a total ordering over all the rules together is to use a pairwise priority comparison of the rules. Pairwise priority comparisons are generally easy for human experts to provide. A pairwise priority comparison leads to the establishment of a binary relationship over the space of rules. If this binary relationship is well behaved we can use results from preference theory to construct an ordering over the set of all rules. However, the problem of incomparability often requires the introduction of some additional meta knowledge to complete the information. Before describing the process for constructing this ordering from pairwise comparison we shall briefly review some ideas from preference theory [18].

Assume $X$ is a collection of elements, a relationship $S$ on $X$ is a subset on the cartesian space $X \times X$. Thus $S$ consists of pairs $(x, y)$ where $x, y \in X$. If $(x, y) \in S$ we shall denote this as $x \, S \, y$. We can also associate with $S$ a membership function such that $S(x, y) = 1$ if $x \, S \, y$ and $S(x, y) = 0$ if $x \, \mathcal{S} \, y$. Three cases can be identified regarding any two pairs of elements from $X$. In the first case we have $x \, S \, y$ and $y \mathcal{S} x$, we say that $x$ has a **strictly higher priority** then $y$, we denote this $x \, P \, y$. In the second case we have $x \, S \, y$ and $y \, S \, x$, here we say that $x$ and $y$ **have the same priority** and denote this $x \, I \, y$. In the third case we have $x \mathcal{S} y$ and $y \mathcal{S} x$, here we say that $x$ and $y$ are **incomparable** and denote this as $x \, T \, y$.

A number of basic properties can be associated with binary relationships. $S$ is called reflexive if $x \, S \, x$, $S(x, x) = 1$, for all $x$. $S$ is called complete if $x \, S \, y$ or $y \, S \, x$ for all $x$ and y, $S(x, y) + S(y, x) \geq 1$. $S$ is called transitive if $x \, S \, y$ and $y \, S \, z$ implies that $x \, S \, z$, formally we can express this as $S(x, y) + S(y, z) - S(x, z) \leq 1$. Relationships possessing combinations of these properties are given special name. A relationship $S$ is called a **weak ordering** if it is reflexive, complete and transitive. It is called a **quasi-ordering** if it is reflexive and transitive.

If $S$ is a weak ordering for all pairs $x$ and $y$ one of the following is always true; $x \, P \, y$, $y \, P \, x$ or $x \, I \, y$. We can associated with any weak ordering a function

$$g(x) = \sum_{x_j \in X} S(x, x_j),$$

called the scoring function of $S$. It can be shown that this function has the following properties:

$x \, P \, y \ \text{if} \ g(x) > g(y)$

$y \, I \, y \ \text{if} \ g(x) = g(y).$

If the binary relationship $S$ resulting from a pairwise priority comparison is a weak ordering we easily obtain the overall ordering of the objects by using the scoring function.

In order to provide a method for obtaining the overall ordering from the pairwise relationship we need one further idea, relational composition. This will help us assure transitivity. Assume $S_1$ and $S_2$ are two relationships on $X$ we denote the composition of $S_1$ and $S_2$ as $S_3 = S_1 \circ S_2$. $S_3$ is also a relationship on $X$ where $S_3(x,z) = \underset{y \in X}{Max}[S_1(x,y) \wedge (y,z)]$. For any relationship on $S$ we denote $S^2 = S \circ S$ and more generally $S^i = S^{i-1} \circ S$. Using this idea of composition we can define the transitive closure of any relationship. Assume $X$ has cardinality $n$ then the transitive closure of $S$ is defined as $\hat{S}$ where

$$\hat{S} = S \cup S^2 \cup S^3 ... \cup S^n.$$

Two properties about transitive closures are important here. The first is that $\hat{S}$ is always a transitive relationship and the second is that if $S$ is transitive than $S = \hat{S}$.

We are now are in a position to describe the procedure for obtaining the HPS from a collection of rules. Assume we have a collection of rules $H = \{h^1, h^2, \ldots, h^n\}$. We first obtain a relationship $S$ on $H$ as follows.

**Algorithm I:**

1. We initiate $S^*$ as the empty set
2. For each $h_i \in H$ we add the tuple $(h_i, h_i)$ to $S^*$
3. For each pair of rules $h_i$ and $h_j$ in $H$ are proceed as follows:
   i. If we know $h_i$ to have a higher priority than $h_j$ we add the tuple $(h_i, h_j)$ to $S^*$
   ii. If we know $h_i$ and $h_j$ to have the same priority we add the tuples $(h_i, h_j)$ and $(h_j, h_i)$ to $S^*$
   iii. If we can't make a comparison between $h_i$ and $h_j$ regarding the priorities we do nothing
4. If $S^*$ is transitive we stop and set $S = S^*$
5. If $S^*$ is not transitive we calculate its transitive closure and set this equal to $S$.

As a result of this algorithm we have a relationship $S$ on $H$ which is a quasi-ordering, reflexive and transitive. We next test whether $S$ is complete, $S(h_i, h_j) + S(h_j, h_i) = 1$ for all pairs $h_i$ and $h_j$. If $S$ is complete then $S$ is a weak ordering. If $S$ is a weak ordering we can then construct the HPS as follows. For each $h_i \in H$ we calculate its score $g(h_i) = \sum_{j=1}^{n} S(h_i, h_j)$. Using $g(h_i)$ we construct the HPS by assigning those rules with the highest score to the highest priority level of the HPS. The rules with the second highest score get assigned to the second level. We continue in this manner until all rules are assigned

**Example:** Let $H = \{h_1, h_2, h_3, h_4\}$. Assume we have the following information regarding the priorities of these rules:

| | |
|---|---|
| *$h_1$ has priority over $h_2$* | *$h_1 \, S \, h_2$* |
| *$h_2$ and $h_3$ are of the same priority* | *$h_2 \, S \, h_3$ & $h_3 \, S \, h_2$* |
| *$h_3$ has priority over $h_4$* | *$h_3 \, S \, h_4$* |

From this we can obtain $S^* = \begin{array}{c} h_1 \\ h_2 \\ h_3 \\ h_4 \end{array} \begin{bmatrix} 1 & 1 & 0 & 0 \\ 0 & 1 & 1 & 0 \\ 0 & 1 & 1 & 1 \\ 0 & 0 & 0 & 1 \end{bmatrix}$ . It can be seen that this is not transitive.

We now apply our transitive closure procedure on $S^*$ to obtain $S = \begin{array}{c} h_1 \\ h_2 \\ h_3 \\ h_4 \end{array} \begin{bmatrix} 1 & 1 & 1 & 1 \\ 0 & 1 & 1 & 1 \\ 0 & 1 & 1 & 1 \\ 0 & 0 & 0 & 1 \end{bmatrix}$ . It is

easy to see that S is complete and hence a weak ordering. Applying our scoring function to $S$ we get $g(h_1) = 4$; $g(h_2) = 3$; $g(h_3) = 3$ and $g(h_4) = 1$. From this we get the HPS structure shown in figure 6.

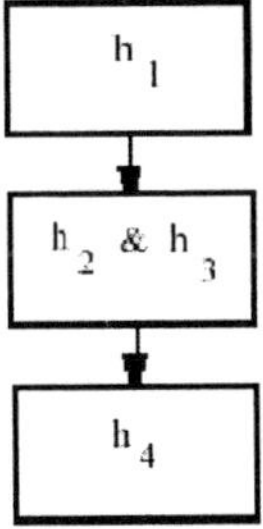


Figure 6. HPS from priority relationship

## 7. Completion of Quasi-Ordering by Maximal Buoyancy

In situations in which some of the rules in our knowledge base are incomparable with each other the relationship $S$ is not complete. In this case the relationship $S$ resulting from the application of algorithm $I$ is not a weak ordering, it is only a quasi-ordering, and we can't use a scoring function to construct the HPS. In order to enable us to use a scoring function to construct the HPS we must obtain from the quasi-ordering a weak ordering by completing $S$. We now look at the process of completing quasi-ordering. The completion of this quasi-ordering will be based upon the principle of maximal buoyancy introduced in [19][20][21]. The use of this principle leads to a completion which introduces the least unjustified information. The principle of maximal buoyancy is very much in the spirit of the principle of maximal entropy.

**Definition**: Assume $S_1$ is a quasi-ordering a weak ordering $S_2$ is said to be a completion of $S_1$:

**I.** If for all pairs $h_i$ and $h_j$ in $H$ we make $S_1$ complete by

　　　*1. if $h_i\, P_1\, h_j$ we assign $h_i\, P_2\, h_j$*

　　　*2. if $h_i\, I_1\, h_j$ we assign $h_i\, I_2\, h_j$*

　　　*3. if $h_i\, T_1\, h_j$ (incomparable) we assign either $h_i\, I_2\, h_j$ or $h_i\, P_2\, h_j$.*

**II.** This completion is done in a way to retain the transitivity.

Essentially we complete a quasi-ordering by turning all incomparable pairs into either strict preference or equality, while leaving all strict preference and identity relationships as they are.

While any quasi-ordering can be completed and turned into a weak ordering, unfortunately, as the following example illustrates, there can be many ways to complete a quasi-ordering.

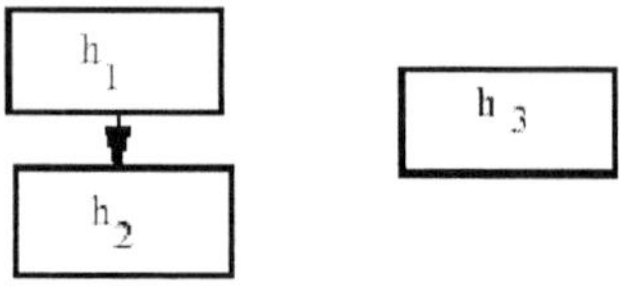

Figure 7. A quasi-ordering

**Example:** Consider the quasi-ordering shown in figure 7.
Acceptable completions of these quasi-ordering are shown in figure 8.

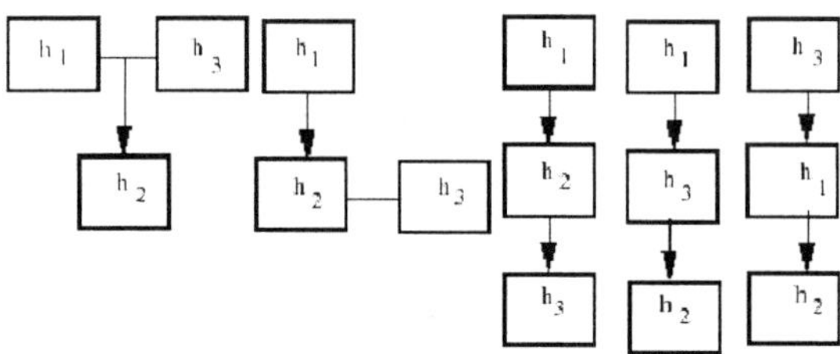

Figure 8. Possible completions of quasi-ordering of figure 6.

Some additional external criteria must be used to select from the multiple possible completions of a quasi-ordering an appropriate one. In [19][20][21] Yager has suggested such an approach based upon the principle of maximum buoyancy. We shall now describe this process.

We first must introduce the measure of buoyancy associated with a weak ordering. Assume $S$ is a weak ordering and let $g$ be the scoring function $g(h_i) = \sum_{j=1}^{n} S(h_i, h_j)$. Let $V_i$ be a normalized scored, $V_i = \dfrac{g(h_i)}{n}$. The measure of buoyancy of $S$ is defined as $Buo(s) = \sum_{j=1}^{n} w_j a_j$, where $a_j$ is the $j$th largest of the $V_i$ and $w_j$ are a set of weights such that: **1.** $w_i \in [0,1]$, **2.** $\sum_i w_i = 1$ and **3.** $w_i > w_j$ if $i < j$. An appropriate set of weights is $w_i = (0.5)^i$ for $i = 1$ to $n - 1$ and $w_n = (0.5)^{n-1}$. Using this measure the process we use for completing a quasi-ordering is as follows. Let $Q$ be a quasi-ordering and let $S_1, \ldots, S_q$ be the set of all weak orderings that are completions of $Q$. Let $S^*$ be the weak ordering in this set such that

$$Buo(S^*) = Max_i\left[Buo(S_i)\right].$$

$S^*$ is the completion of $Q$ with the maximum buoyancy. We use this $S^*$ as our completion. Once having determined $S^*$ we use the scoring function associated with $S^*$ to order the rules in the HPS. Yager [19][20][21] discusses the justification of the principle of maximum buoyancy, which is very much in the spirit of maximum entropy. Essentially the basis of this method is as follows. In selecting a completion of a quasi-order to be used in an HPS structure we desire to pick one that introduces the least possible unjustified information. The information is related to the specificity of any resulting inference. It can be shown that in using the principle of maximum buoyancy we are essentially selecting weak ordering introducing the least information.

## 8. Mathematical Programming for Completion

If the dimension of $H$ is not small it becomes infeasible to just test all the possible completions of our quasi-ordering and select the one with the maximum buoyancy. In this case we need some help in determining the best one In this section we shall describe a mathematical programming approach to determine the weak order that is the best completion.

Assume $S$ is a quasi-ordering which we desire to complete. Let $R$ indicate the desired completed ordering based on the principle of maximal buoyancy. In the following we shall let $R_{ij}$ indicate the membership function of $R$, $R_{ij} = R(h_i, h_j)$. Since $R_{ij}$ must be either 1 or 0, we note that $R_{ij}$ must be a binary integer variable.

To find $R$ we can solve an integer programming problem whose objective is to maximize the buoyancy of $R$. There are six sets of constraints that we must impose upon our problem:

1. *Reflexivity Constraints*
2. *Faithfulness to S Constraints*
3. *Completion Constraints*
4. *Transitivity Constraints*
5. *Scoring Constraints*
6. *Range of R*

The reflexivity constraints are n constraints of the form $R_{ii} = 1$ for $i = 1$ to $n$.

The next set of constraints assure us that $R$ is an extension of $S$. For each pair $h_i$ and $h_j$ for which we have $h_i \, P \, h_j$ in $S$, $S(h_i, h_j) = 1$ and $S(h_j, h_i) = 0$, we add the two constraints $R_{ij} = 1$ and $R_{ji} = 0$. For each pair $h_i$ and $h_j$ where $h_i \, I \, h_j$ in $S$, $S(h_i, h_j) = S(h_j, h_i) = 1$, we add the constraint $R_{ij} + R_{ji} = 2$. With these constraints we assure ourselves that the resulting relationship will be faithful to the relationship $S$ with respect to already established preferences and equalities.

We next add a collection of constraints that assure us that the resulting $R$ is a complete ordering. For each pair $h_i$ and $h_j$ which is not complete in $S$, not covered by the above two conditions, we add a constraint $R_{ij} + R_{ji} \geq 1$. Next we add a collection of constraints that assure us that $R$ is transitive. For each pair $h_i$ and $h_j$, we add a collection of constraints of the form

$$R_{ik} + R_{kj} - R_{ij} \leq 1 \text{ for } k \text{ equal to all } 1, \dots, n \text{ except } j \text{ and } i.$$

From this we see that for any $k$ if $R_{ik} = 1$ and $R_{kj} = 1$ we must have $R_{ij} = 1$ to satisfy the condition and thus these conditions guarantee transitivity. We next include conditions defining the scoring function of $R$. In particular for each $h_i$ we have a constraint $v_i = \dfrac{1}{n}\left(\sum_{j=1}^{n} R_{ij}\right)$. Finally we require that each $R_{ij}$ must be a binary integer variable, $R_{ij} \in \{0, 1\}$.

The above collection of constraints assures us that $R$ is a weak ordering which is a completion of our original $S$.

The objective function, which we desire to maximize, is the $Buo(R)$. However, we must recall that the calculation of the buoyancy function requires an ordering over the set of scores associated with the $v_i$. Thus the objective function is not a simple linear calculation. In order to implement this approach we use a method suggested by Yager [22] for converting objective functions involving an ordering of the arguments into one that doesn't. We first introduce a collection of variables $y_1, \ldots, y_n$ where $y_i$ is used to indicate the ith largest of the calculated scores, the $v_j$. We next use as our objective function

$$\sum_{i=1}^{n} w_i y_i$$

where the $w_i$ are the weights associated with our buoyancy measure. We next must introduce some constraints that assure us that the $y_i$'s are in the appropriate order. We first introduce a collection of constraints guarantying the ordering of the $y_i$'s,

$$y_{i+1} - y_i \leq 0, \quad i = 1, \ldots, n-1$$

We next introduce a collection of constraints assigning the $y_i$'s to the appropriate $V$ value. For each $i = 1$ to $n$ we introduce the following set of constraints

$$y_i - v_j - 1000 Z_{ij} \leq 0, \quad j = 1, 2, \ldots, n$$

$$\sum_{j=1}^{n} Z_{ij} \leq n-1$$

We also require that each $Z_{ij}$ must be an integer 0 - 1 variable.

In [22] Yager discusses how the introduction of these constraints works to obtain the ordering.

## 9. Conclusion

We discussed the approach to fuzzy systems modeling known as the hierarchical prioritized structure. An important part of this is the HEU operator that is used to combine information from different levels of the HPS. Some alternative formulations for this operator were suggested. We considered the issues related to the learning of the HPS and discussed two methodologies that can be used in appropriate situations. First we considered the DELTA method which provides an algorithm that can dynamically adapt the HPS model based on observations. We next considered the situation where rules are provided by an expert and focused on obtaining the appropriate ordering of the rules within the HPS. Here we used the principle of maximal buoyancy to help complete quasi–orderings.

## References

[1] R. R. Yager and D. P. Filev, Essentials of Fuzzy Modeling and Control, John Wiley, New York, 1994.

[2] L. A. Zadeh, Toward a theory of fuzzy information granulation and its centrality in human reasoning and fuzzy logic, *Fuzzy Sets and Systems*, **90**, 1997, pp. 111-127.

[3] L. A. Zadeh, The calculus of fuzzy if/then rules, AI Expert, March, 1992, pp. 23- 27.

[4] B. Kosko, Fuzzy Engineering, Prentice Hall, Upper Saddle River, NJ, 1997.

[5] R. R. Yager, On a hierarchical structure for fuzzy modeling and control, *IEEE Transactions on Systems, Man and Cybernetics* **23**, 1993, pp. 1189-1197.

[6] R. R. Yager, Hierarchical representation of fuzzy if-then rules, in *Advanced Methods in Artificial Intelligence*, Bouchon-Meunier, B., Valverde, L., Yager, R. R. (Eds.), Springer-Verlag: Berlin, 1993, pp. 239–247.

[7] R. R. Yager, On the construction of hierarchical fuzzy systems models, *IEEE Transactions on Systems, Man and Cybernetics, Part C: Applications and Reviews* **28**, 1998, 55-66.

[8] R. Rozich, T. Ioerger and R. R. Yager, FURL - A Theory Revision Approach to Learning Fuzzy Rules, Proceedings of Conference on Fuzzy Systems on the IEEE World Congress on Computational Intelligence, Honolulu, 2002, 791-796.

[9] E. H. Mamdani, Application of fuzzy algorithms for control of simple dynamic plant, *Proc. IEEE* **121**, 1974, 1585-1588.

[10] E. H. Mamdani and S. Assilian, An experiment in linguistic synthesis with a fuzzy logic controller, *Int. J. of Man-Machine Studies* **7**, 1975, pp. 1-13.

[11] R. R. Yager and D. P. Filev, On the issue of defuzzification and selection based on a fuzzy set, *Fuzzy Sets and Systems* **55**, 1993, pp. 255-272.

[12] M. Sugeno and T. Takagi, A new approach to design of fuzzy controller, In Advances in Fuzzy Sets, Possibility Theory and Applications, Wang, P. P. (Ed.), Plenum Press, New York, 1983, pp. 325-334.

[13] T. Takagi and M. Sugeno, Fuzzy identification of systems and its application to modeling and control, *IEEE Transactions on Systems, Man and Cybernetics* **15**, 1985, pp. 116-132.

[14] R. R. Yager and D. P. Filev, Approximate clustering via the mountain method, *IEEE Transactions on Systems, Man and Cybernetics* **24**, 1994, pp. 1279-1284.

[15] E. P. Klement, R. Mesiar and E. Pap, Triangular Norms, Kluwer Academic Publishers, Dordrecht, 2000.

[16] R. R. Yager and D. P. Filev, Generation of fuzzy rules by mountain clustering, *Journal of Intelligent and Fuzzy Systems* **2**, 1994, pp. 209-219.

[17] S. L. Chiu, Fuzzy model identification based on cluster estimation, *Journal of Fuzzy and Intelligent Systems* **2**, 1994, pp. 267-278.

[18] M. Roubens and P. Vincke, Preference Modeling, Springer-Verlag, Berlin, 1989.

[19] R. R. Yager, On the completion of priority orderings in nonmonotonic reasoning systems, *International Journal of Uncertainty, Fuzziness and Knowledge Based Systems* **1**, 1993, pp. 139-165.

[20] R. R. Yager, On the completion of qualitative possibility measures, *IEEE Transaction on Fuzzy Systems* **1**, 1993, pp. 184-194.

[21] R. R. Yager, Completing orderings using the principle of maximal buoyancy, In: Uncertainty Modeling and Analysis: Theory and Applications, Ayyub, B. M, Gupta, M. M. (Eds.), North Holland, Amsterdam, 1994, pp. 41-57.

[22] R. R. Yager, Constrained OWA aggregation, *Fuzzy Sets and Systems* **81**, 1996, pp. 89-101.

# Algebraic Aspects of Information Organization

Ioan TOFAN and Aurelian Claudiu VOLF
*Faculty of Mathematics, University "Al. I. Cuza", Iasi 6600, Romania*
*e-mail: tofan@uaic.ro; volf@uaic.ro*

**Abstract.** In what follows, we approach the problem of information organization from the viewpoint of generalized structures (fuzzy structures and hyperstructures). The fuzzy quantitative information can be modeled by *fuzzy numbers*, while the fuzzy qualitative information has its counterpart in *hyperstructures*, in the sense that, for example, two (fuzzy) informations yield a *set* of possible consequences. The significance of information appears most clearly in *structures*; this induces the necessity of studying the *fuzzy algebraic structures* (fuzzy groups, rings, ideals, subfields and so on) as a means towards the better understanding and processing of information. This report presents some recent results and methods in the rapidly growing fields of fuzzy algebraic structures and hyperstructures and some connections between them. Some results on fuzzy groups, fuzzy rings and fuzzy subfields are given. Likewise, the consideration of diverse *sets of fuzzy numbers* and, more notably, of the *structures* that these sets can be endowed with is of utmost importance. In this direction, the *operations* with fuzzy numbers play a major role and a number of questions regarding these operations are still open. A sample of the different notions of fuzzy number and of the operations with fuzzy numbers and their properties is given in this report. The similarity relations (fuzzy generalizations of equivalence relations) are in direct connection with shape (pattern) recognition. Diverse types of similarity classes and partitions are studied. Several notions of $f$-hypergroup, which combine fuzzy structures and hyperstructures, are presented and studied. Some results that put forward a two-way connection between $L$-fuzzy structures and hyperstructures are given.

## 1. Introduction

In what follows, we deal with the problem of information organization from the viewpoint of generalized structures (fuzzy structures and hyperstructures).

Generally speaking, one can accept the fact that "to solve a problem (not necessarily of a mathematical nature)" means "to determine a set" (the set of the solutions), based upon the problem data (that is, upon a set of informations). But, to determine a set means to give a characteristic property, in other words, to obtain an information. In this context, a classification of properties (informations) may be useful. One can distinguish between *qualitative properties* (corresponding to the *linguistic level* of information) and *quantitative properties* (corresponding to the *numerical level* of information). In most cases, the information is not crisp, precise, but vague and imprecise, "fuzzy". The fuzzy quantitative information can be modeled by *fuzzy numbers*, while the fuzzy qualitative information has its counterpart in *hyperstructures*, in the sense that, for example, two (fuzzy) informations yield a *set* of possible consequences.

The significance of information appears most clearly in structures; this induces the necessity of studying the *fuzzy algebraic structures* (fuzzy groups, rings, ideals, subfields and so on) as a means towards the better understanding and processing of information. The theory of algebraic hyperstructures has surprising connections with the fuzzy structures, which can be interpreted as connections between the two types of information described above. The similarity relations (fuzzy generalizations of equivalence relations) are in direct connection with shape (pattern) recognition.

This report presents some recent results and methods in the rapidly growing fields of fuzzy algebraic structures and hyperstructures and some connections between them. Some results on fuzzy groups, fuzzy rings and fuzzy subfields are given. Likewise, the consideration of diverse *sets of fuzzy numbers* and, more notably, of the *structures* that these sets can be endowed with is of utmost importance. In this direction, the *operations* with fuzzy numbers play a major role and a number of questions regarding these operations are still open. A sample of the different notions of fuzzy number and of the operations with fuzzy numbers and their properties is given in this report. Diverse types of similarity classes and partitions are studied. Several notions of *f*-hypergroup, which combine fuzzy structures and hyperstructures, are presented and studied. Some results that put forward a two-way connection between *L*-fuzzy structures and hyperstructures are given.

## 2. Preliminaries

### 2.1. Fuzzy sets

The theory of fuzzy sets extends the area of applicability of mathematics, by building the instruments and the framework for the management of the imprecision inherent to the human language and thinking. The starting point is generalizing the notion of subset of a set. It is well-known that a subset $A'$ of the set $A$ is perfectly determined by its *characteristic function*

$$\chi_{A'} : A' \to \{0,1\}, \; \chi_{A'}(x) = \begin{cases} 1, \; if \; x \in A' \\ 0, otherwise \end{cases} \cdot$$

One generalizes the notion of "belonging to" the subset $A'$ by introducing a gradual transition from "does not belong to" to "belongs to" (L. Zadeh, 1965). L. Zadeh succeeded in imposing the theory of fuzzy sets, by exhibiting applications of the theory. The idea of rejecting the principle "tertium non datur" is directly connected to the generalization above. It goes back to Aristotle and appears in the modal logic (Mac Coll, 1897) or multivalued logics. The generalization of the concept of "characteristic function" was given by H. Weyl (1940) and appears again in a new interpretation in papers by A. Kaplan & H. F. Schott and K. Menger.

1.1. DEFINITION. Let $U$ be a nonempty set. A pair $(U, \mu)$, where $\mu: U \to [0, 1]$ is a mapping, is called a *fuzzy set*. If $x \in U$, $\mu(x)$ is understood as the "degree to which $x$ belongs to the fuzzy set determined by $\mu$". We shall also call $\mu: U \to [0, 1]$ a *fuzzy subset* of $U$ and denote $\mathsf{F}(U) := [0, 1]^U = \{\mu \mid \mu: U \to [0, 1]\}$ the set of fuzzy subsets of $U$.

It is sometimes useful to replace the interval $[0, 1]$ (which is a lattice with respect to the usual order relation) with a lattice $(L, \wedge, \vee)$. Thus, a pair $(U, \mu)$, where $\mu: U \to L$, is called an

*L-fuzzy set* or *L-fuzzy subset* of *U*. Many definitions and results on fuzzy sets can be transferred to *L*-fuzzy sets, provided some conditions on *L* are imposed.

1.2. Definition. Let $(U, \mu)$ be a fuzzy set and $\alpha \in [0, 1]$. The set

$$_\mu U_\alpha := \{x \in U \mid \mu(x) \geq \alpha\}$$

(also denoted $\mu_\alpha$) is called the $\alpha$-*level set* of $(U, \mu)$. Let supp $\mu := \{x \in U \mid \mu(x) \neq 0\}$.

1.3. PROPOSITION. Let $(U_\alpha)_{\alpha \in [0, 1]} \subseteq P(U)$ be a family of subsets of U. Then $(U_\alpha)_{\alpha \in [0, 1]}$ is the family of level sets of a fuzzy subset $\mu: U \rightarrow [0, 1]$ if and only if it satisfies the conditions:
a) $U_0 = U$.
b) $\forall \alpha, \beta \in [0, 1]$, $\alpha \leq \beta$ *implies* $U_\beta \subseteq U_\alpha$.
c) For any increasing sequence $(\alpha_i)_{i \geq 0}$, $\alpha_i \in [0, 1]$, $\forall i \in \mathbb{N}$, having limit $\alpha$, we have

$$U_\alpha = \bigcap_{i \geq 0} U_\alpha.$$

A fuzzy set is completely determined by the family of its level sets:

1.4. Proposition. Let X be a set and let $\mu$ a fuzzy subset of X. Then

$$\mu(x) = \sup\{k \in [0, 1] / x \in {}_\mu X_k\}.$$

1.5. Definition.
*i)* $\mu_\varnothing \in F(U)$ given by $\mu_\varnothing(x) = 0$, $\forall x \in U$, is called the *empty fuzzy subset* of *U*.
*ii)* If $\mu, \tau \in F(U)$, the *inclusion* $\mu \subseteq \tau$ is defined by $\mu(x) \leq \tau(x)$, $\forall x \in U$.
*iii)* If $\mu, \tau \in F(U)$, define $\mu \, \hat{\cup} \, \tau$ (the *union* of the fuzzy subsets $\mu$ and $\tau$) by

$$\mu \hat{\cup} t : U \rightarrow [0,1], \ (\mu \hat{\cup} t)(x) = \max\{\mu(x), t(x)\}.$$

The *intersection* is defined by $\mu \cap \tau : U \rightarrow [0,1]$, $(\mu \cap \tau)(x) = \min\{\mu(x), \tau(x)\}$. These definitions extend to families of fuzzy subsets: if $\{\mu_i\}_{i \in I} \subseteq F(U)$, then we set:

$$\bigcap_{i \in I} \mu_i : U \rightarrow [0,1], \ \left(\bigcap_{i \in I} \mu_i\right)(x) = \inf\{\mu_i(x)\};$$

$$\bigcup_{i \in I} \mu_i : U \rightarrow [0,1], \ \left(\bigcup_{i \in I} \mu_i\right)(x) = \sup\{\mu_i(x)\}.$$

*iv)* For $\mu \in F(U)$, the fuzzy subset $\mu' \in F(U)$ given by $\mu'(x) = 1 - \mu(x)$, $\forall x \in U$, is called the *complement* of $\mu$.

1.6. REMARK. $(F(U), \cap, \hat{\cup}, ')$ is a de Morgan algebra (as opposed to $(P(U), \cap, \cup, \bar{\ })$ which is a Boole algebra). Note that $\{0, 1\}$ has a Boole algebra structure (with respect to min, max, $x' = 1 - x$), while $[0, 1]$ with the same operations is just a de Morgan algebra. On $F(U)$ the following operations can be defined:

"+" by $\mu + \tau: U \to [0, 1]$, $(\mu + \tau)(x) = \mu(x) + \tau(x) - \mu(x)\tau(x)$, $\forall x \in U$.

"·" by $\mu \cdot \tau: U \to [0, 1]$, $(\mu \cdot \tau)(x) = \mu(x) \cdot \tau(x)$, $\forall x \in U$.

"●" by $\mu \bullet \tau: U \to [0, 1]$, $(\mu \bullet \tau)(x) = \min\{1, \mu(x) + \tau(x)\}$, $\forall x \in U$;

"○" by $\mu \bigcirc \tau: U \to [0, 1]$, $(\mu \bigcirc \tau)(x) = \max\{0, \mu(x) + \tau(x) - 1\}$, $\forall x \in U$

### 2.2. Hyperstructures

The concept of hypergroup was introduced in 1934 by F. Marty as a natural generalization of the notion of group. Many applications in geometry, combinatorics, group theory, automata theory etc. have turned hypergroup theory and subsequently hyperstructure theory into a relevant domain of modern algebra.

Let $H$ be a nonempty set. Let $P^*(H) = P(H) \setminus \{\varnothing\} = \{A \mid A \subseteq H, A \neq \varnothing\}$.

2.1. DEFINITION. A *hyperoperation* "$*$" on $H$ is mapping $*: H \times H \to P^*(H)$. For any $a \in H$ and $B \subseteq H, B \neq \varnothing$, we denote by: $a * B = \bigcup_{b \in B} a * b$.

Similarly one defines $B * a$. If $A, B \in P^*(H)$, let $A * B = \bigcup_{\substack{a \in A \\ b \in B}} a * b$.

A nonempty set endowed with a hyperoperation "$*$" on $H$ is called a *hypergroupoid*. If, $\forall a, b, c \in H$, we have $a * (b * c) = (a * b) * c$ (*associativity*), then $H$ is called a *semihypergroup*. If a semihypergroup $(H, *)$ satisfies $a * H = H * a = H$, $\forall a \in H$ (*reproducibility*) then $H$ is called a *hypergroup*. A hypergroup is called *commutative* if, $\forall a, b \in H, a * b = b * a$.

2.2. REMARK. A hyperoperation $*$ defined on a set $H$ induces two hyperoperations "$/$" and "$\backslash$". For every $x, y \in H$, define:

$$x / y = \{a \in H \mid x \in a * y\}, \qquad x \backslash y = \{b \in H \mid x \in y * b\}.$$

If "$*$" is commutative, then $x / y = x \backslash y$, $\forall x, y \in H$. Also, the reproducibility axiom is equivalent to the condition: $\forall x, y \in H, x / y \neq \varnothing$ and $x \backslash y \neq \varnothing$.

2.3. DEFINITION. A commutative hypergroup $(H, *)$ is called a *join space* if, $\forall a, b, c, d \in H$, $a/b \cap c/d \neq \varnothing$ implies $a * d \cap b * c \neq \varnothing$.

### 3. Fuzzy algebraic structures

### 3.1. Fuzzy subgroups

1.1. DEFINITION. Let $(G, \cdot, e)$ be a group and let $\mu: G \to [0, 1]$ be a fuzzy subset of $G$. We say that $\mu$ is a *fuzzy subgroup* of $G$ if:

i) $\mu(xy) \geq \min\{\mu(x), \mu(y)\}$, $\forall x, y \in G$;

ii) $\mu(x^{-1}) \geq \mu(x)$, $\forall x \in G$.

If moreover $\mu(xy) = \mu(yx)$, $\forall x, y \in G$, then $\mu$ is called a *normal fuzzy subgroup* of $G$.

1.2. REMARK. If $\mu$ is a fuzzy subgroup of $G$, then $\mu(x^{-1}) = \mu(x) \leq \mu(e)$, $\forall x \in G$. Moreover, $\mu$ is a normal fuzzy subgroup if and only if $\mu(y^{-1}xy) \geq \mu(x)$, $\forall x, y \in G$.

The next characterization is typical for all "fuzzy substructures".

1.3. PROPOSITION. A fuzzy set $\mu: G \to [0, 1]$ is a (normal) fuzzy subgroup of G if and only if the level subsets $_\mu G_\alpha$ are (normal) subgroups of G for all $\alpha \in \mathrm{Im}\ \mu$.

1.4. DEFINITION. We say the fuzzy set $(F, \mu)$ satisfies the *sup property* if, for every nonempty subset $A$ of Im $\mu$, there exists $x \in \{y \in F / \mu(y) \in A\}$ such that $\mu(x) = \sup A$. In other words, $\mu$ has the sup property if and only if any nonempty subset $A$ of Im $\mu$ has a greatest element.

1.5. PROPOSITION. *Let $(G, \cdot, e)$, $(H, \cdot, e')$ be groups, $f: G \to H$ group homomorphism and $\mu$, $\eta$ fuzzy subgroups of G, respectively H. Then $f^{-1}(\eta)$ is a fuzzy subgroup of G. If $(G, \mu)$ has the sup property, then $f(\mu)$ is a fuzzy subgroup of H.*

### *3.2. Fuzzy ideals*

2.1. DEFINITION. Let $(R, +, \cdot)$ be a unitary commutative ring.

i) A fuzzy subset $\sigma : R \to I$ is called a *fuzzy subring* of R if, $\forall x, y \in R$:

$$\mu(x - y) \geq \min\{\mu(x), \mu(y)\}; \qquad \mu(xy) \geq \min\{\mu(x), \mu(y)\}.$$

ii) A fuzzy subset $\sigma : R \to I$ is called a *fuzzy ideal* of R if, $\forall x, y \in R$:

$$\mu(x - y) \geq \min\{\mu(x), \mu(y)\}; \qquad \mu(xy) \geq \max\{\mu(x), \mu(y)\}.$$

2.2. PROPOSITION. *Let $\mu : R \to [0, 1]$ be a fuzzy ideal of R. Then:*

i) $\mu(1) = \mu(x) = \mu(-x) = \mu(0)$, $\forall x \in R$;

ii) $\mu(x - y) = \mu(0) \Rightarrow \mu(x) = \mu(y)$, $\forall\ x, y \in R$;

iii) $\mu(x) < \mu(y)$, $\forall y \in R \Rightarrow \mu(x - y) = \mu(x) = \mu(y - x)$.

2.3. PROPOSITION. A fuzzy subset $\mu : R \to [0, 1]$ is a fuzzy subring (ideal) of R if and only if all level subsets $_\mu R_\alpha$, $\alpha \in \mathrm{Im}\ \mu$, are subrings (ideals) of R.

2.4. REMARK. The intersection of a family of fuzzy ideals of $R$ is a fuzzy ideal of $R$. This leads to the notion of *fuzzy ideal generated* by a fuzzy subset $\sigma$ of $R$, namely the intersection of all fuzzy ideals that include $\sigma$, denoted $< \sigma >$. We have: $< \sigma >: R \to [0, 1]$ is given by

$$< \sigma > (x) = \sup\{\alpha \in [0, 1] \mid x \in <_\mu R_\alpha>\}.$$

2.5. PROPOSITION. The union of a totally ordered (with respect to the relation $\mu \leq \eta \Leftrightarrow \mu(x) \leq \eta(x)$, $\forall x \in R$) family of fuzzy ideals of R is a fuzzy ideal of R.

2.6. DEFINITION. Let $\mu$, $\theta$ be fuzzy ideals of $R$. The *product of* $\mu$ and $\theta$ is:

$$\mu \cdot \theta : R \to [0,1], (\mu \cdot \theta)(x) = \sup_{\substack{x = \sum y_i z_i \\ i < \infty}} \left\{ \min_i \left\{ \min \left\{ \mu(y_i), \theta(z_i) \right\} \right\} \right\}, \forall x \in R.$$

The *sum* of $\mu$ and $\theta$ is:

$\mu + \theta : R \to [0, 1], (\mu + \theta)(x) = \sup\{\min\{\mu(y), \theta(z)\} \mid y, z \in R, y + z = x\},$
$\forall x \in R.$

2.7. REMARK. In general, for $\mu$, $\theta$ fuzzy subsets of a set $S$ endowed with a binary operation "·", one defines the *product* $\mu\theta : S \to [0, 1]$,

$$(\mu\theta)(x) = \begin{cases} \sup\limits_{yz=x}\{\min\{\mu(y), \theta(z)\}\}, & \textit{if there exist } y, z \in S \textit{ such that } x = yz, \\ 0, & \textit{otherwise} \end{cases}$$

For any $\mu$, $\theta$ fuzzy ideals of $R$, we have $\mu \cdot \theta = {<}\mu\theta{>}$.

2.8. PROPOSITION. Let f: $R \to R'$ be a surjective ring homomorphism and $\mu$ a fuzzy ideal of R, $\mu'$ a fuzzy ideal of R'. Then:
   i) $f(\mu)$ is a fuzzy ideal of R';
   ii) $f^{-1}(\mu')$ is a fuzzy ideal of R.

2.9. DEFINITION. A nonconstant fuzzy ideal $\mu$ ($|\mathrm{Im}\ \mu| > 1$) of a ring $R$ is called a *fuzzy prime ideal* if, for any fuzzy ideals $\sigma$, $\theta$ of $R$, $\sigma\theta \subseteq \mu \Rightarrow \sigma \subseteq \mu$ or $\theta \subseteq \mu$.

### *3.3. Fuzzy rings of quotients*

The study of the fuzzy prime ideals of a ring leads naturally to the question of the existence of a "fuzzy localization" device, that is, to the problem of the construction of a fuzzy ring of quotients. Let $R$ be unitary commutative ring. $R^*$ denotes the set of the invertible elements of $R$.

3.1. DEFINITION. A fuzzy subset $\sigma : R \to [0, 1]$ is called a *fuzzy multiplicative subset* (FMS for short) if:
   i) $\sigma(xy) \geq \min(\sigma(x), \sigma(y))$, $\forall x, y \in R$.
   ii) $\sigma(0) = \min\{\sigma(x): x \in R\}$;
   iii) $\sigma(1) = \max\{\sigma(x): x \in R\}$.

3.2. PROPOSITION. The fuzzy subset $\sigma$ of the ring R is a FMS if and only if every level subset $\sigma_t = \{x \in R: \sigma(x) \geq t\}$, $t > \sigma(0)$, is a multiplicative system (in the classical sense).

Recall that a multiplicative subset $S$ of $R$ is *saturated* if $xy \in S$ implies $x, y \in S$.

3.3. DEFINITION. A FMS $\sigma$ of a ring $R$ is called *saturated* if, for any $x, y \in R$,
$$\sigma(xy) = \min(\sigma(x), \sigma(y)).$$

3.4. PROPOSITION. The fuzzy subset $\sigma$ of the ring R is a saturated FMS if and only if every level subset $\sigma_t$ is a saturated multiplicative system, $\forall\ t > \sigma(0)$.

3.5. PROPOSITION. If $\mu$ is a fuzzy prime ideal of a ring R, then $1-\mu$ is a saturated FMS.

3.6. PROPOSITION. *Let* $\sigma$ *be a FMS of the ring R. Then the fuzzy subset* $\overline{\sigma}$ *, defined by*

$$\overline{\sigma}(x) = \sup\{\sigma(xy): y \in R\}$$

is a saturated FMS, with $\sigma \leq \overline{\sigma}$. Moreover, if $\tau$ is a saturated FMS with $\sigma \leq \tau$, then $\overline{\sigma} \leq \tau$. This result entitles us to call $\overline{\sigma}$ above the *saturate* of $\sigma$.

Let $\sigma$ be a FMS of the ring $R$ and $m = \sigma(0)$. For every $t > m$, we may construct the classical ring of fractions $\sigma_t^{-1} R = S_t$ with respect to the multiplicative subset $\sigma_t$. Let $\varphi_t$ denote the canonical ring homomorphism $R \to S_t$. If $s < t$, since $\sigma_t \subseteq \sigma_s$, the universality property of the ring of fractions yields the existence of a unique ring homomorphism $\varphi_{ts} : S_t \to S_s$ such that $\varphi_{ts} \circ \varphi_t = \varphi_s$. The system of rings and homomorphisms $(S_t, \varphi_{ts})$, $t$, $s \in [m, 1]$ is an inductive system (if $[m, 1]$ is endowed with the reverse of the usual order). Let $\sigma^{-1}R$ denote the inductive limit of this system and let $\varphi$ be the canonical ring homomorphism $R \to \sigma^{-1}R$ (the inductive limit of the $\varphi_t$, $t > m$). It is natural to call $\sigma^{-1}R$ the *ring of quotients* relative to the FMS $\sigma$.

3.7. PROPOSITION. With the notations above, $\varphi$ has the following universality property: for every $t > \sigma(0)$, $\varphi(\sigma_t) \subseteq (\sigma^{-1}R)^*$; if $T$ is a ring and $\psi : R \to T$ is a ring homomorphism such that for every $t > \sigma(0)$, $\varphi(\sigma_t) \subseteq T^*$, then it exists a unique ring homomorphism f: $\sigma^{-1}R \to T$ such that $f \circ \varphi = \psi$.

3.8. PROPOSITION. *There is a canonical isomorphism* $\psi : \sigma^{-1}R \to \overline{\sigma}^{-1}R$ .

If $\overline{\varphi} : R \to \overline{\sigma}^{-1}R$ denotes the canonical homomorphism, then $\overline{\varphi} = \psi \circ \varphi$ .

By applying Zorn's Lemma to the set **P**, one proves:

3.9. PROPOSITION. If $\sigma$ is a FMS in R and $\mu$ is a fuzzy ideal such that $\mu \cap \sigma = \varnothing$, then the set $\mathbf{P} = \{\eta: \eta$ is a fuzzy ideal of R, $\eta \cap \sigma = \varnothing$, $\mu \subseteq \eta\}$ has maximal elements and any such element is a fuzzy prime ideal. Thus it exists a fuzzy prime ideal $\pi$ such that $\pi \cap \sigma = \varnothing$.

3.10. PROPOSITION. *Let* $\pi$ *be a fuzzy prime in R and denote by* $R_\pi$ *the ring* $(1 - \pi)^{-1}R$. *Then* $R_\pi$ *is a local ring.*

### *3.4. Fuzzy intermediate fields*

Let $F/K$ be a field extension and let $\mathfrak{I}(F/K) = \{L/ L$ subfield of $F$, $K \subseteq L\}$ be the lattice of its *intermediate fields* (we also called them *subextensions* of $F/K$). If $F/K$ is a field extension and $c \in F$ is algebraic over $K$, then we denote by $Irr(c, K) \in K[X]$ the minimal polynomial of $c$ over $K$.

4.1. DEFINITION. Let $F/K$ be an extension of fields and $\mu: F \to [0, 1]$ a fuzzy subset of $F$. We call $\mu$ a *fuzzy intermediate field* of $F/K$ if, $\forall x, y \in F$:

$\mu(x - y) \geq \min\{\mu(x), \mu(y)\}$;
$\mu(xy^{-1}) \geq \min\{\mu(x), \mu(y)\}$ if $y \neq 0$;
$\mu(x) \leq \mu(k)$, $\forall k \in K$.

Let $\mathcal{FI}(F/K)$ denote the set of all fuzzy intermediate fields of $F/K$. If $\mu \in \mathcal{FI}(F/K)$, then $\mu$ is a constant on $K$.

For any fuzzy subset $\mu: F \to [0, 1]$ and $s \in [0, 1]$, define the *level set:*

$$\mu_s := \{x \in F \,/\, \mu(x) \geq s\}.$$

It is well known that a fuzzy subset $\mu: F \to [0, 1]$ is a fuzzy intermediate field if and only if, $\forall s \in \mathrm{Im}\,\mu$, the level set $\mu_s$ is an intermediate field of F/K.

4.2. THEOREM. Let F/K be a field extension. Then every fuzzy intermediate field of F/K has the sup property iff there are no infinite strictly decreasing sequences of intermediate fields of F/K.

4.3. REMARK. This result can be applied, mutatis mutandis, to *any algebraic structure* for which is defined a notion of "fuzzy substructure". For instance, let $(G, \cdot)$ be a group and 1 is its neutral element. By replacing in Theorem 4. 2 "intermediate field" with "subgroup" and $K$ (the base field) with the trivial subgroup $\{1\}$, one obtains the following fact:

4.4. PROPOSITION. Let G be a group. Then every fuzzy subgroup of G has the sup property if and only if there are no infinite strictly decreasing sequences of subgroups of G.

Similarly, in the case of *fuzzy ideals*, we have:

4.5. PROPOSITION. Let R be a unitary commutative ring. Then every fuzzy ideal of R has the sup property if and only if R is Artinian (there are no infinite strictly decreasing sequences of ideals of R).

4.6. DEFINITION. [2] Let $F/K$ be an extension of fields and $\mu \in \mathcal{FI}(F/K)$. Then $\mu$ is called a *fuzzy chain subfield* of $F/K$ if $\forall x, y \in F$, $\mu(x) = \mu(y) \Leftrightarrow K(x) = K(y)$.

Here is a fuzzy characterization of the fact that $\mathcal{I}(F/K)$ is a chain.

4.7. THEOREM. [[2], Th. 3. 3]. The intermediate fields of F/K are chained if and only if F/K has a fuzzy chain subfield.

4.8. THEOREM. Let F/K be an extension such that the intermediate fields of F/K are chained. Then:

a) F/K is algebraic.

b) Any intermediate field L of F/K with $L \neq F$ is a finite simple extension of K.

c) $(\mathcal{I}(F/K), \subseteq)$ satisfies the descending chain condition (there is no strictly decreasing sequence of intermediate fields of F/K). Thus, $(\mathcal{I})F/K), \subseteq)$ is well ordered.

4.9. Corollary. Let F/K be a field extension.

a) Assume that any proper intermediate field of F/K is a finite extension of K. Then every $\mu \in \mathcal{FJ}(F/K)$ has the sup property.

b) If the intermediate fields of F/K are chained, then every $\mu \in \mathcal{FJ}(F/K)$ has the sup property.

c) If every $\mu \in \mathcal{FJ}(F/K)$ has the sup property, then F/K is algebraic.

## 4. Applications and connections

### *4.1. Fuzzy numbers*

1.1. DEFINITION. Let $(G, \cdot)$ be a set endowed with a binary operation "$\cdot$" (usually a group). Let $\mu, \theta \in \mathsf{F}\,(G)$. We use the definition 3.2.7 for $\mu * \theta \in \mathsf{F}\,(G)$,

$$(\mu * \theta)(x) = \begin{cases} \sup_{yz=x}\{\min\{\mu(y),\theta\{z\}\}\}, & \text{if there exist } y,z \in S \text{ such that } x = yz, \\ 0, & \text{otherwise} \end{cases}$$

Thus, "$*$" is a binary operation on $\mathsf{F}\,(G)$.

If $G$ is a group and $e$ is its neutral element, we denote $\chi_{\{e\}}$ by $\varepsilon$. For any $\mu \in \mathsf{F}\,(G)$, let

$$\tilde{\mu} : G \to [0, 1], \ \tilde{\mu}(x) = \mu(x^{-1}), \ \forall x \in G.$$

1.2. Proposition. *Let G be a group.*
i) The operation "$*$" on $\mathsf{F}\,(G)$ is associative;
ii) If $G$ is commutative, then "$*$" is commutative.
iii) $\forall \mu \in \mathsf{F}\,(G), \ \mu * \varepsilon = \varepsilon * \mu = \mu$;
iv) $\varepsilon \subseteq \mu * \tilde{\mu}, \ \varepsilon \subseteq \tilde{\mu} * \mu.$

1.3. Proposition. *Let* $\mu, \tau, \nu \in \mathsf{F}\,(G)$. *Then:*
i) $\mu \subseteq \tau \Rightarrow \mu * \nu \subseteq \tau * \nu, \nu * \mu \subseteq \nu * \tau$;
ii) $\mu * (\tau \,\hat{\cup}\, \nu) = (\mu * \tau) \,\hat{\cup}\, (\mu * \nu); (\tau \,\hat{\cup}\, \nu) * \mu = (\tau * \mu) \,\hat{\cup}\, (\nu * \mu)$;
iii) $\mu * (\tau \cap \nu) = (\mu * \tau) \cap (\mu * \nu); (\tau \cap \nu) * \mu = (\tau * \mu) \cap (\nu * \mu).$

1.4. DEFINITION. *A fuzzy number* is a mapping $\mu : \mathbb{R} \to [0, 1]$ (where $\mathbb{R}$ is the field of real numbers) such that there exists $x_\mu \in \mathbb{R}$ with $\mu(x_\mu) = 1$, the set $\{x \mid \mu(x) \neq 0\}$ is bounded and the level sets $_\mu\mathbb{R}_\alpha$ are closed intervals ($\alpha \in [0, 1]$).

For any $r \in \mathbb{R}$, the mappings $\tilde{r} : \mathbb{R} \to [0, 1]$,

$$\tilde{r}(x) = \begin{cases} 1, & x = r \\ 0, & \text{otherwise} \end{cases}, \text{ are called } \textit{degenerate fuzzy numbers.}$$

One usually takes the fuzzy numbers of the following type:

$$\mu(x) = \begin{cases} 0 & ,x < a \quad ; \\ \pi_1(x), & x \in [a,b); \\ 1 & ,x \in [b,c] ; \\ \pi_2(x), & x \in (c,d] ; \\ 0 & ,x > d. \end{cases} \tag{1}$$

where $a \leq b \leq c \leq d$ are reals, and $\pi_1, \pi_2 : \mathbb{R} \to \mathbb{R}$ satisfy the conditions that turn $\mu$ in a fuzzy number as in the definition. For:

$$\pi_1(x) = \frac{x-a}{b-a}, \quad \pi_2(x) = \frac{d-x}{d-c}$$

one gets *trapezoidal fuzzy numbers*. If $b = c$, *triangular fuzzy numbers* are obtained. A trapezoidal fuzzy number as above is denoted by $A = (a, b, c, d)$, respectively $A = (a, b, d)$ for triangular fuzzy numbers.

The operations with fuzzy numbers $\mu, \eta$ are defined as in the case of $\mathbf{F}(G)$ above:

$$\mu * \eta : R \to [0,1], \quad (\mu * \eta)(z) = \sup_{x \circ y = z} \{ \min \{ \mu(x), \eta(y) \} \},$$

By replacing "$\circ$" with "+", ".", "−", ": ", one obtains the operations "●", respectively "○", "⊖", "⊘".

We use the following notations:
- $\mathcal{R}$ is the set of nondegenerate fuzzy numbers;
- $\mathcal{R}_+ = \{ \mu \in \mathcal{R} \mid \mu(x) > 0 \Rightarrow x > 0 \}$, $\mathcal{R}_- = \{ \mu \in \mathcal{R} \mid \mu(x) > 0 \Rightarrow x < 0 \}$,
- $\mathcal{R}^* = \mathcal{R}_+ \cup \mathcal{R}_-$;

1.5. REMARK. For any $\mu, \eta \in \mathcal{R}$ and $r \in \mathbb{R}$, we have, $\forall x \in \mathbb{R}$:

$$(\tilde{r} ● \mu)(x) = \mu(r - x); \qquad (\mu ● \eta)(x) = \sup_{y \in \mathbb{R}} \{ \min \{ \mu(y), \eta(x - y) \} \};$$

$$(\tilde{r} ○ \mu)(x) = \begin{cases} \mu\left(\dfrac{x}{r}\right), & r \neq 0 \\ \begin{cases} 1, & x = 0 \\ 0, & x \neq 0 \end{cases}, & r = 0 \end{cases} ;$$

$$(\mu ○ \eta)(x) = \begin{cases} \sup_{y \neq 0} \left\{ \min \left\{ \mu(y), \mu\left(\dfrac{x}{y}\right) \right\} \right\}, & x \neq 0 \\ \max \{ \mu(0), \eta(0) \}, & x = 0 \end{cases} .$$

1.6. REMARK. Fuzzy numbers can be characterized by a family of intervals (*intervals of confidence*). Let $\mu \in \mathcal{R}$, $\alpha \in [0, 1]$. Define $\mu_\alpha = [\underline{x}_\alpha, \overline{x}_\alpha]$, where $\underline{x}_\alpha = \inf\{x \mid \mu(x) = \alpha\}$, and $\overline{x}_\alpha = \sup\{x \mid \mu(x) = \alpha\}$. If $\mu$ is of the type (1), we get:

$$[\underline{x}_\alpha, \overline{x}_\alpha] = \begin{cases} \left[\pi_1^{-1}(\{\alpha\}), \pi_2^{-1}(\{\alpha\})\right] & , \alpha \neq 1 \\ [b, c] & , \alpha = 1 \end{cases}.$$

The conditions $\pi_1$ strictly increasing and $\pi_2$ strictly decreasing determine the fuzzy number if the confidence intervals are given. For the numbers of the type $\tilde{r}$ $(r \in \mathbb{R})$ the use of confidence intervals is superfluous. In this context the operations with fuzzy numbers can be defined as follows: $\forall \mu, \eta \in \mathcal{R}$, with $\mu_\alpha = [\underline{x}_\alpha, \overline{x}_\alpha]$, $\eta_\alpha = [\underline{y}_\alpha, \overline{y}_\alpha]$, we define:

$$(\mu \bullet \eta)_\alpha = [\underline{x}_\alpha + \underline{y}_\alpha, \overline{x}_\alpha + \overline{y}_\alpha]; \quad (\mu \ominus \eta)_\alpha = [\underline{x}_\alpha - \overline{y}_\alpha, \overline{x}_\alpha - \underline{y}_\alpha];$$

$$(\mu \bigcirc \eta)_\alpha = [\min\{\underline{x}_\alpha \underline{y}_\alpha, \underline{x}_\alpha \overline{y}_\alpha, \overline{x}_\alpha \underline{y}_\alpha, \overline{x}_\alpha \overline{y}_\alpha\}, \max\{\underline{x}_\alpha \underline{y}_\alpha, \underline{x}_\alpha \overline{y}_\alpha, \overline{x}_\alpha \underline{y}_\alpha, \overline{x}_\alpha \overline{y}_\alpha\}];$$

$$(\mu \oslash \eta)_\alpha = \left[\min\left\{\frac{\underline{x}_\alpha}{\underline{y}_\alpha}, \frac{\overline{x}_\alpha}{\underline{y}_\alpha}, \frac{\underline{x}_\alpha}{\overline{y}_\alpha}, \frac{\overline{x}_\alpha}{\overline{y}_a}\right\}, \max\left\{\frac{\underline{x}_\alpha}{\underline{y}_\alpha}, \frac{\overline{x}_\alpha}{\underline{y}_\alpha}, \frac{\underline{x}_\alpha}{\overline{y}_\alpha}, \frac{\overline{x}_\alpha}{\overline{y}_a}\right\}\right], \text{if } 0 \notin \eta_\alpha, (\alpha \in (0, 1)).$$

1.7. REMARK. For trapezoidal or triangular fuzzy numbers, $A = (a, b, c, d)$, respectively $A = (a, b, c)$, the confidence intervals are $A_\alpha = [(b - a)\alpha + a, (c - d)\alpha + d]$, respectively:

$$A_\alpha = [(b - a)\alpha + a, (b - c)\alpha + c], \alpha \in [0, 1].$$

In these cases, $A_0 = [a, c]$, respectively $A_1 = [a, b]$.

Since $A_0$ and $A_1$ determine completely the triangular fuzzy number $\mu$, sometimes it is taken the following definition (for $A = (a_1, b_1, c_1)$, $B = (a_2, b_2, c_2)$):

$$A \bullet B = (a_1 + a_2, b_1 + b_2, c_1 + c_2); \quad A \bigcirc B = (a_1 a_2, b_1 b_2, c_1 c_2), \text{ for } a_1, a_2 \geq 0;$$

$$A \ominus B = (a_1 - a_2, b_1 - b_2, c_1 - c_2); \quad A \oslash B = \left(\frac{a_1}{c_2}, \frac{b_1}{b_2}, \frac{c_1}{a_2}\right), \text{ for } a_1, a_2 > 0.$$

If $A$ is a triangular fuzzy number, $A = (a, b, c)$, we denote also $-A = (-a, -b, -c)$ and $A^{-1} = (c^{-1}, b^{-1}, a^{-1})$ if $a > 0$. For any $\alpha \in \mathbb{R}$, let $0_\alpha = (-\alpha, 0, \alpha)$ and for any $\alpha \geq 1$, let $1_\alpha = (\alpha^{-1}, 1, \alpha)$. We have $0_\alpha \bullet 0_\beta = 0_{\alpha+\beta}$, $1_\alpha \bigcirc 1_\beta = 1_{\alpha\beta}$.

1.8. DEFINITION. We define on the set of triangular fuzzy numbers $\mathcal{R}_t$ the following relations: $A_1 = (a_1, b_1, c_1)$ and $A_2 = (a_2, b_2, c_2)$ are $\bullet$-*equivalent* (and write $A_1 \sim_\bullet A_2$) if there exist $0_\alpha$, $0_\beta$ such that $A_1 \bullet 0_\alpha = A_2 \bullet 0_\beta$.

We say that $A_1 = (a_1, b_1, c_1)$ and $A_2 = (a_2, b_2, c_2)$ are $\bigcirc$-*equivalent* (we write $A_1 \sim_\bigcirc A_2$) if there exist $1_\alpha$, $1_\beta$ such that $A_1 \bigcirc 1_\alpha = A_2 \bigcirc 1_\beta$.

It is easy to see that $0_\alpha \sim_\bullet 0_\beta$ for every $\alpha$, $\beta \in \mathbb{R}$ and $1_\alpha \sim_\bigcirc 1_\beta$ for every $\alpha$, $\beta \geq 1$.

1.9. PROPOSITION. *The relations* $\sim_\bullet$, $\sim_\bigcirc$ *are equivalence relations.*

Let $\mathcal{R}_\bullet = \mathcal{R}_t / \sim_\bullet$ and for every $A \in \mathcal{R}_t$ denote $\overline{A} \in \mathcal{R}_\bullet$ the equivalence class of $A$.

For $\overline{A}$, $\overline{B} \in \mathcal{R}_\bullet$, we define $\overline{A}\ [+]\ \overline{B} = \overline{A[+]B}$.

1.10. PROPOSITION. The operation $[+]$ is well defined and $(\mathcal{R}_\bullet, [+])$ is an abelian group, $\overline{0_\alpha}$ being its neutral element $(\forall \alpha \in \mathbb{R})$; $\overline{-A}$ is the symmetrical element of $\overline{A}$.

1.11. REMARK. A similar result can be obtained for $\mathcal{R}_\bigcirc = {}_+\mathcal{R}_t / \sim_\bigcirc$, where ${}_+\mathcal{R}_t$ is the set of triangular fuzzy numbers $(a, b, c)$ with $a > 0$.

We note the fact that the operations "$\bigcirc$" or "$\oslash$" defined before (using $\mu * \eta$ or confidence intervals) do not necessarily lead to triangular numbers if one starts with triangular numbers. For instance, $\tilde{1} \bigcirc \tilde{1} = \begin{cases} \sqrt{t}, t \in [0,1]; \\ 0, \ otherwise \end{cases}$ ($\tilde{1} = (1, 1, 1)$). This justifies somehow the operations defined above ("component-wise"), but the deviations for the variant given by "$*$" for product and quotient are considerable. On the other hand, one obtains for usual real numbers (considered as fuzzy numbers) the usual operations. The problem of building an acceptable arithmetic for fuzzy number is still open.

### 4.2. Similarity relations and partitions

The role played by the notion of *relation* in the structure of mathematical concepts is well known. We review known results on the introduction of this notion in the framework of fuzzy set theory.

2.1. DEFINITION. Let $X$ and $Y$ be sets. We call a *fuzzy relation* between $X$ and $Y$ any fuzzy subset $\rho: X \times Y \to [0, 1]$ of the (usual) cartesian product $X \times Y$. If $X = Y$, we say that $\rho$ is a *fuzzy relation on X*.

Let $\mathcal{R}_f(X)$ be the set of all fuzzy relations on $X$.

The *diagonal* fuzzy relation on $X$ is $\Delta: X \times X \to [0, 1]$, $\Delta(x, y) = \begin{cases} 1 & if\ x = y \\ 0 & if\ x \neq y \end{cases}$.

If $\rho: X \times Y \to [0, 1]$ is a fuzzy relation between $X$ and $Y$, then $\rho^{-1}: Y \times X \to [0, 1]$ defined by $\rho^{-1}(y, x) = \rho(x, y)$ is called the *inverse* of $\rho$.

In the same manner as in the classical case, since the fuzzy relations are, in fact, fuzzy subsets, one may introduce the operations $\hat{\cup}$ and $\cap$ with fuzzy relations, as well as defining the inclusion between the fuzzy relations. Among the many possibilities of *composing* the fuzzy relations, we present the definition due to Zadeh:

Let $X$, $Y$, $Z$ be sets and $\rho: X \times Y \to [0, 1]$, $\xi: Y \times Z \to [0, 1]$ fuzzy relations. The *composition of the fuzzy relations* $\rho$ and $\xi$ is the fuzzy relation $\rho \circ \xi: X \times Z \to [0, 1]$, defined by

$$\rho \circ \xi(x, z) = \sup_{y \in Y} \inf \{\rho(x, y), \xi(y, z)\}.$$

For $\rho \in \mathcal{R}_f(X)$, we set $\rho^0 = \Delta$ and $\rho^{n+1} = \rho^n \circ \rho$, $\forall n \in \mathbb{N}$.

2.2. PROPOSITION. *i) If $\rho_1: X \times Y \to [0, 1]$, $\rho_2: Y \times Z \to [0, 1]$, $\rho_3: Z \times U \to [0, 1]$ are fuzzy relations, then $(\rho_1 \circ \rho_2) \circ \rho_3 = \rho_1 \circ (\rho_2 \circ \rho_3)$.*

*ii) Let $\rho: Y \times Z \to [0, 1]$, $\rho_1$ and $\rho_2: X \times Y \to [0, 1]$ be fuzzy relations such that $\rho_1 \subseteq \rho_2$. Then $\rho_1 \circ \rho \subseteq \rho_2 \circ \rho$.*

*iii) Let $\rho: Y \times Z \to [0, 1]$, $\rho_1$ and $\rho_2: X \times Y \to [0, 1]$ be fuzzy relations. Then*
$$(\rho_1 \cup \rho_2) \circ \rho = (\rho_2 \circ \rho) \cup (\rho_2 \circ \rho) \text{ and } (\rho_1 \cap \rho_2) \circ \rho \subseteq (\rho_2 \circ \rho) \cap (\rho_2 \circ \rho).$$

2.3. DEFINITION. Let $\rho$ be a fuzzy relation on a fuzzy set $(X, \mu)$.
– $\rho$ is called *reflexive* if $\rho(x, x) = \mu(x)$, for any $x \in X$ ($\rho(x, x) = 1$ for an usual set);
– $\rho$ is called *symmetric* if $\rho(x, y) = \rho(y, x)$, for any $(x, y) \in X \times X$;
– $\rho$ is called *Z-transitive* if $\rho(x, z) \geq \sup_{y \in X} \min\{\rho(x, y), \rho(y, z)\}$, for any $x, z \in X$;

The fuzzy counterpart of the classical equivalence relation is the *similarity relation*.

2.4. DEFINITION. A relation $\rho : X \times X \to [0, 1]$ is called a *similarity relation* on $X$ if it is reflexive, symmetric and Z-transitive.

2.5. PROPOSITION. *Let $\rho : X \times X \to [0, 1]$ be a similarity relation and $x, y, z \in X$. Then:*
$$\rho(x, y) = \rho(y, z) \text{ or } \rho(x, z) = \rho(y, z) \text{ or } \rho(x, z) = \rho(x, y).$$

By using the level subsets, one obtains:

2.6. PROPOSITION. The relation $\rho: X \times X \to [0, 1]$ is a similarity relation if and only if, for any $\alpha \in [0, 1]$, $_\rho(X \times X)_\alpha$ is an equivalence relation on X.

2.7. PROPOSITION. *Let $\rho: X \times X \to [0, 1]$ be a fuzzy relation on X. The smallest similarity relation $\rho_s$ with $\rho \subseteq \rho_s$ is $\rho_s(x, y) = \sup \{(\rho \, \hat{\cup} \, \Delta \, \hat{\cup} \, \rho^{-1})^n(x, y)| \, n \in \mathbb{N}\}$.* The notion of equivalence class leads, in this setting, to the notion of similarity class. Let $\rho: X \times X \to [0, 1]$ be a similarity relation and $x \in X$. The *similarity class of representative x* is $\rho_x: X \to [0, 1]$, $\rho_x(y) = \rho(x, y)$, for any $y \in X$. Unlike the equivalence classes, the similarity classes are not necessarily disjoint (with respect to fuzzy intersection). We point out some connections with the *fuzzy partitions*.

Let $X$ be a set and $J = \{1, 2, ..., n\}$. The symbols "·", "●", "○" denote the operations on F $(X)$ defined at 2. 1. 6.

2.8. DEFINITION. The fuzzy sets $\mu_1, \mu_2, ..., \mu_n \in$ F $(X)$ are called:
- *s-disjoint*, if, $\forall k \in J$, $(● \,_{i \in J - \{k\}} \mu_j) \,○\, \mu_k = \varnothing$;
- *w-disjoint*, if $○ \,_{1 \leq i \leq n} \mu_j = \varnothing$;
- *i-disjoint*, if, $\forall r, s \in J, r \neq s, \mu_r \cap \mu_s = \varnothing$;
- *t-disjoint*, if, $\forall r, s \in J, r \neq s, \mu_r \cdot \mu_s = \varnothing$.

We say that the letters *s, w, i, t* are associated, respectively, to the operations "●", "○", "∩", ".".

2.9. REMARK. The above definitions can be extended in a natural manner to a *countable family* of fuzzy sets of $\mathsf{F}(X)$: $\forall \alpha \in \{s, w, i, t\}$, $\mu_1, \mu_2, \ldots, \mu_n, \ldots \in \mathsf{F}(X)$ are called $\alpha$-*disjoint* if, for any $n \in \mathbb{N}$, $\mu_1, \mu_2, \ldots, \mu_n$ are $\alpha$-disjoint.

2.10. REMARK. a) If $\mu_1 \cap \mu_2 = \varnothing$ then $\mu_1 \bigcirc \mu_2 = \varnothing$. The converse is not generally true. It is true if $\mu_1$ and $\mu_2$ are characteristic functions.

b) $\mu_1 \bigcirc \mu_2 = \varnothing \Leftrightarrow (\mu_1 \bullet \mu_2)(x) = \mu_1(x) + \mu_2(x)$, $\forall x \in X$.

c) Let $(A_i)_{i \in J}$ a family of *n* subsets of *X* and let $\chi_i$ the characteristic function of $A_i$, $\forall i \in J$. Then $\chi_i$, $i \in J$, are *s*-disjoint if and only if, $\forall i, j \in J$, $i \neq j$ implies $\chi_i \bigcirc \chi_j = \varnothing$.

d) $\mu_1 \cap \mu_2 = \varnothing$ if and only if $\mu_1 \cdot \mu_2 = \varnothing$.

e) $\mu_1, \mu_2, \ldots, \mu_n$ are *s*-disjoint $\Rightarrow \mu_1, \mu_2, \ldots, \mu_n$ are *w*-disjoint.

2.11. Proposition. We have:

$\mu_1, \mu_2, \ldots, \mu_n$ *are s-disjoint* $\Leftrightarrow \forall x \in X$, $\mu_1(x) + \mu_2(x) + \ldots + \mu_n(x) \leq 1$;

$\mu_1, \mu_2, \ldots, \mu_n$ *are s-disjoint* $\Leftrightarrow \forall x \in X$, $\sum_{i \in J} \mu_i(x) = \bullet_{i \in J} \mu_i(x)$;

$\mu_1, \mu_2, \ldots, \mu_n$ *are w-disjoint* $\Leftrightarrow \forall x \in X$, $\mu'_1(x) + \mu'_2(x) + \ldots + \mu'_n(x) \leq 1$;

$\mu_1, \mu_2, \ldots, \mu_n$ *are w-disjoint* $\Leftrightarrow \forall x \in X$, $\mu_1(x) + \mu_2(x) + \ldots + \mu_n(x) \leq n - 1$.

Correspondingly, we obtain the notion of $\sigma$-*partition* with $\sigma \in \{s, w, i, t\}$.

2.12. DEFINITION. Let $\sigma$ be an element of $\{s, w, i, t\}$ and let $\omega$ be the associated operation. The family $\{\mu_i\}_{i \in J} \subseteq \mathsf{F}(X)$ is called a *fuzzy $\sigma$-partition* of $\mu \in \mathsf{F}(X)$ if $\mu_1, \mu_2, \ldots, \mu_n$ are $\sigma$-disjoint and $\omega_{i \in J} \mu_i = \mu$. Similarly, one can define the *countable partitions* of a fuzzy subset of *X*. When $\mu = \chi_A$, with *A* subset of *X*, the $\sigma$-partition is called a *fuzzy $\sigma$-partition* of *A*.

2.13. REMARK. If $\{\mu_1, \mu_2, \ldots, \mu_n\}$ is an *s*-partition of $\mu$ and $\nu \leq \mu$, then:

$\{\nu \cdot \mu_1, \nu \cdot \mu_2, \ldots, \nu \cdot \mu_n\}$ is an *s*-partition for $\nu \cdot \mu$.

Let $\rho: X \times X \to [0, 1]$ be a non-degenerate similarity relation (there exist $x, y \in X$, $x \neq y$, such that $\rho(x, y) = 1$). In the following we consider that *X* is a finite or countable set. For any $x \in X$ we denote $\mu_x : X \to [0, 1]$ the function such that $\mu_x(y) = 1$ if $\rho(x, y) = 1$ and $\mu_x(y) = 0$ if $\rho(x, y) \neq 1$.

2.14. PROPOSITION. *In the conditions above, if* $\exists z \in X$ *such that* $\mu_x(z) = \mu_y(y) = 1$, *then* $\mu_x = \mu_y$. The relation on *X*, defined by $x \sim y$ if and only if $\mu_x = \mu_y$, is an equivalence relation on *X*. Let $K = X/\sim$ and denote by $[x]$ the class of $x$, $\forall x \in X$. Define $\mu_{[x]} = \mu_x$.

2.15. PROPOSITION. The set $H = \{\mu_{[x]} \mid x \in X\}$ is a fuzzy w-partition and a fuzzy i-partition of X.

### 4.3. Connections between hyperstructures and fuzzy sets

The connections between algebraic hyperstructures and the fuzzy sets may take into account the following variants:

Let $H$ be a nonempty set. One may replace (in the definition 2. 2. 1 of a hyperoperation on $H$) $\mathsf{P}^*(H)$ with $\mathsf{F}^*(H)$, where $\mathsf{F}^*(H) = \{\mu\colon H \to [0, 1]\colon \exists x \in H$ such that $\mu(x) \neq 0\}$

(the "family of nonempty fuzzy subsets of $H$").

**B.** For a given hyperstructure, define a *fuzzy subhyperstructure* in an analogous manner to the one used to introduce the fuzzy subgroups.

**C.** Associating a hyperstructure to a fuzzy set (and conversely).

Concerning the variant **A** above, we have:

3.1. DEFINITION. Let $H$ be a nonempty set. An application $\bullet\colon H \times H \to \mathsf{F}^*(H)$ is called an $f$-*hyperoperation* on $H$.

For $a,\, b \in H,\, K \in \mathsf{P}^*(H),\, \mu \in \mathsf{F}^*(H)$, we define:

$a \boxslash b = \{x \in H \mid (a \bullet b)(x) \not\eqslantless 0\},\, a \boxslash K = \bigcup_{k \in K} a \boxslash k,\, k \boxslash a = \bigcup_{k \in K} k \boxslash a;$

$a \boxtimes b = \{x \in H \mid (a \bullet b)(x) = 1\},\, a \boxtimes K = \bigcup_{k \in K} a \boxtimes k,\, k \boxtimes a = \bigcup_{k \in K} k \boxtimes a;$

$a \bullet K \in \mathsf{F}^*(H),\, (a \bullet K)(x) = \sup\{(a \bullet k)(x) \mid k \in K\},\, \forall x \in H.$

$K \bullet a \in \mathsf{F}^*(H),\, (K \bullet a)(x) = \sup\{(k \bullet a)(x) \mid k \in K\},\, \forall x \in H.$

$a \bullet \mu = a \bullet \mathrm{supp}(\mu);\, \mu \bullet a = \mathrm{supp}(\mu) \bullet a,$ where $\mathrm{supp}(\mu)\colon = \{x \in H \mid \mu(x) \neq 0\}.$

We introduce some conditions related to *reproducibility*. We say that the $f$-hyperoperation "$\bullet$" on $H$ satisfies the condition:

$(R_1)$ if: $a \bullet H = \chi_H = H \bullet a,\, \forall a \in H;$

$(R_2)$ if: $a \boxslash H = H = H \boxslash a,\, \forall a \in H;$

$(R_3)$ if: $a \boxtimes H = H = H \boxtimes a,\, \forall a \in H.$

3.2. DEFINITION. A nonempty set $H$ endowed with an $f$-hyperoperation $\bullet$ is called an $f_i$-*hypergroup* ($i \in \{1, 2, 3\}$) if "$\bullet$" is associative ($a \bullet (b \bullet c) = (a \bullet b) \bullet c,\, \forall a,\, b,\, c \in H$) and satisfies the condition $R_i$.

3.3. PROPOSITION. *a)* $(H, \bullet)$ *is a $f_3$-hypergroup* $\Rightarrow$ $(H, \bullet)$ *is a $f_1$-hypergroup* $\Rightarrow$ $(H, \bullet)$ *is a $f_2$-hypergroup.*

b) For any $i \in \{1, 2, 3\}$, if $(H, \bullet)$ is a $f_i$-hypergroup, then $(H, \boxslash)$ is a hypergroup.

c) If $(H, *)$ is a hypergroup, then $(H, \bullet)$ is a $f_i$-hypergroup, for any $i \in \{1, 2, 3\}$, where $\bullet\colon$

$$H \times H \to \mathsf{F}^*(H) \text{ is given by } (a \bullet b)(x) = \begin{cases} 1 & \text{if } x \in a * b \\ 0 & \text{otherwise} \end{cases}.$$

The variant **C** above can be used in the following manner: if $\mu\colon A \to L$ is an $L$-fuzzy set, where $(L, \wedge, \vee)$ is a lattice, define the following *hyperoperation* on $A$:

(1) $a * b = \{x \in A\colon \mu(a) \wedge \mu(b) \leq \mu(x) \leq \mu(a) \vee \mu(b)\}$, where "$\leq$" is the order relation on $L$.

3.4. Proposition. *In the conditions above, for every $a,\, b,\, c \in A$, we have:*

i) $a \in a * b;$

ii) $a * b = b * a;$

iii) $a * (a * b) = a * b = (a * a) * b = (a * a) * (b * b) = (a * b) * b.$

3.5. PROPOSITION. If $\mu(L)$ is a distributive sublattice in L (it is stable with respect to the operations $\wedge$ and $\vee$ and $a\wedge(b\vee c) = (a\wedge b)\vee (a\wedge c)$, for any a, b, c $\in \mu(L)$), then:

iv) $(a*b)*c = a*(b*c)$,  for every a, b, c $\in$ A.

From 3. 4. *i*) it follows at once that $a*A = A*a$, for any *a* in *A*. Together with 3. 4. *ii*) and 3. 5. *iv*), this allows us to say that $(A, *)$ is a *commutative hypergroup* if $\mu(L)$ *is a distributive sublattice in L*. Moreover, 3. 4. *iii*) shows that $(a*a)*(b*b) = a*b$; the set $a*b$ depends only on $a*a$ and $b*b$.

3.6. QUESTION. A natural problem arises: characterize the lattices L (e.g. by means of identities) with the property that, the hyperoperation induced on L – viewed as an L-fuzzy set by:

$1_L: L \rightarrow L$ – as in (1) is associative. The result 3. 4. iv) says that the class of lattices with this property includes the distributive lattices. In the case L = [0, 1] (or, more generally, a totally ordered set), the hypergroup obtained above is even a join space.

Suppose now that $\mu(L)$ is a sublattice which possesses a greatest element denoted 1 (that is, $x \leq 1$ for any $x$ in $\mu(L)$). We then have the additional properties:

3.7. Proposition. *In the conditions above, there exists* $\omega \in$ *A, such that:*

v) For any a, b $\in$ A, the condition $a*\omega = b*\omega$ implies $a*a = b*b$;

vi) For any a, b $\in$ A, there exist m, M $\in$ A such that

$$M * \omega = \bigcap \{x * \omega : x \in a * b\} \text{ and}$$

$$\bigcap \{x * \omega : x * \omega \supseteq \{a,b\}\} = m * \omega.$$

Let us consider the reverse problem: given a hyperstructure (H, $*$) satisfying the properties i)-iii) and v)–vi) above, can one find a lattice L and a mapping $\mu$: H $\rightarrow$ L such that "$*$" is the hyperoperation induced by $\mu$, as in (1)? In order to answer this, let H satisfy the properties above. Define a relation "~" on H by:

$$a\sim b \text{ iff } a*a = b*b.$$

One readily checks that this is an *equivalence relation* on A. Let L be the factor set $H/\sim$ (the set $\{\hat{a} : a \in H\}$, where $\hat{a} = \{x \in H: x\sim a\}$ is the equivalence class of a). Define a relation $\rho$ on L by:

$$\textit{for any } a,b \in L, \hat{a}\rho\hat{b} \textit{ iff } b*b \subseteq a*\omega$$

The relation $\rho$ is *well-defined* (does not depend on the representatives *a* and *b*). This is in fact an *order relation on L* and the ordered set $(L, \rho)$ is a *lattice*. Define now the application $\mu: H \rightarrow L$ as the canonical projection: $\mu(a) = \hat{a}$, for any $a \in H$; define the hyperoperation "$*$" in *H* as in (1).

3.8. Proposition. *In the conditions above, for any a and b in H,*

$$a*b = \{x \in A: \ \mu(a)\wedge\mu(b) \leq \mu(x) \leq \mu(a)\vee\mu(b)\}.$$

**Acknowledgements.** The authors wish to thank Professor H. N. Teodorescu for his support and competent advice given during the preparation of this paper.

## References


[1] Alkhamees, Y., Mordeson, J., Fuzzy principal ideals and fuzzy simple extensions, *Fuzzy Sets and Systems,* **3** (1995), 911-919.

[2] Alkhamees, Y., Mordeson, J., Reduced fields, primitive and fuzzy Galois theory, *J. Fuzzy Math.*, vol. **8**, no. 1 (2000), 157-173.

[3] Corsini, P., *Prolegomena on Hypergroup Theory*, Aviani Ed., 1993.

[4] Corsini, P., Tofan, I., On fuzzy hypergroups, *PUMA*, vol. **8**, no. 1, 1997, 29-37.

[5] Erceq, M. A., Functions, Equivalence Relations, Quotient Spaces and Subsets in Fuzzy Set Theory, *Fuzzy Sets and Systems*, **3** (1980), 75-92.

[6] Kerre, E. E., Introduction to the Basic principles of the Set Theory and some of its Applications, *Comm. & Cognition*, Gent, Belgium, 1991.

[7] Kumar, R., *Fuzzy Algebra* I, Univ. Delhi Publ. Division, 1993.

[8] Liu, Wang-Jiu: , *Fuzzy Invariant Subgroups and Fuzzy Ideals*, F. S. S., vol. **8** (1982), 133-139.

[9] Liu, Wang-Jiu, Operations on Fuzzy Ideals, *F. S. S.,* vol. **8** (1983), 31-41.

[10] Maturo, A., Tofan, I., Similitude relations and applications in architecture communication, *SIGEF Conference*, Napoli 2001.

[11] Mordeson, J. N., Fuzzy Galois theory, *J. Fuzzy Math.* (1993), 659-671.

[12] Mukherjee, T. K., Sen, M. K., On Fuzzy Ideals of a Ring (1), *F. S. S.*, vol. **21** (1987), 99-104.

[13] Nistor, S., Tofan, I., Volf, A. C., Fuzzy algebraic structures, *Fuzzy Systems & A. I.,* Acad. Româna, vol. **5**, no. 1 (1996), 5-15.

[14] Ovchinnikov, S. V., Structure of Fuzzy Binary Relations, *Fuzzy Sets and Systems* **6** (1981), 169-197.

[15] Teodorescu, H. N., Binary-like codes for trapezoidal fuzzy numbers, vol. *Proc. 3rd IFSA Congress*, Washington, 1989, 548-551.

[16] Teodorescu, H. N., Congruencies of fuzzy numbers, *Proc. Int. Poznan Seminar on Fuzzy Sets and Interval Analysis*, 1988.

[17] Teodorescu, H. N., in cooperation with S. Nistor, I. Tofan, *Fuzzy Sets, part. 2: Fuzzy arithmetic, R. R.,* Ecole Pol. Féd. Lausanne 1994.

[18] Tofan, I., Toposes of totally fuzzy sets, *BUSEFAL* **58** (1994), 60-65.

[19] Tofan, I., Volf, A. C., Fuzzy rings of quotients, *Italian Journal of Pure and Applied Mathematics* no. **8** (2000), 83-90.

[20] Tofan, I., Volf, A. C., On some connections between hyperstructures and fuzzy sets, *Italian Journal of Pure and Applied Mathematics*, no. **7** (2000), 63–68.

[21] Tofan, I., Volf, A. C., Some remarks on fuzzy relations, *Fuzzy Systems & A. I.,* Romanian Academy, vol. **4**, no. 2 (1995), 1-4.

[22] Zadeh, L. A., Fuzzy sets, *Inform. and Control,* vol. **8** (1965), 338-353.

[23] Zahedi, M. M., Bolurian, M., Hasankhani, A., On polygroups and fuzzy subpolygroups, *J. of Fuzzy Math.*, vol. **2**, no. 4, 1994, 1-15.

# Automated Quality Assurance of Continuous Data

Mark LAST[1] and Abraham KANDEL[2]

[1]*Ben-Gurion University of the Negev, Department of Information Systems Engineering, Beer-Sheva 84105, Israel*
[2]*Department of Computer Science and Engineering, ENB 118, University of South Florida, Tampa, Florida 33620, USA.*

**Abstract**

Most real-world databases contain some amount of inaccurate data. Reliability of critical attributes can be evaluated from the values of other attributes in the same data table. This paper presents a new fuzzy-based measure of data reliability in continuous attributes. We partition the relational schema of a database into a subset of input (predicting) and a subset of target (dependent) attributes. A data mining model, called information-theoretic connectionist network, is constructed for predicting the values of a continuous target attribute. The network calculates the degree of reliability of the actual target values in each record by using their distance from the predicted values. The approach is demonstrated on the voting data from the 2000 Presidential Elections in the US.

## 1. Introduction

Modern database systems are designed to store accurate and reliable data only. However, the assumption of zero defect data (ZDD) is far from being true in most real-world databases, especially when their data comes from multiple sources. The issue of data reliability has been in the focus of the recent controversy regarding the results of the Year 2000 presidential elections in the State of Florida. After the elections, the leaders of the Democratic Party questioned the accuracy of the official voting results in certain Florida counties, based on the demographic characteristics of the voters in those counties. Their suspicions have led to a manual re-count of the votes, which was aimed at improving the reliability of the results. Though the accuracy of the punch card counting machines, used in some counties, was known to be limited, a complete manual re-count of all Florida votes was not feasible within a several weeks time frame. Eventually, the courts stopped the manual re-count process and Mr. George W. Bush was declared as the new President of the United States.

In small databases the users have enough time to check manually every record "suspected" of poor data quality and correct data, if necessary. In a large database, like the data on Florida voting results, this approach is certainly impractical. The task of assuring data reliability and data quality, known as "data cleaning", becomes even more acute in rapidly emerging *Data Warehouses*. Thus, there is a strong need for an efficient automated tool, capable of detecting, filtering, representing and analyzing poor quality data in large databases.

In our previous work (see [18] - [20]), we have introduced an information-theoretic fuzzy method for evaluating reliability of discrete (nominal) attributes. Our methodology for data quality assurance includes three main stages: modification of the database schema, induction of a data mining model (information-theoretic network), and using the constructed network to calculate reliability degrees of attribute values. In this paper, we present a new fuzzy-based measure for evaluating the reliability of continuous attributes and demonstrate it on a set of real voting data from the US presidential elections.

Our paper is organized as follows. In Section 2 we present an overview of existing approaches to various aspects of data quality and data reliability. Section 3 briefly describes the algorithm for building an information-theoretic connectionist network from relational data. In Section 4, we present the fuzzy-based approach to evaluating data reliability of continuous attributes. In Section 5 we apply the info-fuzzy methodology presented in Sections 3 and 4 to a set of real voting data. Section 6 concludes the paper with summarizing the benefits of our approach to data reliability and representing a number of issues for the future research.

## 2. Data Quality and Data Reliability

As indicated by Wang *et al.* [32], data reliability is one of *data quality dimensions*. Other data quality dimensions include ([31] - [33]): accuracy, timeliness, relevance, completeness, consistency, precision, etc. Various definitions of these and other dimensions can be found in [33]. Ahituv *et al.* [1] refer to accuracy and relevance as *content attributes* of an information system. According to Wand and Wang [31], the reliability "indicates whether the data can be counted on to convey the right information". Unreliable (deficient) data represents an inconformity between the state of the information system and the state of the real-world system. The process of mapping a real-world state to a wrong state in an information system is termed by [31] as "garbling". Two cases of garbling are considered: the mapping to a meaningless state and the mapping to a meaningful, but wrong state. In the first case the user knows that the data is unreliable, while in the second case he relies upon an incorrect data. Wand and Wang suggest to solve the garbling problem by adding data entry controls, like check digits and control totals, methods that are not applicable to qualitative data. The paper follows a "Boolean" approach to data reliability: the information system states are assumed to be either correct or incorrect. No "medium" degree of reliability is provided.

An attribute-based approach to data quality is introduced by Wang *et al.* in [33]. It is based on the entity-relationship (ER) model (see [13]) and assumes that some attributes (called *quality indicators*) provide objective information (metadata) about data quality of other attributes. The data quality is expressed in terms of *quality parameters* (e.g., believability, reliability, and timeliness). Thus, if some sources are less reliable than the others, an attribute *data source* may be an indicator of data reliability. Each quality parameter has one or more quality indicators attached to it via *quality keys*. A quality indicator may have quality indicators of its own, leading easily to an exponential total number of quality indicators. Wang *et al.* [33] suggest integration of quality indicators, to eliminate redundancy and inconsistency, but no methodological approach to this problem (crucial for dimensionality reduction) is presented.

An extended database, storing quality indicators along with data, is defined as a *quality database*. The quality indicator values are stored in quality indicator relations. The quality database is strictly deterministic: once the values of quality indicators are given, the values of quality parameters are uniquely defined by the database structure. The values of quality parameters are often qualitative and subjective (like "highly reliable" vs. "unreliable"). Wang *et al.* [33] warn that quality parameters and quality indicators are strongly user-dependent and application-dependent. The database structure described by [33] enables an experienced user to infer *manually* from values of quality indicators about the quality of relation attributes, but their work provides no method for automated evaluation of data quality in large databases.

Kandel *et al.* [11] mention unreliable information as one of sources of data uncertainty, other sources including fuzziness of human concepts, incomplete data, contradicting sources of information, and partial matching between facts and events. According to Kandel *et al.* (1996), the main drawback of the probabilistic approaches to uncertainty (e.g., the Bayesian approach) is their limited ability to represent human reasoning, since humans are not Bayesian when reasoning under uncertainty.

Kurutach [14] discusses three types of data imperfection in databases: *vagueness*, or *fuzziness* (the attribute value is given, but its meaning is not well-defined), *imprecision* (the attribute value is given as a set of possible items), and *uncertainty* (the attribute value is given along with its degree of confidence). All these types of imperfection are defined by users themselves during the data entry process. The author suggests a unified approach, based on fuzzy set theory, to incorporating these aspects of imperfection in an extended relational database containing, primarily, discretely-valued, qualitative data. In addition to imprecision and uncertainty, Motro [22] defines a third kind of imperfect data: erroneous information. Database information is erroneous, when it is different from the true information. Motro [22] follows the binary approach to errors: both "small" and "large" errors in a database should not be tolerated. He also mentions inconsistency as one of the important kinds of erroneous information.

Since, in a general case, data reliability is a *linguistic variable* (the data can be considered "very reliable", "not so reliable", "quite unreliable", etc.), the models of fuzzy databases seem to be helpful for treating reliability of database attributes. As indicated by Zemankova and Kandel [35], the main problem of fuzzy databases is to propagate the level of uncertainty associated with the data (reliability degree in our case) to the level of uncertainty associated with answers or conclusions based on the data. The fuzzy relational algebra proposed by Klir and Yuan [12] enables to check similarity between values of fuzzy attributes by using a similarity relation matrix and a pre-defined threshold level of minimum acceptable similarity degree.

Zemankova and Kandel [35], Kandel [10] propose a Fuzzy Relational Data-Base (FRDB) model which enables to evaluate fuzzy queries from relational databases. The attribute values in the FRDB can represent membership or possibility distributions defined on the unit interval [0, 1]. According to this model, a single value of a membership distribution can be used as a value of a fuzzy attribute. Another model of fuzzy querying from regular relational databases (called SQLf) is presented by Bosc and Pivert [2]. The main purpose of this model is to define imprecise answers based on precise data and on fuzzy conditions (which contain fuzzy predicates and fuzzy quantifiers).

The Fuzzy Data model developed by Takahashi [29] assumes that some nonkey attributes may have values defined by fuzzy predicates (e.g., "very reliable"). All key attributes and some other attributes are assumed to have nonfuzzy values only. Any tuple in Takahashi data model has a *truth value z* defined over the unit interval [0, 1]. The value of $z$ is interpreted as a degree to which the tuple is true, with two special cases: $z = 0$ when the tuple is completely false and $z = 1$ when the tuple is completely true. This approach treats a tuple as a set of attribute values, all having the same truth-value. The case of different truth-values associated with values of different attributes in the same tuple is not covered by the model of [29]. A similar idea of associating a single truth value (a *weight)* with each tuple is described by Petri [25]. Petri terms such tuples as *weighted tuples* and defines their weight as a membership degree expressing the extent to which a tuple belongs to a fuzzy relation. Three possible meanings of tuple weights are proposed. One of them is "the certainty of information stored in the tuple", i.e. the reliability of all tuple attributes. The concept of reliability degree associated with every column in a fuzzy spreadsheet table is used by [23]. According to their definition, the degree of reliability can take any continuous value between 0 and 1, but no explicit interpretation of this variable is provided.

All the above-mentioned models assume that both crisp and fuzzy quality dimensions of database attributes are available from the database users. Obviously, this assumption may not be realistic for large and dynamically changing databases. Consequently, there is a need for methods that perform automated assessment of data quality. An information theoretic approach to automated data cleaning is presented by Guyon *et al.* [8]. The paper assumes that erroneous ("garbage") data has a high information gain. The information gain is defined by [8] as a self-information (logarithm of probability) of predicting the correct data value. This means that the most "surprising" patterns (having the lowest probability to be predicted correctly) are suspicious to be unreliable. The authors propose a computer-aided cleaning method where a human operator must check only those patterns that have the highest information gain and remove from the database patterns, which are truly corrupted, while keeping all the rest. The prediction itself is performed in [8] by using a neural network trained with a "cross-entropy" cost function. One can easily accept the approach of [8] that values having lower probability are more likely to be erroneous. However, the values having the *same* probability (and, accordingly, the same information gain) cannot be treated alike in different databases. Reliability may also depend on the inherent distributions of database attributes and some other, user-related factors. Thus, the approach of [8] should be enhanced to cope with real-world problems of data quality.

In [19], we have presented a fuzzy-based approach to automated evaluation of data reliability. The method of [19] is aimed at detecting unreliable nominal data by integrating objective (information-theoretic) and subjective (user-specific) aspects of data quality. In this paper, we extend the method of [19] to handle partially reliable continuous attributes.

## 3. Information-Theoretic Connectionist Networks

Uncertainty is an inherent part of our life. Delivery time of manufactured products is not constant, stock prices go up and down, and people vote according to their personal beliefs. Most real-world phenomena cannot be predicted with perfect accuracy. The reasons for that may include limited understanding of the true causes for a given phenomenon (e.g., detailed considerations of each specific voter), as well as missing and erroneous data (e.g., incomplete or inaccurate voting results).

Data mining methods (see [4], [5], [17], [21], [26], and [27]) are aimed at reducing the amount of uncertainty, or gaining *information*, about the data. More information means higher prediction accuracy for future cases. If a model is useless, it does not provide us with any new information and its prediction accuracy is not higher than just a random guess. On the other hand, the maximum amount of information transferred by a model is limited: in the best case, we have an accurate prediction for every new case. Intuitively, we need more information to predict a multi-valued outcome (e.g., percentage of votes for a certain candidate) than to predict a binary outcome (e.g., customer credibility).

The above characteristics of the data mining problem resemble the communication task: predictive attributes can be seen as input messages and each value of the system output as an output message. If we have a model with a perfect accuracy, each output value can be predicted correctly from the values of input attributes. In terms of the Information Theory (see [3]), this means that the entropy of the output $Y$, given the input $X$ is zero, i.e., the mutual information between $Y$ and $X$ is maximal.

The information-theoretic approach to data mining (see [6], [7], [15], [16], [18], [19], and [20]) is a powerful methodology for inducing information patterns from large sets of imperfect data, since it uses meaningful network structure, called *information-theoretic connectionist network*. The measures of information content, expressed by the network connection weights, include mutual information, conditional mutual information, and divergence. The connection weights can incorporate prior knowledge on probability distributions of database values. Information-theoretic connectionist techniques have been successfully applied to the problems of extracting probabilistic rules from pairs of interdependent attributes [6], speech recognition [7], feature selection [15], and rule induction [16]. The procedure for constructing a multi-layer information-theoretic network is briefly described in the next sub-sections. Complete details can be found in [20].

### 3. 1 Extended Relational Model

We use the following formal notation of the relational model [13]:
- $R = (A_1,..., A_N)$ - a schema of a relation (data table) containing $N$ attributes
- $D_i$ - the domain of an attribute $A_i$.
- $V_{ij}$- the value $j$ in the domain $D_i$.
- $t_k[A_i]$ - value of an attribute $A_i$ in a tuple $k$, $t_k[A_i] \in D_i$.

To build an information-theoretic network, we define the following types of attributes in a relation schema:

1. A subset $O \subset R$ of *target* ("output") attributes ($|O| \geq 1$). This is a subset of attributes, which can be predicted by the information-theoretic network. If the values of these attributes are already available, we can evaluate their reliability by using the method of Section 4 below.

2. A subset $C \subset R$ of *candidate input* attributes ($|C| \geq 1$). These attributes can be used to predict the values of target attributes.

The following constraints are imposed on the above partition of the relation schema:

1. $C \cap O = \varnothing$, i.e. the same attribute cannot be both a candidate input and a target.

2. $C \cup O \subseteq R$, i.e. some attributes are allowed to be neither candidate inputs nor targets. Usually, these will be the key (identifying) attributes.

Now we proceed with describing the structure of a connectionist network designed to predict the values of target attributes.

### 3. 2 Connectionist Network Structure

An information-theoretic connectionist network has the following components:

1. $I$ - a subset of input (predicting) attributes selected by the network construction algorithm from the set C of candidate input attributes.

2. $|I|$ - total number of hidden layers (levels) in a network. Unlike the standard decision tree structure [27], where the nodes of the same tree level are independent of each other, all nodes of a given network layer are labeled by the same input attribute associated with that layer. Consequently, the number of network layers is equal to the number of input attributes. In layers associated with continuous attributes, an information network uses multiple splits, which are identical at all nodes of the corresponding layer. The first layer in the network (Layer 0) includes only the root node and is not associated with any input attribute.

3. $L_l$ - a subset of nodes $z$ in a hidden layer $l$. Each node represents an attribute-based test, similarly to a standard decision tree. If a hidden layer $l$ is associated with a nominal input attribute, each outgoing edge of a non-terminal node corresponds to an attribute distinct value. For continuous features, the outgoing edges represent the intervals obtained from the discretization process. If a node has no outgoing edges, it is called a terminal node. Otherwise, it is connected by its edges to the nodes of the next layer, which correspond to the same subset of input values.

4. $K$ - a subset of target nodes representing distinct values in the domain of the target attribute. For continuous target attributes (e.g., percentage of votes for certain candidate), the target nodes represent the user-specified intervals of the attribute range. The target layer does not exist in the standard decision-tree structure. The connections between terminal nodes and the nodes of the target layer may be used for predicting the values of the target attributes and extracting information-theoretic rules (see [16]).

### 3. 3 The Network Construction Procedure

The network construction algorithm starts with defining the target layer, where each node stands for a distinct target value, and the "root" node representing an empty set of input attributes. The connections between the root node and the target nodes represent unconditional (prior) probabilities of the target values. The network is built only in one direction (top-down). After the construction process is stopped, there is no bottom-up post-pruning of the network branches. The process of *pre-pruning* the network is explained below.

A node is split on the values of an input attribute if it provides a statistically significant increase in the *mutual information* of the node and the target attribute. Mutual information, or information gain, is defined as a decrease in the conditional entropy of the target attribute (see [3]). If the tested attribute is nominal, the splits correspond to the attribute values. Splits on continuous attributes represent thresholds, which maximize an increase in mutual information.

At each iteration, the algorithm re-computes the best threshold splits of continuously-valued candidate input attributes and chooses an attribute (either discrete, or continuous), which provides the maximum overall increase in mutual information across all nodes of the current final layer.

The maximum increase in mutual information is tested for statistical significance by using the Likelihood-Ratio Test [28]. This is a general-purpose method for testing the null hypothesis $H_0$ that two discrete random variables are statistically independent. If $H_0$ is rejected, a new hidden layer is added to the network and a new attribute is added to the set $I$ of input attributes.

The nodes of a new layer are defined for a Cartesian product of split nodes of the previous final layer and the values of a new input attribute. According to the chain rule (see [3]), the mutual information between a set of input attributes and the target (defined as the overall decrease in the conditional entropy) is equal to the sum of drops in conditional entropy across all the layers.

If there is no candidate input attribute significantly decreasing the conditional entropy of the target attribute, no more layers are added and the network construction stops.

The main steps of the construction procedure for a single target attribute are summarized in Table 1. If a data table contains several target attributes, a separate network is built, by using the same procedure, for each target attribute. Complete details are provided in [20].

### 3. 4 Predicting Continuous Target Values

Like in decision trees, a predicted target value is assigned to every terminal node of an information-theoretic network. Each record of a training set is associated with one and only one terminal node, which can be found by the procedure described in

Table 2 below.

The *predicted value* $Pred_{iz}$ of a continuous target attribute $A_i$ at a terminal node $z$ is calculated as the expected value of $A_i$ over all the training records associated with the node $z$.

Table 1. Network Construction Algorithm

| | |
|---|---|
| Input: | The set of $n$ training instances; the set $C$ of candidate input attributes (discrete and continuous); the target (classification) attribute $A_j$; the minimum significance level sign for splitting a network node (default: $sign = 0.1\%$). |
| Output: | A set $I$ of selected input attributes and an information-theoretic network. Each input attribute has a corresponding hidden layer in the network. |
| Step 1 | Initialize the information-theoretic network (single root node representing all records, no hidden layers, and a target layer for the values of the target attribute). |
| Step 2 | While the number of layers $|I| < |C|$ (number of candidate input attributes) do |
| Step 2.1 | For each candidate input attribute $A_i' \notin I$ do<br>If $A_i'$ is continuous then<br>Return the best threshold splits of $A_i'$.<br>Return the conditional mutual information $cond\_MI_i'$ between $A_i'$ and the target attribute $A_j$.<br>End Do |
| Step 2.2 | Find the candidate input attribute $A_i'^*$ maximizing $cond\_MI_i'$ |
| Step 2.3 | If $cond\_MI_i'^* = 0$, then<br>End Do.<br>Else<br>Expand the network by a new hidden layer associated with the attribute $A_i'$, and add $A_i'$ to the set $I$ of selected input attributes. |
| Step 2.4 | End Do |
| Step 3 | Return the set of selected input attributes $I$ and the network structure |

Table 2. Associating Record with a Terminal Node

| | |
|---|---|
| Input: | The set $I$ of selected input attributes; the values of input attributes in a tuple (record) $k$; the information-theoretic network |
| Output: | The ID of a terminal node corresponding to the tuple $k$: Node_$F_k$ |
| Step 1 | Initialize the current node ID: $z = 0$ |
| Step 2 | Initialize the layer number: $m = 0$ |
| Step 3 | **If** a node $z$ is terminal, **then** go to Step 7<br>**Else**, go to the next step |
| Step 4 | Increment the number of layers: $m = m+1$ |
| Step 5 | Find the next hidden node $z$ by following the edge corresponding to the value of the input attribute $m$ in the tuple $k$ |
| Step 6 | Go to Step 3 |
| Step 7 | Return $Node\_F_k = z$ |

## 4. Evaluating Reliability of Target Attributes

### *4. 1 Fuzzy Approach to Data Reliability*

The main cause of having unreliable data in a database are the errors committed by an information source, which may be a human user, an automated measuring device, or just another database. In the case of the Year 2000 elections in the State of Florida, the Democrats have argued that the votes were not counted properly. The legal controversy was focused on the so-called "undervotes", votes not tabulated by the counting machine due to apparent defects in the punch cards. The claim of the Democrats was that the undervotes have biased the results in favor of their opponent, the Republican Candidate George W. Bush. For example, they have questioned the voting results of Palm Beach County, which seemed particularly unreliable based on the demographic characteristics and the voting traditions of people in that specific county [24].

An expert user examining a familiar database can estimate quickly, and with a high degree of confidence, the reliability of stored information. He, or she, would define some records as "highly reliable", "not so reliable", "doubtful", "absolutely unreliable", etc. However, what is the exact definition of "data reliability"?

The most common "crisp" approach to data reliability is associated with data validity: some attribute values are valid while others are not. For example, if the valid range of a numeric attribute is [50, 100], the value of 100.1 is considered invalid and will be rejected during the data entry process. This is similar to the statistical concept of confidence intervals: any observation outside the interval boundaries is rejected, which means that its statistical validity is zero. The limitations of this approach are obvious: a real validity range may have "soft" boundaries.

It seems reasonable to define the reliability of an attribute value as a mean frequency (or probability) of that particular value, since values of low probability may be assumed less reliable than the most common values. This is similar to the information gain approach of [8]: the most surprising patterns are suspicious as unreliable. However, the information gain approach is not readily applicable to evaluating reliability of continuous attributes, which can take an infinite number of distinct values, each having a very low probability of occurrence.

Noisy data is not necessarily unreliable data, and vice versa. In some areas, like the stock market, the data may be inherently noisy (having a high variance and a high entropy), because the real-world phenomenon, it represents, depends on many independent and dynamic, mostly unknown, factors. Still, the source of noisy data may be completely reliable. On the other hand, the information on a very stable phenomenon (having a low variance) may be corrupted during the data entry process.

Statistical information, obtained from training data, is certainly not sufficient for distinguishing between reliable and unreliable values. People use their intuition, background knowledge, and short-time memory, rather than any probabilistic criteria, for detecting lowly reliable data. Moreover, as indicated by Kandel *et al.* [10], the probabilistic approach seems to be against the nature of human reasoning. Thus, we turn to the fuzzy set theory, which is a well-known approach to catching different aspects of human perception and making use of available prior knowledge.

The fuzzy set theory provides a mathematical tool for representing imprecise, subjective knowledge: the fuzzy membership functions. These functions are used for mapping precise values of numeric variables to vague terms like "low", high", "reliable", etc. The form of a specific membership function can be adjusted by a set of parameters. For example, a triangular membership function is defined by its prototype, minimum, and maximum values. For modeling human perception of reliability, the non-linear, sigmoid function seems more appropriate, since more probable values are usually perceived as more reliable, though all lowly reliable values are considered unreliable to nearly the same degree. The shape of this membership function depends on user perception of unexpected data, ranging from a "step function" (the crisp approach: only values in a specific range are reliable) to a continuous membership grade, giving a non-zero reliability degree even to very distant and unexpected values.

Thus, adopting the fuzzy logic theory and looking at the reliability degree as a fuzzy measure seems an appropriate approach to automating the human perception of data reliability. In [19], we have proposed the following definition for the degree of data reliability:

**Definition 1**. *Degree of Reliability of an attribute A in a tuple k is defined on a unit interval [0, 1] as the degree of certainty that the value of attribute A stored in a tuple k is correct from user's point of view.*

This definition is consistent with the definition of fuzzy measures in Klir and Yuan [12], since a set of correct attribute values can be viewed as a "crisp" set, and we are concerned with the certainty that a particular attribute belongs to that set. It is also related to the fuzzy concept of "usuality" [34], where the fuzzy set of normal (or regular) values is considered the complement of a set of exceptions. Two special cases of Definition 1 are: degree of reliability = 0 (the data is clearly erroneous) and degree of reliability = 1 (the data is completely reliable, which is the implicit assumption of most database systems).

According to Definition 1, the degree of reliability is an attribute-dependent, tuple-dependent and user-dependent measure. It may vary for different attributes of the same tuple, for the same attribute in different tuples and for different users who have distinct views and purposes with respect to the same data. The subjectiveness of data reliability was best demonstrated in the 2000 election controversy. While the Democrats complained about the unreliable voting results, the same numbers seemed perfectly accurate to their political opponents.

Data correctness does not imply precision. It just means that if a user could know the exact state of the real-world system, his or her decision, based on that data, would not be changed. After the 2000 elections, the real controversy was not about the exact number of votes for each candidate, which could be determined only by a tedious hand count. Both parties were just eager to know who won the *majority* of votes in the State of Florida.

### 4. 2 Calculating Degree of Reliability

After finding a *predicted* value of the target attribute $A_i$ in a tuple $k$, we compute the degree of reliability of the *actual* target value by the following formula:

$$t_k[R_i] = \frac{2}{1+e^{\alpha \bullet d_{ik}}} \tag{1}$$

where:

$\alpha$ - exponential coefficient expressing the user perception of "unexpected" data. Low values of $\alpha$ (about 1) make it a sigmoid function providing a gradual change of reliability degree between 0 and 1 within the attribute range. Higher values of $\alpha$ (like 10 or 20) make it a step function assigning a reliability degree of zero to any value, which is different from the expected one.

$d_{ik}$ – a measure of distance between the actual value $t_k[A_i]$ and the predicted value $Pred_{iz^*}$ ($z^* = Node\_F_k$) of a target attribute $A_i$ in a tuple $k$. For continuous target attributes, the distance measure is calculated by:

$$d_{ik} = \frac{abs\left(t_k[A_i] - Pred_{iz*}\right)}{Range_i} \tag{2}$$

where *Range_i* is the difference between the maximum and the minimum values of the attribute $A_i$. According to Equation 2, $d_{ik}$ is a linear measure of the difference between predicted and actual values, which is normalized to the [0, 1] range. The reliability degree in Equation 1 is defined on the same range, but it represents the *non-linearity* of reliability perception as a function of data deviation from the expected value, which can be predicted from the information-theoretic network.

In

Figure 1, we show the reliability degree $t_k[R_i]$ as a function of the distance $d_{ik}$ for two different values of α: α = 1 and α = 5. Equation 1 satisfies the four requirements of a fuzzy measure (see [12], p. 178): boundary conditions, monotonicity, continuity from below and continuity from above. The way to verify that is to look at the proximity to the predicted value as a reciprocal of the distance $d_{ik}$. Then the reliability of the empty set (zero proximity, or infinite distance) is zero and the reliability of the complete set (infinite proximity, or zero distance) is one. Reliability degree is a continuous monotonic function of proximity by its mathematical definition in Equation 1.

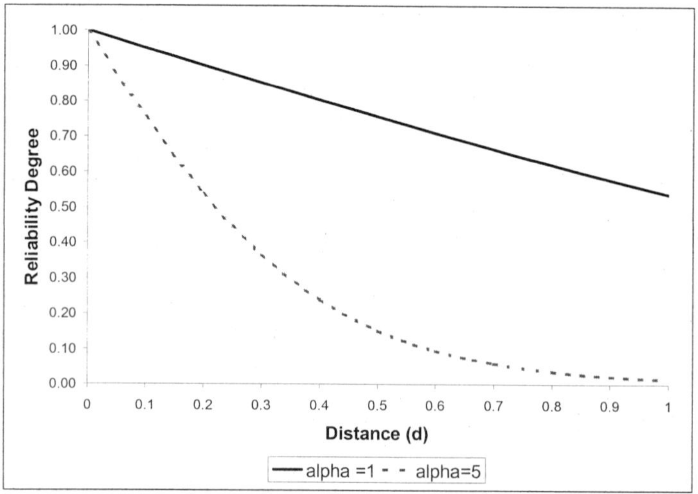


Figure 1. Reliability perceptions for different values of alpha

## 5. Case Study: Palm Beach Election Data

We have applied the information-theoretic fuzzy approach to the precinct-level voting data of the 2000 Presidential Election in Palm Beach County, Florida. The results of the initial count and the demographic data on each precinct (including voter registration information) have been downloaded from the web page of Dr. Bruce E. Hansen [9] in November 2000. The original source for the data was Palm Beach County web page. The list of attributes in the Palm Beach dataset is presented in Table 3 below). The raw data included absolute numbers (number of votes and number of voters). We have normalized these numbers to the percentage out of the total number of votes / voters in the corresponding precinct. Since there is a strong dependency between the percentages of votes for each major candidate in the same precinct, we have arbitrarily chosen the percentage of votes for Bush as the target attribute. After normalization, the values of the target attribute have been discretized to five intervals of approximately equal frequency. The Palm Beach dataset includes 14 candidate-input attributes, representing the percentage of registered voters in each party and the distribution of the voting population across several age groups. The dataset has 494 records referring to all the voting precincts of Palm Beach County. The 106 absentee precincts were excluded from the analysis due to the lack of demographic information.

Table 3 Palm Beach Dataset - List of Attributes

| Ser No | Attribute Name | Meaning | Type | Use in Network |
|---|---|---|---|---|
| 1 | Precinct | Precinct No | Nominal | None |
| 2 | Bush | Percentage of Votes | Continuous | Target |
| 3 | Gore | Percentage of Votes | Continuous | None |
| 4 | Nader | Percentage of Votes | Continuous | None |
| 5 | Buchanan | Percentage of Votes | Continuous | None |
| 6 | Total_Vote | Percentage of Votes | Continuous | None |
| 7 | McCollum | Percentage of Votes | Continuous | None |
| 8 | Nelson | Percentage of Votes | Continuous | None |
| 9 | DEM_PTY | Percentage of Registered Voters | Continuous | Candidate Input |
| 10 | REP_PTY | Percentage of Registered Voters | Continuous | Candidate Input |
| 11 | OTHER_PTY | Percentage of Registered Voters | Continuous | Candidate Input |
| 12 | WHITE | Percentage of Registered Voters | Continuous | Candidate Input |
| 13 | BLACK | Percentage of Registered Voters | Continuous | Candidate Input |
| 14 | HISPANIC | Percentage of Registered Voters | Continuous | Candidate Input |
| 15 | OTHER_RACE | Percentage of Registered Voters | Continuous | Candidate Input |
| 16 | MALE | Percentage of Registered Voters | Continuous | Candidate Input |
| 17 | FEMALE | Percentage of Registered Voters | Continuous | Candidate Input |
| 18 | AGE_18-20 | Percentage of Registered Voters | Continuous | Candidate Input |
| 19 | AGE_21-29 | Percentage of Registered Voters | Continuous | Candidate Input |
| 20 | AGE_30-55 | Percentage of Registered Voters | Continuous | Candidate Input |
| 21 | AGE_56-64 | Percentage of Registered Voters | Continuous | Candidate Input |
| 22 | AGE_65&UP | Percentage of Registered Voters | Continuous | Candidate Input |

The results of applying the information-theoretic procedure of sub-section 3.3 above to the Palm Beach Dataset are shown in

Table 4. Only three out of 14 candidate input attributes (*REP_PTY, WHITE,* and *DEM_PTY*) have been identified as statistically significant and included in the Information-Theoretic Network. The column "Conditional MI" in Table 4 shows the net decrease in the entropy of the target attribute *"Bush"* due to adding each input attribute. The first input attribute (*REP_PTY*) alone contributes nearly 90% of the overall mutual information (1. 435

bits). This attribute is shown in bold. The next two input attributes (*WHITE* and *DEM_PTY*) contribute about 8% and 2% respectively. The first and the third input attributes are not surprising, since people tend to vote by their political association. The input attribute No. 2 (White) is an indicator of some weak relationship between the racial origin of the voters and their votes.

Table 4 Palm Beach Dataset - Summary of Results

| Iteration | Attribute Name | Mutual Information | Conditional MI | Percentage Of MI | Conditional Entropy |
|:---:|:---:|:---:|:---:|:---:|:---:|
| **0** | **REP_PTY** | **1.282** | **1.282** | **89.3%** | **1.04** |
| 1 | WHITE | 1.4 | 0.118 | 8.2% | 0.922 |
| 2 | DEM_PTY | 1.435 | 0.035 | 2.4% | 0.887 |

The constructed information-theoretic network has been used for evaluating the reliability of the target attribute (percentage of Bush votes in each precinct) by the fuzzy-based method of Section 4 above. We have calculated the degrees of reliability with $\alpha = 1.00$. The resulting reliability degrees range between 0.617 and 1.000. As indicated above, these reliability degrees refer to the initial voting results certified by the Palm Beach County after the Election Day. During the following weeks, these results were in the center of a legal controversy until the US Supreme Court halted the vote recount on December 12, 2000. However, the public was still interested to know the "ground truth": who would be the actual winner of the Election in Florida, if the hand recount could be continued to its completion? For this reason, the Miami Herald and other media organizations have conducted a complete review of the "undervote" ballots in all Florida counties. The precinct-level results have been posted on the Miami Herald web site [30]. To evaluate the usefulness of the data reliability calculations, we have examined the number of undervotes and the resulting change in the gap between the candidates for the precincts having the highest and the lowest reliability degrees (see Tables 5 and 6).

The total number of undervotes in 20 precincts having the lowest reliability degrees (Table 5) is much larger than the number of undervotes in 20 precincts with highest reliability (Table 6). In other words, starting the count of undervotes in low reliability precincts would help to detect significant gaps, like the one in Precinct No. 191, as early as possible. From a close look at the data of this precinct, one can see that the predicted percentage of Bush votes is high (51.5%) due to high percentages of Republicans and whites among the voters. However, Mr. Bush has got only 37.2% of votes in this precinct. The low reliability of this result (0.838) has been confirmed by the count of undervotes, which has added the net amount of 28 votes to Bush.

## Table 5 Low Reliability Precincts

| Precinct | Dem | Rep | White | Pred. Vote | Act. Vote | Reliability | Under votes- Bush | Under votes-Gore | Total Under votes | Abs. Gain |
|---|---|---|---|---|---|---|---|---|---|---|
| 154C | 4.12 | 84.02 | 97.94 | 0.515 | 0.866 | 0.617 | 4 | 0 | 4 | 4 |
| 33 | 13.17 | 77.72 | 98.37 | 0.515 | 0.846 | 0.637 | 0 | 1 | 1 | 1 |
| 154B | 13.18 | 74.00 | 99.63 | 0.515 | 0.797 | 0.687 | 2 | 1 | 3 | 1 |
| 167 | 15.25 | 69.50 | 97.34 | 0.515 | 0.761 | 0.724 | 1 | 0 | 1 | 1 |
| 162I | 8.07 | 78.60 | 97.54 | 0.515 | 0.758 | 0.727 | 0 | 0 | 0 | 0 |
| 37 | 52.20 | 37.11 | 59.21 | 0.295 | 0.519 | 0.749 | 0 | 5 | 5 | 5 |
| 122A | 49.66 | 24.16 | 73.83 | 0.251 | 0.452 | 0.773 | 0 | 1 | 1 | 1 |
| 001A | 15.79 | 67.64 | 97.68 | 0.515 | 0.679 | 0.814 | 6 | 2 | 8 | 4 |
| 36 | 46.83 | 38.03 | 78.52 | 0.404 | 0.562 | 0.820 | 0 | 0 | 0 | 0 |
| 148E | 37.43 | 45.99 | 76.47 | 0.515 | 0.667 | 0.827 | 0 | 1 | 1 | 1 |
| 151 | 17.68 | 60.10 | 91.92 | 0.515 | 0.659 | 0.836 | 0 | 1 | 1 | 1 |
| 163 | 30.87 | 48.23 | 94.86 | 0.515 | 0.372 | 0.837 | 0 | 1 | 1 | 1 |
| 191 | 29.73 | 50.23 | 96.77 | 0.515 | 0.372 | 0.838 | 84 | 56 | 140 | 28 |
| 121A | 36.14 | 42.57 | 89.11 | 0.398 | 0.53 | 0.850 | 0 | 0 | 0 | 0 |
| 158 | 35.34 | 48.54 | 98.25 | 0.515 | 0.394 | 0.863 | 2 | 1 | 3 | 1 |
| 225 | 32.62 | 46.34 | 92.34 | 0.515 | 0.398 | 0.867 | 23 | 21 | 44 | 2 |
| 90 | 17.31 | 64.24 | 96.24 | 0.515 | 0.631 | 0.867 | 1 | 1 | 2 | 0 |
| 045A | 33.19 | 46.36 | 93.51 | 0.515 | 0.399 | 0.868 | 5 | 11 | 16 | 6 |
| 49 | 35.22 | 45.34 | 97.03 | 0.333 | 0.448 | 0.869 | 0 | 4 | 4 | 4 |
| 093A | 39.78 | 40.37 | 94.50 | 0.398 | 0.511 | 0.871 | 1 | 0 | 1 | 1 |
| Total | | | | | | | | | 236 | 62 |

Table 6 High Reliability Precincts

| Precinct | Dem | Rep | White | Pred. Vote | Act. Vote | Reliability | Under votes- Bush | Under votes- Gore | Total Under votes | Abs. Gain |
|---|---|---|---|---|---|---|---|---|---|---|
| 156C | 41.76 | 31.87 | 79.12 | 0. 326 | 0.322 | 0.996 | 0 | 0 | 0 | 0 |
| 38 | 82.69 | 8.20 | 2.96 | 0. 072 | 0.069 | 0.997 | 0 | 0 | 0 | 0 |
| 78 | 42.18 | 40.47 | 83.97 | 0. 404 | 0.401 | 0.997 | 0 | 1 | 1 | 1 |
| 110 | 43.39 | 39.31 | 91.10 | 0. 398 | 0.4 | 0.997 | 6 | 9 | 15 | 3 |
| 128G | 43.26 | 33.80 | 80.26 | 0. 326 | 0.328 | 0.997 | 1 | 1 | 2 | 0 |
| 159J | 36.03 | 40.24 | 84.63 | 0. 404 | 0.406 | 0.997 | 2 | 1 | 3 | 1 |
| 201 | 35.63 | 43.19 | 90.48 | 0. 398 | 0.396 | 0.997 | 3 | 0 | 3 | 3 |
| 003B | 32.08 | 42.64 | 95.00 | 0. 398 | 0.4 | 0.998 | 2 | 2 | 4 | 0 |
| 114 | 68.21 | 19.44 | 26.54 | 0. 253 | 0.255 | 0.998 | 0 | 2 | 2 | 2 |
| 119 | 40.12 | 41.93 | 95.18 | 0. 398 | 0.397 | 0.998 | 2 | 3 | 5 | 1 |
| 120 | 45.00 | 27.50 | 84.53 | 0. 251 | 0.249 | 0.998 | 0 | 3 | 3 | 3 |
| 144E | 44.19 | 34.99 | 75.35 | 0. 326 | 0.324 | 0.998 | 33 | 43 | 76 | 10 |
| 162A | 80.42 | 9.88 | 98.19 | 0. 072 | 0.074 | 0.998 | 0 | 2 | 2 | 2 |
| 205E | 41.51 | 35.82 | 92.55 | 0.28 | 0.279 | 0.998 | 4 | 1 | 5 | 3 |
| 007A | 31.71 | 60.98 | 95.12 | 0.515 | 0.515 | 0.999 | 0 | 0 | 0 | 0 |
| 018J | 50.45 | 33.32 | 93.99 | 0.28 | 0.281 | 0.999 | 1 | 7 | 8 | 6 |
| 88 | 37.09 | 42.29 | 92.81 | 0.398 | 0.397 | 0.999 | 5 | 1 | 6 | 4 |
| 115 | 45.53 | 41.30 | 93.12 | 0.398 | 0.397 | 0.999 | 4 | 5 | 9 | 1 |
| 203 | 29.88 | 52.74 | 96.34 | 0.515 | 0.515 | 1.000 | 0 | 0 | 0 | 0 |
| Total | | | | | | | | | 144 | 40 |

## 6. Conclusion

In this paper, we have presented a novel fuzzy-based approach to evaluating reliability of continuous attributes in a relational database. The approach includes partition of a data table into input and target attributes, induction of a data mining model (information-theoretic network) from a set of training data, and calculation of reliability degrees for target values based on their distance from the values predicted by the network.

The proposed approach combines objective information about the data, which is represented by an information-theoretic network, with a subjective, user-specific perception of data quality. In our case study, we have shown that the method can be an efficient tool for detection of inaccurate information in a real-world database.

Related issues, to be further studied, include: integrating the method with other data mining models, evaluating reliability of input attributes, and detecting unreliable information in non-relational data.

*Acknowledgment.* This work was supported in part by the National Institute for Systems Test and Productivity at USF under the USA Space and Naval Warfare Systems Command Grant No. N00039-01-1-2248 and in part by the USF Center for Software Testing under grant No. 2108-004-00.

# References

[1] N. Ahituv, S. Neumann and H. N. Riley, Principles of Information Systems for Management, B & E Tech, Dubuque, Iowa, 1994.

[2] P. Bosc and O. Pivert, SQLf: A Relational Database Language for Fuzzy Querying, *IEEE Transactions on Fuzzy Systems*, **3**, no. 1, pp. 1-17, 1995.

[3] T. M. Cover, Elements of Information Theory, Wiley, New York, 1991.

[4] J. F. Elder IV and D. Pregibon, A Statistical Perspective on Knowledge Discovery in Databases. In: *Advances in Knowledge Discovery and Data Mining*, U. Fayyad, G. Piatetsky-Shapiro, P., Smyth, R. Uthurusamy, Eds. AAAI/MIT Press, Menlo Park, CA, pp. 83-113, 1996.

[5] U. Fayyad, G. Piatetsky-Shapiro and P. Smith, From Data Mining to Knowledge Discovery in Databases, AI Magazine, vol. **3**, pp. 37-54, 1996.

[6] R. M. Goodman, J. W. Miller and P. Smyth, An Information Theoretic Approach to Rule-Based Connectionist Expert Systems. In: *Advances in Neural Information Processing Systems*, Touretzky, D. S. (Ed.), Morgan Kaufmann, San Mateo, CA, pp. 256-263, 1988.

[7] A. L. Gorin, S. E. Levinson, A. N. Gertner and E. Goldman, Adaptive Acquisition of Language, *Computer Speech and Language*, vol. **5**, no. 2, pp. 101-132, 1991.

[8] N. Guyon, N. Matic and V. Vapnik, Discovering Informative Patterns and Data Cleaning. In: *Advances in Knowledge Discovery and Data Mining*, Fayyad, U., Piatetsky-Shapiro, G., Smyth, P., Uthurusamy, R. (Eds.), AAAI/MIT Press, Menlo Park, CA, pp. 181-203, 1996.

[9] B. E. Hansen, Florida Data Page [http: //www. ssc. wisc. edu/~bhansen/vote/data. html]

[10] A. Kandel, Fuzzy Mathematical Techniques with Applications, Addison-Wesley, Reading, MA, 1986.

[11] A. Kandel, R. Pacheco, A. Martins and S. Khator, The Foundations of Rule-Based Computations in Fuzzy Models. In: *Fuzzy Modeling, Paradigms and Practice*, Pedrycz, W. (Ed.), Kluwer, Boston, pp. 231-263, 1996.

[12] G. J. Klir, B. Yuan, Fuzzy Sets and Fuzzy Logic: Theory and Applications, Prentice-Hall Inc., Upper Saddle River, CA, 1995.

[13] H. F. Korth and A. Silberschatz, Database System Concepts, McGraw-Hill, Inc., New York, 1991.

[14] W. Kurutach, Managing Different Aspects of Imperfect Data in Databases, *Proc. 1995 IEEE Int'l Conf. on SMC*, Vancouver, BC, Canada, pp. 2812-2817, 1995.

[15] M. Last, A. Kandel and O. Maimon, Information-Theoretic Algorithm for Feature Selection, Pattern Recognition Letters, vol. **22**, no. 6, pp. 799-811, 2001.

[16] M. Last, Y. Klein and A. Kandel, Knowledge Discovery in Time Series Databases, *IEEE Transactions on Systems, Man, and Cybernetics*, vol. **31**, part B, no. 1, pp. 160-169, 2001.

[17] H. Lu, R. Setiono and H. Liu, Effective Data Mining Using Neural Networks, *IEEE Transactions on Knowledge and Data Engineering*, vol. **8**, no. 6, pp. 957-961, 1996.

[18] O. Maimon, A. Kandel and M. Last, Information-Theoretic Fuzzy Approach to Knowledge Discovery in Databases. In: *Advances in Soft Computing - Engineering Design and Manufacturing*, R. Roy, T. Furuhashi, P. K. Chawdhry (Eds.), Springer-Verlag, London, pp. 315-326, 1999.

[19] O. Maimon, A. Kandel and M. Last, Fuzzy Approach to Data Reliability. In: *Knowledge Management in Fuzzy Databases*, O. Pons, A. Vila, J. Kacprzyk (Eds.), Physica-Verlag, pp. 89-101, 2000.

[20] O. Maimon and M. Last, Knowledge Discovery and Data Mining, The Info Fuzzy Network (IFN) Methodology, Kluwer, Boston, 2000.

[21] T. M. Mitchell, *Machine Learning*, McGraw-Hill, New York, 1997.

[22] Motro, Imprecision and Uncertainty in Database Systems. In: *Fuzziness in Database Management Systems*, P. Bosc, J. Kacprzyk (Eds.), Springer-Verlag, pp. 3-22, 1995.

[23] H. Nakajim and Y. Senoh, A Spreadsheet-Based Fuzzy Retrieval System, International Journal of Intelligent Systems, vol. **11**, no. 1, pp. 661-670, 1996.

[24] Palm Beach County Elections - Election Results [http: //www. pbcelections. org/eresults. htm]

[25] F. E. Petry, Fuzzy Databases, Principles and Applications, Kluwer, Boston, MA, 1996.

[26] J. R. Quinlan, Induction of Decision Trees, *Machine Learning*, vol. **1**, no. 1, pp. 81-106, 1986.

[27] J. R. Quinlan, C4. 5: Programs for Machine Learning, Morgan Kaufmann, San Mateo, CA, 1993.

[28] C. R. Rao and H. Toutenburg, *Linear Models: Least Squares and Alternatives*, Springer-Verlag, 1995.

[29] Y. Takahashi, Fuzzy Database Query Languages and Their Relational Completeness Theorem, *IEEE Transactions on Knowledge and Data Engineering*, vol. **5**, no. 1, pp. 122-125, 1993.

[30] The Miami Herald Ballot Review [http: //advapps. herald. com/election_results/default. asp]

[31] Y. Wand and R. Y. Wang, Anchoring Data Quality Dimensions in Ontological Foundations, Communications of the ACM, vol. **39**, no. 11, pp. 86-95, 1996.

[32] R. Y. Wang and V. C. Storey, C. P. Firth, A Framework for Analysis of Data Quality Research, *IEEE Transactions on Knowledge and Data Engineering*, vol. **7**, no. 4, pp. 623-639, 1995.

[33] R. Y. Wang, M. P. Reddy and H. B. Kon, Toward Quality Data: An Attribute-based Approach, Decision Support Systems, vol. **13**, pp. 349-372, 1995.

[34] L. A. Zadeh, Syllogistic Reasoning in Fuzzy Logic and its Application to Usuality and Reasoning with Dispositions, *IEEE Transactions on Systems, Man, and Cybernetics*, vol. **15**, no. 6, pp. 754-763, 1985.

[35] M. Zemankova-Leech, A. Kandel, Fuzzy Relational Databases - a Key to Expert Systems, Verlag TUV, Koln, Germany, 1984.

# Relevance of the fuzzy sets and fuzzy systems

Paulo SALGADO
*Universidade de Trás-os-Montes e Alto Douro, 5000-911 Vila Real, Portugal*

**Abstract.** The readability of the fuzzy models is related to its organizational structure and the correspondent rules base. In order to define methodologies for organizing the information describing a system, it is important to specify metrics that define the relative importance of a set of rules in the description of a given region of the input/output space. The concept of relevance, landmarked by a set of intuitive axioms, enables this measurement. Considering this, a new methodology for organizing the information, *Separation of Linguistic Information Methodology* (SLIM), was developed. Based on these results, different algorithms were proposed for different structures: the *Parallel Collaborative Structure* (PCS) – SLIM-PCS algorithm and the *Hierarchical Prioritized Structure* (HPS), SLIM-HPS algorithm. Finally, a new Fuzzy Clustering of Fuzzy Rules Algorithm (FCFRA) is proposed. Typically, the FCM (Fuzzy C-means) algorithms organize clusters of points with same similarity. The FCFRA organize the rules of a fuzzy system in various fuzzy sub-systems, interconnected in a structure. Its application in the organization of information of a fuzzy system in HPS and CPS structures are demonstrated as well.

## 1. Introduction

Fuzzy modeling is a very important and active research field in fuzzy logic systems. Compared to traditional mathematical modeling and pure neural network modeling, fuzzy modeling possesses some distinctive advantages, such as the mechanism of reasoning in human understandable terms, the capacity of taking linguistic information from human experts and combining it with numerical data, and the ability of approximating complicated non-linear functions with simpler models. In recent years, a variety of different fuzzy modeling approaches have been developed and applied in engineering practice [1][2][3][4][5][6]. These approaches provided powerful tools to solve complex non-linear system modeling and control problems. However, most existing fuzzy modeling approaches concentrate on model accuracy that simply fit the data with the highest possible accuracy, paying little attention to simplicity and interpretability of the obtained models, which is considered a primary merit of fuzzy rule-based systems. Often, users require the model to not only predict the system's output accurately but also to provide useful description of the system that generated the data. Such a description can be elicited and possibly combined with the knowledge of experts, helping to understand the system and validate the model acquired from data. Thus, it is desired to establish a fuzzy model with satisfactory accuracy and good interpretation capability.

In order to organize the fuzzy rules and reduce its number, it is of utmost importance to define metrics to quantify each one of the fuzzy rules that describes the process. It should be noted that the relative importance of different sets of rules describing a region of the input/output space in a fuzzy system might not be directly related to the contribution for the minimization of the error in the output. Possibly, the various rules of the system will assume different contexts, with some rules covering large regions of the space, while others may be

---

This work was supported by FCT – POSI/SRI/33574/1999.

related to regions where the gradient is higher. Moreover, one rule can be relevant in a particular aspect of the model.

This work addresses this fundamental aim of fuzzy modeling. As result of a new concept recently proposed by Paulo Salgado [7], namely the relevance of the rule set, this objective is near at hand. The relevance is a measure of the relative importance of sets of rules, in the description of a given region of the input/output space. Depending on the context where the relevance is to be measured, different metrics may be defined. The spread of definition of relevance in the boundary regions is proposed here. These new concepts bounded by a set of intuitive axioms open the doors for new types of fuzzy systems. These axioms lead to a set of properties that are analyzed in some detail.

In order to corroborate the validity of the new concept, a new methodology is reviewed. It has been called SLIM, Separation of Linguistic Information Methodology [7][8]. It is useful for organizing the information in a fuzzy system: a system $f(x)$ is organized as a set of $n$ fuzzy sub-systems $f_1(x)$, $f_2(x)$,..., $f_n(x)$. Each of these systems may contain information related with particular aspects of the system $f(x)$.

Two main structures are introduced: firstly, HPS (*Hierarchical Prioritized Structure*), which allows organizing the information in the prioritized fashion, Yager [9][10][11]. Contrarily, PCS (*Parallel Collaborative Structure*), where each model collaborates equally with the other models, is presented. With the HPS structure, Yager has introduced a new perspective. Instead of a fuzzy system consisting of a set of rules with no ordering, apparently all with the same relative importance, priorities are defined, these priorities being connected to the importance of the rules in the description of the process being modeled. The method suggested by Yager for ordering the rules is based on the comparison of every pair of rules in the system. This leads to the establishment of binary relations between the rules. Using some results from preference theory [11], and if the relations are well behaved, it is possible to define a ranking order for the rules. However, the proposed method is not adequate for situations where the number of rules is large. On the contrary, in the PCS structure, each system works independently and collaborates with the others, without any order or inhibition factor. However, different sub-systems and fuzzy rules will have different relevance values.

The application of the SLIM methodology in these structures is here used to organize the information, by exchanging information among various layers of the structure. It is also possible to reduce the number of rules representing the original system by discarding rules with lower relevance values.

Finally, two algorithms that implement Fuzzy Clustering of Fuzzy Rules are presented. By using the proposed algorithm, it is possible to group a set of rules in $c$ subgroups (clusters) of similar rules. It is a generalization of the fuzzy $c$-means clustering algorithm, here applied to rules instead of points in $R^n$ [12]. With this algorithm, the system obtained from the data is transformed into a new system, organized into subsystems, in PCS or HPS structures.

In order to corroborate the proposed concepts, experimental results are presented for the organization of information, namely in the fuzzy identification and fuzzy clustering of fuzzy rules. Practical experiments have been conducted in the identification of environmental variables in agricultural greenhouse (temperature and humidity).

The paper is organized as follows. The concept of relevance of a set of rules is defined in the section 2. The SLIM methodology and different structures (HPS and PCS) is discussed in section 3. In section 4 a new Fuzzy Clustering of Fuzzy Rules (FCFR) strategy is proposed. Different examples and experimental tests are presented in section 5. Finally, the main conclusions are outlined in section 6.

## 2. The concept of relevance

### 2.1. Relevance in well-defined support region S

Fuzzy systems are based on a set of rules that map regions in input space, $U$, to regions in output space, $V$, describing a region in product space $S = U \times V$. For this relationship, the contributions of the different rules will be unequal. One main question will be formulated:

"How to measure the relative importance of the rules that describe the region $S$?". Moreover, the fuzzy system can't completely describe the region $S$: "How to measure the quality of fuzzy system in describing the $S$ region?" or "Is the region $S$ perfectly described by fuzzy system?". In any case, the concept of *relevance* is expected to help clarifying those questions.

Thus, adopting the fuzzy logic theory and looking at the relevance degree as an extended fuzzy measure seems to be an appropriate approach to automating relevance perception of the rules and fuzzy systems. In [7], it was proposed the following definition for the relevance of a set of rules.

**Definition 1:** Consider $\Im$ a set of rules from $U$ into $V$, covering the region $S = U \times V$ in the product space. Any function defined as a measure of relevance must be of the form $\Re_S : P(\Im) \to [0 , 1]$, where $P(\Im)$ is the power set of $\Im$ (the set of all subsets of rules in $\Im$).

Contrarily of a fuzzy measure, the relevance measures involve the relativity of a support region. Only in the case where the support of rules agrees with region $S$, the fuzzy measure is a relevance measure. So, the relevance is a generalization of a fuzzy measure.

Given $A$, a set of rules defined in $P(\Im)$ $(A \subset P(\Im))$, the function $\Re_S$ associates a value $\Re_S(A)$ to each subset of rules $A$ in $\Im$. This value is the measure of the relevance of the subset of rules in the description of the space $S$.

The concept is defined by a set of axioms, which are illustrated, by a set of properties.　　In order to qualify as a measure of relevance, any function $\Re_S (A)$ must obey the following five axioms.

**Axiom 1** (border conditions): $\Re_S(\varnothing) = 0$ and $\Re_S(\Im) = 1$.

**Axiom 2** (monotony): For all $A, B \in P(\Im)$, if $A \subseteq B$, then $\Re_S(A) \le \Re_S(B)$.

**Axiom 3**: Given the space $S$ partitioned in $n$ regions $S = S_1 \cup \ldots \cup S_n$ and $A \in P(\Im)$, then:

$$\Re_S (A) = \max \left\{ \Re_{S_i} (A), \ i = 1, \cdots, n \right\}.$$

**Axiom 4** (continuity): Consider a sequence of sub-regions in $S$, $S_1 \supseteq S_2 \supseteq \ldots$ or $S_1 \subseteq S_2 \subseteq .$, and let $B \in P(\Im)$ be a set of rules that describe the region of the space $T \subset S$. If the sub-region $S^* = \lim_{i \to \infty} S_i$ exists then $\Re_{S^*} (B) = \lim_{i \to \infty} \Re_{S_i} (B)$.

**Axiom 5**: Consider the sequence of sub-regions in $S$, $T \supseteq T_1 \supseteq T_2 \supseteq \ldots$ associated to the sequence of sets of rules $B, B_1, B_2, \ldots$ so that the rules in $B_i$ cover the region $T_i$ and only this region. Then $\lim_{i \to \infty} \Re_S (B_i) = \Re_S \left( \lim_{i \to \infty} B_i \right)$.

The following properties may be derived from axioms and its proof founded in [7][8].

**Property 1:** Let $A, B \in P(\mathfrak{I})$. If $C = A \cup B$, then $\mathfrak{R}_S(C) \geq \max(\mathfrak{R}_S(A), \mathfrak{R}_S(B))$, i.e., $\mathfrak{R}_S(C) = S(\mathfrak{R}_S(A), \mathfrak{R}_S(B))$, where S represents any *S-norm* operation.

**Property 2:** Let $A, B \in P(\mathfrak{I})$. If $C = A \cap B$, then $\mathfrak{R}_S(C) \leq \min(\mathfrak{R}_S(A), \mathfrak{R}_S(B))$, i.e., $\mathfrak{R}_S(C) = T(\mathfrak{R}_S(A), \mathfrak{R}_S(B))$, where T represent any *T-norm* operation.

**Property 3:** If $S_i$ is a sub-region of $S$, $S_i \subseteq S$, and $A \in P(\mathfrak{I})$ is a subset of the rules that describe the region $S$, then $\mathfrak{R}_{Si}(A) \leq \mathfrak{R}_S(A)$.

**Property 4:** If $A$ is a set of rules which only covers the region $T \subseteq S$, then $\mathfrak{R}_S(A) = \mathfrak{R}_T(A)$.

**Property 5:** Consider a sequence of sub-regions in $S$, $S_1 \supset S_2 \supset ..$, and let $B \in P(\mathfrak{I})$ be the set of rules defining the $T \subset S$ region. If the sub-region $S^* = \lim\limits_{i \to \infty} S_i$ exists, and if:

$T \cap S_i \to \varnothing$ when $i \to \infty$ (the intersection of the two regions tends to zero, as $i$ tends to infinity), then $\mathfrak{R}_{S^*}(B) = \lim\limits_{i \to \infty} \mathfrak{R}_{S_i}(B) = 0$.

In determined conditions, given by theorem 1, it's possible to exchange the support spaces $S$, without changing the relevance of a set of rules.

**Theorem 1:** Let $\mathfrak{I}$ a set of rules covering the region $S$. Let, also, the function $f$ that maps a region $S$ in $R$.

$$f : S \mapsto R$$

If the relation map $f$ is a bijective function, the relevance values of rules $\mathfrak{I}$ in space S have the same values in space $R$, $\mathfrak{R}_S(\mathfrak{I}) \equiv \mathfrak{R}_R(\mathfrak{I})$.

Next, we review the definition of relevance for a rule in a *single point* of the product space and the measure of relevance for a rule in a *region of the space*.

**Definition 2:** Let $\mathfrak{I}$ be a set of rules that map $U$ into $V$, describing a region $S = U \times V$ in the product space. The relevance of a rule $l \in \mathfrak{I}$ in a point of the product space $(x, y) \in S$ is defined as

$$\mathfrak{R}_S\big(l,(x,y)\big) = \max_y \big(G_l / G\big) \tag{1}$$

i.e., the relevance in $(x, y)$ is the maximum of the ratio between the value of the output membership function of rule $l$ in $(x\ y)$, and the value of the membership of the union of all the functions in $(x, y)$.

**Definition 3:** Let $\mathfrak{I}$ be a set of rules that map $U$ into $V$, describing a region $S = U \times V$ in the product space. The *relevance* of a rule $l \in \mathfrak{I}$ in the product space $S$ is here defined as

$$\mathfrak{R}_S(l) = \max_{x,y} \mathfrak{R}_S\big(l,(x,y)\big) \quad \forall (x,y) \in S \tag{2}$$

i.e., the maximum of the value for all points $(x, y) \in S$, of the ratio between the membership output function of rule $l$, and the value of the union of all the output membership functions.

In many situations, e.g. in control and modeling applications, it is desirable to have a crisp value $y^*$ for the output of a fuzzy system, instead of a fuzzy value $y$. As example, the center of area defuzzification method, applied to fuzzy sets obtained by using the arithmetic inference mechanism, result:

$$f(x) = \sum_{l=1}^{M} p^l(x) \cdot \theta^l \qquad (3)$$

where $p^l(x) = \mu^l(x) \bigg/ \sum_{l=1}^{M} \mu^l(x)$ is the *fuzzy basis functions* (FBF), $M$ represent the number of rules, $\theta^l$ is the point at which the output fuzzy set $l$ achieves its maximum value, and $\mu^l$ is the membership of antecedent of rule $l$.

This result can be interpreted as the sum of the output membership centroids [13][14], weighted by the relevance values of each rule in the point $(x, y) \in S$.

$$y^* = \sum_{l=1}^{M} \theta^l \cdot \Re_S\big(l,(x,y)\big) \qquad (4)$$

This result and equation (3) lead us to a definition of a type of relevance of a rule.

**Definition 4:** Let $\mathfrak{I}$ be a set of rules that map $U$ into $V$, describing completely the region $S$. The *relevance* of a rule $R_l \in \mathfrak{I}$, of fuzzy system (3) in $S$ space is defined as

$$\Re_l(x_k) = \frac{\mu^l(x_k) \cdot \delta^l}{\sum_{l=1}^{M} \mu^l(x_k) \cdot \delta^l} \qquad (5)$$

i.e., the relevance in $(x, y)$ is the maximum of the ratio between the value of the output membership function of rule $l$ in $(x, y)$, and the value of the membership of the union (sum) of all the functions in $(x, y)$. From axiom 1, it is easy to see that the relevance of all rules is the sum of the relevance of each rule in the point $x_k \in S$ and equal to one:

$$\Re_{\mathfrak{I}}(x_k) = \sum_{l=1}^{M} \Re_l(x_k) = 1 \qquad (6)$$

In the traditional systems, as equation (3), all the rules are considered as having the same contribution in the characterization of the fuzzy system. Other times, the rules were weighed with different weights ($\delta$), which express our faith or conviction on its validity. In any way, this didn't implicate any differentiation in the fuzzy system, even if a given area of the space was described by a single rule with low weight. Now, for the existence of a function of relevance, the weight of the rules has impact in the characterization of the system through its relevance.

### 2.2. The relevance in the transition region

In the previous section, the $S$ region has been considered as a crisp set, i.e., its boundary has been well defined in the *Universe of Space X*. Definitions 2 and 3 have been made for non-fuzzy region $S$. However, in the generality of the fuzzy systems, the validity of fuzzy system outside the region of support $S$ is not guaranteed. In other situations, the behavior of the fuzzy system inside the region $S$ leads us to disbelief of its real support. Otherwise, it is not possible to have an exact knowledge of region $S$. This result can be expressed as a belief measure.

The functionality of a fuzzy system containing various sub-systems with different supports and interconnected in different structures is only possible by a definition of relevance of fuzzy systems, to take into account the boundary regions. The firing fuzzy rules in the boundary regions decaying strongly to zero, in so far as the characterization of regions by fuzzy rules is smaller. This result can be expressed by a membership function, which varies in the range [0,1]. So, the previous definitions of relevance will be now incorporating its approaches. The definition 5 introduces one possible approach.

**Definition 5:** Let $S$ be a region of the universe space $X$ defined by a set of rules $\Im$. The membership function characterization in the space $S$ is defined as

$$\mu_S(x,y) = \begin{cases} 1 & ; if \ \max_y G(y) \geq Sl \\ \dfrac{\max_y G(y)}{Sl} & ; outherwise \end{cases} \tag{7}$$

where $Sl$ is the threshold level that discriminates regions $S$.

As $\max_y G(y) \to 0$, the value of the membership function of the space $S$ tends to zero. From property 2, the jointly relevance can be defined as follows.

**Definition 6:** Let $\Im$ be a set of rules that map $U$ into $V$, describing a region $S$, a sub-region of Universe $X$. The *relevance* of a rule $l \in \Im$ in a point of the product space $(x, y) \in S$ is defined as

$$\Re_S\big(l,(x,y)\big) = \max_y \Big(T\big(\Re^l(x,y),\mu_S\big)\Big) \tag{8}$$

where $\Re^l(x,y)$ is the relevance of $l$ rule.

The relevance of fuzzy rule $l$ in the point $(x, y)$ of space $S$ is obtained by intersecting the relevance of rule $l$ with its relevance covering of region. $\Re^l(x,y)$ is done by definition 3.

**Definition 7:** Let $\Im$ be a set of rules that map $U$ into $V$, describing a region $S$, a sub-region of Universe $X$. The *relevance* of a rule $l \in \Im$ in $S$ region is defined as

$$\Re_S(l) = \max_{x,y} \Big(T\big(\Re^l(x,y),\mu_S\big)\Big) \tag{9}$$

Therefore, the relevance of one rule or a set of rules characterizes simultaneously the relevance importance of the rule in its region and the region cover degree by the fuzzy system. This new proposal enables the aggregation of different fuzzy systems, with a different set of rules and different or shared cover regions.

This result can be now incorporated in the fuzzy system (3). The relevance of equation (4) will contain the relevance of system in the transition region.

**Definition 8:** Let $S$ be a region of the universe space $X$ defined by a set of rules $\mathfrak{I}$, of fuzzy system (3). The relevance of rule $l$ in describing the region $S$ in point $x_k$ is defined as

$$\mathfrak{R}_l(x_k) = \begin{cases} \dfrac{\mu^l(x_k)\cdot\delta^l}{\displaystyle\sum_{l=1}^{M}\mu^l(x_k)\cdot\delta^l} & ; \ if \ G(x_k) \geq Sl \\[4ex] \dfrac{\mu^l(x_k)\cdot\delta^l}{Sl} & ; \ outherwise \end{cases} \tag{10}$$

where $G(x_k) = \displaystyle\sum_{l=1}^{M}\mu^l(x_k)\cdot\delta^l$ and $Sl$ is the threshold level that discriminates regions $S$. The expression is the result of the defuzzification process of a fuzzy relevance.

## 3. SLIM, Separation of Linguistic Information Methodology

### 3.1. SLIM methodology

In order to corroborate the validity of the concept of relevance, a new methodology based on this concept is presented. It has been named SLIM, Separation of Linguistic Information Methodology. The SLIM methodology may be used to rank the rules in a system and/or to distribute them among the various layers of an organized structure.

When the number of rules in a system is large, the hierarchical organization of the rules and the reduction of their number become one important scope [15]. It is render possible by discarding rules with a lower measure of relevance.

Let $R=\{R^1, .., R^M\}$ be a set of rules, which describes a region $S$. The idea is to separate the information of this set in $n$ sets, $R_1, R_2,..., R_n$, making sure that the description of $S$ given by the new sets will be identical to the description given by $R$.

A generic algorithmic description for implementing the SLIM methodology, using two sets, follows.

*SLIM methodology*

***Step 1*:** Let $R_1 = R$.

***Step 2*:** Choose an appropriate number $n$ of rules for $R_2$. Initiate $R_2$ with a set of $n$ rules in $S$, with the restriction of these rules having no relevance in $S$, i.e. $\mathfrak{R}_S(R_2) = 0$. That is to say, the transfer function of the new system is the transfer function of the previous system.

***Step 3*:** Diminish the relevance of every rule in $R_1$, and compensate this effect by increasing the relevance, and possibly tuning, rules in $R_2$, under the condition of maintaining the transfer function of the system invariant.

***Step 4*:** Eliminate the rules in $R_1$ whose values of relevance are considered sufficiently low.

The *steps 3* and *4* may be repeated.

SLIM methodology, involving more than two steps levels, running in same way, i.e., by using the behind process, applied to all combinational pairs of levels.

As referred to before, the SLIM methodology may be used with different structures.

### 3.2. SLIM structures

The SLIM methodology organizes a fuzzy system $f(x)$ as a set of $n$ fuzzy systems $f_1(x)$, $f_2(x)$, ..., $f_n(x)$. Each of these subsystems may contain information related with particular aspects of the system $f(x)$. Different interconnections will be made in order to build complex fuzzy systems structures. We follow with a description of the PCS and HPS structures in the SLIM context.

### A. PCS structure

The figure 1 shows a PCS structure with two fuzzy subsystems. It's possible to observe that sub-system have an extra output, that indicate the relevance of its output, y.

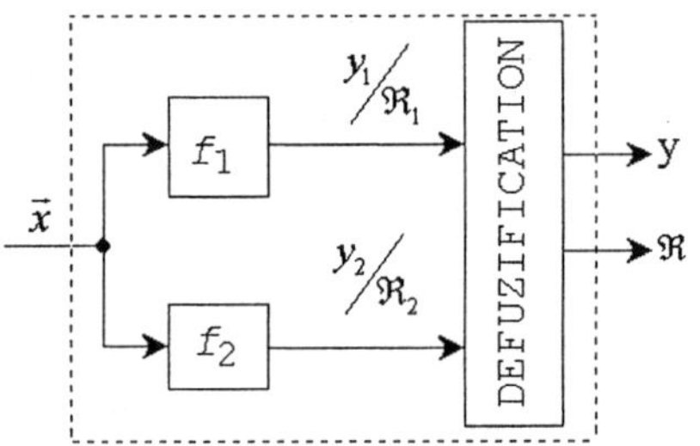


Figure 1: A practical implementation of PCS.

Therefore, the output of the SLIM model is the integral of the individual contributions of fuzzy subsystems:

$$f(x) = \int_{i=1}^{n} f_i(x) \cdot \mathfrak{R}_i(x) \qquad (11)$$

where $\mathfrak{R}_i(x)$ represents the relevance function of the $i^{th}$ fuzzy subsystem covering the point x, of the Universe of Discourse, and the $\int$ an aggregation operator.

The relevance of aggregated system can be done by:

$$\mathfrak{R}_i(x) = \mathfrak{R}_i(x) \cup \cdots \cup \mathfrak{R}_n(x) \qquad (12)$$

Naturally, if the $i^{th}$ fuzzy subsystem covers appropriately the region of the point $x$, its relevance value is high (very close to one), otherwise the relevance value is low (near zero or zero). The separation of information may be used to organize the information of a system. Consequently, it's possible to reduce the number of rules representing the original system by discarding sets of rules with lower relevance values.

Afterwards, there are going propose a SLIM algorithm, for PCS structure of figure 1. The generalizations of this algorithm to a PCS structure with more than two levels are overcome without any difficulty.

Let $f_2$ a fuzzy subsystem with early null output values for all domain (or relevance null). $M_1$ and $M_2$, are the number of rules of the $f_1$ and $f_2$, respectively. The first subsystems can be expressed by:

$$f_{\Re,1}(x) = q_{\Re}^T(x) \cdot Y \tag{13}$$

where $q_{\Re}(x) = q(x) \otimes \Re_l$, i.e., $q_{\Re}(x) = \left[ q^{l_1}(x) \cdot \alpha^{l_1}, q^{l_2}(x) \cdot \alpha^{l_2}, \cdots, q^{l_{M_1}}(x) \cdot \alpha^{l_{M_1}} \right]^T$ is the inner product between the FBF vector and the relevance vector $Y = \left[ \bar{y}^{l_1}, \bar{y}^{l_2}, \cdots, \bar{y}^{l_M} \right]^T$ is the vector (or matrix) of all the centers of output membership functions.

The parameter $\alpha^{l_1}$ is closely connected to the relevance of the rule in the fuzzy system. When $\alpha^{l_1}$ is equal to unity, rule $l_1$ has maximum relevance, while for null $\alpha^{l_1}$, the rule loses its relevance. If $\alpha^{l_1} = 1$ for all $l_1 = 1, \ldots, M_1$, then $f^*_1 = f_1$. If parameter $\alpha^{l_1}$, associated to rule $l_1$, converges to null, rule $l_1$ is eliminated from function $f_{\Re,1}$. If this is possible for all rules of $f_1$ then $f_1$ is eliminated ($\lim\limits_{\alpha \to \infty} f_1 = 0$).

Similarly, the second fuzzy system is expressed by:

$$f_{\Re,2}(x) = p^T(x) \cdot \theta_{\Re} \tag{14}$$

where $\theta_{\Re}$ is the inner product between the $\theta$ vector and the relevance vector. Initially, $f_1$ contains all the information, $\Re_l(l_1) = 1$, $\forall l_1 \in \{1, 2, \cdots, M_1\}$, while $f_2$ is empty, i.e., $\Re_l(l_2) = 0$ $\left( \theta_{\Re}^{l_2} = 0 \right)$, $\forall l_2 \in \{1, 2, \cdots, M_2\}$, $f_{\Re,2}(x_k) = 0$.

The relevance of the rules of $f_1$ decreases at the same proportion that $f_2$ assumes a greater importance. Thus, during this transfer of process information there is no change in the sum of the models. By the end of this process, all or part of the rules of model $f_1$ may have null or almost null relevance and under these circumstances they should be eliminated. Those who keep a significant relevance index should not be eliminated, as they still contain relevant information.

Considering what was stated, the problem consists on the optimization of the cost function $J$: $\min J(\alpha) = \min (\alpha)^T \cdot \alpha$ *subject to*

$$f(x) = f_{\Re,1}^i(x) + f_{\Re,2}^i(x) \tag{15}$$

The purpose is then to keep the invariability of the identification model ($i = 1, 2, \ldots$, iteration) and, simultaneously, to reduce the importance of model $f_1$ in favor of $f_2$. In order to achieve this, the Lagrange multipliers technique has been used. For optimization problem purposed, the Lagrange multipliers technique has been used. So, the previous problem become transformed in the optimization of function $L = 1/2 \cdot \alpha^T \cdot \alpha - \lambda^T \left( Q \cdot (1-\alpha) - P \cdot \theta \right)$, where $Q$ and $P$ are the matrix, which rows are the $q^T(x_k)$ and $p^T(x_k)$ vectors, respectively.

*SLIM-PCS Algorithms*

**Step 1** and **Step 2** are as explained before.

**Step 3:** Diminish the relevance of every rule in $R_1$, and compensate this effect by increasing the relevance, and possibly tuning, rules in $R_2$, under the condition of maintaining the transfer function of the system invariant, equation (15).

**Step 4:** Eliminate the rules in $R_1$ whose values of relevance are considered sufficiently low.

Obviously, if the radius of the rules of $f_1$ is so small that the center of the rule is a point representative of the region covered by the rule. Then, the resolution of the problem has been constrained to the set of $M_1$ points. Under these circumstances, the solution of the problem is obtained by solving the following system of non-linear equations:

$$Q \cdot \theta + e = R \tag{16}$$

where $R = \sum_{k=1}^{M_1} p(x_k) \big/ \overline{y}^k$ is a vector, $Q = \sum_{k=1}^{M_1} \left( p(x_k) \cdot (p(x_k))^T \right) \big/ \overline{y}^k$ is a symmetric matrix with $M_1 {\times} M_1$ dimension and $\vec{e}$ is a vector error.

$$\alpha_k = 1 - \frac{p^T \cdot \theta}{\overline{y}^k}, \quad \text{for } k = 1, \cdots, M_1 \tag{17}$$

If the $\alpha_k$ value for rule $k$ is a null value or near zero then the rule can be discharged.

The singular value decomposition (SVD) of $Q$ matrix is there used to solve the equation (16). Results a factorization of the matrix into a product of three matrices $Q = U \Sigma V$, where $U, V \in \Re^{M_1 \times M_1}$ are orthogonal matrices and $\Sigma = diag\left( \sigma_1, \sigma_2, \cdots, \sigma_{M_1} \right)$ is a diagonal matrix, which elements are called the *singular values* of $Q$.

In practice, the minimum 2-norm solution of equation (16) is usually approximated by:

$$\theta = \sum_{i=1}^{r} \frac{u_i R v_i}{\sigma_i} \tag{18}$$

where $r \leq M_1$ is some numerically determined value. The existence of small singular values implies the presence of redundant or less important rules among the rules that comprise the underlying model. If the input membership functions are static then equations (17)-(18) provide a systematic way of transferring the information from system $f_1$ to system $f_2$ in one step.

Otherwise, the tuning of membership parameters will improve the minimization of $J$ cost function.

An iterative procedure can be used:

$$a_p^{i+1} = a_p^i - \eta \cdot \frac{\partial J}{\partial a} \tag{19}$$

where $a_p$, with $p = 1, \cdots, np$, are parameters of input memberships function, with $np$ the total number of parameters liable to optimization procedure, and $i$ the current number of the iteration optimization step.

The SLIM-PCS algorithm results in the association of tuning of the input memberships functions, eq. (19), with the tuning of the center of fuzzy outputs sets, eqs. (17)-(18). Finally, if the $\alpha_k$ value for rule $k$ is a null value or near zero, then the rule can be discharged. In the section 6, this algorithm is used to modeling the environmental climate variable of an agriculture greenhouse. This algorithm was so been used in modeling of various real systems [16][17].

*B. HPS structure*

The HPS structure, illustrated in figure 2, allows for prioritization of the rules by using a hierarchical representation, as defined by Yager [11]. If $i < j$ the rules in level $i$ will have a higher priority than those in level $j$.

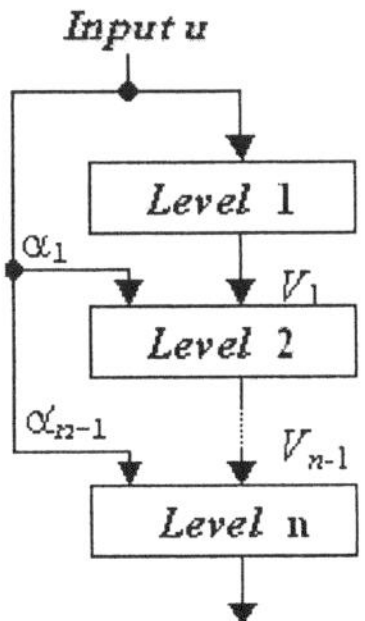


Figure 2. Hierarchical Prioritized Structure (HPS)

Consider a system with $i$ levels, $i$=1,..., $n$-1, each level with $M_i$ rules:

I) If $U$ is $A_{ij}$ and $\hat{V}_{i-1}$ is low, then $V_i$ is $B_{ij}$ and rule II is used

II) $V_1$ is $\hat{V}_{i-1}$

Rule I is activated if two conditions are satisfied: $U$ is $A_{ij}$ and $\hat{V}_{i-1}$ is low. $\hat{V}_{i-1}$, which is the maximum value of the output membership function of $V_{i-1}$, may be interpreted as a measure of satisfaction of the rules in the previous levels. If these rules are relevant, i.e. $\hat{V}_{i-1}$ is not low, the information conveyed by these rules will not be used. On the other hand, if the rules in the previous levels are not relevant, i.e. $\hat{V}_{i-1}$ is low, this information is used.

Rule II states that the output of level $i$ is the union of the output of the previous level with the output of level $i$.

The output of a generic level $i$ is given by the expression

$$G_i = \left( (1-\alpha_{i-1}) \wedge \bigcup_{l=1}^{M_i} F_i^l \right) \cup G_{i-1} \tag{20}$$

where $F_i^l = A_{il}\left(x^*\right) \wedge B_{il}$ , $l = 1, \cdots, M_i$ is the output membership function of rule $l$ in level $i$.

Equation (20) can be interpreted as an aggregation operation, for the hierarchical structure. The coefficient $\alpha_i$ translates the *relevance* of the set of rules in level $i$. Level 1 gets from level 0: $\alpha_0 = 0$, $G_0 = \varnothing$

Yager proposes [11]:

$$\alpha_i = \max_y G_i(y) \tag{21}$$

The relevance, as is defined by equation (21), restricts the contribution of one level, $i$, to the maximum value of its membership $G_i(y)$, in the output domain. The relevance contributions of a significant number of rules are disregarded. However, each level contains a fuzzy system, in which doesn't exist a prioritized relevance of rules.

Moreover, it's our intuition that the information containing in each rule can be complemented by the others rules, increasing the relevance of the level's fuzzy system. Using the axiomatic presented in this paper, other functions of relevance may be defined.

**Definition 9** (*Relevance of fuzzy system just i level*): Let $S_i$ be the input-output region covered by the set of rules of level $i$. The relevance of the set of rules in level $i$ is defined as:

$$\alpha_i = S\left( \alpha_{i-1} , T\left[ (1-\alpha_{i-1}), S\left( \left\{ \Re_{S_i}\left(F_i^l\right), l=1,\cdots,M_i \right\} \right) \right] \right) \tag{22}$$

where $S$ and $T$ are, respectively, *S-norm* e *T-norm* operations, and $\Re_{Si}$ is restricted by the proposed axiomatic.

Taking $\alpha_{i-1} = \Re(G_{i-1})$, and applying *Property 1* and *Property 2*, we verify that $\alpha_i$ is a measure of relevance. $\Re_{S_i}\left(F_i^l\right)$ represent the relevance of rule $l$ in level $i$, defined in (15). Using the product implication rule, and considering $B_{il}$ a centroid of amplitude $\delta_{il}$ centered in $y=\ddot{y}$, then

$$\Re_{S_i}\left(F_i^l\right) = A_{il}\left(x^*\right) \cdot \delta_{il} \tag{23}$$

When the relevance of a level $i$ is 1, the relevance of all the levels below is null.

If the *S-norm* and *T-norm* operations used in (22) are continuous and derivable, and given a cost function, it is possible to develop an optimization process for tuning the parameters of the membership functions of the rules in a HPS system.

An algorithm implementing the SLIM methodology for the HPS structure is presented. For simplicity, the algorithm is presented for a particular HPS structure (with three levels, $n=3$, two inputs, $ni=2$, and one output, $no=1$). All the rules are of the type:

I) If $U$ is $A_{ij}$ and $\hat{V}_{i-1}$ is low, then $V_i$ is $\theta_{ij}$ and rule II is used

II) $V_1$ is $\hat{V}_{i-1}$

The consequent rules are of the Sugeno type, with singleton membership functions, $\theta_{ij}$. The antecedent sub-sets $A_{ij}$ (product aggregation of the input membership functions) are of the Gaussian type. The system output is obtained by a defuzzification center of area:

$$y^* = \frac{\sum_{i=1}^{n}\sum_{j=1}^{M_i} G_n\left(b_{ij}\right)\cdot\theta_{ij}}{\sum_{i=1}^{n}\sum_{j=1}^{M_i} G_n\left(b_{ij}\right)} \tag{24}$$

$F_i$ may be written as

$$F_i = \bigcup_{j=1}^{M_i}\left\{ A_{ij}\Big/\theta_{ij}\right\} \tag{25}$$

and $G_n$ [11]:

$$G_n\left(b_{ij}\right) = \left(1-\alpha_{i-1}\right)\cdot A_{ij}\left(x\right) \tag{26}$$

Equation (18) can be rewritten as:

$$y^* = \frac{\sum_{j=1}^{M_i}\left(1-\alpha_1^j\right)\cdot\theta_1^j\cdot\mu_{A_{1j}}\left(x\right)+\left(1-\alpha_1\right)\cdot\sum_{j=1}^{M_2}\theta_2^j\cdot\mu_{A_{2j}}\left(x\right)+\left(1-\alpha_2\right)\cdot\sum_{j=1}^{M_3}\theta_3^j\cdot\mu_{A_{3j}}\left(x\right)}{\sum_{j=1}^{M_i}\left(1-\alpha_1^j\right)\cdot\mu_{A_{1j}}\left(x\right)+\left(1-\alpha_1\right)\cdot\sum_{j=1}^{M_2}\mu_{A_{2j}}\left(x\right)+\left(1-\alpha_2\right)\cdot\sum_{j=1}^{M_3}\mu_{A_{3j}}\left(x\right)} \tag{27}$$

where $y^* = f\left(x\right)$ and $\alpha_1^j$ is associated to the relevance of rule $j$ of level 1.

When $\alpha_1^j = 0$ the rule $j$ of level 1 has the original relevance value. If $\alpha_1^j = 1$, the relevance of rule $j$ is null and so its contribution to the model is very low.

Parameters $\alpha_1$ and $\alpha_2$ are associated with the relevance of the sets of rules in levels 1 and 2, respectively. For a given point in the input space, the associated relevance value can be obtained using equation (22), which is here considered as continuous and derivable. It is clear that, when the relevance of a given rule becomes sufficiently low, the rule can be discarded.

This problem can be formalized as follows:

$$Min\ J = \sum_{j=1}^{M_i}\left(1-\alpha_1^j\right)^2,\ subject\ to:\ f\left(x\right)\ is\ invariant,\ \forall\alpha_1^j \in \left[0,1\right] \tag{28}$$

The solution to this problem with restrictions can be obtained using the Lagrange method, which converts the previous problem in a dual problem without restrictions:

$$W\left(X,\lambda\right)=\sum_{j=1}^{M_1}\left(1-\alpha_1^j\right)^2+\lambda\sum_{k=1}^{p}\left(f\left(\boldsymbol{x}_k\right)-f^*\left(\boldsymbol{x}_k\right)\right)^2 \tag{29}$$

where $\lambda$ is the Lagrange multiplier, $X$ is the vector with the adjustable parameters of $f$, $f^*$ is the original fuzzy function and $p$ is the number of significant points that cover the region of interest.

Formulated in this way, the problem can be solved using the generic SLIM algorithm presented in section 0. Some details for this particular case are given below.

*SLIM-HPS Algorithms*

**Step 1** and **Step 2** are as explained before.

**Step 3** is divided in two parts *Step 3.1* and *Step 3.2*.

**Step 3.1**: Diminish the relevance of every rule in $R_1$ and compensate this effect by increasing the relevance, and possibly tuning, rules in $R_3$, by solving equation (29). The relevance of the rules in level 2 is considered null.

**Step 3.2**: Diminish the relevance of every rule in $R_1$ and compensate this effect by increasing the relevance, and possibly tuning, rules in $R_2$, by solving equation (29).

**Step 4**: Eliminate the rules in $R_1$ whose values of relevance are considered sufficiently low.

Equation (29) can be solved by using an iterative optimization method, e.g. gradient descendent; $W$ may be minimized in two steps by solving iteratively a system of non-linear equations [18].

The interconnection of HPS and PCS structures will make a General Structure (GS). The resulting structure has necessarily merged the characteristics of both HPS and PCS. The generalization of SLIM methodology at a GS is an extended case.

## 4. Fuzzy rules clustering algorithms

### 4.1. Probabilistic and possibilistic clustering algorithm

Clustering numerical data forms the basis of many modeling and pattern classification algorithms. The purpose of clustering is to find natural groupings of data in a large data set revealing patterns that can provide a concise representation of data behavior [5].

The classification problem consists on the separation of a set of objects $\boldsymbol{O}=\left\{o_1,o_2,\cdots,o_k\right\}$ in $c$ clusters, according to a similarity criterion, using all available data, $\boldsymbol{X}=\left\{x_1,x_2,\cdots,x_n\right\}$, where $x_k$ can be any element (for example, $\boldsymbol{x}_k\in R^p$ ).

A real matrix $U$ of dimensions $c{\times}n$, called the fuzzy partition matrix, can be used to show the results of classifying the data $\boldsymbol{X}$ into clusters. This is achieved interpreting each element $[u_{ik}]$ as a measure representing the degree of membership of a data vector $\boldsymbol{x}_k$ belonging to the cluster $i$.

The Fuzzy C-Means Clustering Algorithm, FCM is based on the calculation of the fuzzy partition matrix $U$ under the following constraints:

$$u_{ik}\in[0,1],\ 1\le i\le c,\ 1\le k\le n \tag{30}$$

$$\sum_{i=1}^{c} u_{ik} = 1, \quad 1 \le k \le n \tag{31}$$

$$0 < \sum_{k=1}^{n} u_{ik} < n, \quad \forall i \in \{1, 2, \cdots, c\} \tag{32}$$

A central issue in FCM, also namely *probabilistic algorithm*, is the constraint of equation (31) that is strongly influenced by probability theory: the sum of the membership values for all clusters of a data vector equals one. This constraint prevents the trivial solution $U$=**0** and provides reasonable results when the degree of membership of each vector to a cluster is interpreted as a probability, or as the degree of sharing an attribute. On the contrary, the fuzzy partition matrix that results from the application of FCM should not be used in cases when the degree of membership implies typicality or compatibility. Having these facts in mind, Krishnapuram and Keller [19] reformulated the fuzzy clustering problem in a way that the fuzzy partition matrix value can be considered as deviations from typicality. So, in the Possibilistic Clustering Algorithm and its c-means algorithm (P-FCM), the condition (31) is dropped.

For the FCM and P-FCM algorithms, the objective is then to find a $U$=[$u_{ik}$] and $V = [v_1, v_2, \cdots, v_C]$ with $v_i \in R^p$ where:

$$J(U,V) = \sum_{k=1}^{n} \sum_{i=1}^{c} u_{ik}^m \cdot D(x_i, v_k), \quad 1 < m \le \infty \tag{33}$$

is minimized [20].

The fuzzifier that controls the fuzziness of the final partition is $m > 1$, $c \ge 2$ is the number of clusters of the final partition, and $n$ is the number of the available data points. $D(\cdot)$ is a measure of the distance between the data vector $x_k$ and the prototype vector $v_i$ of cluster $i$.

It can be shown that the following algorithm may lead the pair $(U^*, V^*)$ to a minimum [21].

The fuzzifier that controls the fuzziness of the final partition is $m > 1$, $c \ge 2$ is the number of clusters of the final partition, and $n$ is the number of the available data points. $D(\cdot)$ is a measure of the distance between the data vector $x_k$ and the prototype vector $v_i$ of cluster $i$.　　It can be shown that the following algorithm may lead the pair $(U^*, V^*)$ to a minimum[21]:

*Probabilistic and Possibilistic Fuzzy C-Means Algorithm*

**Step 1**– For a set of points $X$={$x_1$, $x_2$,..., $x_n$}, with $x_i \in R^p$, keep $c$, $2 \le c < n$, and initialize $U^{(0)} \in M_{cf}$.

**Step 2** – On the $r^{th}$ iteration, with $r = 0, 1, 2,...$, compute the $c$ mean vectors.

$$v_i^{(r)} = \frac{\sum_{k=1}^{n} \left( u_{ik}^{(r)} \right)^m \cdot x_k}{\sum_{k=1}^{n} \left( u_{ik}^{(r)} \right)^m} \tag{34}$$

where $\left[ u_{ik}^{(r)} \right] = U^{(r)}$, $i$=1, 2,..., $c$.

**Step 3**– Compute the new partition matrix $U^{(r+1)}$ using the expression:

$$u_{ik}^{(r+1)} = \cfrac{1}{\displaystyle\sum_{j=1}^{c} \left( \cfrac{d_{ik}^{(r)}}{d_{ij}^{(r)}} \right)^{\frac{2}{m-1}}} \tag{35}$$

$$u_{ik}^{(r+1)} = \cfrac{1}{1 + \left( \cfrac{d_{ik}^{2(r)}}{\eta_k} \right)^{\frac{1}{m-1}}} \tag{36}$$

for $1 \leq i \leq c$, $1 \leq k \leq n$, where $d_{ik}^{(r)}$ denotes $D\left(x_k, v_i^{(r)}\right)$ and $\eta_k$ belongs to **R**.

**Step 4** – Compare $U^{(r)}$ with $U^{(r+1)}$: If $\| U^{(r+1)}\text{-}U^{(r)}\| < \varepsilon$ then the process ends. Otherwise let $r=r+1$ and go to step 2. $\varepsilon$ is a small real positive constant.

Equations (35) and (36) defines, respectively, the probability (FCM) and the possibility (P-FCM) membership function for cluster $i$ in the universe of discourse of all data vectors $X$. The values of $\eta_k$ control the distance from the prototype vector of cluster $i$ where the membership function becomes 0.5, i.e., the point(s) with the maximum fuzziness.

### 4.2. Fuzzy Clustering of Fuzzy Rules

If the fuzzy system entirely characterizes one region of product space $U \times V$, for any existing characterization (operation) of the input-output space map, there will be one correspondence in the rules domain (Theorem 1). The fuzzy clustering of fuzzy rules here proposed, as well as one clustering of data or region of space, corroborates this idea.

The objective of the fuzzy clustering partition is the separation of a set of fuzzy rules $\Im=\{R_1, R_2,..., R_M\}$ in $c$ clusters, according to a "similarity" criterion, finding the optimal clusters center, $V$, and the partition matrix, $U$. Each value $u_{ik}$ represents the membership degree of the $k^{th}$ rule, $R_k$, belonging to the $i^{th}$ cluster $i$, $A_i$, and obeys to equations (30)-(32).

Let $x_k \in S$ be a point covered by one or more fuzzy rules. Naturally, the membership degree of point $x_k$ belonging to $i^{th}$ cluster is the sum of products between the relevance of the rules $l$ in $x_k$ point and the membership degree of the rule $l$ belong to cluster $i$, $u_{il}$, for all rules, i.e.:

$$\sum_{i=1}^{c} \sum_{l=1}^{M} u_{il} \cdot \mathfrak{R}_l\left(x_k\right) = 1, \quad \forall x_k \in S \tag{37}$$

For the fuzzy clustering proposes, each rule and $x_k$ point, will obey simultaneously to equations (30)-(32). These requirements and the relevance condition of equation (6) are completely satisfied in equation (37).

So, for the Fuzzy Clustering of Fuzzy Rules Algorithm, FCFRA, the objective is to find a $U=[u_{ik}]$ and $V = [v_1, v_2, \cdots, v_C]$ with $v_i \in R^p$ where:

$$J(U,V) = \sum_{k=1}^{n}\left[\sum_{i=1}^{c}\sum_{l=1}^{M}\left(u_{il}\Re_{l}(x_{k})\right)^{m}\|x_{k},v_{i}\|^{2}\right] \tag{38}$$

is minimized, with a weighting constant $m>1$, with constrain of equation (37). It can be shown that the following algorithm may lead the pair $(U^{*},V^{*})$ to a minimum:

Also in previous section, the models specified by the objective function (38) were minimized using alternating optimization. The results can be express in fellow algorithm:

*Fuzzy Clustering algorithms of fuzzy rules – FCAFR*

**Step 1**– For a set of points $X=\{x_{1}, x_{2},..., x_{n}\}$, with $x_{i}{\in}S$, and a set of rules $\Im=\{R_{1}, R_{2},..., R_{M}\}$, with relevance $\Re_{l}(x_{k})$, $k=1, .., M$, keep $c$, $2 \leq c < n$, and initialize $U^{(0)}{\in} M_{cf}$.

**Step 2**– On the $r^{th}$ iteration, with $r= 0, 1, 2,...$, compute the $c$ mean vectors.

$$v_{i}^{(r)} = \frac{\displaystyle\sum_{l=1}^{M}\left[\left(u_{il}^{(r)}\right)^{m}\cdot\sum_{k=1}^{np}\left(\Re_{l}(x_{k})\right)^{m}\cdot x_{k}\right]}{\displaystyle\sum_{l=1}^{M}\left[\left(u_{il}^{(r)}\right)^{m}\cdot\sum_{k=1}^{np}\left(\Re_{l}(x_{k})\right)^{m}\right]} \tag{39}$$

where $\left[u_{il}^{(r)}\right]=U^{(r)}$, $i=1, 2,..., c$.

**Step 3**– Compute the new partition matrix $U^{(r+1)}$ using the expression:

$$u_{il}^{(r+1)} = \frac{1}{\displaystyle\sum_{j=1}^{c}\left(\frac{\displaystyle\sum_{k=1}^{np}\left(\Re_{l}(x_{k})\right)^{m}\cdot\|x_{k}-v_{i}^{(r)}\|}{\displaystyle\sum_{k=1}^{np}\left(\Re_{l}(x_{k})\right)^{m}\cdot\|x_{k}-v_{j}^{(r)}\|}\right)^{\frac{2}{m-1}}} \text{, with } 1\leq i\leq c, 1\leq l\leq M \tag{40}$$

**Step 4**– Compare $U^{(r)}$ with $U^{(r+1)}$: If $\| U^{(r+1)}-U^{(r)}\| < \varepsilon$ then the process ends. Otherwise let $r=r+1$ and go to step 2. $\varepsilon$ is a small real positive constant.

If the rules describe one region $S$, instead of a set of points, the equation (39) and (40) will be exchange the sum by integral:

$$v_{i}^{(r)} = \frac{\displaystyle\sum_{l=1}^{M}\left[\left(u_{il}^{(r)}\right)^{m}\cdot\int_{S}\left(\Re_{l}(x)\right)^{m}\cdot x\cdot dx\right]}{\displaystyle\sum_{l=1}^{M}\left[\left(u_{il}^{(r)}\right)^{m}\cdot\int_{S}\left(\Re_{l}(x)\right)^{m}\cdot dx\right]} \tag{41}$$

$$u_{il}^{(r+1)} = \frac{1}{\displaystyle\sum_{j=1}^{c} \left( \frac{\int_S \left( \Re_l \left( x \right) \right)^m \cdot \left\| x - v_i^{(r)} \right\| \cdot dx}{\int_S \left( \Re_l \left( x \right) \right)^m \cdot \left\| x - v_j^{(r)} \right\| \cdot dx} \right)^{\frac{2}{m-1}}} \tag{42}$$

It's possible to proof that: for the condition where the membership functions of fuzzy rules are symmetrical, and in the same way, the relevance functions of the rules, the equation (41) and (42) can be rewrite as:

$$v_i^{(r)} = \frac{\displaystyle\sum_{l=1}^{M} \left[ \left( u_{il}^{(r)} \right)^m \cdot \alpha_l \cdot \overline{x}_l \right]}{\displaystyle\sum_{l=1}^{M} \left[ \left( u_{il}^{(r)} \right)^m \cdot \alpha_l \right]} \tag{43}$$

$$u_{il}^{(r+1)} = \frac{1}{\displaystyle\sum_{j=1}^{c} \left( \frac{\left\| \overline{x}_l - v_i^{(r)} \right\|}{\left\| \overline{x}_l - v_j^{(r)} \right\|} \right)^{\frac{2}{m-1}}} \tag{44}$$

where $\overline{x}_l$ is the center of rule $l$ and $\alpha_l = \int_S \left( \Re_l \left( x \right) \right)^m dx$ is one form factor.

If all rules have the same shape, the fuzzy clustering of fuzzy can be obtaining by considering only the center of rules, as a normal *fuzzy c-means algorithm* process.

*Possibilistic Fuzzy Clustering algorithms of fuzzy rules – P-FCAFR*

**Step 1** and **Step 2** are as explained in FCAFR.
**Step 3**– Compute the new partition matrix $U^{(r+1)}$ using the expression:

$$u_{il}^{(r+1)} = \frac{1}{\displaystyle\sum_{j=1}^{c} \left( \sum_{k=1}^{np} \left( \Re_l \left( x_k \right) \right)^m \cdot \left\| x_k - v_i^{(r)} \right\| \Big/ \eta_l \right)^{\frac{2}{m-1}}} , \tag{45}$$

with $1 \leq i \leq c$, $1 \leq l \leq M$.

**Step 4**– Compare $U^{(r)}$ with $U^{(r+1)}$: If $\| U^{(r+1)} - U^{(r)} \| < \varepsilon$ then the process ends. Otherwise let $r = r+1$ and go to step 2. $\varepsilon$ is a small real positive constant.

This methodology allows separating the original fuzzy system in the sum of $c$ fuzzy systems, in form of equation (11), where each system (subsystem) represents one cluster.

$$f_{cluster}^{i}\left(x\right)=\frac{\sum_{l=1}^{M}\theta^{l}\cdot\mu^{l}\left(x\right)\cdot u_{il}}{\sum_{l=1}^{M}\mu^{l}\left(x\right)} \tag{46}$$

The Rule Clustering Algorithm, organize the rules of fuzzy systems in the PCS structure. So, the FCAFR take part of the SLIM methodology, where the partition matrix can be interpreted as containing the values of the relevance of the sets of rules in each cluster.

As same manner as in FCAFR algorithms, if the rules describe one region $S$, instead of a set of points, the equation (45) will be reformulate to:

$$u_{il}^{(r+1)}=\frac{1}{\sum_{j=1}^{c}\left(\int_{S}\left(\Re_{l}\left(x\right)\right)^{m}\cdot\left\|x-v_{i}^{(r)}\right\|\cdot dx\Big/\eta_{l}\right)^{\frac{2}{m-1}}} \tag{47}$$

and still more the membership of relevance function of the rules was symmetrical, the last equation can be rewrite as:

$$u_{il}^{(r+1)}=\frac{1}{\sum_{j=1}^{c}\left(\left\|\overline{x}_{l}-v_{i}^{(r)}\right\|\Big/\eta_{l}^{*}\right)^{\frac{2}{m-1}}} \tag{48}$$

where $\overline{x}_{l}$ is the center of rule $l$.

The same conclusion of FCAFR algorithms are obtaining: If all rules have the same shape, the possibilistic fuzzy clustering of fuzzy can be determined by only considering the center of rules.

This methodology allows expand the original fuzzy system in the HPS structure. For fuzzy systems (11), the P-FCM result in following HPS structure, of the equation form(27):

$$f\left(x\right)=\frac{\sum_{i=1}^{c}\left(\sum_{l=1}^{M}\theta^{l}\cdot\mu^{l}\left(x\right)\cdot u_{il}\right)}{\sum_{i=1}^{c}\left(\sum_{l=1}^{M}\mu^{l}\left(x\right)\cdot u_{il}\right)} \tag{49}$$

In conclusion, the FCAFR organize the information of fuzzy systems in PCS structures, while the P-FCAFR algorithm organizes fuzzy systems in the HPS structure form.

## 5. Experimental results

In this section it was tested the SLIM methodology in two cases. The first one, it's the modeling of a volcano surface with the hierarchical model. The second case, the SLIM methodology is applied to modeling the greenhouse environmental variables with real data.

**Example 1:** Volcano Description as in HPS structure.

The figure 3a shown a volcano surface, generated with 40×40 data points. The exercise is to capture in an HPS system the description of the function. First, we identified the raw data using the nearest neighborhood identification method, with a radius of 0.4, with a negligible error. A set of 400 fuzzy rules has been generated.

It is general perception that the volcano function, $W=F(U,V)$, can been generated by the following three level hierarchic HPS structure, with one rule in each level:

*Level 1*: IF $(U,V)$ is **very close** to (5,5) THEN $W$ is **quasi null** (Rule 1)

*Level 2*: IF $(U,V)$ is **close** to (5,5) THEN $W$ is **high** (Rule 2)

*Level 3*: IF $U$ and $V$ are **anything** THEN $W$ is **low** (Rule 3)

The fuzzy sets membership functions **very close**, **close**, **anything**, **quasi null**, **high** and **low** can be defined as follows as two-dimensional Gaussian functions:

$$\mu_{I(i)}(\vec{x}) = e^{-\left(\frac{x_1-\bar{x}_{i1}}{\sigma_{i1}}\right)^2} \times e^{-\left(\frac{x_1-\bar{x}_{i2}}{\sigma_{i2}}\right)^2} \quad \text{with } \vec{x}=(x_1,x_2) \text{ and } I=\{\text{"very close", "close", "anything"}\}$$

$$\mu_{J(i)}(w) = \left\{\frac{1}{\theta_i}\right\} = \left\{\frac{1}{0}\right\} \quad \text{with } J=\{\text{"quasi null", "high", "low"}\};$$

where $\bar{x}_{ij}$ and $\sigma_{ij}$ are the central value and the width of the input membership functions and $\theta$ is the central value of the singleton output membership function for the $i$ level. Because there is only a rule in each level its subscript was omitted.

Now, we begin building the HPS structure as mentioned in the SLIM-HPS Algorithm. At first, all 400 rules are placed in the first level. As the membership functions of the other levels are unknown, we will assume arbitrary values as starting values. The next step consists in diminishing the relevance of the rules in level 1, in favor of the rules in level 3. This is accomplished by tuning the membership functions of level 3 to compensate for the alterations in level 1. This has been done minimizing the equation

$$W(X,\lambda) = \sum_{j=1}^{M_1}\left(1-\alpha_1^j\right)^2 + \lambda\sum_{k=1}^{p}\left(f(x_k)-f^*(x_k)\right)^2 \tag{50}$$

using the gradient-descendent method [8]. We repeat the process for diminishing the relevance of the rules in level 1, in favor of the rules in level 2.

The output of the resulting HPS system, after these two processes, can be seen in figure 4b). The end values $\alpha_1^j$ (at this stage), associated to the relevance of the rules, are shown in figure 2c). The value of each $\alpha_1^j$ is presented located at the center of the associated rule.

One should note that the majority of the rules were assimilated by the levels 2 and 3, with the exception of the ones that represent the volcano opening, where the coefficients $\alpha_1^j$ are close to null. These are the rules that, as a whole, constitute the level with the highest priority. Next we suppress, from level 1, the rules with low relevance ($\alpha_1^j > 0.9$). This results in keeping 49 rules in level 1, without significant loss of the model quality, as we can observe in figures 4d).

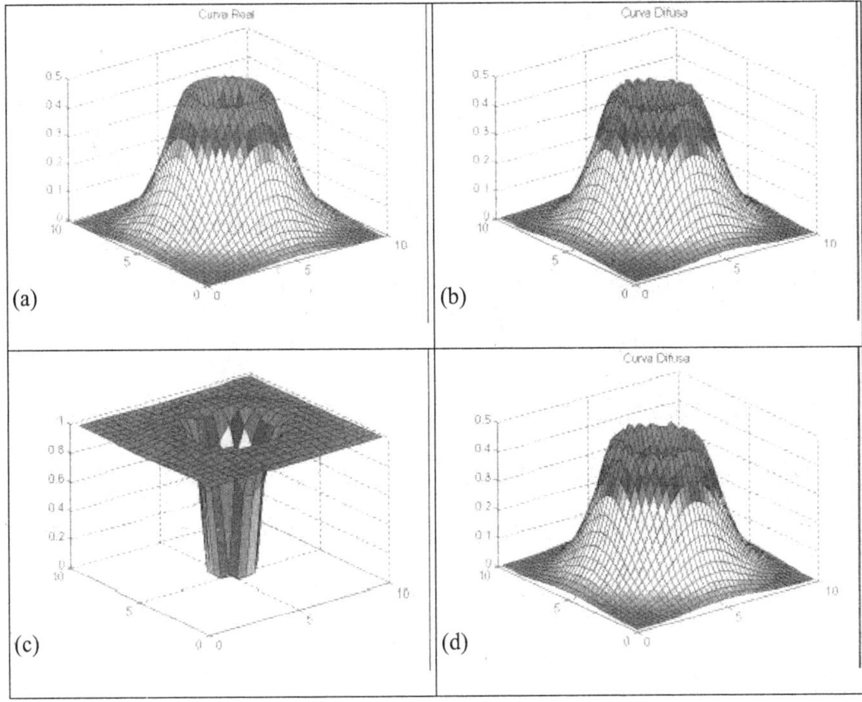


(a)

(b)

(c)

(d)

Figure 3 - SLIM methodology applied to Volcano description

**Example 2:** Environmental Greenhouse model, in PCS structure

The greenhouse climate model describes the dynamic behavior of the state variables by the use of differential equations of temperature, humidity and $CO_2$ concentrations. In this paper the $CO_2$ concentration model is not considered. The model equations can be written by deriving the appropriate energy and mass balances

$$\frac{dT_{est}}{dt} = \frac{1}{C_{Temp}}\left(Q_{T,h} - Q_{T,out} + Q_{soil} + Q_{T,rad}\right) \tag{51}$$

$$\frac{dH_{est}}{dt} = \frac{1}{C_{cap,h}}\left(\Phi_{h,C,Al} - \Phi_{h,Al,AE}\right) \tag{52}$$

where, $C_{Temp}$ [J m$^{-2}$ °C$^{-1}$], $C_{CO2}$ [Kg m$^{-2}$], $C_{cap,h}$ [m] are the heat and mass capacities per square meter.

The energy balance in the greenhouse air is affected by the energy supplied by heating system, $Q_{T,h}$ [W m$^{-2}$], by energy losses to outside air due to transmission through the greenhouse cover and forced ventilation exchange, $Q_{T,out}$ [W m$^{-2}$], by the energy exchange with the soil $Q_{soil}$ [W m$^{-2}$], and by the heat supply by sun radiation, $Q_{T,rad}$ [W m$^{-2}$].

The humidity balance in the greenhouse air is determined by canopy and soil transpiration, $\Phi_{h,CAI}$ [kg m$^{-2}$ s$^{-1}$], and by the free and forced ventilation exchange with the outside air $\Phi_{h,AI,AE}$ [kg m$^{-2}$ s$^{-1}$].

Energy and mass transport phenomena at greenhouse cover and the contribution of ventilation, induced by temperature differences between the inside and outside air is only significant at very low wind speeds [22], and consequently they are neglected in this model.

This model requires a great domain of the physical process and measurement accuracy on the process variables. Fuzzy modeling can be an alternative representation for describing the process and is easily interpretable by someone.

The model can be significantly improved if it is divided in sub models. The temperature and humidity models are breaking in two parts: the daylight and night sub models for the temperature and with and without forced ventilations for the humidity sub models.

| Daylight Temperature (DT) sub model ($RAD>0$) | Night temperature (NT) sub-model |
|---|---|
| $$\frac{dT_{est}}{dt} = f_{Temp}\left(\Delta T, Rad, Q_{aquec}, Q_{vent}\right) \quad (53)$$ | $$\frac{dT_{est}}{dt} = f_{Temp}\left(\Delta T, Q_{solo}, Q_{aquec}\right) \quad (54)$$ |

where $\Delta T(t) = T_{est}(t) - T_{ext}(t)$ is the inside and outside temperature difference; $RAD$ is the sunlight radiation intensity; $Q_{aquec}(t) = U_{aquec}(t) \cdot \left(T_p - T_{est}(t)\right)$ is the heat flux from the heating system. $Tp$ is the temperature of the coil water of the heating (about 60 °C); $Q_{vent}(t) = U_{vent}(t) \cdot \left(T_{ext}(t) - T_{est}(t)\right)$ is the heat flux exchange with the outside air and $Q_{solo}(t) = \left(T_{solo}(t) - T_{est}(t)\right)$ is the heat exchange with the greenhouse soil.

| Low Dynamical Humidity (LDH) sub model ($U_{vent}=0$) | High Dynamical Humidity (HDH) sub model ($U_{vent}>0$) |
|---|---|
| $$\frac{dH_{est}}{dt} = f_H\left(\Delta HS, \Delta H\right) \quad (55)$$ | $$\frac{dH_{est}}{dt} = f_H\left(\phi_{vent}\right) \quad (56)$$ |

where $HS(T)$ is the dewpoint humidity of the air at $T$ (°C) temperature; $\Delta H(t) = H_{est}(t) - H_{ext}(t)$ is the difference between inside and outside absolute humidity's; $\phi_{vent}(t) = U_{vent}(t) \cdot \left(H_{est}(t) - H_{ext}(t)\right)$ is the water mass exchange between the inside and outside air and $\Delta HS(t) = HS\left(T_{est}(t)\right) - H_{est}(t)$ is the difference between the dewpoint humidity of the air, at temperature $Test(t)$, and the absolute humidity in the greenhouse air.

Here, the required task is to develop the above fuzzy systems that can match all the $N$ pairs of collected data (temperature, humidity,...) to any given accuracy.

The SLIM methodology has been applied to the identification of the dynamic behavior of temperature and humidity in an agricultural greenhouse. Three different fuzzy identification techniques have been applied to model the data: back propagation [2], table lookup [23] and the RLS [16]. The idea is to compare a system organized by the SLIM methodology against the systems produced by the reference methods: The structure for the organization of information adopted in this case was a PCS structure with two fuzzy subsystems. The number of rules chosen for the level $f_2$ was 10, for each one temperature and humidity sub-models. The identification of the above models was realized by using daily input-output data points collected from the process, between January 15 and February 4, 1998, at 1-minute sample time. Two others periods of data were been used for test the models: 8 to 14 of January and 5 to 11 of April.

The first step is to capture the greenhouse models in a PCS system from real data. The identification process was performed with the RLS identification method, with a radius of 0.4 and $4 \times 10^{-4}$, respectively for temperature and humidity model.

In order to proceed, the information starts to be transferred from the sub-system $f_1$ to $f_2$ in the PCS structure. Previously, all the original rules are placed in the first level. The next step consists in diminishing the relevance of the rules in level 1, in favor of the rules in level 2. This is accomplished by tuning the membership functions of level 2 to compensate for the relevance rules in level 1.

The optimization procedure is achieved by using equation (17) and equation (18).

The output of the resulting PCS system after applying the SLIM methodology is represented in Figure 4 for temperature model and in Figure 5 for the absolute humidity model: The results are obtained in the test period 5 to 11 of April.

Real data curves are also represented for comparison purposes. The variances of the errors between the simulations and the experimental data for temperature and humidity, for each one of Fuzzy Methods, are indicated in Table 1 and 2 respectively.

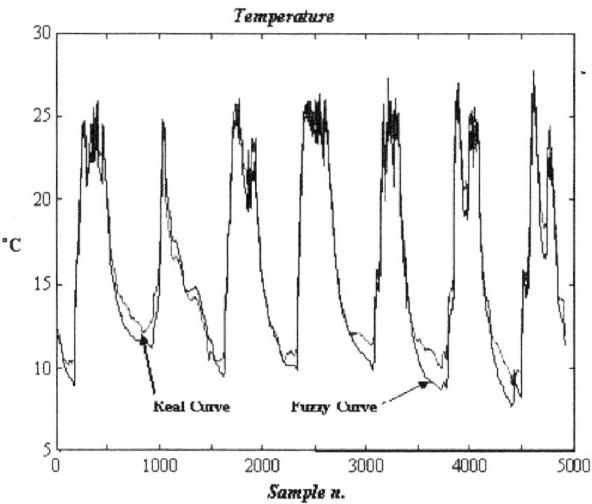


Figure 4 - Temperature Fuzzy model

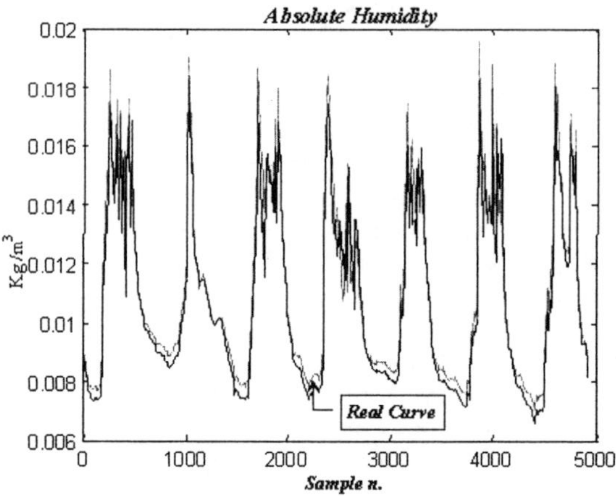


Figure 5 – Humidity Fuzzy model.

Experimental work shows that the fuzzy identification systems created with the reference methods are made of a very high number of IF…THEN rules. The SLIM methodology, applied to the systems produced with the RLS method, has produced systems with a number of rules that varies between 2% and 20% of the number of rules of the other systems, with negligible differences in the behavior, as illustrated in Table 1 and 2. The "slimed" system, with a superior organization, contains the same level of information, and uses a lower number of rules.

Table 1: Errors, number of rules and iterations of the different temperature fuzzy system models

| Fuzzy Method | n. of rules | n. of iteration | Temperature Error - $\hat{E}$ ( $\bar{E}$ ) [*] (°C) | | |
| --- | --- | --- | --- | --- | --- |
| | | | 8 to 14 Jan. | 15 of Jan to 4 of Feb | 5 at 11 of April |
| Back-Propagation | 50 (day) | 1000 | 0,98 (0,79) | 0,81 (0,56) | 1,01 (0,77) |
| | 50 (night) | | | | |
| RLS | 777 (day) | 1 | 0,88 (0,70) | 0,74 (0,53) | 1,07 (0,86) |
| | 349 (night) | | | | |
| Table-lookup | 596 (day) | 1 | 1,00 (0.83) | 0.55 (0.54) | 1.15 (0.85) |
| | 201 (night) | | | | |
| MSIL-PCS | 10 (day) | 1 | 0.88 (0.73) | 0.82 (0.64) | 1.18 (1.01) |
| | 10 (night) | | | | |

(*) $\hat{E}$ and $\bar{E}$ are the mean square error and the means error, respectively.

Table 2: Errors, number of rules and iterations of the different humidity's fuzzy system models

| | | | Absolute Humidity Error - $\hat{E}$ ( $\bar{E}$ )×10$^{-4}$ kg m$^{-3}$ (*) | | |
|---|---|---|---|---|---|
| Fuzzy Method | n. of rules | n. of iterations | 8 to 14 Jan. | 15 of Jan to 4 of Feb. | 5 at 11 de April |
| Back-Propagation | 50 ($U_{vent}$ =0) | 1000 | 3.56 (2.83) | 4.47 (3.13) | 7.17 (4.56) |
| | 10 ($U_{vent}$ >0) | | | | |
| RLS | 143 ($U_{vent}$ =0) | 1 | 3.04 (2.07) | 3.84 (2.46) | 7.90 (5.53) |
| | 134 ($U_{vent}$ >0) | | | | |
| Table-lookup | 175 ($U_{vent}$ =0) | 1 | 4.52 (3.41) | 4.58 (3.14) | 5.58 (3.79) |
| | 52 ($U_{vent}$ >0) | | | | |
| MSIL-PCS | 10 ($U_{vent}$ =0) | 1 | 3.62 (2.18) | 4.81 (2.94) | 6.80 (4.91) |
| | 10 ($U_{vent}$ >0) | | | | |

(*) $\hat{E}$ and $\bar{E}$ are the mean square error and the means error, respectively.

## 6. Conclusions and future work

The concept of relevance, presented in the paper, responds to some important questions in the field of fuzzy systems. Its applicability has been demonstrated in a methodology for the separation of information (SLIM), in HPS and PCS structures.

Also in this work, the mathematical fundaments for fuzzy clustering of fuzzy rules were presented, where the relevance concept has a significant importance. Based in this concept, it is possible to make a fuzzy clustering algorithm of fuzzy rules, which is naturally a generalization of fuzzy clustering algorithms. Moreover, it was proved that different clusters strategies lead it's to the SLIM methodology and of a structured organization of information.

The definition of new relevance functions opens new perspectives in other fields, namely in pattern recognition. Work is this field is under way, in image pattern recognition [24]. Another promising line of work is the stability analysis of fuzzy systems.

**References**

[1] Y. Klir, Fuzzy Sets and Fuzzy Logic, Prentice Hall PTR, N. J., 1995.
[2] L. X. Wang, Adaptive Fuzzy Systems and Control, Design and Stability Analysis, Prentice-Hall, 1994.
[3] L. X. Wang, A course in fuzzy systems and control, Prentice-Hall, 1997.
[4] M. Jamshidi, A. Titli, L. A. Zadeh and S. Bovorie, (Eds.), Applications of Fuzzy Logic, Towards Quotient Systems, Prentice Hall, 1997.
[5] F. Höppner, F. Klawonn, R. Kruse and T. Runkler, Fuzzy Cluster Analysis, methods for classification, data analysis and image recognition, John Wiley & Sons, 1999.
[6] S. K. Pal and S. Mitra, Neuro-Fuzzy Pattern Recognition, methods in soft computing, John Wiley & Sons, 1999.
[7] P. Salgado, Novos métodos de Identificação Difusa, Ph.D. Thesis, UTAD, July 1999, Portugal.
[8] P. Salgado, C. Couto, P. Melo-Pinto and J. Bulas-Cruz, Relevance as a new measure of relative importance of sets of rules, *Proceedings of the IEEE Conf. on Syst., Man and Cybernetics*, Nashville, USA, pp. 3770-3777, 2000.
[9] R. Yager, On a Hierarchical Structure for Fuzzy Modeling and Control, *IEEE Trans. On Syst., Man, and Cybernetics*, vol. **23**, no. 4, pp. 1189-1197, 1993.

[10] R. Yager, Hierarchical representation of fuzzy if-then rules, In: Advanced Methods in Artificial Intelligence, Bouchon-Meunier, B., Valverde, L., Yager, R. R. (Eds.), Berlin, Germany: Springer-Verlag, pp. 239-247, 1993.

[11] R. Yager, On the Construction of Hierarchical Fuzzy Systems Models, *IEEE Trans. On Syst., Man, and Cybernetics – Part C: Applications and reviews*, vol. **28**, no. 1, pp. 55-66, 1998.

[12] P. Salgado, Fuzzy Rule Clustering, *Proceedings of the IEEE Conference on System, Man and Cybernetics 2001*, Arizona, Tucson, AZ, USA, pp. 2421-2426, October 7-10, 2001.

[13] B. Kosko, Additive Fuzzy Systems: from functions approximation to learning, In: Fuzzy Logic and Neural Network Handbook, Chen, C. H. (Ed.), pp. 9.1-9.22, McGraw-Hill, Inc., 1996.

[14] B. Kosko, Global Stability of Generalized Additive Fuzzy Systems, *IEEE Trans. on Syst., Man, and Cybernetics*, vol. **28**, pp. 441-452, August 1998.

[15] M. Jamshidi, Large-scale systems: modeling, control and fuzzy logic, Prentice Hall PTR, Upper Saddle River, N. J. 07458, 1997.

[16] P. Salgado, P. Afonso and J. Castro, Separation of Linguistic Information Methodology in a Fuzzy Model, *Proceedings of the IEEE Conference on System, Man and Cybernetics 2001*, Arizona, Tucson, AZ, USA, pp. 2133-2139, October 7-10, 2001.

[17] P. Salgado and P. Afonso, Parallel Collaborative fuzzy model, accepted for publication.

[18] D. G. Luenberger, Introduction to linear and non-linear programming, Addison Wesley, 1973.

[19] R. Krishnapuram and J. M. Keller, A possibilistic approach to clustering, *IEEE Trans. Fuzzy Syst.*, vol. **1**, pp. 85-110, May 1993.

[20] J. C. Bezdek and S. K. Pal, (Eds.), Fuzzy Models for Pattern Recognition: Methods that Search for Patterns in Data, IEEE Press, New York, 1992.

[21] J. C. Bezdek, A Convergence Theorem for Fuzzy ISODATA Clustering Algorithms, *IEEE Trans. Pattern Analysis and Machine Intelligence*, vol. **2**, pp. 1-8, 1980.

[22] G. P. A. Bot, Greenhouse climate: from physical process to a dynamic model, Ph.D. Thesis, Wageningen Agricultural University, Wageningen, 1983.

[23] Li Xin Wang and J. M. Mendel, Generating fuzzy rules by learning from examples, *IEEE Trans. System Man Cybernetics*, vol. **22**(6), pp. 1414-1427, 1992.

[24] P. Salgado, J. Bulas- Cruz and P. Melo-Pinto, A New Paradigm for the Description of Image Patterns – From Pixels to Fuzzy Sets of Rules, *RECPAD2000*, Porto, 2000.

# Self-Organizing
# Uncertainty-Based Networks

Horia-Nicolai TEODORESCU
*Romanian Academy, Calea Victoriei 125, Bucharest, Romania*
*and*
*Technical University of Iasi, Faculty Electronics and Tc., Carol 1, 6600 Iasi, Romania*
*e-mail hteodor@etc.tuiasi.ro*

**Abstract**. The first section is introductory and is devoted to clarify the meanings
we use in this paper for "information aggregation." In the second section, fuzzy-
coupled map networks and other uncertainty-carrying networks are introduced
and analyzed. Several examples are discussed throughout this chapter.
Simulation results, a discussion and conclusions are presented in the next
sections.

## 1. Introduction

Recently, it has been a large interest in the scientific community in the structures that self-
organize. Many volumes (see, for instance, [1], [2], [3]) and a large number of papers (e.g.,
[4]-[11]) have been published on the theory and applications of such systems in physics
biology, chemistry, and in almost any other domain.

Self-organizing structures are structures that dynamically produce clustering, patterns, or
other form of structuring during their temporal or spatial evolution. There are numerous
examples of self-organization in natural processes, from crystals to people aggregating into
communities. Some of these processes are entirely due to the laws of physics, like the
crystals, while other processes involve information, utility and knowledge to produce the
aggregation. We are interested here in the last category, especially emphasizing the case of
systems involving uncertainty. We aim to build models for such processes.

In this chapter, the meanings of "information aggregation" is according to the subsequent
description, which partially departs from the usual definitions [12]. We shall mean by
information (data) aggregation any technique that:

i)   relates a dynamical behavior to another dynamical behavior and possibly
     produces a measure of relationship, or
ii)  relates two or several parameters in some specific way, for example as
     representing the coordinates of the elements of a class, or
iii) relates the dynamic behavior of a time-dependent membership function to the
     dynamics of other time-dependent membership functions, or
iv)  forces data in a system to self-aggregate in some way or another, i.e., to self
     organize. Self-organization is seen in this chapter as the most interesting way of
     information aggregation, and will be given considerable emphasis.

Throughout this chapter, we make little distinction between certainty factors, truth-values in the frame of fuzzy logic, and belief degrees. In fact, replacing fuzzy systems with certainty factors based systems or similarly with other uncertainty-based systems does not change the essence of our discussion, nor the general formal framework. However, for brevity, we will exemplify all processes for fuzzy systems only.

We introduce and discuss in this chapter a class of entities that can be used in information aggregation, namely the **uncertain** information-based **coupled map networks** (UCMNs). We are especially interested in the dynamics of information aggregation when a large number of players behave in a specified manner and how the informations they manipulate aggregate the players. Here, the players are represented by "cells" (nodes) in a network. The analysis of these structures is by no means exhaustive, but aims to provide illustrative results.

## 2. Coupled Map Networks and information aggregation

### 2.1. Coupled Map Networks

A Coupled Map Network (CMN), also named Coupled Lattice Map, or Coupled Map Lattice (CML) is an entity defined by:

i) a lattice topology, typically isomorphic with $\mathbf{Z}$, $\mathbf{Z}^2$, or more generally, with $\mathbf{Z}^k$, or a subset of $\mathbf{Z}^k$,

ii) the characteristic function(s) of a set of "cells" laying in the nodes of the lattice, the cells performing operations on the signals it receives, while the state space of the cells is either continuous, or, at least, uncountable.

A coupled map lattice with all cells having the same characteristic function is named *homogeneous* CMNs. We will discuss only homogeneous CMNs in this chapter. The propagation of the information in the lattice is performed by various mechanisms, for example by shifting, diffusion, and one-way mechanisms.

CMLs are produced as a result of the discretization of partial differential equations (PDE), see Annex 1 and Annex 2, or as a result of discrete modeling of processes. Examples, namely an infinite "linear" lattice and the finite "ring topology" 1D-lattice, are shown in Figure 1.

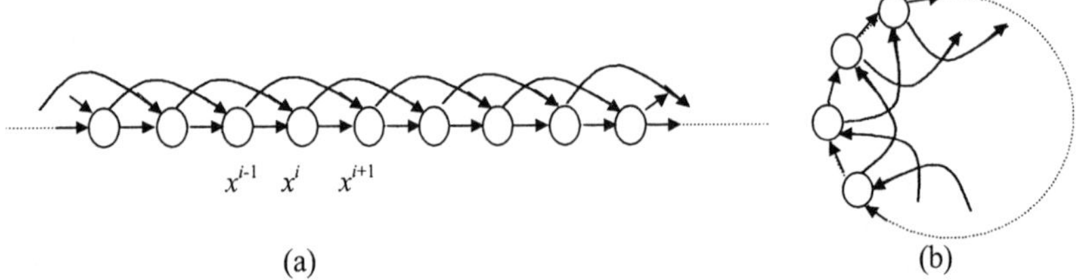


Figure 1. Examples of 1D lattices: (a) infinite linear lattice and (b) ring lattice, connected in one direction only, both with two lags

A linear infinite CMN is described by the initial state (initial condition) $\left\{x^i[0]\right\}_{i\in \mathbf{N}}$ and the equation of the lattice. The equation describing the lattice in Figure 1 a) is

$$x_{n+1}^{i+1} = f\left(x_n^{i-1}, x_n^i\right)$$

where the index $n$ stands for the number of times the map $f$ has been iterated. The function $f(\cdot)$ defines the nodes. In the linear case,

$$x_{n+1}^{i+1} = a \cdot x_n^{i-1} + b \cdot x_n^i$$

A set of selected nodes can be source nodes, i.e., they are given values not related to the recurrence relation. For example, if $x_n^0 = \sin(\pi \cdot n)$, the $0^{th}$ node is a source node. Source nodes are frequently named boundary conditions, because in physical applications, the boundary has valued assumed independent on the process (here, independent on the iteration process). If the boundary is "rigid", then $x_n^0 = $ constant. Frequently, the evolution of the linear CMN is represented as an image, with successive lines standing for successive time moments.

A (crisp), planar, CML, consists of rows of elements connected in one direction only, from one row to the next. The rows can extend to indefinitely, or be finite; in the last case, boundary conditions should be stated. In another version, CMNs have finite rows on the surface of a cylinder, i.e., the rows are placed along parallel closed curves (e.g., circles.) This translates by the conditions: $x^T = x^0$, $x^{T-1} = x^{-1}$, ..., for a CMN with rows including $T$ nodes.

The information signal is introduced as initial condition of the system; also, information can be input continuously in time, as excitation to the cells in the first row. The data propagate from one row to the next. Cells in a neighborhood act on a cell on the next row (Figure 2). This propagation can be interpreted as space-domain or as time-domain propagation, depending on the application. Notice that the elements in the networks are also named nodes, cells, agents, players etc.

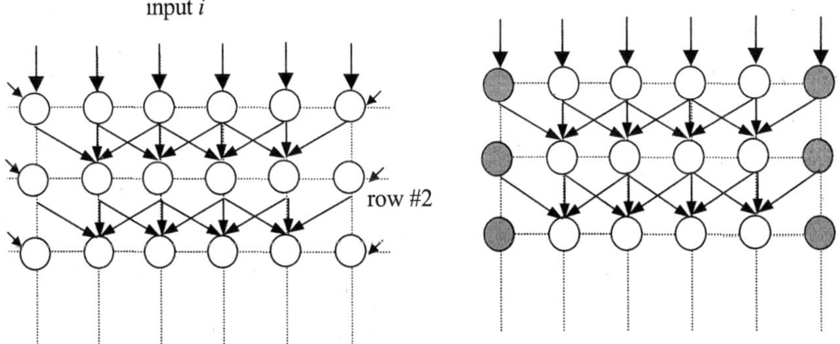


Figure 2. A semi-planar coupled map network (a) with infinite rows, and (b) with boundary conditions. (The shaded nodes may have fixed values or simply reproduce the values of the corresponding boundary nodes on the first row.)

Every cell is defined by the same characteristic function, $f$. For convenience, we may choose $f:[0,1]^{2Q+1} \to [0,1]$ or $f:[-1,1]^{2Q+1} \to [-1,1]$. Thus, the output of a cell is:

$$x_i^{j+1} = f\left(x_{i-Q}^j, x_{i-Q+1}^j, \ldots, x_{i+Q}^j\right) \tag{1}$$

where the neighborhood is centered on the current column and is $2Q+1$ wide, $j$ is the rank of the row, and $i$ is the index of the current cell (node) in the network. The initial condition is $\left\{x_i^0\right\}_i$, $x_i^0 \in \mathbf{R}$.

A typical class of coupled maps is governed by the equation:

$$x_i^{j+1} = f\left(\sum_{k=-Q}^{Q} w_k \cdot x_{i+k}^j\right), \quad w_k \in [0,1], \quad \sum_k w_k = 1 \tag{2}$$

where $w_k$ are the weights and are generally assumed symmetric ($w_k = w_{-k}$). One of the best-known examples of a similar case is [13]:

$$x_i^{j+1} = k \cdot f\left(x_i^j\right) + \frac{1-k}{2Q} \sum_{\substack{k=-Q \\ k \neq 0}}^{Q} f\left(x_{i+k}^j\right), \quad 0 \leq k \leq 1 \tag{3}$$

which becomes, for $Q = 1$,

$$x_i^{j+1} = k \cdot f\left(x_i^j\right) + \frac{1-k}{2}\left(f\left(x_{i-1}^j\right) + f\left(x_{i+1}^j\right)\right)$$

and after linearization,

$$x_i^{j+1} = k \cdot x_i^j + \frac{1-k}{2}\left(x_{i-1}^j + x_{i+1}^j\right) \tag{3'}$$

The relationship of the equations (3), (3') with the numerical integration of the diffusion process is briefly presented in Annex 1.

Another frequently used coupled map equation is:

$$x_i^{j+1} = k \cdot f\left(x_i^j\right) + \frac{k}{N} \sum_{k=1}^{N} f\left(w_{ki} \cdot x_i^j\right) + \xi_i^j$$

where $\xi_i^j$ ($\left|\xi_i^j\right| \ll 1$) represents a noise term. With less generality,

$$x_i^{j+1} = f\left(x_i^j\right) + \sum_{k=1}^{N} w_{ki} \cdot x_i^j \tag{4}$$

where $N$ is the number of cells in the row. Quite often, the function $f$ represents a process that can generate chaos by iteration, for example $f(x) = \eta \cdot x \cdot (1-x)$. For a symmetric vicinity coupling, and applying the function $f$ only to the current $j$ node:

$$x_i^{j+1} = f\left(x_i^j\right) + \sum_{k=-Q}^{Q} w_k \cdot x_{i+k}^j \tag{5}$$

Symmetric ($w_{i+k} = w_{i-k}$) and anti-symmetric ($w_{i+k} = -w_{i-k}$) coupling are frequently used. Notice that in all the above CMNs, the neighborhood includes elements on both sides. This type is named diffusion-type CMN. In contrast, if the neighborhood includes only elements on one side only, like in Figure 1, the CMN has one-way coupling. Such networks are also named unidirectionally coupled map lattice. In case of one-way coupling, the equations should be rewritten accordingly; for example, equation (3') becomes:

$$x_i^{j+1} = k \cdot x_i^j + (1-k) \cdot x_{i-1}^j \tag{3''}$$

A time-delayed map is defined by equations involving time lags larger than 1, for example:

$$x_i^j = k \cdot f\left(x_i^{j-p}\right) + \frac{(1-k)}{2} \cdot \left(f\left(x_{i-1}^{j-p}\right) + f\left(x_{i+1}^{j-p}\right)\right) \tag{6}$$

In the above description of the planar CMNs, the time variable has not been present. Notice that a planar CMN with no time variable involved is an interesting object *per se*. For example, if the initial condition represent the pixels of an original image, the CMN is equivalent to an image transform. Also notice that a planar CMN with no time involved should not be confused with a 1D, time-dependent CMN, the former having a 2D vector of initial conditions, while the latter has a 1D vector of initial data.

The above-described planar structures can be extended to three dimensions, like in Figure 3, with an appropriate definition of the vicinity. Propagation takes place from one plane to the next plane, and spatial or spatio-temporal aggregation may be produced.

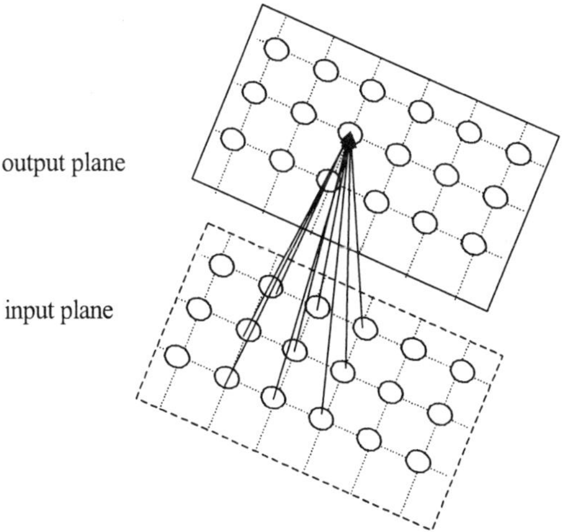


Figure 3. A bi-planar coupled map

The operation of the 3D structure in Figure 3 may be described by the equation:

$$x_{ij}^{t+1} = f\left(x_{hk}^{t}\right)\Big|_{(h,k)\in Vicinity\,of\,(i,j)}$$

where $t$ denotes the number of the plane in the layered structure.

Coupled map networks have been applied in physics, biology, and economics. They are good models for space-time propagation of epidemics and spreading of populations, for flow in fluid layers etc. CMNs have also been proposed as means to encrypt images and in several other engineering applications.

The behavior of the CMNs is depending on both the topology of interconnection (type of vicinity) and the operation of the nodes, i.e., the characteristic function of the nodes. Specifically, the synchronization and patterns produced are dependent on the graph-theoretical properties of the underlying coupling topology and the type of nonlinearity of the nodes. The spatio-temporal pattern formation in networks of oscillating or chaotic systems can be derived by nonlinear stability analysis ([3].)

Linear and finite length, non-cylindrical CMNs (i.e., CMNs like in equations (3') and (3''), with finite number of cells on a row and with no boundary conditions transforming them into cylindrical CMNs) are apparently of little interest. Nevertheless, they are very popular under the name of "cascaded linear filters". Indeed, because of the linearity, we can apply the $Z$ transform to obtain the $z$-space transfer function corresponding to the transformation of one row in the next one. For example, for the CMN described by equation (3''):

$$H(z) = \frac{Z\left(x_n^{t+1}\right)}{Z\left(x_n^{t}\right)} = c + (1-c)\cdot z^{-1}$$

Thus, the CMN is equivalent to repeatedly applying a linear transform (filter, in terms of signal processing) to the initial values. Let us denote by $H(\omega)$ the corresponding Fourier transfer function of the transform in equ. (3''). Then, the spectrums of the rows are obtained as:

$$S^{(n+1)}(\omega) = H(\omega)\cdot S^{(n)}(\omega) = H(\omega)\cdot H(\omega)\cdot S^{(n-1)}(\omega) = ... = H(\omega)\cdot H(\omega)...\, H(\omega)\cdot S^{(0)}(\omega)$$

where $S^{(n)}(\omega)$ denotes the spectrum of the $n^{th}$ row. Notice that the filter is a low-pass one, because in $H(z)$ both coefficients are positive. Therefore, high-frequency components will be eliminated.

Because the initial sequence $\left\{x_n^0\right\}$ is finite, the spectrum $S^{(n)}(\omega)$ includes important low-frequency components due to the finitude (the spectrum is dominated by the spectrum of the "window".) Denoting by $H^{\infty}(\omega) = \lim_{n\to\infty} H^n(\omega)$, we conclude that after a reasonable number of iterations, the spectrums of the rows tend to be identical, possibly dominated by the windowing effect.

There are several extensions of the CMNs, namely the Cellular Automata (CAs), Cellular Neural Networks, CNNs, and various certaity/fuzzy coupled map networks (U/FCMNs) that we will introduce in this chapter. The cellular automata represent networks of cells situated according to a grid on a line, in the plane or space, like CMNs; every cell has connections in a 1D, 2D, or 3D vicinity. What differentiate CAs and CMNs is that in CAs, the states of the cells in the nodes belong to a countable set (discrete, countable state space). In many cases of well-known CAs, like "the game of life", the state space includes just two states ("alive" and "dead", for the game of life.) Because of the finite number of states, the operation of the nodes in a CAs is easily described by rules. Consequently, CAs have many applications in fields related to biology and human activities, like traffic control, which are effectively described by rules. The CNNs are specific, continuous-time networks, representing an analog version of CMNs and being easily implemented in analog electronic circuits. The U/FCMNs are introduced below.

### 2.2. Uncertain Information-based Coupled Map Networks (UCMNs) [14-17]

In this section, we introduce several types of structures with lattice topologies inherited from the CMNs, but using uncertain information-based systems and propagating uncertain information through the network. For exemplification in this chapter, the uncertainty will be fuzzy-type uncertainty. Specifically, in Fuzzy CMNs (FCMNs), the crisp cells are replaced by fuzzy systems. The nodes transmit to their neighbors either crisp (defuzzified) values, or both crisp and fuzzy values.

Consider first fuzzy systems with defuzzified outputs and consider that only the defuzzified output values are used in the coupling. Then, the FCMN acts in a similar manner to the crisp CMNs. Thus, these coupled fuzzy map networks are actually crisp CMNs, with the cell law implemented by a fuzzy system with defuzzified output. Therefore, such CMNs do not really deserve the "fuzzy" attribute in their name; we will denote them as CFMNs.

Subsequently, we introduce the concept of fuzzy-coupled (fuzzy-) map networks (F-CMNs). These CMNs make use of the intrinsic properties of the fuzzy cells, by carrying fuzzy information from one cell to the cell on the next row (see Figure 4). The type of fuzzy information sent to the next row and the way of aggregating this information with other pieces of information will define the particularities and the class of FCMNs.

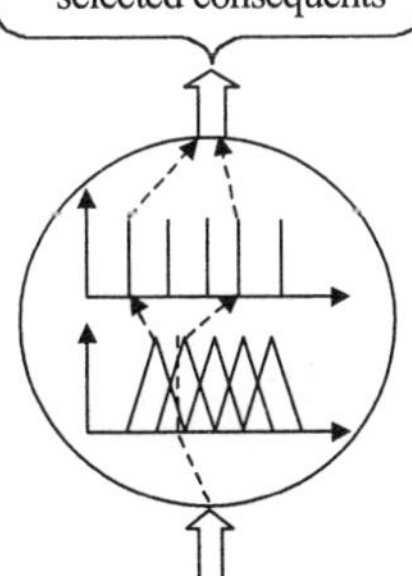


Figure 4. A fuzzy cell in a F-CMN may transfer to its neighbors the numerical (defuzzified) output value, moreover information on the values of the output membership functions. Here, a Sugeno-type fuzzy system is implementing a node in a F-CMN

Under a general framework, in a fuzzy-coupled map network, the values of the membership functions of a cell:

- are functions of the crisp (defuzzified) output values of the neighboring fuzzy cells on the previous row, moreover
- are functions of the values of the membership functions of the neighboring fuzzy cells on the previous row.

The general equation of a F-CMN is:

$$\mu^{k}_{x_i^{j+1}} = f_k\left( \left[\mu^{k}_{x_{i-Q}^{j}}\right], \dots, \left[\mu^{k}_{x_{i-Q+h}^{j}}\right], \dots, \left[\mu^{k}_{x_{i+Q}^{j}}\right], \left[y^{j}_{i-Q+h}\right]\right) \tag{7}$$

where:

- $k = 1, \dots, p$ is the index of the membership functions defining a cell;
- $\mu^{k}_{x_i^{j+1}}$ is the $k^{th}$ membership function value of the $i^{th}$ fuzzy cell on the $j^{th}$ row, for the current set of input values;
- $\left[\mu^{k}_{x_{i-Q}^{j}}\right], \dots, \left[\mu^{k}_{x_{i-Q+h}^{j}}\right], \dots, \left[\mu^{k}_{x_{i+Q}^{j}}\right]$ are the vectors of the values of the membership functions of the neighboring cells;
- $\left[y^{j}_{i-Q+h}\right]$ is the vector of crisp values at the output of the neighboring cells, and
- $f_k(\cdot)$ is one of the $k$ functions defining the cells.

**Example 1.**
Consider a Sugeno system having three Gaussian input membership functions (m.f.):

$$\mu^{k}(u) = \exp\left(-\frac{(x-a_k)^2}{s_k}\right), \qquad k=1,2,3; \quad a_k, \ s_k \in \mathbf{R}, s_k > 0 \tag{8}$$

The output singletons $b_k$ have certainty factors $\mu_k(u)$, where $u$ is the crisp input value. Then, for a 3-neighbors case and assuming identical nodes,

$$\mu^{k}_{x_i^{j+1}} = f\left( \left[\mu^{k}_{x_{i-1}^{j}}\right], \left[\mu^{k}_{x_i^{j}}\right], \left[\mu^{k}_{x_{i+1}^{j}}\right]\right)$$

Here, we have dropped the crisp output vector, $\left[y^{j}_{i-Q+h}\right]$, (see equ. (7)) because, in case of identical nodes, the output is a function of the values of the membership functions only:

$$y(u) = \frac{\mu_1 b_1 + \mu_2 b_2 + \mu_3 b_3}{\mu_1 + \mu_2 + \mu_3} \tag{9}$$

(we have dropped the indices $_i^j$). Therefore, we can see the F-CMN as operating solely with the values of the membership function.

The meaning of equation (7) will be further presented in the subsequent sections.

### 2.3. Interpretation and examples

What is specific in the manner we build the coupled map network is that we take into account the membership functions values of the cells in the vicinity, beyond considering the defuzzified outputs. In this way, the (fuzzy) information contained in the previous row cells is aggregated in a direct manner, not just mediated by the defuzzified values. This can be paralleled with the way of operation of an expert system-user tandem: the user needs not only the result, but the *HOW* information too ("How have you arrived at this conclusion?"). It is typical that a user changes his/her decision by contrasting the degree of belief in partial results she/he has and the belief degrees in the partial results of the adviser. Indeed, if the adviser has a strong belief in a partial conclusion, this can change our reasoning way. This shows that only the defuzzified output is not enough for a fuzzy cell to aggregate fuzzy information and that information on the values of the membership functions leading to the defuzzified values is also needed. Hence, the use of the equation (7) for a fuzzy-coupled map network.

Such networks can be also models for opinion dissemination and rumor spreading. For example, the first row can be the rumor generating persons, while the propagation takes place from friends-to-friend, based on the degrees of confidence people have in the opinion, authority and judgements of their friends, neighbors, or advisors. Buying goods and services like insurance policies is frequently based on such opinion and rumor spreading.

A special way to transfer the fuzzy information is to allow a cell on the previous row to exert its influence *on a specific membership function* of the cell on the next row. For example, the fuzzy aggregation may work as follows:

*IF　　The next left neighbor has the value of the membership function "LOW" significantly LARGER than that of the current cell, THEN it exerts its influence on the output by lowering it through domination on the LOW membership function,*
　　*ELSE　the neighbor exerts no influence.*

This can be written as:

$$\mu_{x_i^{j+1}}^{LOW} = \max\left( \mu_{x_{i-1}^{j}}^{LOW}\left(input\ to\ x_{i-1}^{j}\right), \mu_{x_i^{j+1}}^{LOW}\left(input\ to\ x_i^{j+1}\right)\right)$$

The strategy in the above rule can be summarized as:

*Every neighbor exerts its influence (dominates) on its expertise domain.*

As a matter of example:

*IF　　next left neighbor is RED and redder than the current cell, THEN it exerts its influence on the output through domination on the RED color spectrum,*
　　*ELSE　the neighbor exerts no influence.*

These strategies can be further elaborated using extra rules like:

*IF        the second left neighbor is GREEN, moreover GREENER than the current element,*
*THEN     it exerts its influence on the output through the domination on the GREEN color membership function*
*ELSE      the second-to-left neighbor exerts no influence*

### Example 2.
Another example of rules for uncertain information propagation is:

*IF        Difference Current-Neighbor is Negative_Big,*
*THEN     Neighbor dominates the LOWEST membership function (m.f.)*

*IF        Difference Current-Neighbor is Negative,*
*THEN     Neighbor dominates the SECOND_LOWEST m.f.*

*IF        Difference Current-Neighbor is ZERO,*
*THEN     Neighbor dominates NO m.f.*

### Example 3. Model of a panel
Consider a panel sitting in a line at the panel table. Every panelist is identified by a number $j$, when counted from left to right. Every panelist can discuss with his close neighbors only. They have to discuss some issue related to three choices and should determine an averaged optimum of the three choices. For example, the question to be decided could be: "What would be the best guess of the inflation rate (IR) next year and the years to come?" The panelists start the decision process with some initial guess for the IR next year, $u$ (the "initial opinion", "initial condition".)

We assume that all panelists are modeled (behave) in the same "rational" manner, except the initial guess, which is based on experience or intuition. Precisely, each panelist uses three linguistic attributes for the IR, namely, #1 "*low*", #2 "*average*", and #3 "*high*", then they all determine in the same way the belief degrees for the attributes, $\mu_1, \mu_2$, and $\mu_3$. Based on these, they determine the "best guess" for the inflation rate, applying the formula (see equ. (9)):

$$u_{n+1} = \frac{\mu_1 b_1 + \mu_2 b_2 + \mu_3 b_3}{\mu_1 + \mu_2 + \mu_3}$$

where $b_1, b_2$ and $b_3$ are constants and $\mu_1, \mu_2$, and $\mu_3$ are derived as explained below. Notice that this is equivalent to assuming that the panelists aggregate the three choices according to a Sugeno system (see the method presented in Example 1.)

We assume that, for some reason or another, every panelist believes that his left-side neighbor is an expert in the matter covered by the attribute #1. Moreover, every panelist believes that the right-side neighbor is an expert in matter covered by the attribute #3, while she/he thinks to be the expert in matters covered by attribute #2.

The rules of the panel require that the panelists write down their opinion in two copies and pass them to their close neighbors, i.e., to the panelist sitting at left, and respectively at right of her/him. They write down the overall result of their guesses, moreover the three values $\mu_1, \mu_2$, and $\mu_3$. The panelists at the ends of the row receive only information from a single neighbor.

After receiving the information, they aggregate the information and form their guess. The guess is computed according to the rules:

i)  Aggregate the values guessed by himself and the neighbors as:

$$x_{n+1}^{j} = k \cdot u_n^{j} + \frac{(1-k)}{2} \cdot u_n^{j-1} + \frac{(1-k)}{2} \cdot u_n^{j+1}$$

ii)  Determine $\mu_{1,2,3}^{j}(x_{n+1}^{j})$.

iii)  If $\mu_1^{j}(x_{n+1}^{j}) < \mu_1^{j-1}(x_{n+1}^{j-1})$ then assign $\mu_1^{j}(x_{n+1}^{j}) \leftarrow \mu_1^{j-1}(x_{n+1}^{j-1})$.

iv)  If $\mu_3^{j}(x_{n+1}^{j}) < \mu_3^{j+1}(x_{n+1}^{j+1})$ then assign $\mu_3^{j}(x_{n+1}^{j}) \leftarrow \mu_1^{j+1}(x_{n+1}^{j+1})$.

v)  Use the aggregation formula (9) for the guess:

$$u_{n+1} = \frac{\mu_1 b_1 + \mu_2 b_2 + \mu_3 b_3}{\mu_1 + \mu_2 + \mu_3}$$

The guess, together with the three values $\mu_1, \mu_2$, and $\mu_3$ will be communicated to their close neighbors at the next time moment. All opinions are communicated simultaneously. Then, the opinion formation process and the opinion communication restarts, as above.

The described process can be modeled by a F-CMN; indeed, the panelists transmit both crisp values and information on the membership degrees. The problem is to determine the dynamics of the guesses for successive years and how patterns of opinions, if any, form in the panel.

Most of the comments in sections 2.2 and 2.3 are valid for any type of network, including CMNs and fuzzy ganglions [18].

### 2.4. A broader perspective on information aggregation in networks

In this sub-section, we draw a parallel between some aggregation methods for uncertain information and Sugeno-type systems. A certainty factor- (belief degree-) aggregator is a system performing some operation on the belief degrees. The operation may be arithmetic, for instance a weighting of the belief degrees, or a selection-type operation, for example picking the highest or lowest belief degree.

For example, if $\{h[k]\}_{k \in S}$ denotes a set of belief degrees (values of the membership functions), the following represent aggregations:

$$h = \sum_{k \in S} h[k], \, h = \sum_{k \in S} w_k \cdot h[k], \, \max_{k \in S}\{h[k]\} \, , \min_{k \in S}\{h[k]\}$$

First, notice that a Sugeno system can be seen from a different perspective as an aggregator of belief degrees.

Indeed, the formula providing the output value[1]:

$$output = \sum_k h[k] \cdot b_k$$

can be seen as a weighted value of the belief degrees. If $\alpha_k \in [-1,1]$, then the weighting interpretation is more clear. Indeed, this provides the idea of yet another version of aggregating fuzzy information, namely, to use values of the coefficients $\alpha_k$ in the $[-1,1]$ interval, or even $\alpha_k \in \{-1,0,1\}$, to aggregate information on the belief degrees. Two "paired systems", one with the classical interpretation as Sugeno system, and the other one in the new interpretation, can be associated to produce a hybrid formal cell to model such behaviors.

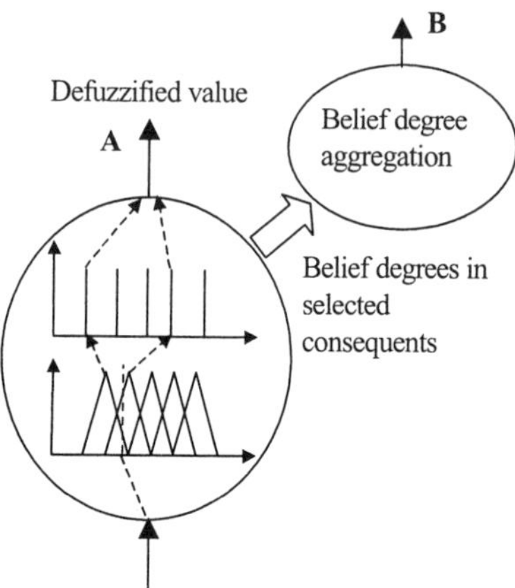


Figure 5. A couple of Sugeno systems

The model is shown in Figure 5, where the output B of the node is the result of the belief degree aggregation. Using these notations, the equations of a F-CMNs with 3 node-vicinity are:

$$A_i^{j+1} = f\left(A_{i-1}^j, A_i^j, A_{i+1}^j, B_{i-1}^j, B_i^j, B_{i+1}^j\right)$$

$$B_i^{j+1} = g\left(A_{i-1}^j, A_i^j, A_{i+1}^j, B_{i-1}^j, B_i^j, B_{i+1}^j\right)$$

---

[1] This formula is valid when the input membership functions are isoscelous triangular and equal, and they superpose two by two. In this case, it is easy to prove that $\sum_k h[k] = 1$. Therefore, equation (9) becomes $output = \sum_k h[k] \cdot b_k$ and allows us the above interpretation of the Sugeno system.

**Example 4.**

Consider the situation in Example 3, except that here, the panelists communicate to their neighbors an aggregated belief index, for example the maximal value of the three membership functions. The rules of the panel, including the rules determining the guess should be slightly modified accordingly. We leave this task to the reader.

### 2.5. Information aggregation and decision mechanisms

Aggregation performed at the level of a node determines the output of the node. For a version of the aggregation operation, the computation of the membership function values to be used in determining the output is performed in the following way:

$$h_i^{t+1}[k] = f\left(h_j^t[k], y_j^t\right)\Big|_{j \in Vicinity}$$

where $h_i^{t+1}[k]$ stands for the $k^{th}$ belief degree of the node $i$ in a linear F-CMN, at time moment $t+1$, and $y_j^t$ stands for the crisp output of the node $j$ at time moment $t$, $j$ being any index of a node in the vicinity of the current node.

Alternatively, when the system yields a weighted sum of belief degrees, the formula providing the values of the membership degrees of the predecessor is:

$$h_i^{t+1}[k] = f\left(h_j^t[k]\right)\Big|_{j \in Vicinity}$$

as $y_j^t$ is a function of the membership degrees; indeed, $y_j^t = y\left(h_j^t[k]\right)$ in a Sugeno system (see equ. (9)). The successor may take into account in an "optimistic" way the belief degrees expressed by its predecessors. The "optimistic" way means that the successor increases its own belief degrees depending on the values of the belief degrees of the predecessors, if the last ones are higher:

$$h_i^{t+1}[k] = \max_{j \in Vicinity}\left(h_{i+j}^t[k]\right)$$

where the index $j$ is according to the vicinity. If the vicinity is symmetric and includes $2Q+1$ neighbors:

$$h_i^{t+1}[k] = \max_{j \in \{-Q,\dots,Q\}}\left(h_{i+j}^t[k]\right)$$

Alternatively, the successor may take into account in a "pessimistic" way the belief degrees expressed by its predecessors. The "pessimistic" way means that the successor decreases its own belief degrees to the values of the belief degrees of the predecessors, if the last ones are lower:

$$h_i^{t+1}[k] = \min_{j \in \{-Q,\dots,Q\}}\left(h_{i+j}^t[k]\right)$$

The following mechanisms are examples of mechanisms that can be used in aggregation:

a) Single dominance, single influence, "optimistic" strategy. Compare only certainty factors with a selected rank, $k$, i.e., compare $h^j_{i+r}[k]$ ($r$ for the predecessors) with $h^{j+1}_i[k]$. If one of the predecessors has the selected certainty factor *higher* than the current cell, then force $h^{j+1}_i[k] = h^j_{i+r}[k]$. (In case several predecessors satisfy the condition, choose the one with the higher value of the selected certainty factor.) The other membership functions are not changed.

b) Single dominance, single influence, "pessimistic" strategy. Compare only certainty factors with a selected rank, $k$, i.e., compare $h^j_{i+r}[k]$ ($r$ for the predecessors) with $h^{j+1}_i[k]$. If one of the predecessors has the selected certainty factor *lower* than the current cell, then force $h^{j+1}_i[k] = h^j_{i+r}[k]$. When several predecessors satisfy the condition, choose the one with the lower value of the selected certainty factor.) The other membership functions are not changed.

c) Single dominance, global influence, "optimistic" strategy. Compare only certainty factors with a selected rank, $k$, i.e., compare $h^j_{i+r}[k]$ with $h^{j+1}_i[k]$. If one of the predecessors has the selected certainty factor higher than the current cell, then force all the $h^{j+1}_i[l] = h^j_{i+r}[l]$ (for all $l$.)

d) Single dominance, global influence, "pessimistic" strategy. Compare only certainty factors with a selected rank, $k$, i.e., compare $h^j_{i+r}[k]$ with $h^{j+1}_i[k]$. If one of the predecessors has the selected certainty factor lower than the current cell, then force all the $h^{j+1}_i[l] = h^j_{i+r}[l]$ (for all $l$.)

e) Global dominance, single influence, "optimistic" strategy. Compare *all* certainty factors, i.e., compare for all $l$, $h^j_{i+r}[l]$ with the corresponding $h^{j+1}_i[l]$. If one of the predecessors has all the certainty factors higher than the current cell, then force only certainty factors with a selected rank, $k$, i.e., $h^{j+1}_i[k] = h^j_{i+r}[k]$.

f) Global dominance, single influence, "pessimistic" strategy. Similarly to the above.

g) Aggregated dominance, single influence, "optimistic" strategy. Compare an aggregated index of certainty factors; for example, compare $h^j_{i+r}[1] + h^j_{i+r}[2] - h^j_{i+r}[3]$, with the corresponding $h^{j+1}_{i+r}[1] + h^{j+1}_{i+r}[2] - h^{j+1}_{i+r}[3]$. If one of the predecessors has the aggregated index higher than the current cell, then force only certainty factors with a selected rank, $k$, i.e., $h^{j+1}_i[k] = h^j_{i+r}[k]$.

h) Aggregated dominance, single influence, "pessimistic" strategy. Similarly to the above.

In summary, combinations of the following choices, among others, can be used to build uncertain information-based cellular map networks:

- Choices for the input $u$ aggregation:

a) Classic aggregation way, even if the output represents certainty factors, for example:

```
u← kapa*x_prev[i] + (1.-kapa)/2)*(x_prev[i-1]+x_prev[i+1])
```

b)Transfer dominated by a rule of domination, as explained in the text.

- Choices for the output **A** function:

(a) $A = \sum a_k \cdot h_k^j \Big/ \sum h_k^j$ $a_k \in \mathbf{R}$. No restriction on the $\alpha_k$ values (output singletons).

(b) Other defuzzification rules.

- Choices for the output **B** function:

a) $B = \sum w_k \cdot h_k^j$ , $w_k \in [0,1]$, $\sum w_k = 1$

b) $B = \sum w_k \cdot h_k^j$ , $w_k \in \{-1,0,1\}$

c) $B = h_k^{j+r}$

- Choices for the way of transferring information on the membership functions:

a) None
b) Compare a single certainty factor for the current and the neighboring cells
c) Compare some weighted sum of certainty factors for the current and the neighboring cells
d) Compare each certainty factor for the current and the neighboring cells

- Choices for the dominance of the certainty factor:

a)   None
b)   If one of the predecessors has the selected certainty factor (with a selected rank, $k$) **higher** than the current cell, $h_i^{j+1}[k] \leftarrow h_{i+r}^j[k]$. The other membership functions are unchanged.
c)   If one of the predecessors has the selected certainty factor **lower** than the current cell, $h_i^{j+1}[k] \leftarrow h_{i+r}^j[k]$.

We will name *strategy* or *behavior* any combination of the above; for example, *acdb* is a strategy. For a given optimization problem, the strategies can be evolved to find the best one solving the problem in a societal game, i.e. to optimize the societal behavior in a game framework. One of the purposes of the game may be to obtain the best opinion aggregation in an expert panel like in Example 3.

For a sketch of an information propagation theory leading to models represented by CMNs as presented in this chapter, see Annex 3.

### 3. Examples and algorithms

In this section, we use Sugeno 0-type fuzzy systems, with the singletons $\alpha_i, i = 1,...,q$ and their truth-degree denoted by $h_i$. Let be the rules describing the system given by the couples $(k, i)$, where $k$ is the index of the input interval associated with the $i^{th}$ output singleton:

*IF        input is $A_k$        THEN  output is $\alpha_i \in \mathbf{R}$*

We associate the neighbor #$j$ ( $j = -q...q$) to the $i_q$ membership function of the systems (cells), i.e., the $j$ neighbor causes the $\alpha_{j_q}$ to appear with the belief degree:

$$h_{j_q} = h(\text{output of neighbor } \# j).$$

Define a fuzzy coupled map by:

*Every neighbor exerts its influence (dominates) on the corresponding domain (interval): partial result #1 is* $\alpha_{j_q}$ *occurs with* $h_{j_q} = h(\text{output of neighbor } \# j)$

*MOREOVER,*

*IF        input is* $A_k$ *with* $\mu_{A_k}(x_0)$

*THEN    partial result #2 is* $\alpha_i$ *with* $h_i' = \mu_{A_k}(x_0)$

*The final result is*

$$\alpha_i \text{ with } h_i = \max\left( h_i', h_{j_q} \mid j_q = i \right)$$

The corresponding algorithm is:

```
1. input prev_x [i]
2. for j=0 to N
3.     for i = 0 to Nr_cells_in_rows
4.        x[i] ← fuz_cell_X(i, prev_x [ ], Q, a, b, alpha[ ], w[ ])
5.        plot (i, j, color(x[i]))
6.     end i
7.     for i = 0 to Nr_cells_in_rows
8.        prev_x [i] ← x[i]
9. end j
```

The procedure fuzz_cell_X ( ), where X stands for the version number (1,2, …), has various versions used to implement various strategies, as previously discussed.

The procedure fuzz_cell_1 ( ), for the case of aggregating when the neighbor # $i+h-Q$ in the $2Q+1$ vicinity acts on the membership function $h$ only.

```
1. fuz_cell_X(i, prev_x[ ], Q, a, b, alpha[ ], w[ ])
                     // actually, in this procedure, w[ ] has no effect
2.     for h = 0 to 2Q+1
3.        if h < > 0
4.           u ← prev_x[i]
5.           r ← compute_m_f(h, u, a,b)
6.           u ← x[i+h-Q]
7.           s ← compute_m_f(h, u, a,b)
8.           height[h] ← max(r,s)
9.        else
10.          height[h] ← r
11.       out ← out + height[h]*alpha[h]
12.       sum_height ← sum_height + height[h]
13.    end for h
14     out ← out/sum
15.    return out
```

The line 8 can be replaced by a weighted maximum version:

8.        height[h] ← min(1, max(r, s_weigt*s) )

where s_weight can be, for example, 1.5, to give a higher emphasis to the neighbors.

The procedure compute_m_f(h, u, a, b) is used when the input membership functions are triangular. This procedure computes the membership function value, for the $h$-th input membership function, for the $u$ input value, assuming the parameters $a$, $b$ and $h$ are enough to determine the membership function. (This is true for equal, equally spaced triangular membership functions). The procedure is written depending on the type of the input membership functions used. In most cases, in this chapter we use Gauss membership function at the input.

In another version of the algorithm for the F-CMN, the aggregation is based on the comparison of an aggregated result of the neighbors with the result of the current cell:

```
1. fuz_cell_2 (i, previous_x[ ], Q, a, b, alpha[ ], w[ ])
2.        for h = 0 to 2Q+1
3.                    y ← aggregate_neighbors (i, previous_x[ ], Q, w[ ])
4.        for h = 0 to 2Q+1
5.                        u ←previous_x[i]
6.                        r ←compute_m_f(h, u, a,b)
7.                        u ←y
8.                        s ←compute_m_f(h, u, a,b)
9.                height[h] ← max(r,s)
10.               out ← out + height[h]*alpha[h]
11.               sum_height ← sum_height + height[h]
12.       end for h
13        out ←out/sum
14.       return out
```

In all the examples below, the Gauss membership functions associated to the cells have the centers in the set singleton = {- 0.4, 0.5, 0.1, - 0.5, 0.9}, and the spreading is spread = 0.5. The input to the network is a random sequence (shown for convenience in every figure and represented by the left-side bar.)

In the next examples, we use aggregation formulas with integer weights {-1, 0, 1} to compute the output of the Sugeno system with Gauss input membership functions. We use the input aggregation law (see equ. (5)):

u← kappa*x_prev[i] +  (1.-kappa)/2)*(x_prev[i-1]+x_prev[i+1])

For various combinations of weights, the results yielded by the fuzzy coupled map networks are shown in Figures 6 a) – e).

In all figures below, the left-hand side pictures represent, by colors (gray shades), the output values of the cells in the network, while the right-hand side represents the variation of the values on the last computed row in the cell. (In most figures, the last row has the number 800, whatever the settings of the display are.)

Of course, the way of choosing the value-color correspondence may play an essential part in showing the self-organization of the outputs of the cells.

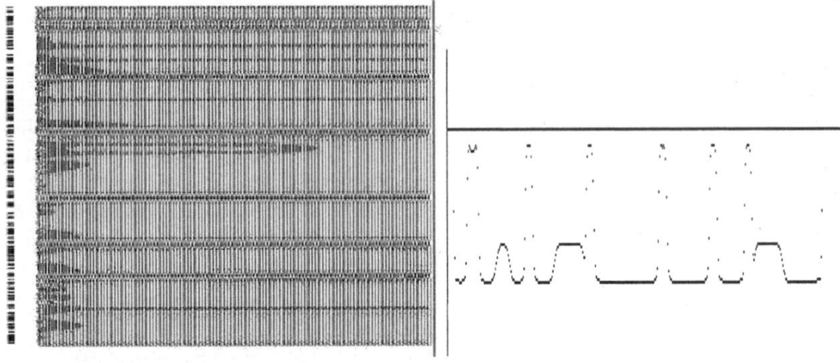

a) Output generated by a system returning the value h[3]-h[0]+h[2]+h[5].
Graph drawn time samples # 0 to 800.

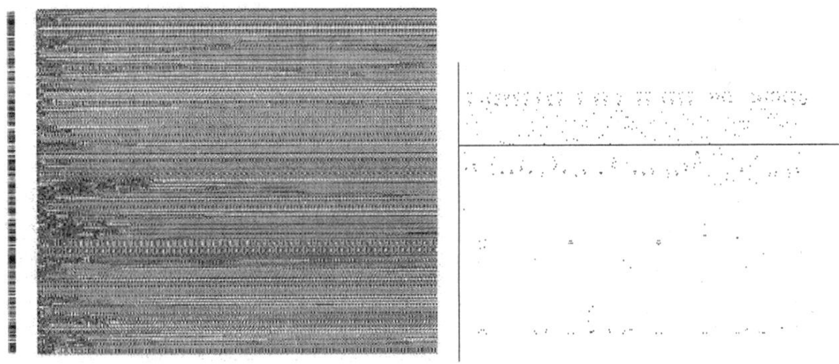

b) Conditions as above; the returned value is h[3]-h[4]+h[2].

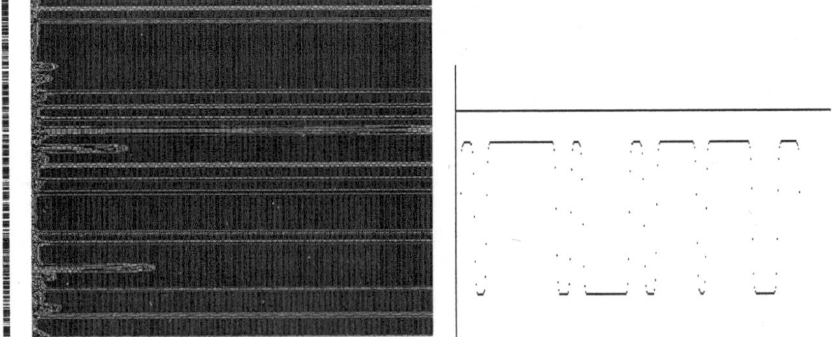

d) Conditions as above; the returned value is h[1].

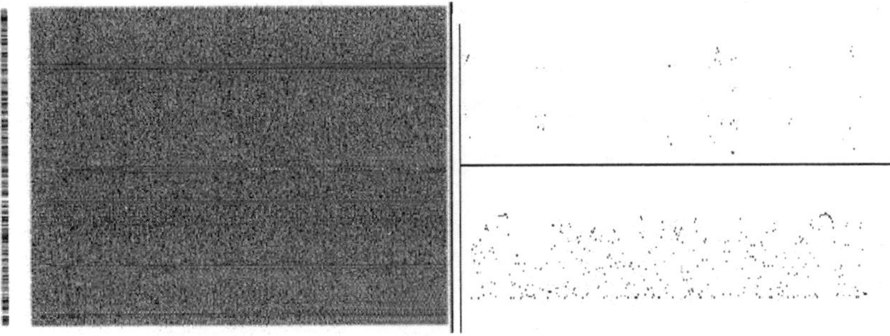

e) Conditions as above; the returned value is h[1] – h[4].

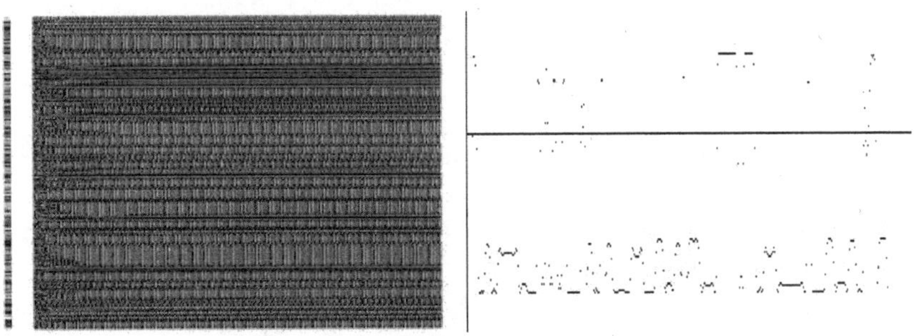

e) Conditions as above; the returned value is h[2] – h[4]

Figure 6. Examples of results obtained with F-CMNs.

The graphs in Figure 7 a), b) are obtained with a non-adaptive but information-sensitive system based on the conditions:

```
if (i>0) AND (i<N2-1)
    if h[1]>hh_prev[i-1][1]
      h[K] ←hh_prev[i+1][K]
return h[2]
```

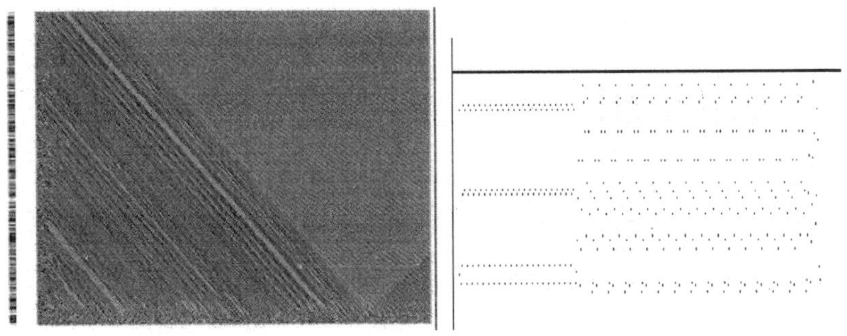

a) Results obtained for the first 800 time samples ($i = 0..800$)

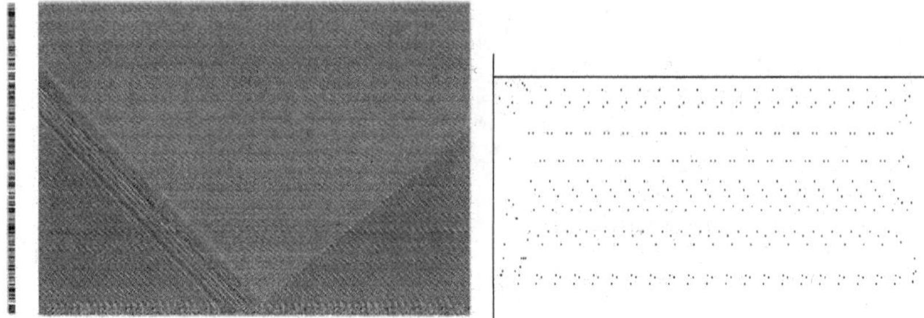

b) Results obtained for the time samples $i = 2000..2800$

Figure 7. Results obtained with an information-sensitive system

Finally, as a matter of example, we notice that geometric shapes can also be obtained, like in Figure 8.

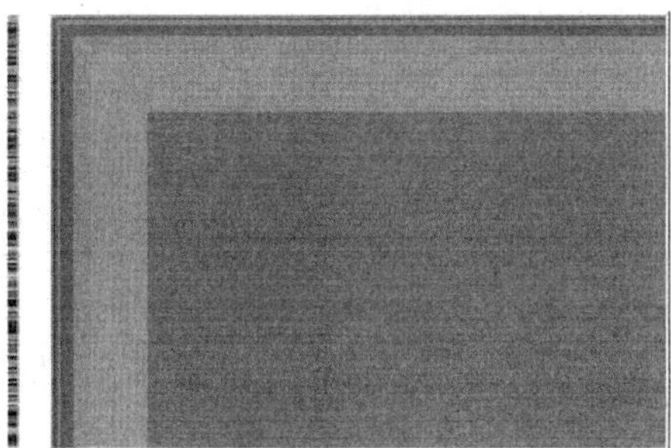

Figure 8. Example of simple geometric image obtained using a F-CMN

An example of procedure used to take into account previous values of the membership functions is provided below.

```
Procedure pseudo-Sugeno
1. out ← 0, out_sum ← 0.
2. for K= 0 to M
3.      if v = u-c[k])^2/spread[k] <1000.
4.          h[k] ← exp (-v)
5.      else
6.          abort
7.      if h[0] < h_previous [i −1][0] AND i>0
                    // The comparison of only the first m.f. of the current and the previous
                    //nodes decide on the change (inheritance) of all the membership functions
8.          h[k] ← h_previous[i][k-1]
                        // Notice that a shifting in the k index is enforced here (k ←k-1)
9.      out ← out+h[k]*singleton[k]
10.     out_sum ← out_sum +h[k]
11.     float vv←out/out_sum
12.     return h[3]-h[0]+h[2]+h[5]
```

## 4. Discussions

### *4.1. Discussion of the adaptation issues*

An "adaptation" procedure can be envisaged either to model an adaptation process of the individual players, or to perform an adaptation of the network to a specific task. Here, we discuss only the first case.

For modeling purposes, we may assume that a "player" changes its strategy when one of its predecessors provides a different answer than the player. For example, the adaptation can be performed by changing the input by a constant bias, whenever the third membership function of the left-side neighbor has higher value than the third membership function of the current cell:

```
if  h[3]>hh_prev[i-1][3]
     u←u+0.5;
```

Above, the first membership function value of the left-side neighbor has been taken into account in a dramatic manner. Namely, if that value was higher then the value of the first membership function of the current cell, then the behavior of the current cell is replaced by that of the left-side neighbor:

```
if  h[0]<hh_prev[i-1][0]) AND i>0
     h[K] ←hh_prev[i][K]
   out_Sugeno += h[K]*singleton[K]
   out_sum +=h[k]
end K
```

The output of the cell is the weighted-sum of four of its membership functions:

```
   return h[3]-h[0]+h[2]+h[5]
```

We may choose that, if another membership function of the left neighbor dominates, other types of "adaptation" occur, for instance, if h[0]<hh_prev[i-1][0] then h[K]=hh_prev[i][K].

The aggregation may use a "prudent" strategy, copying the behavior of the neighbors when they have belief degrees that are less then the belief degrees of the current player (cell). Moreover, a single neighbor having a single belief degree less then the corresponding belief degree of the current player may trigger the decrease of the "optimism" (belief degree) of the current player. For example:

```
for all K
     if h[2]>hh_prev[i-1][2]
        then h[K] ←hh_prev[i-2][K]
```

while another strategy, asymmetric to the first, may be used for some other membership function:

```
   IF h[3]>hh_prev[i-1][3] THEN u = u+0.5
```

To date, there is no systematical way to choose the adaptation (decision mechanism), except the modeling requirements – when the method is used in creating models.

### 4.2. Clustering, fuzzy clustering, and texture definition

In the previous sections, we have adopted an empirical manner of visualization the clustering and texturing properties. This manner, which is generally used in the literature, let the clustering and texturing be visualized by an arbitrary choice of colors for values of the output falling in some interval of the output space.

Formally, the clustering can be defined as follows.

Let $U$ be the output space of the cells in the network, $U \subset \mathbf{R}$; $U$ is considered bounded. Consider a finite partition $\Omega = \{I_k\}_{k=1,2,\dots,r}$ of $U$, i.e., a finite set of intervals satisfying the conditions:

$$I_k \subset U \,, \ I_k \cap I_j = \phi \ \forall k \neq j \,, \ U = \bigcup_k I_k$$

We need another concept, which it looks to be undefined and dealt with intuitively (empirically) in the literature, namely the way we use to geometrically arrange the nodes for visualizing them.

We define the "visualization topology" the map $V \to \{(x, y)\}$, i.e., the way we place the nodes in the plane. For sake of convenience, we choose the visualization topology obtained by applying the rules:

i) the nodes with no in-edge are placed all in the same row, on a line, equally spaced, at distance (arbitrarily) equal to 1 in the plane; the line should be parallel to the $Ox$ axis;

ii) all nodes having in-edges from the previously drawn row are placed in the same row, on a line, equally spaced, at distance equal to 1; moreover, the distance between one line and the previous one should be 1; also, the nodes connected by the edges corresponding to the indices $j_i$ and $j_i^{+1}$ should be placed at distance equal to 1, along a parallel to the $Oy$ axis.

The above recursive procedure guarantees the visualization "as usually" of the rectangular-shape CMNs.

Define on the set of nodes of the network, $V$, a distance $d(\ )$. For example, the usual distance on a graph may be used. We define a *visual cluster* in $V$ a subset $C$ of $V$ with the properties:

a) all nodes in the cluster belong to the same interval in the partition $\Omega$, $\upsilon \in I_k \ \forall \upsilon \in C$;

b) every node in the cluster has a neighbor in the cluster, i.e. a node in the cluster at distance 1;

c) no "foreign nodes" are admitted in a cluster, i.e., no node with a different color or sets of nodes of different colors should be surrounded by the class, except when those nodes form together themselves a class[2];

d) the dimension of a cluster, along both plane directions, should be at least a threshold value (for example, 3).

The above definitions allow us to create algorithms for automatically checking the existence of clusters in a network. What is disputable in the above definitions is the arbitrary way of choosing the partition for coloring. We may find it preferable to use an adaptive partition, with the intervals determined by conditions like "maximize the total number of nodes belonging to a cluster, while preserving the number of clusters."

---

[2] Notice that a class can be included into another class.

Fuzzy clusters may also be defined, by allowing nodes to belong with some membership degree to fuzzy intervals and coloring the nodes with the color of the interval to which the nodes "mainly" belongs.

The formation of repetitive patterns (textures) is also an interesting behavior of the CMNs. The automatic recognition of the patterns requires statistical techniques as applied in image processing.

## 5. Conclusions

In this chapter, we have introduced several new types of structures processing information with various types of uncertainty. We have been interested into the propagation of information and the self-organization processes driven by various types of local information aggregation procedures, while the adaptations are driven by information and uncertainty.

The discussed networks may be the basis for modeling a large variety of processes in social sciences, macro-economics, and biology. The interpretation of the various local functions and propagation procedures discussed in this chapter should be given in relation to the modeled process.

We presented preliminary arguments on the power of this new theoretical and modeling tool and on the possible use of uncertain-information driven networks in several domains. We have argued that the dynamic process should include local decision based on uncertain information in the network, to account for the adaptive behavior of real agents in economic, social and biologic processes. The presented results demonstrate that the methodology we have introduced produces much richer results than the classic similar methods. The rational of using elaborated methods of transferring the uncertain information from one node to the others, and especially the asymmetric and "preference based" ways suggested consists in the observation of such processes in economy and sociology. It is well known, as a matter of example, that the asymmetric information concept and related economic theories have won two times in 5 years the Nobel Prize in economy [19].

The CMNs are frequently related to ODEs solving and specifically to the flow problems solving. On the other side, the understanding we have of the Navier-Stokes equations is not satisfactory (see the Clay Mathematics Institute description of the problem, [20], whose solution is 1 million worth.)

Further developments should establish a more detailed methodology for transferring uncertain information to a node from its vicinity and develop a complete theory of dynamic behavior for such networks. Extensions to cellular automata and other types of networks should also be carried on.

**Acknowledgments**. Most part of this research has been supported by Techniques and Technologies Ltd., to whom the copyright for the algorithms, software and the pictures produced with the software remains. The software used in the analysis above can be obtained from the author or from Techniques and Technologies Ltd.



## References

[1]. K. Kaneko. Simulating Physics with Coupled Map Lattice. Formation, Dynamics and Statistics of Patterns 1, World Scientific, Singapore, 1990

[2]. P. Thiran, Dynamics and Self-Organization of Locally-Coupled Neural Networks. Presses Polytechniques et Universitaires Romandes, Collection META, Lausanne, Switzerland, 1997

[3]. Chai Wah Wu, Synchronization in Coupled Chaotic Circuits and Systems, World Scientific Series on Nonlinear Science, Series A, Vol. **41**, 188 pp., 2002

[4]. K. Kaneko. Overview of Coupled Map Lattices. Chaos vol. **2**, nr. 3, pp. 279-282, 1992.

[5]. R. Carretero-Gonzalez, D.K. Arrowsmith, and F. Vivaldi, Mode-locking in Coupled Map Lattices. Physica **D 103** (1997), pp. 381-403

[6]. O. Rudzick, A. Pikovsky, Ch. Scheffczyk, and J. Kurths, Dynamics of Chaos-order Interface in Coupled Map Lattices, Physica **D 103**, 1-4, p. 330-347 (1997)

[7]. S.C. Manrubia and A.S. Mikhailov, Mutual Synchronization and Clustering in Randomly Coupled Chaotic Dynamical Networks. Physical Review **E 60**, 1579-1589 (1999)

[8]. S.C. Manrubia and A.S. Mikhailov, Very Long Transients in Globally Coupled Maps. Europhysics Letters **50**, 580-586 (2000)

[9]. V.I. Sbitnev, Chaos Structure and Spiral Wave Self-Organization in 2D Coupled Map Lattice, International Journal of Bifurcation and Chaos, Vol. **8**, Number 12, December 1998, pp. 2341- 2352

[10].H. Chaté, A. Pikovsky, and O. Rudzick, Forcing Oscillatory Media: Phase Kinks vs. Synchronization, Physica **D 131**, 1-4, p. 17-30 (1999)

[11].O. Rudzick and A. Pikovsky, Unidirectionally Coupled Map Lattice as a Model for Open Flow Systems. Physical Review E November 1996 Volume **54**, Number 5, pp. 5107- 5115

[12].G.J. Klir, Tina A. Folger, Fuzzy Sets, Uncertainty, and Information. Prentice-Hall International Editions, Englewood Cliffs, N.J., 1988

[13].J.C. Sprott, G. Rowlands, Chaos Demonstrations. User's Manual. American Institute of Physics, New York, 1995

[14].H.N. Teodorescu, Coupled Map Networks and Uncertainty propagation. Research Report, 2000, Techniques and Technologies Ltd., Iasi, Romania

[15].H.N. Teodorescu, Coupled Map Networks and Uncertainty propagation. Int. J. Chaos Theory and Applications, vol. **8**, (2002), nr. 1, pp. 3-16

[16].H.N. Teodorescu, Uncertainty Aggregation and Uncertainty Complexity in Coupled Map Networks. Applications. Proc. ECIT2002 Conference, 17-20 July, 2002, Iasi, Romania

[17].H.N. Teodorescu, Uncertainty Propagation through Fuzzy-Coupled Map Networks. Research Report, 2001, Techniques and Technologies Ltd., Iasi, Romania

[18].H.N. Teodorescu, Fuzzy- and t-Ganglions. Research Report, 2001, Techniques and Technologies Ltd., Iasi, Romania

[19].J. Barkley Rosser, Jr., A Nobel Prize for Asymmetric Information: Economic Contributions of George Akerlof, Michael Spence, and Joseph Stiglitz http://cob.jmu.edu/rosserjb/

[20].C.L. Fefferman, Existence & Smoothness of the Navier–Stokes Equation. http://www.claymath.org/prizeproblems/navierstokes.htm

[21].N. Racoveanu, Gh. Dodescu, I. Mincu, Numerical Methods for Hyperbolic Partial Differential Equations. (In Romanian). Editura Tehnica, Bucharest, 1976

## Annex 1. Coupled Map Lattice for diffusion modeling

The diffusion equation for homogeneous media is a parabolic partial derivative equation:

$$\frac{\partial f}{\partial t} = D \cdot \sum_i \frac{\partial^2 f}{\partial x_i^2} = D \cdot \Delta_x f(x,t) \tag{A1.1}$$

where D is the diffusion coefficient. We will consider a single spatial variable:

$$\frac{\partial f}{\partial t} = D \cdot \frac{\partial^2 f}{\partial x^2} \tag{A1.2}$$

We use the following notation for discretization: $f(x,t) \to f(x_i, t_j) \to f_i^j$. The equation can be discretized according to various schemes. According to a common scheme,

$$\frac{\partial f}{\partial t} \to f_i^{j+1} - f_i^j$$

$$\frac{\partial^2}{\partial x^2} f \to \frac{1}{2}\left[f_{i+1}^j - 2f_i^j + f_{i-1}^j\right]$$

The diffusion equation becomes:

$$f_i^{j+1} - f_i^j = D\frac{1}{2}\left[f_{i+1}^j - 2f_i^j + f_{i-1}^j\right]$$

or

$$f_i^{j+1} = (1-D)\cdot f_i^j + D\frac{1}{2}\left[f_{i+1}^j + f_{i-1}^j\right] \tag{A1.3}$$

The numerical integration is stable if $D < 1$. Notice that the above is equivalent to a convolution operation with the coefficients in the window: $\left[\frac{D}{2}, 1-D, \frac{D}{2}\right]$, the window being centered on the current point. Compare equation (A1.3) with equation (3) and the subsequent ones. The physical interpretation of the equation (A1.3) is that the amount of material at time moment $j+1$ at location $i$ is composed by:

- the amount at that location at previous time moment minus the amount that diffused toward other locations (the term $-D\cdot f_i^j$), plus
- the amount of material that diffused into the respective location from the neighboring locations, namely the term $D\frac{1}{2}\left[f_{i+1}^j + f_{i-1}^j\right]$.

## Annex 2. Coupled Map Lattice for flow modeling

The steady flow equation in a tube without losses, fluid generation or transfer is[3]:

$$\frac{\partial f}{\partial t} + c\cdot\frac{\partial f}{\partial x} = 0 \tag{A2.1}$$

where $c$ is a coefficient. Based on the Taylor series expansion, we obtain:

---

[3] This is a very simple case. Navier-Stokes equations governing the general flow processes can be found in the literature, for example [20].

$$f_i^{\,j+1} = \left(1 + u\frac{\partial}{\partial t} + \frac{1}{2}u^2\frac{\partial^2}{\partial t^2} + \ldots\right)\cdot f_i^{\,j}$$

where $u$ is the discretization step. Neglecting terms of order higher than 1:

$$f_i^{\,j+1} = \left(1 + u\frac{\partial}{\partial t}\right)\cdot f_i^{\,j}$$

and taking into account (A2.1), we obtain:

$$f_i^{\,j+1} = \left(1 - uc\frac{\partial}{\partial x}\right)\cdot f_i^{\,j}$$

Using the left-hand side derivation formula, after computation [21 p. 84], we obtain:

$$f_i^{\,j+1} = (1 - c)\cdot f_i^{\,j} + c\cdot f_{i-1}^{\,j}$$

This is the equation of a unidirectionally coupled map network.

### Annex 3. Modeling the information propagation in societal systems

In this Annex, we briefly propose an outline for a theory of information propagation theory, to model information spreading in societal systems. At our best knowledge, such a theory is lacking, although it is obviously needed in social sciences, behavioral sciences, economics, and other human-related sciences. We borough much from mechanist paradigms to build our model. The result is that CMNs and F-CMNs naturally appear as models of information propagation.

Information may be characterized by several attributes: quantity, flux (quantity of transmitted information per time unit), entropy, relevance, and credibility, while the sources of information are characterized by attributes like credibility, expertise etc. All these attributed should be used to model the information propagation in societal systems. In what follows, we sketch a set of rules to build such a model.

First, we notice that information is transmitted from individual to individual, i.e., between discrete partners. Therefore, while time may be continuous, space should be discrete in information transmission.

Second, we notice that at least two different types of information exist, namely, linguistic and numerical. For example "this cake is sweet", "this cake is fresh" are linguistic information, while "this cake is 27 hour old" is numerical information.

**Rule 1.** The source of information transmits information without loss of information. Denoting by $\Xi_s$ the quantity of information at the source,

$$\left.\frac{\partial \Xi_s}{\partial t}\right|_{\Xi_{in}=0} = 0 \qquad\qquad\qquad\qquad \text{(A3.1)}$$

where $\Xi_{in}$ is the received information.

Indeed, the information transmitted is not lost at the source, as it is with physical quantities like mass, heat, or energy, when transmitted from a source of mass, heat, or energy to the surrounding environment.

**Rule 2.** We assume that the flux of transmitted (received) information is proportional to the relevance of information. The relevance of information is considered proportional to the difference between the levels of information at the source and at the destination. Namely, denoting by $\Xi_r$ the information at the receptor:

$$\frac{\partial}{\partial t}\Xi_r = c \cdot (\Xi_s - \Xi_r) \tag{A3.2}$$

In discrete time, this equation reads:

$$\Xi_{r,t+1} - \Xi_{r,t} = c \cdot (\Xi_{s,t} - \Xi_{r,t})$$

or

$$\Xi_{r,t+1} = c \cdot \Xi_{s,t} + (1-c) \cdot \Xi_{r,t} \tag{A3.2'}$$

where $t$ is the time moment the information is transmitted.

The relevance of information is the counterpart of the gradient in physical laws. The reason for the formula (A3.2) is that when the receptor is poorly informed (in a specific field), any new information is "absorbed" with a low rejection ratio. (It may be the case of adolescents, who a keen to get new information in various domains.)

**Rule 3 (law of optimistic receptor).** Information on a specific subject arriving in a receptor from two different sources aggregate in information that has the credibility equal to the level of the higher source credibility. **Namely,**

$$\Xi_{r \leftarrow s1}(h_1) \hat{+} \Xi_{r \leftarrow s2}(h_2) = \Xi_{r \leftarrow s}(\max(h_1, h_2)) \tag{A3.3}$$

where "$\hat{+}$" means an aggregation operation, with result as in the right-hand side, $h$ is the belief degree associated to the information, and the indices $s_1$, $s_2$ indicate the information sources, while the index in $\Xi_{r \leftarrow s}$ indicates that information is transmitted from the source $s$ toward the receptor $r$.

Combining equations (A3.2') and (A3.3), we obtain for the case of multiple sources, single receptor:

$$\Xi_{r,t+1} = c\sum_{s}\Xi_{s,t}\left(\max_{s}(h_s)\right) + (1-c)\cdot\Xi_{r,t} \qquad\text{(A3.2'')}$$

**Rule 4 (law of pessimistic receptor).** Information arriving in a receptor from two different sources aggregate in an information that has the credibility equal to the level of the lower source credibility:

$$\Xi_{s1}(h_1) + \Xi_{s2}(h_2) = \Xi_s(\min(h_1,h_2)) \qquad\text{(A3.3')}$$

**Rule 5 (law of aggregation at the receptor).** Information arriving at the receptor is added to the existing information at the receptor site whenever non-contradictory, while it annihilates the existing information, when contradictory. In case of linguistic information, this can be written as:

$$\Xi_r = \Xi_{r0} + \Xi_s \qquad\text{(A3.4)}$$

In case of numerical information, aggregation operations that combine the numerical data should be used, according to the application in hand, for example a weighted sum, as:

$$n_r^{t+1} = \alpha\cdot n_r^t + (1-\alpha)\cdot n_s^t$$

Other numerical aggregations may be conceived as well.

### Annex 4. Statistical CMNs

In this annex, we introduce another class of CMNs, whose nodes operation is based on statistical operation. For brevity, we name a statistical CMN any CMN obeying the general equation:

$$x_i^{j+1} = f\left(x_{i-Q}^j, x_{i-Q+1}^j, \ldots, x_{i+Q}^j\right) \qquad\text{(A4.1)}$$

where $f(\ )$ is a statistical function. Notice that the equ. (A4.1) is identical with equ. (1), but we impose a restrictive condition to the function $f(\ )$. For example, the operation described by

$$x_i^{j+1} = \begin{cases} x_i^j & if \quad \left|x_i^j - \bar{x}\right| < \sigma \\ \bar{x} & else \end{cases} \qquad\text{(A4.2)}$$

where $\bar{x}$ is the average value in a vicinity of $x_i^j$, and $\sigma$ is the corresponding spreading, is a statistical operation. Such a CMN acts as a "sigma-filter".

**Example.** This is a good model for some societal games, when the actors keep their own opinions unchanged, except if the average opinion significantly differs from theirs. In the last case, many people tend to adopt the average opinion, according to (A4.2) or according to more intricate laws. More complex adaptations may be represented by:

$$x_i^{j+1} = \begin{cases} x_i^j & if \quad \left| x_i^j - \bar{x} \right| < \sigma \\ x_i^j + a \cdot \left( x_i^j - \bar{x} \right) & if \quad \sigma < \left| x_i^j - \bar{x} \right| < 2\sigma \\ \bar{x} & else \end{cases} \tag{A4.2}$$

where $a$ is a constant.

Also, the node operation described by:

$$x_i^{j+1} = median\left( x_{i+k}^j \right)\Big|_{k=-2Q..+2Q} \tag{A4.3}$$

represents a statistical operation. Notice that the weighted-average can also be seen as a statistical operation, along with formulas (A4.1-3).

*Systematic Organisation of Information in Fuzzy Systems*
*P. Melo-Pinto et al. (Eds.)*
*IOS Press, 2003*

161

# Intuitionistic Fuzzy Generalized Nets. Definitions, Properties, Applications

Krassimir T. ATANASSOV and Nikolai G. NIKOLOV
*Center for Biomedical Engineering, Bulgarian Academy of Sciences,*
*Bl. 105, Acad. G. Bonchev Str., Sofia-1113, BULGARIA*
*e-mails: {krat, shte}@bgcict. acad. bg*

**Abstract**. Generalized Nets (GNs) and intuitionistic fuzzy logic are briefly reminded. The concepts of Third and Fourth types of intuitionistic fuzzy GNs are defined and some of their properties and applications in abstract systems theory are discussed. It is stated that they are conservative extensions of the ordinary GNs.

## 1. Introduction

Generalized Nets (GNs) are extensions of Petri nets and Petri net modifications and extensions. They were defined in 1982 (see [1]). Their transitions have two temporal components (moment of transition firing and its duration), two indexed matrices (the $(i, j)$-th element of the first one is a predicate that determines whether a token from the $i$-th input place can be transferred to the $j$-th output place; the $(i, j)$-th element of the second one determines the capacity of the arc between these two places). GNs have three global time-components: moment of GN-activation, elementary time-step and duration of the GN-functioning. The GN-tokens enter the GN with initial characteristics and at the time of their transfer in the net they obtain next (current and final) characteristics. A large number of operations, relations and operators (global, local, dynamical, and others) are defined over the GNs.

Intuitionistic Fuzzy Sets (IFSs), defined in 1983, are extensions of fuzzy sets (see [2]). They have two degrees – degree of membership ($\mu$) and degree of non-membership ($\nu$) such that their sum can be smaller that 1, i.e., a third degree – of uncertainty ($\pi = 1 - \mu - \nu$) – can be defined, too. A variety of operations, relations and operators (from modal, topological and other types) are defined over the IFSs. These ideas are transferred also to Intuitionistic Fuzzy Logics (IFLs), which we shall discuss shortly below.

GNs have so far over 20 extensions. The first one, proposed in 1985 (see [3]), was called Intuitionistic Fuzzy GN (IFGN). The transition condition predicates of these nets are estimated in intuitionistic fuzzy sense. Later, this extension was called IFGN of type 1, because IFGN of a second type was defined. In it, the tokens were replaced by "quantities" that flew throughout the net. Now it was places, instead of tokens those obtain characteristics. For both extensions of the GNs (as it is done for all other GN-extensions) it is proved that they are conservative extensions of the ordinary GNs (for instance, e.g., of the GNs, which themselves are not conservative extensions of the ordinary Petri nets). Here a third and a fourth types of IFGNs will be defined and their properties will be discussed. Now, in addition to transition condition predicates and /or the form of the tokens being intuitionistic fuzzy, the tokens characteristics can be intuitionistic fuzzy, too. Below we discuss applications of the IFGNs from the three types in the areas of Artificial Intelligence, abstract systems theory, medicine, chemical industry and others.

## 2. Short introduction to Generalized Nets

Since 1983, more than 400 papers related to the concept of the Generalized Nets (GNs) have been published. A large part of them is included in the Petri Nets database, which can be consulted at the Internet address *www. daimi. aau. dk/PetriNets/bibl/aboutpnbibl. html.*

GNs are defined as extensions of the ordinary Petri nets and their modifications, but in a way that is in principle different from the ways of defining the other types of Petri nets. The additional components in the GN-definition provide more and greater modeling possibilities and determine the place of the GNs among the individual types of Petri nets, similar to the place of the Turing machine among the finite automata.

The first basic difference between GNs and the ordinary Petri nets is the "place – transition" relation [4]. Here, the transitions are objects of a more complex nature. A transition may contain $m$ input and $n$ output places where $m, n \geq 1$.

Formally, every transition is described by a seven-tuple (Figure 1):

$$Z = <L', L'', t_1, t_2, r, M, \square >$$

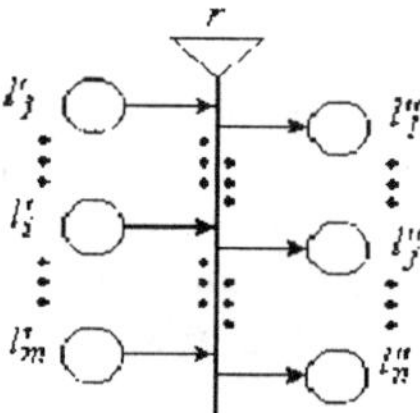

Figure 1.

where:

**(a)** $L'$ and $L''$ are finite, non-empty sets of places (the transition's input and output places, respectively); for the transition in Figure 1 these are $L' = \{l'_1, l'_2, ..., l'_m\}$ and $L'' = \{l''_1, l''_2, ..., l''_n\}$;

**(b)** $t_1$ is the current time-moment of the transition's firing;

**(c)** $t_2$ is the current value of the duration of its active state;

**(d)** $r$ is the transition's *condition* determining which tokens will pass (or *transfer*) from the transition's inputs to its outputs; it has the form of an Index Matrix (IM; see [5]):

$$R = \begin{array}{c|c} & L''_1 \ldots l''_j \ldots l''_n \\ \hline Ll'_1 & \\ \ldots & r_{i,j} \\ Ll'_i & (r_{i,j} - predicate) \\ \ldots & (1 \leq i \leq m, 1 \leq j \leq n) \\ Ll'_m & \end{array}$$

$r_{i,j}$ is the predicate that corresponds to the $i$-th input and $j$-th output places. When its truth value is **true**, a token from $i$-th input place transfers to $j$-th output place; otherwise, this is not possible;

**(e)** $M$ is an IM of the capacities of transition's arcs:

$$M = \begin{array}{c|c} & l''_1 \dots l''_j \dots l''_n \\ \hline L\,l'_1 & \\ l'i & \begin{array}{l} m_{i,j} \\ (m_{i,j} \quad 0 \text{ - } natural\ number) \\ (1 \quad i \quad m,\ 1 \quad j \quad n) \end{array} \\ l'_m & \end{array}$$

**(f)** $\square$ is an object of a form similar to a Boolean expression. It may contain as variables the symbols that serve as labels for transition's input places, and $\square$ is an expression built up from variables and the Boolean connectives $\wedge$ and $\vee$. When the value of a type (calculated as a Boolean expression) is **true**, the transition can become active, otherwise it cannot.

The ordered four-tuple

$$E = \ll A, \pi_A, \pi_L, c, f, \theta_1, \theta_2 >, < K, \pi_K, \theta_K >, < T, t^0, t^* >, < X, \Phi, b \gg$$

is called a *Generalized Net* (GN) if:

**(a)** $A$ is a set of transitions;

**(b)** $\pi_A$ is a function giving the priorities of the transitions, i. e., $\pi_A : A \to N$, where $N = \{0, 1, 2, \dots\} \cup \{1\}$

**(c)** $\pi_L$ is a function giving the priorities of the places, i.e., $\pi_L : L \to N$, where and $pr_i X$ is the $i$-th projection of the $n$-dimensional set, where $n \in N$, $n \geq 1$ and $1 \leq k \geq n$ (obviously, $L$ is the set of all GN-places);

**(d)** $c$ is a function giving the capacities of the places, i.e., $c : L \to N$;

**(e)** $f$ is a function that calculates the truth values of the predicates of the transition's conditions (for the GN described here let the function f have the value **false** or **true**, i.e., a value from the set $\{0, 1\}$;

**(f)** $\theta_1$ is a function giving the next time-moment when a given transition $Z$ can be activated, i.e., $\theta_1(t) = t'$, where $pr_3 Z = t, t' \in [T, T + t^*]$ and $t \leq t'$. The value of this function is calculated at the moment when the transition terminates its functioning;

**(g)** $\theta_2$ is a function giving the duration of the active state of a given transition $Z$, i.e., $\theta_2(t) = t'$, where $pr_4 Z = t \in [T, T + t^*]$ and $t' \geq 0$. The value of this function is calculated at the moment when the transition starts its functioning;

**(h)** $K$ is the set of the GN's tokens;

**(i)** $\pi_K$ is a function giving the priorities of the tokens, i.e., $\pi_K : K \to N$;

**(j)** $\theta_K$ is a function giving the time-moment when a given token can enter the net, i.e., $\theta_K(\alpha) = t$, where $\alpha \in K$ and $t \in [T, T + t^*]$;

**(k)** $T$ is the time-moment when the GN starts functioning. This moment is determined with respect to a fixed (global) time-scale;

**(l)** $t^0$ is an elementary time-step, related to the fixed (global) time-scale;

**(m)** $t^*$ is the duration of the GN functioning;

**(n)** $X$ is the set of all initial characteristics the tokens can receive on entering the net;

**(o)** $\Phi$ is a characteristic function that assigns new characteristics to every token when it makes the transfer from an input to an output place of a given transition;

**(p)** $b$ is a function giving the maximum number of characteristics a given token can receive, i.e., $b : K \to N$.

For example, if $b(\alpha) = 1$ for some token $\alpha$, then this token will enter the net with some initial characteristic (marked as its zero-characteristic) and subsequently it will keep only its current characteristic.

When $b(\alpha) = 1$, the token $\alpha$ will keep all its characteristics.

When $b(\alpha) = k < 1$, except its zero-characteristic, the token $\alpha$ will keep its last $k$ characteristics (characteristics older than the last k will be "forgotten").

Hence, in the general case, every token $\alpha$ has $b(\alpha) + 1$ characteristics on leaving the net.

A given GN may lack some of the above components. In these cases, any missing component will be omitted. GNs of this kind form a special class of GNs called *reduced GNs*.

The definition of a GN is more complex than the definition of a Petri net. Thus the algorithms of the tokens' transfer in the GNs are also more complex. On the other hand, as the GNs are more general, the algorithms for movement of tokens in the GN are more general than those of Petri nets. In a Petri net implementation, parallelism is reduced to a sequential firing of its transitions and the order of their activation in the general case is probabilistic or dependent on the transitions' priorities, if such exist. The GN's algorithms provide a means for a more detailed modeling of the described process. The algorithms for the token's transfers take into account the priorities of the places, transitions and tokens, i.e., they are more precise.

Operations and relations are defined as over the transitions, as well as over the GNs in general.

The operations, defined over the GNs – "union", "intersection", "composition" and "iteration" (see [1]) do not exist anywhere else in the Petri net theory. They can be transferred to virtually all other types of Petri nets (obviously with some modifications concerning the structure of the corresponding nets). These operations are useful for constructing GN models of real processes.

In [1] different properties of the operations over transitions and GNs are formulated and proved. Certain relations over transitions and GNs are also introduced there.

The idea of defining operators over the set of GNs in the form suggested below dates back to 1982 (see [1]). It is a proper extension of the idea of self-modifying Petri nets.

Now, the operator aspect has an important place in the theory of GNs. Six types of operators are defined in its framework. Every operator assigns to a given GN a new GN with some desired properties. The comprised groups of operators are:

- global ($G$-) operators,
- local ($P$-) operators,
- hierarchical ($H$-) operators,
- reducing ($R$-) operators,
- extending ($O$-) operators,
- dynamic ($D$-) operators.

The *global operators* transform, according to a definite procedure, a whole given net or all its components of a given type. There are operators that alter the form and structure of the transitions ($G_1$, $G_2$, $G_3$, $G_4$, $G_6$), temporal components of the net ($G_7$, $G_8$); the duration of its functioning ($G_9$), the set of tokens ($G_{10}$), the set of the initial characteristics ($G_{11}$); the characteristic function of the net ($G_{12}$) (this function is the union of all places' characteristic functions); the evaluation function ($G_{13}$), or other net's functions ($G_5$, $G_{14}$, ..., $G_{20}$).

One of the global operators can collapse a given GN to a GN-transition ($G_2$). Another operator ($G_4$) adds two special places and, connected to two special transitions of the GN: a *general input place*, where all tokens enter the net and are later distributed among the net's actual input places; and a *general output place* that collects all tokens leaving the GN from their respective output places.

Another global operator ($G_3$) transforms a given GN after its functioning so that it removes all tokens which have not participated in the process and all places which have not been visited by tokens. The new net has the same functional behavior as the original one, however, all its tokens and places are actually involved in the modeled process.

Some global operators ($G_5$, $G_{12}$, ..., $G_{20}$) alter the different (global) functions defined on the net.

The second type of operators is *local operators*. They transform single components of some of the transitions of a given GN. There are 3 types of local operators:

- temporal ($P_1$, $P_2$, $P_3$, $P_4$), that change the temporal components of a given transition,
- matrix ($P_5$, $P_6$), that change some of the index matrices of a given transition,
- other operators: these alter the transition's type ($P_7$), the capacity of some of the places in the net ($P_8$) or the characteristic function of an output place ($P_9$), or the evaluation function associated with the transition condition predicates of the given transition ($P_{10}$).

For any of these operators, a continuation ($P_i, 1 \leq i \leq 10$) to a global one ($\overline{P_i}, 1 \leq i \leq 10$) can be made by defining the corresponding operator in such a way that it would transform all components of a specified type in every transition of the net.

The third type of operators are the *hierarchical operators*. These are of 5 different types and fall into two groups according to their way of action:

- expanding a given GN ($H_1$, $H_3$ and $H_5$),
- shrinking a given GN ($H_2$, $H_4$ and $H_5$).

The $H_5$ operator can be expanding as well as shrinking, depending on its form;

According to their object of action the operators fall again into two groups:

- acting upon or giving as a result a place ($H_1$ and $H_2$),
- acting upon or giving as a result a transition ($H_3$, $H_4$ and $H_5$).

The hierarchical operators $H_1$ and $H_3$ replace a given place or transition, respectively, of a given GN with a whole new GN. Conversely, operators $H_2$ and $H_4$ replace a part of a given GN with a single place ($H_2$) or transition ($H_4$). Finally, operator $H_5$ changes a subnet of a given GN with another subnet. Expanding operators can be viewed as tools for magnifying the modeled process' structure; while shrinking operators – as a means of integration and ignoring the irrelevant details of the process.

The next (fourth) group of operators defined over the GNs produces a new, reduced GN from a given net. They would allow the construction of elements of the classes of reduced GNs. To find the place of a given Petri net modification among the classes of reduced GNs, it must be compared to some reduced GN obtained by an operator of this type. These operators are called *reducing operators*.

Operators from the fifth group extend a given GN. These operators are called *extending operators*. The extending operators are associated with every one of the GN extensions.

Finally, the operators from the last sixth group are related to the ways the GN functions, so that they are called *dynamic operators*. These are the following:

- operators $D(1, i)$ that determine the procedure of evaluating the transition condition predicates ($1 \leq i \leq 18$).

- operators governing token splitting: one that allows ($D(2,1)$) and one that prohibits splitting ($D(2,2)$), respectively; and operators governing the union of tokens having a common predecessor: an allowing one ($D(2,4)$) and a prohibiting one ($D(2,...3)$);

- operators that determine the strategies of the tokens transfer: one by one at a time vs. all in groups (the operator $D(3,2)$; the operator $D(3,1)$ does not allow this);

- operators related to the ways of evaluating the transition condition predicates: predicate checking ($D(4,1)$); changing the predicates by probability functions with corresponding forms ($D(4,2)$); expert estimations of predicate values ($D(4,3)$); predicates depending on solutions of optimization problems (e.g., transportation problem) ($D(4,4)$).

The operators of different types, as well as the others that can be defined, have a major theoretical and practical value. On the one hand, they help us study the properties and the behavior of GNs. On the other hand, they facilitate the modeling of many real processes. The basic properties of the operators are discussed in [1][6]. "*A Self-Modifying GN*" (SMGN) is constructed and described in details in [1]. The SMGNs has the property of being able to alter its structure (number of transitions, places, tokens, transition condition predicates, token characteristics, place and arc capacities, etc.) and the token transfer strategy during the time of the GN-functioning. These changes are done by operators that can be defined over the GN. Of course, not all operators can be applied over a GN during the time of its functioning. Some of them are only applicable before, and others – after the GN-functioning. The necessary conditions for an operator to be applicable during the time of the GN-functioning are discussed in [1].

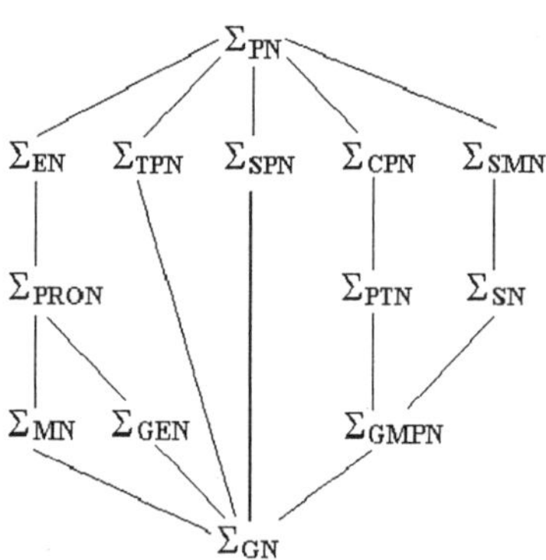

Figure 2.

The relations between the GNs and the most important classes of Petri net modifications are shown on Figure 2, where $\Sigma_{PN}$, $\Sigma_{EN}$, $\Sigma_{TPN}$, $\Sigma_{SPN}$, $\Sigma_{CPN}$, $\Sigma_{SMN}$, $\Sigma_{PRON}$, $\Sigma_{PTN}$, $\Sigma_{SN}$, $\Sigma_{MN}$, $\Sigma_{GEN}$, $\Sigma_{GMPN}$ and $\Sigma_{GN}$ are the respective classes of all ordinary Petri nets, all E-nets, all time Petri nets, all stochastic Petri nets, all color Petri nets, all self-modifying Petri nets, all PRO-nets, all predicative-transition nets, all super nets, all M-nets, all generalized E-nets, all generalized modified Petri nets and all GNs.

## 3. The Basics of Intuitionistic Fuzzy Logic

To each proposition (in the classical sense) we can assign its truth-value: truth – denoted by 1, or falsity – 0. In the case of fuzzy logic this truth-value is a real number in the interval [0, 1] and may be called "truth degree" of a particular proposition. Here we add one more value – "falsity degree" – that will be in the interval $[0,1]$ as well. Thus two real numbers, $\mu(p)$ and $v(p)$, are assigned to the proposition $p$ with the following constraint to hold (see [2]):

$$\mu(p)+v(p)\le 1.$$

Let this assignment be provided by an evaluation function $V$ defined over a set of propositions $S$ in such a way that:

$$V(p)=<\mu(p),v(p)>.$$

The evaluation of the negation $\neg p$ of the proposition $p$ will be defined as:

$$\neg V(p)=V(\neg p)=<v(p),\mu(p)>.$$

When the values $V(p)$ and $V(q)$ of the propositions $p$ and $q$ are known, the evaluation function V can be extended also for the operations "& ", "$v$" through the definition:

$$V(p)\&V(q)=V(p\&q)=<\min(\mu(p),\mu(q)),\max(v(p),v(q))>,$$

$$V(p)\lor V(q)=V(p\lor q)=<\max(\mu(p),\mu(q)),\min(v(p),v(q))>$$

$$V(p)\to V(q)=V(p\to q)=<\max(v(p),\mu(q)),\min(\mu(p),v(q))>.$$

For the needs of the discussion below we shall define the notion of *intuitionistic fuzzy tautology* (IFT) through:

$$\text{"}A \text{ is an IFT" } \textbf{iff} \ a\ge b, \text{ where } V(A)=<a,b>.$$

Shortly, the quantifiers are represented by

$$V(\forall xA)=<\min_{x}\mu(A),\ \max_{x}v(A)>,$$

$$V(\exists xA)=<\max_{x}\mu(A),\ \min_{x}v(A)>$$

It is very interesting to note that the intuitionistic fuzzy quantifier interpretations coincide with the intuitionistic fuzzy interpretations of the two topological operators "interior" and "closure", respectively.

Let $A$ be a fixed propositional form for which

$$V(A) = <a,b>.$$

The two basic modal operators can be defined as follows:

$$V(\Box A) = <a, 1-a>,$$

$$V(\Diamond A) = <1-b, b>$$

and they can be extended for every two á, â $\in$ [0, 1]:

$$V(D_\alpha(A)) = <a+\alpha\cdot(1-a-b), b+(1-\alpha)\cdot(1-a-b)>,$$

$$V(F_{\alpha,\beta}(A)) = <a+\alpha\cdot(1-a-b), b+\beta\cdot(1-a-b)>, \text{ for } \alpha+\beta\leq 1,$$

$$V(G_{\alpha,\beta}(A)) = <\alpha\cdot a, \beta\cdot b>,$$

$$V(H_{\alpha,\beta}(A)) = <\alpha\cdot a, b+\beta\cdot(1-a-b)>,$$

$$V(H_{\alpha,\beta}*(A)) = <\alpha\cdot a, b+\beta\cdot(1-\alpha\cdot a-b)>,$$

$$V(J_{\alpha,\beta}(A)) = <a+\alpha\cdot(1-a-b), \beta\cdot b>,$$

$$V(J_{\alpha,\beta}*(A)) = <a+\alpha\cdot(1-a-\beta\cdot b), \beta\cdot b>.$$

Level, temporal and other operators are defined in IFL, too (see [2]).

### 4. Short Remarks on the Extensions of Generalized Nets

First, we shall note that the GNs have more than 20 conservative extensions (i.e., extensions, the functioning and the results of working of each of which can be represented by an ordinary GN).

The most important of them are:

- Intuitionistic fuzzy GNs of type 1: their transition condition predicates are evaluated into the set $[0,1]^2$ with a degree of truth ($\mu$) and a degree of falsity ($\nu$) for which $\mu+\nu\leq 1$ (see [3]);

- Intuitionistic fuzzy GNs of type 2 are intuitionistic fuzzy GNs of type 1, which have "quantities" instead of tokens, which "flow" from input to output GN places;

- Color GNs are GNs whose tokens and transition arcs are painted in different colors and the tokens transfer depends on these colors;

- GNs with global memory have a global tool (i.e., common for all GN-components) for keeping data (during the functioning of the GN) or for determining of the values of various parameters related to the modeled processes;

- GNs with optimization components use optimization procedures for determining of the way tokens transfer from transition input to output places;

- GNs with additional clocks comprise tools for taking into account the duration of the predicates' truth-values evaluation; should this process take up more time than a specified limit, it is stopped and the predicate is assigned its previous value determining the future functioning of the net;

- GNs with complex transition types have additional conditions for activation of the GN transition types;

- GNs with stop-conditions have additional conditions for termination of the GN functioning;

- Opposite GNs – their tokens transfer (within a separate GN area) in the opposite direction, i.e., from output to input places. This enables searching for necessary conditions for the desired flow of a given process;

- GNs where tokens can obtain variables as characteristics can be used for solving optimization problems related to the processes modeled by the GNs.

## 5. Definitions of Two New Types of Intuitionistic Fuzzy Generalized Nets

Here we shall introduce two new types of IFGNs, to be called **Third type of IFGNs (IFGN3s)** and **Fourth type of IFGNs (IFGN4s)**. They will be extensions of IFGN1s and IFGN2s, respectively. The idea is generated on the basis of the idea of an abstract system with properties (see next chapter). Now, the token characteristics will be estimated in intuitionistic fuzzy sense and only when their estimations (intuitionistic fuzzy tuples) satisfy the definition of intuitionistic fuzzy tautology, the tokens will obtain characteristics. Therefore, the two new types of GNs allow for describing situations where the model determines its status with degrees of validity and of non-validity.

Every IFGN3 has the form:

$$E = << A, \mu_A, \mu_L, c, f, \theta_1, \theta_2 >, < K, \pi_K, \theta_K >, < T, t^0, t^* >, < X, \Phi, b >>$$

where the elements of the set $A$ (of the IFGN3-transitions) are the same as the GN-transitions. All other components, without the components $f$ and $\Phi$ are also the same. The function $\Phi$ (as in the case of the IFGN1s) gives to each token (let us denote it by $\alpha$) as a current characteristic (let us mark it as $x_{CU}^\alpha$ two values: the first coincides with the token characteristic in the sense of the GN's (let us denote it by $x_{CU}^\alpha$; the second value is an ordered tuple with real numbers, each of which is an element of the set $[0, 1]$. They are equal to the truth values of the predicate of the transition condition between the place whence the token $\alpha$ transfer starts and the place where this transfer ends. The function f calculates these two values of the corresponding predicate $r_{ij}$ of the transition condition index matrix in the form

$$f(r_{ij}) = < \mu(r_{ij}), \nu(r_{ij}) >,$$

where $\mu(r_{ij})$ is the degree of truth of the predicate $r_{ij}$, $\nu(r_{ij})$ is its degree of false, and:

$$\mu\left(r_{ij}\right)+\nu\left(r_{ij}\right)\leq 1 .$$

Therefore, $x_{CU}^{\alpha}=<\overline{x}_{CU}^{\alpha},\mu\left(r_{ij}\right),\nu\left(r_{ij}\right)>$ .

For instance in IFGN1s, function $\Phi$ is estimated too, i.e., now there are two real numbers $\mu\left(x_{CU}^{\alpha}\right)$ and $\nu\left(x_{CU}^{\alpha}\right)$, such that $\mu\left(x_{CU}^{\alpha}\right)$, $\nu\left(x_{CU}^{\alpha}\right)\in[0,1]$ and $\mu\left(x_{CU}^{\alpha}\right)+\nu\left(x_{CU}^{\alpha}\right)\leq 1$ .

The following two theorems hold.

**THEOREM 1:** The functioning and the results of the work of every IFGN3 can be described by an ordinary GN.

**THEOREM 2:** The set of all IFGN3 $\Sigma_{IFGN3}$ is a conservative extension of the set of all ordinary GNs $\Sigma_{GN}$ .

Now, we proceed to define the concept of "**Fourth type of IFGN**" (**IFGN4**). In this net, tokens are some "quantities" moving throughout the net.

The values of the transition condition's predicates can be intuitionistic fuzzy, i.e. they can have degrees of truth and of false, as in the cases of IFGN1s and IFGN3s. The IFGN2 has the form

$$E=<< A,\mu_{A},\mu_{L},c,f,\theta_{1},\theta_{2}>,< K,\pi_{K},\theta_{K}>,< T,t^{0},t^{*}>,< \Phi,b >>$$

where $A$ is the set of the net's transitions which have the ordinary GN-form and there exists only one difference: here the index matrix $M$ contains as elements real numbers – capacities of the transition's arcs.

The functions in the first component are similar to those of IFGN1 and they satisfy the same conditions.

The essential difference between IFGN2s and IFGN4s on the one hand, and the rest of GNs, on the other, is the set $K$ and the functions related to it. Now the elements of $K$ are some "quantities", that do not receive characteristics. Function $è_{K}$ gives the time-moment when a given token will enter the net as in the ordinary GNs.

The temporal components are also as in the other GN-types.

For both types of GNs – IFGN2s and IFGN4s, function $\underline{\Phi}$ has new meaning. Now it is assigns the places characteristics (the quantities of the tokens from the different times in the corresponding places). As in GNs, it can be extended: it can give also other data about the modeled process (e.g. the time-moments for the entry of the "quantities" in the places).

For instance, in IFGN2s and similarly to the IFGN3s, function $\Phi$ in the IFGN4s is estimated too, i.e., as above, for each token $\alpha$ this function gives two real numbers $\mu\left(x_{CU}^{\alpha}\right)$ and $\nu\left(x_{CU}^{\alpha}\right)$, such that $\mu\left(x_{CU}^{\alpha}\right)$, $\nu\left(x_{CU}^{\alpha}\right)\in[0,1]$ and $\mu\left(x_{CU}^{\alpha}\right)+\nu\left(x_{CU}^{\alpha}\right)\leq 1$ .

The following two theorems hold:

**THEOREM 3:** The functioning and the results of the work of every IFGN4 can be described by an ordinary GN.

**THEOREM 4:** The set of all IFGN4 $\Sigma_{IFGN4}$ is a conservative extension of the set of all ordinary GNs $\Sigma_{GN}$.

## 6. On the Concept of Intuitionistic Fuzzy Abstract System

In a series of eight papers, collected in [6], the author shows the possibilities for GN-interpretations of abstract systems theory elements, using notations from [7, 8, 9, 10].

Let a family $\overline{V} = \{V_i \mid i \in I\}$ of sets be given, where $I$ is an index set. A (general) system is a relation on nonempty (abstract) sets (cf. [9]):

$$S \subset \prod_{i \in I} V_i$$

A component set $V_i (i \in I)$ is referred to as a system object. Let $I_x \subset I$, $I_y \subset I$, $I_x \cap I_y = \varnothing$, $I_x \cup I_y = I$. The sets

$$X = \prod_{i \in I_x} V_i \text{ and } Y = \prod_{i \in I_y} V_i$$

are named the input and the output objects of the system, respectively. The system $S$ is then

$$S \subset X \times Y$$

and will be referred to as an input-output system. If $S$ is interpreted as a function

$$S : X \to Y$$

it is referred to as a function-type (or functional) system.

Given a general system S, let C be an arbitrary set and R a function,

$$R : (C \times X) \to Y$$

such that

$$(x, y) \in S \ \textit{iff} \ (\exists c)(R(c, x) = y),$$

$C$ is then a global state object or set, its elements being global states, while $R$ is a global (systems)-response function (for $S$).

Some GN-interpretations of such described system are discussed in [6].

In [11] two new extensions of the concept of an abstract system were introduced. The first of them is called *an Abstract System with Properties (ASP)*. It is an abstract system, which input(s), output(s), global state object(s), time component(s) and response function(s) have additional properties.

Let $X$ be one of the above system components (inputs, outputs, etc.) and let $P(X)$ be the list of its properties. Of course, this list may be empty or have just one element (property). Let $CP(X)$ be the cardinality of the set $P(X)$. Let **P** be the set of the different properties the systems components may have. If we want to discuss the form of this set from the standpoint of the NBG-axiomatic set theory, **P** may be a class.

Let a family $\overline{V} = \{V_i \mid i \in I\}$ of sets be given, where $I$ is an index set. An ASP is a relation on nonempty (abstract) sets

$$S \subset \left( \prod_{i \in I} \left( V_i \times P^{CP(V_i)} \right) \right) \times P^{CP(\sigma)} \quad (*)$$

The last member of (*) shows that the system $S$ can have specific properties, not representable by the other properties, related to the individual system components. Of course, for the system, these properties will have a global nature.

As above, let $C$ be a global state object for a given ASP $S$ and $R$ be a global (system)-response function (for $S$). Now, the form of $R$ is: $R : (C \times (P \times X \times P) \to Y \times P$, such that

$$(x, P(x), y, P(y)) \in S \quad iff \quad (\exists c)(c, P(c), x, P(x)) = (y, P(y)) \,\&\, P(R).$$

Now, the set of time-moments $T$ can also be characterized with some properties.

Component $P(T)$ can be interpreted, e.g., as the condition that set $T$ be discrete or continuous, finite or infinite, etc.

In [11], a GN is constructed that describes the functioning of the results of the work of an ASP. Now we can use the additional conveniences that the IFGN3 gives.

In [11], the concept of an Intuitionistic Fuzzy Abstract Systems with Properties (IFASP) is being introduced, too. Now we have intuitionistic fuzzy estimations not only of the system components behavior, but also of the validity of the system components properties. Now, the form of the new system is

$$S \subset \left( \prod_{i \in I} \left( V_i \times [0,1]^2 \times \left( P \times [0,1]^2 \right)^{CP(V_i)} \right) \right) \times [0,1]^2 \times \left( P \times [0,1]^2 \right)^{CP(\sigma)}$$

We can extend the above model in the following sense. The ASP works on input data $X$ with properties $P(X)$ and with global state $C$ with properties $P(C)$ at time moment $t$ with properties $P(t)$. The global results of the work of system $S$ can be estimated with the help of two degrees – for example, a degree of "goodness" $g_S$ and a degree of "badness" $b_S$, so that $g_S, b_S \in [0, 1]$ and $g_S + b_S \leq 1$. It is suitable to define that the truth-value of $P(t)$ is

$$V(P(t)) = < g_S b_S >.$$

The behavior of the system will depend on the values of $g_S$ and $b_S$. We can determine two intervals $[\alpha_S, \beta_S] \in [0,1]$ and $[\gamma_S, \delta_S] \in [0,1]$, so that the system behavior will be:

- *passable* if $g_S \geq \beta_S$ and $b_S \leq \delta_S$
- *insufficient* if $g_S \leq \alpha_S$ and $b_S \geq \delta_S$
- *allowable* if $g_S \geq \alpha_S$ or $b_S \leq \delta_S$.

In some particular case, the values of $\alpha$, $\beta$, $\Upsilon$ and $\delta$ will be determined by different ways.

For example, we can assert that the system works *surely* if $g_S > \dfrac{1}{2} > b_S$.

**THEOREM 5:** There exists an IFGN3 that is universal for the set of all IFASPs. i.e., that can represent the functioning and the results of the work of each IFASP.

**THEOREM 6:** There exists a GN universal for the set of all IFASPs, i.e., that can represent the functioning and the results of the work of each IFASP.

## 7. On the Applications of GN, IFGN3, IFGN4, IFASP

Up to now, Generalized nets (ordinary GNs and IFGNs) were applied to modeling of real processes in the medical diagnosis, chemical engineering, economics, computer architectures and others (see, e.g. [12, 13, 14, 15, 16, 17]). Already, the possibility for using the GNs for modeling of Artificial Intelligence objects (e.g., expert systems, neural networks, semantic networks, genetic algorithms and others) or processes (e.g., of machine learning, decision making and others) was discussed (see, e.g., [18, 19, 20, 21, 22, 23, 24, 25, 26]).

In [18] it is shown that the functioning and the results of the work of each production-based expert system can be represented by a GN. The discussed expert systems can have facts and rules with priorities, metafacts, temporal fact parameters, the estimations of the truth values of the hypotheses can be represented in intuitionistic fuzzy forms. There it is shown that there exists a GN, universal for all expert system from each type, which represents these expert systems.

Also, the processes of machine learning of abstract objects, neural networks, genetic algorithms, intellectual games, GNs, IFGNs, abstract systems and others, and the results of these processes can be represented by universal GNs (see [19]).

There are separate attempts for GN-description of the functioning and the results of the work of neural networks, processes related to pattern and speech recognition and others (see e.g., [27, 28, 29]).

In [30, 31] some processes related to human body and brain are discussed.

What is common in all these research is the fact that in each of the so constructed GN-models there are subnets that model the informational processes of the modeled objects. The GNs from their separate types (ordinary GNs, IFGN1s, etc.) contain static and dynamic elements as all other types of Petri nets, but for instance of the latter nets, the GN-tokens have characteristics that can be used for modeling of information, related or generated in/by the modeled processes. This information can be represented by the respective characteristic functions associated to the GN-places. They can be:

(a) calculated in the frameworks of the GNs;
(b) obtained from external sensors through suitable interface;
(c) obtained from suitable expert or information systems;
(d) determined by specialists;
(e) calculated by external software (e.g., "Mathematica", "MatLab", "Maple", etc).

Practically, all information that can be calculated in abstract systems can be represented by GN-token characteristics. The real processes, no matters that they are abstract ones (functioning of expert systems, machine learning, human body, etc.) or real ones (chemical plants, medical diagnosis, economics, etc.), submit to logical conditions that determine the ways of their flowing.

These conditions, especially when they are complex enough, cannot be represented by the other types of Petri nets and by ordinary systems in the sense of systems theory. The GN-transition condition predicates can exactly determine the individual logical conditions and therefore, the GN-models can adequately represent both the functional and the logical sides of the modeled processes. In addition, we can construct different GN-models at a single or multiple levels of hierarchy that represent separate parts of the modeled processes. Using the operations and the operators (and especially the hierarchical ones) over GNs, we can represent in suitable level of detailing the objects of modeling. For example, if we wish to model a conversation of two persons by the means of the Artificial Intelligence, we need to use methods from the means of the pattern and speech recognition, pattern and speech generation; in some cases – translation. For each of these areas of the AI there are separate mathematical formalisms. They are close, but different. Therefore, we cannot describe by the means of the AI the complex process of communication. If we construct GN-models of all mentioned four or five processes, we can construct a common GN-model that will represent the modeled process in its entirety. This is an example illustrating the thesis, discussed in [18] that it will be convenient for all areas of the AI to have GN-interpretations.

The example shows the fact that GNs can be used as a suitable means for modeling of information processes. The fact that the items of information can be fuzzy is represented by the means of GNs, IFGN1s and IFGN3s. If we would like to interpret the information of a continued "brook", then we can use IFGN2s and IFGN4s.

### References


[1]  K. Atanassov, Generalized Nets. World Scientific, Singapore, New Jersey, London, 1991.

[2]  K. Atanassov, Intuitionistic Fuzzy Sets. Springer-Verlag, Heidelberg, 1999.

[3]  K. Atanassov, Generalized nets and their fuzzings. *AMSE Review*, **2** (1985), No. 3, 39-49.

[4]  P. Starke, Petri-Netze. Berlin, VEB Deutscher Verlag der Wissenschaften, 1980.

[5]  K. Atanassov, Generalized index matrices. *Compt. Rend. de l'Academie Bulgare des Sciences*, **40**, 1987, No 11, 15-18.

[6]  K. Atanassov, Generalized Nets and Systems Theory. "Prof. M. Drinov" Academic Publishing House, Sofia, 1997.

[7]  M. Mesarovic and Y. Takahara, General system theory: mathematical foundations, Academic Press, New York, 1975.

[8]  M. Mesarovic and Y. Takahara, Abstracts systems theory, *Lecture Notes in Control and Information Sciences*, Vol. **116**, 1989.

[9]  M. Mesarovic, D. Macko and Y. Takahara, Theory of hierarchical, multilevel systems, Academic Press, New York, 1970.

[10] J. Schreider and A. Sharov, Systems and Models, Moskow, Radio I Svjaz, 1982 (in Russian).

[11] K. Atanassov, Generalized net aspects of systems theory. *Advanced Studies in Contemporary Mathematics,* **4**, 2001, No. 1, 73-92.

[12] K. Atanassov, (Ed.), Applications of generalized nets. World Scientific, Singapore, New Jersey, London, 1993.

[13] K. Atanassov, M. Daskalov, P. Georgiev, S. Kim, Y. Kim, N. Nikolov, A. Shannon and J. Sorsich, Generalized Nets in Neurology. "Prof. M. Drinov" Academic Publishing House, Sofia, 1997.

[14] D. Dimitrov, The Paretian liberal in an intuitionistic fuzzy context, Ruhr-Universitat Bochum Discussion Paper No. 02-00, April 2000.

[15] D. Dimitrov, Intuitionistic fuzzy preferences: A factorization, Ruhr-Universitat Bochum Discussion Paper No. 05-00, June 2000.

[16] A. Shannon, J. Sorsich and K. Atanassov, Generalized Nets in Medicine. "Prof. M. Drinov" Academic Publishing House, Sofia, 1996.

[17] A. Shannon, J. Sorsich, K. Atanassov, N. Nikolov and P. Georgiev, *Generalized Nets in General and Internal Medicine.* "Prof. M. Drinov" Academic Publishing House, Sofia, Vol. **1**, 1998; Vol. **2**, 1999; Vol. **3**, 2000.

[18] K. Atanassov, *Generalized Nets in Artificial Intelligence*: Generalized nets and Expert Systems, Vol. **1**, "Prof. M. Drinov" Academic Publishing House, Sofia, 1998.

[19] K. Atanassov and H. Aladjov, *Generalized Nets in Artificial Intelligence*: Generalized nets and Machine Learning, Vol. **2**, "Prof. M. Drinov" Academic Publishing House, Sofia, 2001.

[20] R. K. De Biswas and A. Roy, Multicriteria decision making using intuitionistic fuzzy set theory, *The Journal of Fuzzy Mathematics,* **6**, 1998, No. 4, 837-842.

[21] N. Nikolov, Generalized nets and semantic networks, *Advances in* Modeling *& Analysis,* AMSE Press, **27**, 1995, No. 1, 19-25.

[22] E. Szmidt and J. Kacprzyk, Intuitionistic fuzzy sets for more realistic group decision- making, *Proceedings of International Conference "'Transitions to Advanced Market Institutions and Economies",* Warsaw, June 18-21, 1997, 430-433.

[23] E. Szmidt and Kacprzyk J., Applications of intuitionistic fuzzy sets in decision making, *Actas del VIII Congreso Espanol sobre Technologias y Logica Fuzzy,* Pamplona, 8-10 Sept. 1998, 143-148.

[24] E. Szmidt and J. Kacprzyk, Group decision making under intuitionistic fuzzy preference relations, *Proceedings of Seventh International Conference "Information Processing and Management of Uncertainty in Knowledge-Based Systems",* Paris, July 6-10, 1998, 172-178.

[25] E. Szmidt, J. Kacprzyk, Decision making in an intuitionistic fuzzy environment, *Proceedings of EUROFUSE-SIC'99,* Budapest, 25-28 May 1999, 292-297.

[26] E. Szmidt, Application of Intuitionistic Fuzzy Sets in Decision-Making. DSc. Thesis, Technical University, Sofia, June 2000.

[27] G. Gluhchev, K. Atanassov and S. Hadjitodorov, Handwriting analysis via generalized nets. *Proceedings of the international Scientific Conference on Energy and Information Systems and Technologies.* Vol. **III**, Bitola, June 7-8, 2001, 758-763.

[28] S. Hadjitodorov, An intuitionistic fuzzy version of the nearest prototype classification method, based on a moving-pattern procedure. *Int. J. General Systems,* **30**, 2001, No. 2, 155-165.

[29] S. Hadjitodorov and N. Nikolova, Generalized net model of the self-organizing map of Kohonen classical training procedure, *Advances in* Modeling *& Analysis,* B, **43**, 2000, No. 1-2, 51-60.

[30] V. Sgurev, G. Gluhchev and K. Atanassov, Generalized net interpretation if information process in the brain. *Proc. of the International Conference "Automatics and Informatics",* 31 May - 2 June 2001, Sofia, A-147-A-150.

[31] J. Sorsich, A. Shannon and K. Atanassov, A global generalized net model of the human body. *Proc. of the Conf. "Bioprocess systems'2000",* IV. I-IV, **4**, Sofia, 11-13 Sept. 2000.

# Part III

# Techniques

# Dynamical Fuzzy Systems with Linguistic Information Feedback

X. Z. GAO and S. J. OVASKA
*Institute of Intelligent Power Electronics*
*Helsinki University of Technology*
*Otakaari 5 A, FIN-02150 Espoo, Finland*

Abstract. In this paper, we propose a new dynamical fuzzy system with linguistic information feedback. Instead of crisp system output, the delayed conclusion fuzzy membership function in the consequence is fed back locally with adjustable feedback parameters in order to overcome the static mapping drawback of conventional fuzzy systems. We give a detailed description of the corresponding structure and algorithm. Our novel scheme has the advantage of inherent dynamics, and is therefore well suited for handling temporal problems like dynamical system identification, prediction, control, and filtering. Simulation experiments have been carried out to demonstrate its effectiveness.

## 1. Introduction

Due to its great flexibility in coping with ill-defined processes, fuzzy logic theory has found numerous successful applications in industrial engineering, e.g., pattern recognition [1], process automation [2], and fault diagnosis [3]. Generally, a fuzzy logic-based system with embedded linguistic knowledge maps an input data (feature) vector into a scalar (conclusion) output [4]. Most fuzzy systems applied in practice are *static*. In other words, they lack the internal dynamics, and can thus only realize nonlinear but non-dynamic input-output mappings. This disadvantage hinders their wider employment in such areas as dynamical system modeling, prediction, filtering, and control. In order to introduce advantageous dynamics into regular fuzzy systems, we propose a new kind of dynamical fuzzy model with *linguistic* information feedback in this paper. The goal of the present paper is to introduce the idea of linguistic information feedback and illustrate the principal characteristics of the corresponding dynamic fuzzy system.

Our paper is organized as follows: necessary background knowledge of classic fuzzy logic systems is first presented in Section 2. In the following section, we discuss two conventional recurrent fuzzy systems with *crisp* output feedback and local memory nodes, respectively. Next we introduce our linguistic information feedback-based fuzzy system in Section 4. Both the structure and algorithm of this dynamical fuzzy system are then presented in details. Computer simulations are made to illustrate its characteristics in Section 5. Finally in Section 6, a conclusion and some remarks are given.

## 2. Fuzzy Logic-based Systems

The basic knowledge of fuzzy systems is provided in this section. In general, a fuzzy logic-based system consists of four essential components [5]: a fuzzification interface, a fuzzy rule base, an inference engine, and a defuzzification interface, as shown in Fig. 1. The fuzzification interface converts crisp inputs into corresponding fuzzy values. It is well known that linguistic variable [6] is a core concept in fuzzy logic theory, and it takes fuzzy variables as its values. A linguistic variable can be represented by a quintuple $\{x, T(x), U, G, M\}$, in which $x$ is the name of the variable; $T(x)$ is the term set of $x$, i.e., the set of names of linguistic values of $x$ with each value being a fuzzy number defined in $U$—universe of discourse; $G$ is a syntactic rule for generating the names of the values of $x$; and $M$ is a semantic rule for associating each value with its meaning. Actually, a fuzzy number is a normal and convex fuzzy set in $U$. It can be characterized by a membership function $\mu_F$, which usually takes values within the interval of [0, 1]:

$$\mu_F : U \rightarrow [0,1]. \tag{1}$$

Thus, a fuzzy set $F$ in $U$ may be described as a group of ordered pairs of a generic element $u$ and its grade of membership function $\mu_F(u)$ :

$$F = \{u, \mu_F(u)\}, \quad u \in U. \tag{2}$$

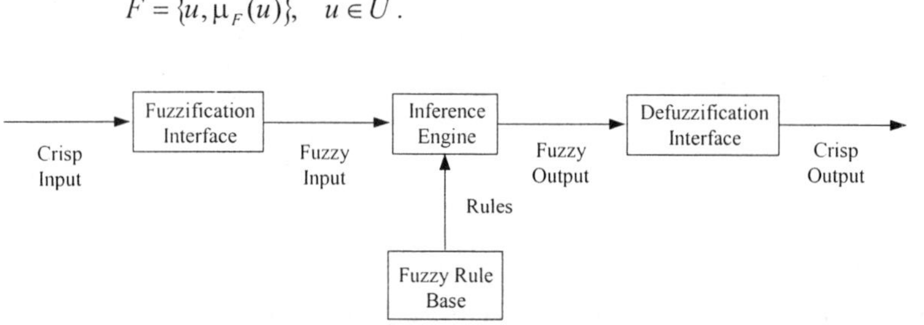


Figure 1. Basic architecture of a fuzzy logic system.

The fuzzy rule base is a collection of embedded fuzzy rules that are normally expressed by the statements of *IF-THEN*. The inference engine is the kernel of fuzzy systems. It has the capability of performing human-like decision-making, and inferring conclusion output by employing a fuzzy implication operation, such as the well-known Max-Min operator [6]. Since the output of fuzzy reasoning is usually a membership function, in most cases, the defuzzification interface is needed to extract a crisp value best representing the possibility distribution of the inferred fuzzy action for driving external plants. There are three commonly applied defuzzification methods in engineering: the Max Criterion, Centroid of Area (COA), and Mean of Maximum (MOM). More details of the working principles and performance comparisons of these defuzzification strategies can be found in [5].

In a fuzzy logic-based system, its behavior is characterized by a set of linguistic rules, created often from expert knowledge. These IF-THEN fuzzy rules usually have the following forms (a two-input-single-output case):

IF $x$ is $A$ AND (OR) $y$ is $B$ THEN $z$ is $C$,

where $x$, $y$, and $z$ are the input and output linguistic variables represented by fuzzy sets, respectively. The IF and THEN portions of the above fuzzy rules are referred to be the premise as well as consequent parts of a fuzzy system. Since fuzzy rules are formulated in linguistic rather than numerical terms, it is very difficult to acquire them directly from practical measurement data, although there exists already some preliminary approaches [7]. Typically, operator's heuristic experience and engineering knowledge play a pivotal role in the derivation of fuzzy rules. In a word, the design of a typical fuzzy system is quite subjective, and it depends heavily on the *trial and error* procedure. However, this particular feature gives fuzzy systems the advantages of flexibility and robustness, and thus provides them with effectiveness of handling complex ill-defined processes without much prior information, where conventional hard computing methods may fail [2].

Normally, according to the forms of their consequence parts, fuzzy logic systems are classified into two categories: Mamdani-type [8] and Sugeno-type [9]. In a Mamdani-type fuzzy system, the conclusion output consists of pure membership functions, i.e., $C$ is a fuzzy membership function of Gaussian, triangular-, or trapezoidal-shape, etc. The Sugeno-type fuzzy systems, on the other hand, have explicit *function-like* consequences. For example, the consequence part of a Sugeno-type fuzzy system with two inputs and a single output can be expressed as:

$$\text{IF } x \text{ is } A \text{ AND (OR) } y \text{ is } B \text{ THEN } z = f(x,y), \tag{3}$$

where $f$ is a function of input variables $x$ and $y$. Based on (3), we can see that there is no fuzziness in the reasoning output of Sugeno-type fuzzy systems. Therefore, the defuzzification phase in Fig. 1 is not needed here. Is it worth pointing out that neither one of these two fuzzy systems presented so far to possess any internal dynamical characteristics. Next we will discuss some conventional dynamical fuzzy systems with feedback.

## 3. Conventional Dynamical Fuzzy Systems

It has been proved that like neural networks, fuzzy systems are universal approximators [10]. In other words, they can approximate any continuous function with an arbitrary degree of accuracy. Nevertheless, as aforementioned in the previous sections, classical fuzzy logic systems do not have internal dynamics, and can therefore only realize *static* input-output mappings. This disadvantage makes it difficult to employ them in such dynamics-oriented applications as coping with temporal problems of prediction, control, and filtering.

A straightforward approach to introducing dynamics into conventional fuzzy systems is feeding the defuzzified (*crisp*) system output to the input through a unit delay, which is depicted in Fig. 2 [11] [12]. These recurrent fuzzy systems have been shown to occupy distinguished dynamical properties over regular static fuzzy models, and they have found various real-world applications, e.g., long term prediction of time series [11].

However, in such output feedback fuzzy systems, some useful information reflecting the dynamical behavior is unavoidably lost during the defuzzification procedure. In addition, with global feedback structures, it is not so easy to analyze and guarantee their stability.

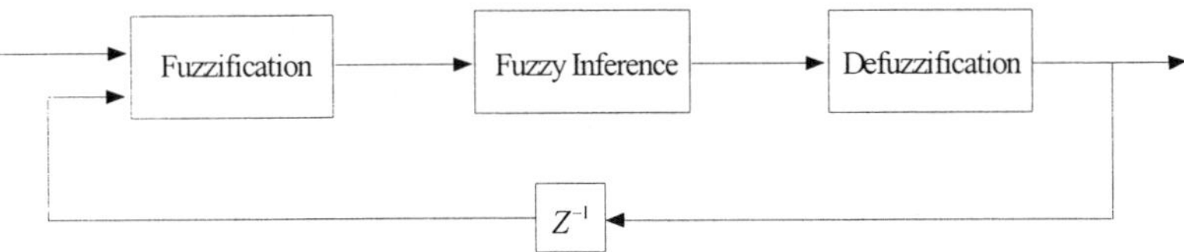


Figure 2. Conventional output feedback fuzzy system.

It is argued that internal memory units are compulsory for feedforward neural networks to deal with temporal problems [13]. Inspired by this idea, Lee and Teng proposed a recurrent fuzzy-neural network model with self-feedback connections [14]. To put it to more details, weighted outputs of the antecedent membership functions are fed back to their own inputs in order to construct *local* closed loops. As a matter of fact, the memory elements can store temporal information in the input signal, and realize dynamical input-output mappings. In [14], the performance of this scheme has been verified using challenging tasks of system identification and adaptive control of nonlinear plants. Unfortunately, like the output feedback fuzzy systems discussed above, it shares the common shortcoming of *crisp* feedback [15]. Different from these conventional dynamical fuzzy systems with greatly limited crisp signal feedback, we propose a new linguistic information feedback-based recurrent fuzzy model in the following section.

## 4. Dynamical Fuzzy Systems with Linguistic Information Feedback

### A. Basic Structure

In this section, we introduce a new dynamical fuzzy system with *linguistic* information feedback, whose conceptual structure is depicted in Fig. 3. Note that *only* the Mamdani-type fuzzy systems are considered here.

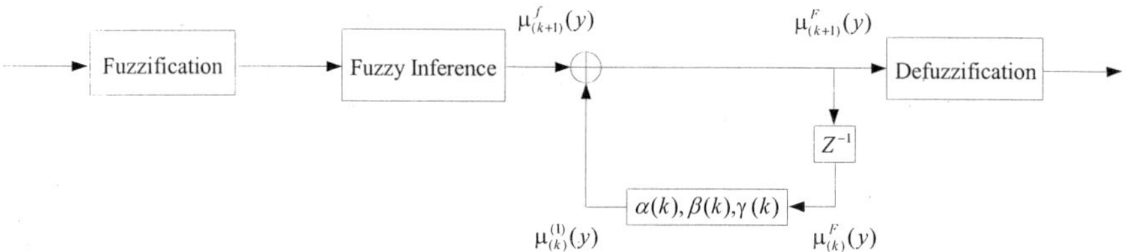


Figure 3. Structure of dynamical fuzzy system with linguistic information feedback.

It is clearly visible in Fig. 3 that instead of the defuzzified system output, the one-step delayed fuzzy output membership function in the consequence part is fed back as a whole to construct a *local* recurrent loop. Since this membership function is the final conclusion of our fuzzy system, it indeed has some practical *linguistic* meaning, such as an inferred fuzzy control action. In practical simulations, the membership function can be first 'sampled' in a discrete universe of discourse and then stored as a vector. To allow more degrees of freedom in our system, we introduce three variable coefficients, $\alpha$, $\beta$, and $\gamma$, in the feedback loop. In fact, $\alpha$ and $\beta$ are considered as shifting and scaling parameters, which can adjust the center location and height of the feedback membership function, respectively. $\gamma$ is, on the other hand, a concentration/dilation parameter. Suppose $\mu_A(y)$ is the *concluded* membership function, the transformed feedback membership functions, $\mu_A^1(y)$, $\mu_A^2(y)$, and $\mu_A^3(y)$, using these three parameters can be described as follows:

$$\mu_A^1(y) = \mu_A(y + \alpha) \, , \tag{4}$$

$$\mu_A^2(y) = \beta \mu_A(y) \, , \tag{5}$$

and

$$\mu_A^3(y) = \mu_A(y)^\gamma . \tag{6}$$

The above four fuzzy membership functions when $\alpha = -0.2$, $\beta = 0.7$, and $\gamma = 10$ are shown in Fig. 4 (a), (b), (c), and (d), respectively. With the joint contribution of $\alpha$, $\beta$, and $\gamma$, the actual feedback membership function is

$$\mu_A^{(1)}(y) = \left[\beta \mu_A(y + \alpha)\right]^\gamma . \tag{7}$$

An illustrative diagram of our fuzzy information feedback scheme is given in Fig. 5. More precisely, let $\alpha(k)$, $\beta(k)$, and $\gamma(k)$ be the three adaptive feedback coefficients at iteration $k$. $\mu_{(k+1)}^F(y)$ is defined as the fuzzy output of our system at $k+1$. Like in (6), we have

$$\mu_{(k)}'(y) = \mu_{(k)}^F\left[y + \alpha(k)\right] , \tag{8}$$

$$\mu_{(k)}''(y) = \beta(k)\mu_{(k)}'(y) , \tag{9}$$

and

$$\mu_{(k)}^{(1)}(y) = \left[\mu_{(k)}''(y)\right]^{\gamma(k)} \tag{10}$$

We denote $\mu_{(k+1)}^f(y)$ as the inference output directly from embedded reasoning rules in this recurrent fuzzy system, i.e., before feedback, at step $k+1$. If the *max* T-conorm is used for aggregation of individual fuzzy rule outputs [4], the resulting conclusion output $\mu_{(k+1)}^F(y)$ is obtained:

$$\mu_{(k+1)}^F(y) = \max\left\{\mu_{(k)}^{(1)}(y), \mu_{(k+1)}^f(y)\right\} . \tag{11}$$

     *X.Z. Gao and S.J. Ovaska / Dynamical Fuzzy Systems*

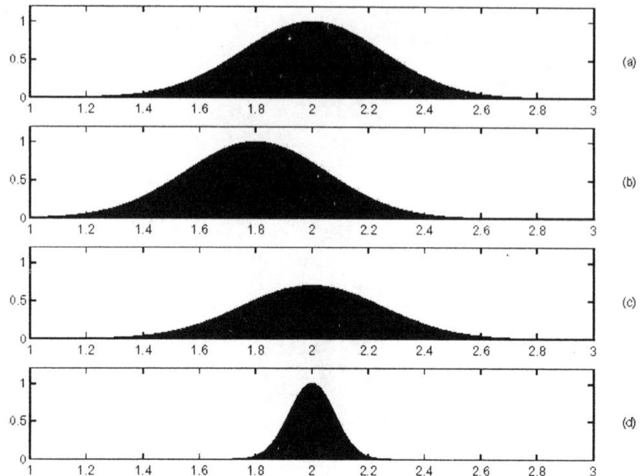

Figure 4. (a) Fuzzy output membership function $\mu_A(y)$.

(b) Fuzzy feedback membership function $\mu_A^1(y)$ with $\alpha = -0.2$ (shifted).

(c) Fuzzy feedback membership function $\mu_A^2(y)$ with $\beta = 0.7$ (scaled).

(d) Fuzzy feedback membership function $\mu_A^3(y)$ with $\gamma = 10$ (concentrated).

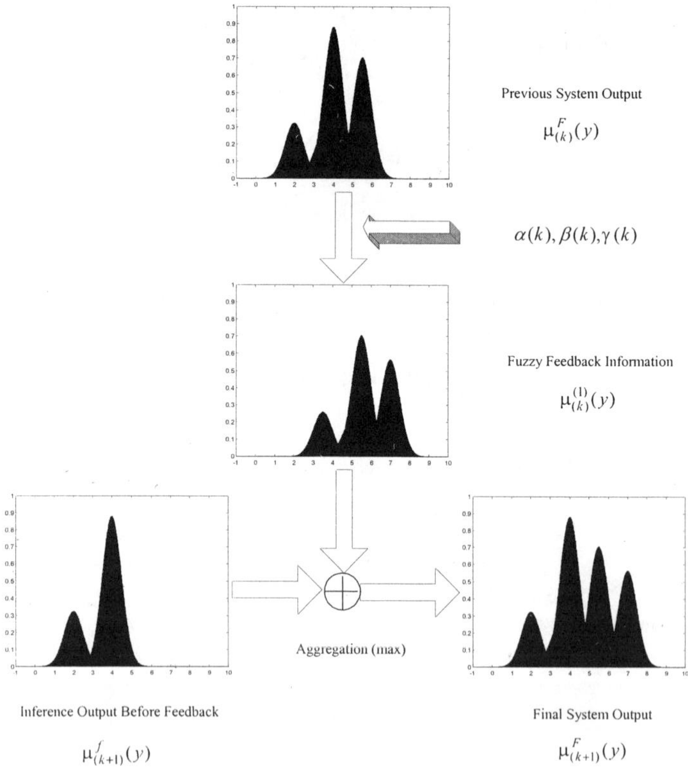


Figure 5. Diagram of fuzzy information feedback scheme

The defuzzification approach has to be applied to acquire a final crisp output from $\mu_{(k+1)}^{F}(y)$. Suppose $z_{(k+1)}^{*}$ is the defuzzified output of our fuzzy system, and we get:

$$z_{(k+1)}^{*} = \text{DEFUZZ}\left\{\mu_{(k+1)}^{F}(y)\right\}, \tag{12}$$

where $\text{DEFUZZ}\{\ \}$ is a defuzzification operator, as mentioned in Section 2. In a representative identification or prediction problem, if we denote $z_{(k+1)}^{d}$ as the desired output of this dynamical fuzzy system at iteration $k+1$, the following approximation error to be minimized is defined:

$$E_{(k+1)} = \frac{1}{2}\left(z_{(k+1)}^{d} - z_{(k+1)}^{*}\right)^{2}. \tag{13}$$

However, due to the inherent strong nonlinearity of our linguistic information feedback as well as the defuzzification phase, it is difficult to derive a closed-form learning algorithm for coefficients $\alpha(k)$, $\beta(k)$, and $\gamma(k)$ with regard to the minimization of $E_{(k+1)}$. Hence, some general-purpose algorithmic optimization method, such as a genetic algorithm (GA) [16], could be applied to optimize them in the first place. We emphasize that $\alpha$, $\beta$, and $\gamma$ can be intuitively considered as 'gains' in the feedback loop of the dynamical fuzzy system, which makes it possible to utilize *a priori* knowledge to choose appropriate initial values for them. The actual tuning procedure is realized, on the other hand, using an algorithmic learning approach.

### B. Structure Extension

As a matter of fact, the idea of linguistic information feedback in fuzzy systems presented above can be generalized to the individual fuzzy rules as well as higher order cases, as shown in Figs. 6 and 7, respectively. In Fig. 6, $m$ is the number of fuzzy rules, each of which has a *local* linguistic feedback. Totally, there are $3 \times m$ adaptive parameters, i.e., $\left\{\alpha_{1}(k), \beta_{1}(k), \gamma_{1}(k), \alpha_{2}(k), \beta_{2}(k), \gamma_{2}(k), \cdots, \alpha_{m}(k), \beta_{m}(k), \gamma_{m}(k)\right\}$, for these fuzzy rules. From the adaptation point of view, more degrees of freedom are introduced in this way.

Besides the first order linguistic information feedback in Figs. 3 and 6, we can also construct higher order feedback structures. Illustrated in Fig. 7, for Rule $i$ the delayed membership functions $\mu_{(k)}^{F}(y), \mu_{(k-1)}^{F}(y), \cdots, \mu_{(k-n)}^{F}(y)$ (up to $n$ steps) are fed back within separate loops. Therefore, long history information in the input signal is stored by these feedback loops, and this approach can significantly enhance the dynamics of our linguistic information feedback-based fuzzy systems.

The number of feedback parameters for a single rule $\left\{\alpha_{i}^{(1)}(k), \beta_{i}^{(1)}(k), \gamma_{i}^{(1)}(k), \alpha_{i}^{(2)}(k), \beta_{i}^{(2)}(k), \gamma_{i}^{(2)}(k), \cdots, \alpha_{i}^{(n)}(k), \beta_{i}^{(n)}(k), \gamma_{i}^{(n)}(k)\right\}$ is now $3 \times n$.

Our dynamical fuzzy system with linguistic information feedback has several remarkable features. First, instead of crisp signals, the 'rich' fuzzy inference output is fed back without any information transformation and loss. Second, the local feedback connections act as internal memory units here to store temporal input information, which is pivotal in identifying dynamical systems.

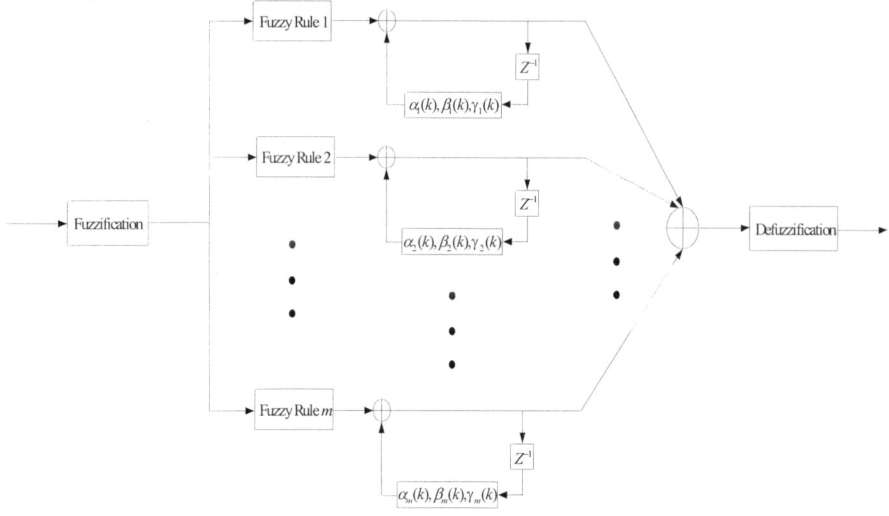


Figure 6. Linguistic information feedback in the consequences of individual fuzzy rules.

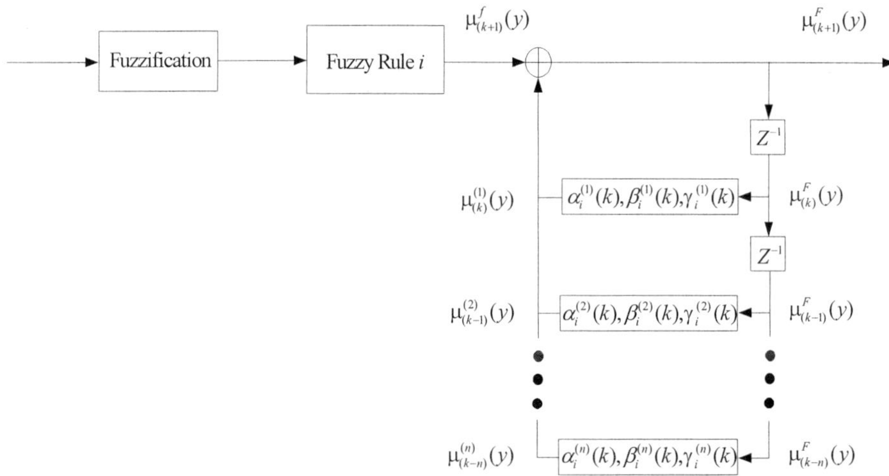


Figure 7. Dynamical fuzzy system with high-order linguistic information feedback.

That is to say, the linguistic information feedback can effectively and accurately capture the dynamics of the nonlinear systems to be modeled. Third, training of the three feedback coefficients, $\alpha$, $\beta$, and $\gamma$, leads to an equivalent update of those membership functions for output variables. Actually, this approach can be regarded as an *implicit* parameter adjustment in the fuzzy systems, and thus adds the advantage of self-adaptation to our model. In the following section, we will illustrate the characteristics of the proposed dynamical fuzzy system using numerical simulations.

## 5. Simulations

### *Simulation I. Step responses*

To gain some preliminary understanding of the dynamical characteristics of our linguistic information feedback, a simple single-input-single-output fuzzy system with only two fuzzy reasoning rules is studied in this part. Let us first assume that $x$ and $y$ are the inputs and output variables, respectively. The two fuzzy rules are given as follows:

IF $x$ is Small THEN $y$ is Small,

and

IF $x$ is Large THEN $y$ is Large.

The Gaussian fuzzy membership functions (input and output) for linguistic terms of 'Small' and 'Large' are illustrated in Fig. 8 (a) and (b).

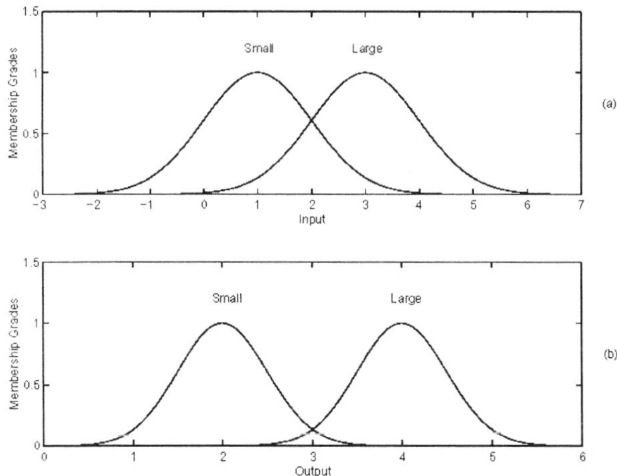

Figure 8. (a) Input fuzzy membership functions for linguistic terms of 'Small' and 'Large'.
(b) Output fuzzy membership functions for linguistic terms of 'Small' and 'Large'.

Note linguistic information feedback is added only at the final conclusion output of our system, i.e., there is no feedback for individual fuzzy rule outputs. Feedback parameters $\alpha$, $\beta$, and $\gamma$ are fixed to be $-1.5$, 0.8, and 1, respectively. We employ both the *max-min* and *sum-product* composition operators [5] as the fuzzy implication here. The COA defuzzification method is used in the simulations. The responses of our dynamical fuzzy system with a step input $x = 0 \rightarrow 2.5$ are shown in Fig. 9, in which the solid line represents the response using the max-min composition while the dotted line indicates that with the sum-product operator. We can conclude from Fig. 9 that concerning the step response these two outputs look very similar with that of a linear feedback system. It should also be emphasized that there is no principal difference in the responses using max-min and sum-product composition, although the latter gives us a slightly larger steady-state output than the former.

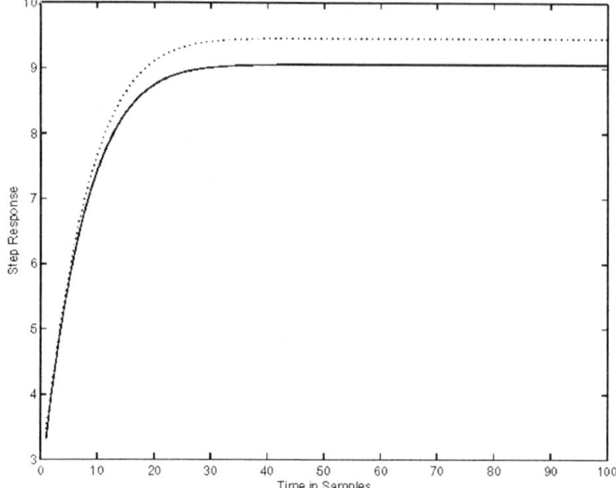

Figure 9. Step responses of recurrent fuzzy system with first order linguistic information feedback. Solid line: max-min composition. Dotted line: sum-product composition.

Next, we investigate the step response of the same fuzzy system but with a second-order feedback at the conclusion output. The feedback parameters are set as the following: $\alpha^1 = 1.5, \beta^1 = 0.2, \gamma^1 = 0.8$ and $\alpha^2 = -1.5, \beta^2 = 0.5, \gamma^2 = 0.6$. Only the sum-product composition is utilized this time. The step response of our second-order recurrent fuzzy system is given in Fig. 10, from which it can be seen that with the second-order linguistic information feedback, high-frequency oscillations occur in the beginning of system output. The response also depicts that high-order fuzzy information feedback can provide us with stronger nonlinearity. In summary, the above two examples demonstrate the internal dynamical property of our new fuzzy system.

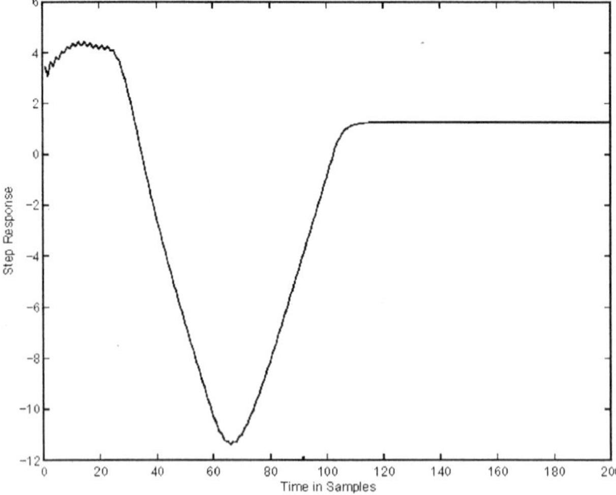

Figure 10. Step response of recurrent fuzzy system with second order linguistic information feedback.

We have to stress that the simulation results shown in this part are solely based on a simple two-rule recurrent fuzzy system with a step input. Comprehensive mathematical analysis and empirical investigations are needed to further explore the complex characteristics of the proposed linguistic information feedback.

### Simulation II. Single-input time sequence prediction

In this part of simulation, we apply the proposed dynamical fuzzy system to predict a simple time sequence $x(k)$, namely Time Sequence I, which consists of just three points: $\{-1\}$, $\{0\}$, and $\{1\}$. The trajectory of these points is shown in Fig. 11. Our goal is to predict the trajectory of time series $x(k)$ as below:

$$\{-1\} \rightarrow \{0\} \rightarrow \{1\} \rightarrow \{0\} \rightarrow \{-1\}.$$

In other words, based on the *current* point $x(k)$ only, the succeeding one $x(k+1)$ has to be predicted. Apparently, it is impossible for static fuzzy systems to make such a prediction, since point $\{0\}$ has two different successors, depending on its predecessors. If the predecessor point is $\{-1\}$, the consequent point should be $\{1\}$, while if the predecessor is $\{1\}$, the successor is $\{-1\}$. Our dynamical fuzzy system employed here has a single input, $x(k)$.

We denote $[-1]$, $[0]$, and $[1]$ as the fuzzy sets, whose membership functions are centered at $x(k) = \{-1\}$, $x(k) = \{0\}$, and $x(k) = \{1\}$, respectively. The three membership functions for input $x(k)$ are shown in Fig. 12.

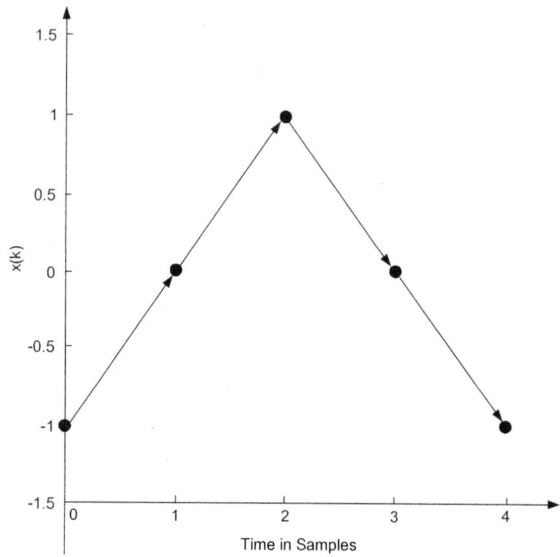

Figure 11. Trajectory of Time Sequence I.

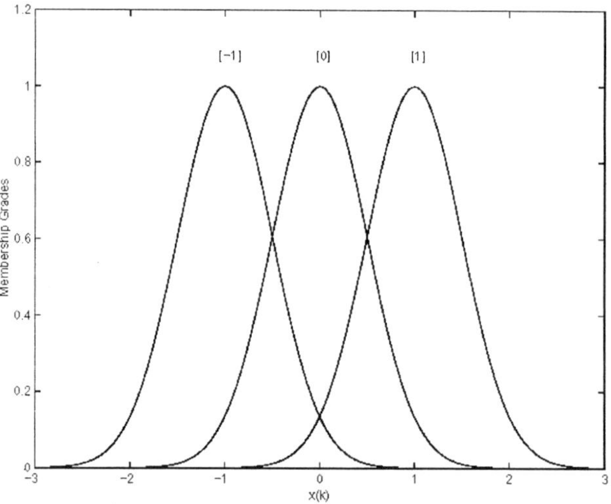


Figure 12. Membership functions of $x(k)$ in prediction of Time Sequence I.

From the above trajectory of $x(k)$, the following four straightforward fuzzy rules can be derived [4]:

Rule 1: IF $x(k)$ is $[-1]$ THEN $x(k+1)$ is $[0]$.
Rule 2: IF $x(k)$ is $[0]$ THEN $x(k+1)$ is $[1]$.

Rule 3: IF $x(k)$ is $[1]$ THEN $x(k+1)$ is $[0]$.

Rule 4: IF $x(k)$ is $[0]$ THEN $x(k+1)$ is $[-1]$.

Notice that sharing the same antecedent, Rule 2 and Rule 4 are completely conflicting. Therefore, when $x(k) = \{0\}$, a static fuzzy system embedded with these four rules will give a false prediction of $x(k+1)$, which is supposed to be an *average* between the different consequences of Rules 2 and 4, i.e., $\{1\}$ and $\{-1\}$.

To solve this problem, in our dynamical fuzzy system we introduce linguistic feedback at the individual conclusion outputs of the two rules. The corresponding two sets of feedback parameters are $\alpha_2$, $\beta_2$, $\gamma_2$ and $\alpha_4$, $\beta_4$, $\gamma_4$, respectively. However, there is no feedback added for the final system output.

With the prediction error as the performance criterion (fitness), a genetic algorithm is used to optimize the aforementioned feedback parameters. The evolution procedure of this genetic algorithm-based optimization is illustrated in Fig. 13. From Fig. 13, we can obtain the following optimized $\alpha_2$, $\beta_2$, $\gamma_2$, $\alpha_4$, $\beta_4$, $\gamma_4$ after about 20 iterations:

$$\alpha_2 = -0.7742, \ \beta_2 = 1.7606, \ \gamma_2 = 0.0606, \ \alpha_4 = -0.0026, \ \beta_4 = 1.5905, \ \gamma_4 = 5.8774.$$

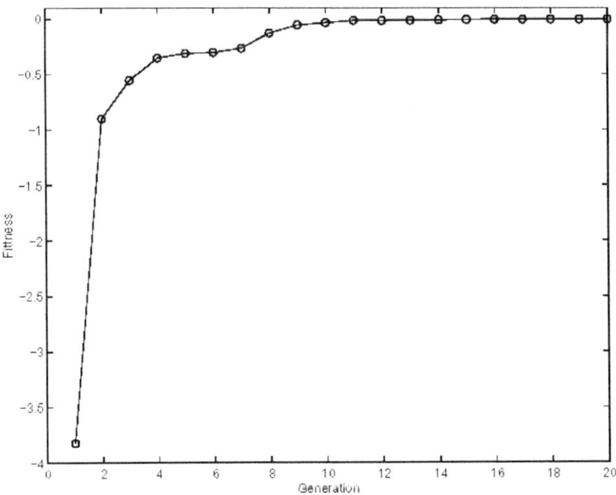

Figure 13. Evolution of a genetic algorithm in optimization of feedback parameters.

The prediction outputs of the static and dynamical fuzzy predictors are shown in Fig. 14, represented by dotted and solid lines, respectively. It is clearly visible that the dynamical fuzzy system can successfully predict the trajectory of Time Sequence I, while static fuzzy predictor fails.

This simulation result demonstrates that the introduced linguistic information feedback can store useful *temporal* input patterns for appropriate prediction of time sequences. Moreover, it is capable of correcting those wrong conclusions caused by the conflicting rules in our dynamical fuzzy system.

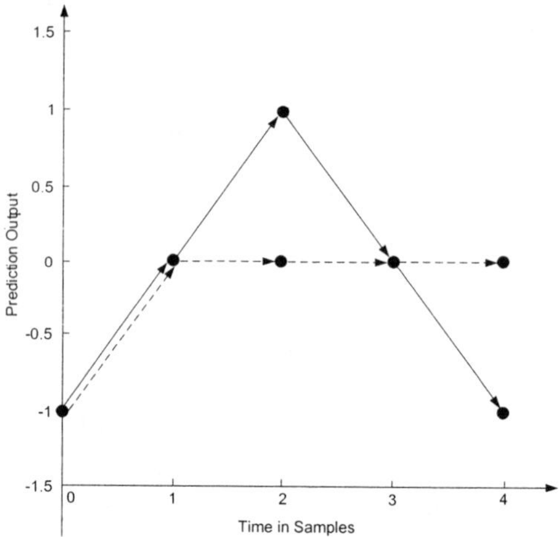

Figure 14. Prediction outputs of two fuzzy systems in Simulation II.
Dotted line: static fuzzy predictor. Solid line: dynamical fuzzy predictor.

### *Simulation III. Two-input time sequence prediction*

Next, we will study a more complex time sequence, Time Sequence II. Actually, it is a simplification of the problem discussed in [14], which can be considered as a 'figure 8' prediction case shown in Fig. 15.

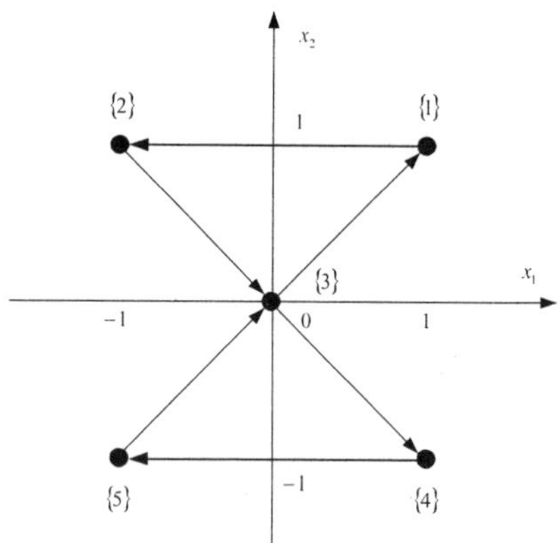

Figure 15. Trajectory of Time Sequence II.

More precisely, each point in Time Sequence II, $y(k)$, consists of two coordinates $\langle x_1(k), x_2(k) \rangle$. Its trajectory is given as follows:

$$\{1\} \rightarrow \{2\} \rightarrow \{3\} \rightarrow \{4\} \rightarrow \{5\} \rightarrow \{3\} \rightarrow \{1\},$$

where $\{1\} = \langle 1,1 \rangle$, $\{2\} = \langle -1,1 \rangle$, $\{3\} = \langle 0,0 \rangle$, $\{4\} = \langle 1,-1 \rangle$, and $\{5\} = \langle -1,-1 \rangle$. Note, similar with Time Sequence I, in this sequence the key point is $\{3\}$, which also has two successors, $\{1\}$ and $\{4\}$. According to different predecessors of $\{3\}$, the fuzzy predictors should give the corresponding prediction of its successors. Therefore, it requires that the predictors have some kind of *internal* memory. To make one-step-ahead prediction of Time Sequence II, we choose two inputs, $x_1(k)$ and $x_2(k)$, for our dynamical fuzzy system. Their membership functions are selected exactly the same as in Simulation II. However, $[1], [2], \cdots, [5]$ are defined as the fuzzy sets for points $y(k) = \{1\}, y(k) = \{2\}, \cdots, y(k) = \{5\}$, respectively. The membership functions of these fuzzy sets are demonstrated in Fig. 16.

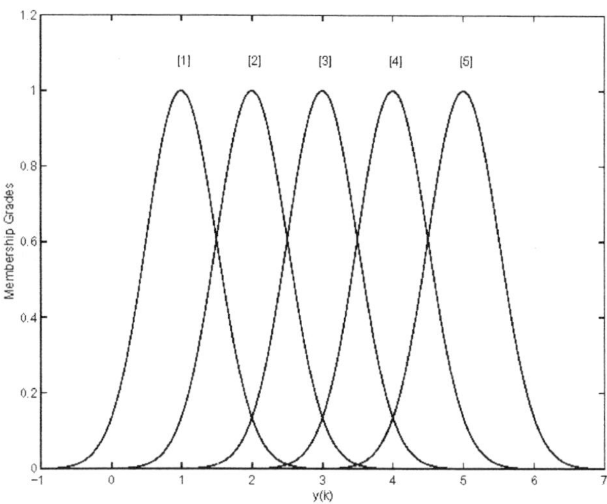


Figure 16. Membership functions of $y(k)$ in prediction of Time Sequence II.

We first derive the following six fuzzy inference rules based on the trajectory of Time Sequence II:

Rule 1: IF $x_1(k)$ is $[1]$ AND $x_2(k)$ is $[1]$ THEN $y(k+1)$ is $[2]$.
Rule 2: IF $x_1(k)$ is $[-1]$ AND $x_2(k)$ is $[1]$ THEN $y(k+1)$ is $[3]$.
Rule 3: IF $x_1(k)$ is $[0]$ AND $x_2(k)$ is $[0]$ THEN $y(k+1)$ is $[4]$.
Rule 4: IF $x_1(k)$ is $[1]$ AND $x_2(k)$ is $[-1]$ THEN $y(k+1)$ is $[5]$.
Rule 5: IF $x_1(k)$ is $[-1]$ AND $x_2(k)$ is $[-1]$ THEN $y(k+1)$ is $[3]$.
Rule 6: IF $x_1(k)$ is $[0]$ AND $x_2(k)$ is $[0]$ THEN $y(k+1)$ is $[1]$.

Obviously, Rule 3 and Rule 6 are not consistent in the above fuzzy rule base. Therefore, linguistic information feedback is introduced in these two rules to deal with this problem. $\alpha_3$, $\beta_3$, $\gamma_3$ and $\alpha_6$, $\beta_6$, $\gamma_6$ are denoted as the feedback parameters for Rules 3 and 6, respectively. Nevertheless, different from Simulation II, we further add fuzzy feedback (with parameters $\alpha$, $\beta$, and $\gamma$) at the final conclusion output. To summarize, totally nine feedback coefficients have been used in our dynamical fuzzy system, and after initialization they are supposed to be optimized by a genetic algorithm.

Those optimized parameters are:

$\alpha_3 = 0.5849$, $\beta_3 = 15.0878$, $\gamma_3 = 2.6088$, $\alpha_6 = 0.6828$, $\beta_6 = 6.2841$, $\gamma_6 = 3.0732$, $\alpha = -0.9955$, $\beta = 19.7197$, $\gamma = 1.2591$.

The prediction outputs from the dynamical as well as static fuzzy predictors (without linguistic information feedback) are illustrated in the following figure, which are represented by solid and dotted lines, respectively. From Fig. 17, we can see that with only the *current* time sequence values available, our dynamical fuzzy system can solve this challenging prediction problem. The regular fuzzy predictor, on the other hand, only gives an 'average' prediction (about 2.5) between the two possible successors, $\{1\}$ and $\{4\}$, at point $\{3\}$.

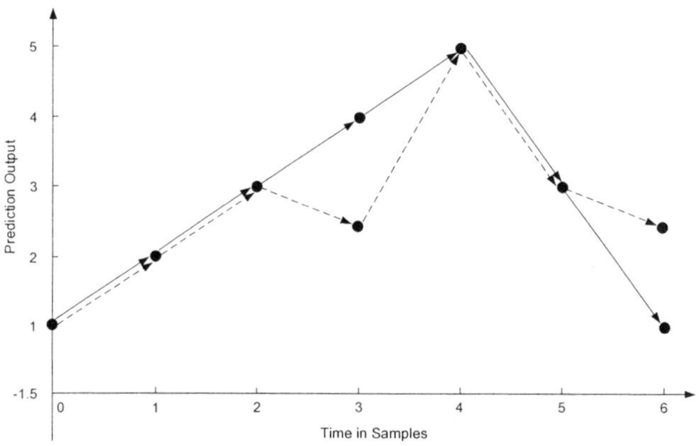


Figure 17. Prediction outputs of two fuzzy systems in Simulation III.
Dotted line: static fuzzy predictor. Solid line: dynamical fuzzy predictor.

Based on the above theoretical analysis and simulation results, we draw the conclusion that the linguistic information feedback embedded in the dynamical fuzzy system can store temporal patterns of input signals for effective prediction. It also corrects the false inference outputs from some conflicting fuzzy rules. Unfortunately, the GA-based adaptation of feedback parameters is a time consuming procedure, and often suffers from heavy computation burden. Fast training algorithms thus should be developed to meet the needs of on-line applications.

## 6. Conclusions

A new dynamical fuzzy system with linguistic information feedback is proposed in our paper. Preliminary simulations have been made to demonstrate its dynamical characteristics and applications in the time series prediction. Note that we only consider the Mamdani-type fuzzy systems here. It is, however, possible to construct dynamical Sugeno-type systems by similarly introducing fuzzy feedback in the antecedent. We also would like to point out that utilizing the internal *linguistic* information of fuzzy systems is not limited to the feedback structure presented in the current paper. Actually, this original idea can be generalized to a general framework including the *feedforward-like* configurations. Applications of our dynamical fuzzy system in other interesting areas such as process modeling are now being investigated. Study on the derivation of a closed-form learning algorithm for the feedback parameters is also under development.

## References


[1] W. Pedrycz, "Fuzzy sets in pattern recognition: Methodology and methods," *Pattern Recognition*, vol. **23**, no. 1, pp. 121-146, 1990.

[2] D. Driankov, M. Reinfrank, and H. Hellendoorn, An Introduction to Fuzzy Control. Berlin, Germany: Springer-Verlag, 1993.

[3] W. A. Kwong, K. M. Passino, E. G. Laukonen, and S. Yurkovich, "Expert supervision of fuzzy learning systems for fault tolerant aircraft control," *Proc. IEEE*, vol. **83**, no. 3, pp. 466-483, Mar. 1995.

[4] J. M. Mendel, "Fuzzy logic systems for engineering: A tutorial," *Proc. IEEE*, vol. **83**, no. 3, pp. 345-377, Mar. 1995.

[5] C. C. Lee, "Fuzzy logic in control systems: Fuzzy logic controller, Parts I and II," *IEEE Trans. on Systems, Man, and Cybernetics*, vol. **20**, no. 2, pp. 404-435, Apr. 1990.

[6] L. A. Zadeh, "Outline of a new approach to the analysis of complex systems and decision process," *IEEE Trans. on Systems, Man, and Cybernetics*, vol. **3**, no. 1, pp. 28-44, 1973.

[7] L.-X. Wang and J. M. Mendel, "Generating fuzzy rules by learning from examples," *IEEE Trans. on Systems, Man, and Cybernetics*, vol. **22**, no. 6, pp. 1414-1427, Nov./Dec. 1992.

[8] E. H. Mamdani, "Application of fuzzy logic to approximate reasoning using linguistic systems," *IEEE Trans. on Computers*, vol. **26**, pp. 1182-1191, 1977.

[9] M. Sugeno and G. T. Kang, "Structure identification of fuzzy model," *Fuzzy Sets and Systems*, vol. **28**, no. 1, pp. 15-22, 1988.

[10] L.-X. Wang, A Course in Fuzzy Systems and Control, Upper Saddle River, NJ: Prentice-Hall, 1997.

[11] J. Zhang and A. J. Morris, "Recurrent neuro-fuzzy networks for nonlinear process modeling," *IEEE Trans. on Neural Networks*, vol. **10**, no. 2, pp. 313-326, Mar. 1999.

[12] G. C. Mouzouris and J. M. Mendel, "Dynamic non-singleton fuzzy logic systems for nonlinear modeling," *IEEE Trans. on Fuzzy Systems*, vol. **5**, no. 2, pp. 199-208, May 1997.

[13] P. S. Sastry, G. Santharam, and K. P. Unnikrishnan, "Memory neuron networks for identification and control of dynamic systems," *IEEE Trans. on Neural Networks*, vol. **5**, no. 2, pp. 306-319, Mar. 1994.

[14] C.-H. Lee and C.-C. Teng, "Identification and control of dynamic systems using recurrent fuzzy neural networks," *IEEE Trans. on Fuzzy Systems*, vol. **8**, no. 4, pp. 349-366, Aug. 2000.

[15] F.-Y. Wang, Y. Lin, and J. B. Pu, "Linguistic dynamic systems and computing with words for complex systems," in *Proc. IEEE International Conference on Systems, Man, and Cybernetics*, Nashville, TN, Oct. 2000, pp. 2399-2404.

[16] T. Bäck, Evolutionary Algorithms in Theory and Practice, New York, NY: Oxford University Press, 1996.

# Fuzzy Bayesian Nets and Prototypes for User Modeling, Message Filtering and Data Mining

Jim F. BALDWIN
*Engineering Mathematics Department*
*University of Bristol*
*Email: jim.baldwin@bristol.ac.uk*

**Abstract.** In this paper, we will discuss the basic theory of Fuzzy Bayesian Nets and their application to user modeling, message filtering and data mining. We can now capture and store large amounts of data that we would like to transform into useful knowledge. We also expect to receive far too many messages to be able to handle without some form of intelligent filtering which acts on our behalf. Our Society feeds on information with data banks web sites, Internet wireless etc. We require a more intelligent way of handling all this data and methods for transforming it into useful knowledge. Summarizing data, finding appropriate rules to discover models, sending only relevant information to any individual, modeling users are all-important tasks for our modern day computer systems. We want to use the available data, not be swamped with it.

## 1. Introduction

Various machine learning techniques such as decision trees, neural nets and Bayesian Nets, [1, 2], have been successful in the general field of data mining. A database is given and used to find rules, determine a neural connectionist model or provide the required probabilities for a Bayesian net to answer queries that cannot be answered directly. Finding the optimal architectures or form of rules is more difficult. A decision tree can be derived to fly an airplane for a given flight path from a database of states and events taken every time a pilot makes a decision. To find a decision tree to fly more generally for any flight plan cannot be done. There are limitations to all these machine-learning methods. Whichever model is used it should not be too complex and must provide good generalization. If it is over complex by over fitting the data then errors in classifications or predictions will be made. Good generalization means sensible interpolation between the cases in the database to match the desired case. It is easier to treat discrete variables than continuous ones. We will let continuous variables take fuzzy values rather than a continuum of values. Thus, a variable representing the height of a person will be allowed to take values such as short, average and tall rather than a range of values. The semantics of short, average, tall will be given by a fuzzy set interpretation. All machine learning methods and algorithms can be modified to allow variables to take these fuzzy values. These modifications will be illustrated in this paper. This provides an example of computing with words. The advantage of using fuzzy words as compared to crisp sets is that better interpolation is provided and this results in more simple models. For example, to classify points belonging to the upper triangle of a square as opposed to those in the lower triangle requires only 2 overlapping fuzzy sets on each of the axes of the square to give 97% accuracy. To obtain the same accuracy in the case of crisp sets we would require 16 divisions of each axis. We would expect the modifications required to move from crisp to fuzzy sets to not be too complex. Certainly, we would not expect a totally different mathematics. For prediction problems, the variable whose value is to be predicted is allowed to take a discrete set of fuzzy values. The prediction from our machine learning method will be a probability distribution over these fuzzy sets. A method of defuzzification will be described which converts this to a single point value. For filtering messages we will use prototypes representing different types of people for given contexts. Each prototype will output a support for receiving the message. Each user will have a neural net to input these supports and provide a command to send the message to the user or reject it. The prototype can be modeled with fuzzy rules, a fuzzy decision tree, a fuzzy Bayesian net or a fuzzy conceptual graph. In this paper we will only consider Bayesian nets and decision trees.

## 2. Prototypes for Message Passing and other Applications

As discussed above a message is sent to prototypes representing prototypical persons and a support of interest is output from each prototype. These are sent to a user's neural net that decides to either send or reject the message. We can illustrate this diagrammatically in figure 1:

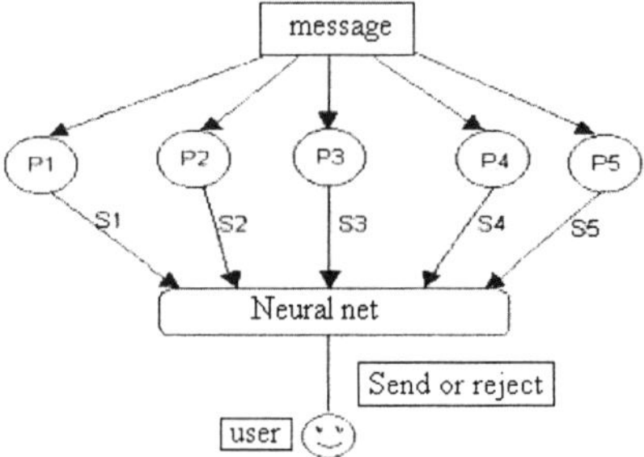


Figure 1.

The representation for the prototypes, P1, …, P5 could be fuzzy rules, fuzzy conceptual graphs, fuzzy decision trees or fuzzy Bayesian nets. We will choose for this paper fuzzy Bayesian nets. A fuzzy Bayesian net is like a Bayesian net, [3], but the variables can take fuzzy values. An example of such a net is depicted in figure 2:

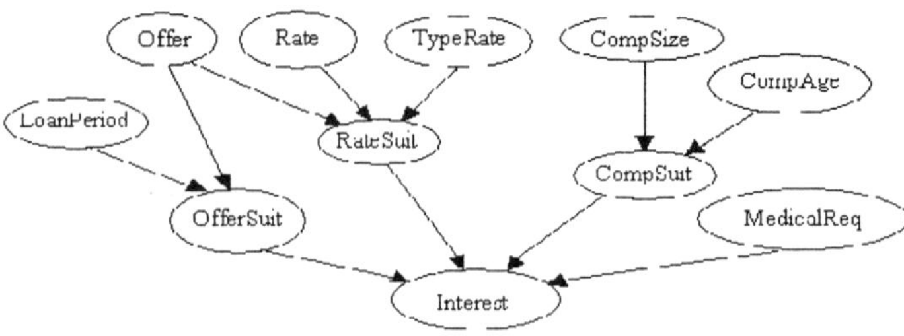


Figure 2.

The values of the variables are:

Offer: {mortgage, personal loan, car loan, car insurance, holiday insurance, credit card, payment protection, charge card, home insurance}

TypeRate:{fixed, variable}
Rate: {good, fair, bad}
CompSize:{large, medium, small}
CompAge:{old, medium, young}
LoanPeriod:{long, medium, short}
RateSuit:{very, fairly, a little}
CompSuit:{very, fairly, a little}
OfferSuit:{good, average, bad}
MedicalReq:{yes, no}
Interest:{high, medium, low}

The meaning of the variables is obvious for this financial application. Some of the values are fuzzy sets such as old, long, very, good, high etc.

Prior probability distributions are given for those nodes without parents and conditional probabilities Pr (Node | Parents) for those nodes with parents. These probabilities define a type of person – risk taking for example. Each prototype would have a different set of conditional probabilities. They could, of course, also have different architectures with different nodes. These probabilities can be determined from appropriate databases. The entries in these databases would more likely be precise values and we would need to convert this database to one where the attributes could take the fuzzy values given above. We will discuss this later.

A message might be as follows. The Robert Building Society is offering long term mortgages at 4% fixed rate. From this message we can extract the some variable instantiations. The Company size and age can be deduced from published details of the Robert Company. There is missing information about the variable medical required. The rate of loan, Company size and age will be given as numbers and we must convert these to probability distributions over the word values of the variables. The words are represented by fuzzy sets and mass assignment theory is used to obtain these distributions as discussed later.

Normally we would compile the net without instantiations and then update this prior representation using the variable instantiations to obtain a final probability distribution over the values of the Interest variable. A clique tree is first formed and a message-passing algorithm used to do the compiling and updating. This method must be modified to take account of the fact that some of the instantiations are probability distributions. Bayesian updating assumes precise values for the evidence and does not handle the situation in which distributions are given. We will later discuss the philosophy of this modification but we will not give all the details of the modification to the message-passing algorithm.

For now we can assume that we use the modified algorithm to give us a final probability distribution over the values of the Interest variable. Suppose this is high: $p_1$, medium: $p_2$, low: $p_3$, where $p_1 + p_2 + p_3 = 1$.

The values high, medium and low are fuzzy sets and the mass assignment theory can give expected values of these fuzzy sets as the expected values of the least prejudiced distributions associated with the fuzzy sets. Let these be $\mu_1$, $\mu_2$ and $\mu_3$ respectively. The Interest variable then takes the defuzzified value s where

$$s = p_1\mu_1 + p_2\mu_2 + p_3\mu_3$$

Each prototype delivers a support in a similar manner. The neural net for a given user is trained on an example set.

This is a toy example to illustrate the general approach. More generally contexts would be defined and the net prototypes given for each defined contexts. Message interpretation would use some form of semantic analysis to extract the relevant variable instantiations.

The approach can be used for many different types of personalization examples. A web page could be personalized for a given user to give relevant and interesting information. Books and films could be chosen to interest the user. Internet wireless will require approaches like this to prevent deadlock in communication traffic.

## 3. The Bayesian Modification Philosophy

Classical Bayesian updating cannot handle updating with distributions. We require a modification and we base the idea of choosing the updated distribution to minimize the relative entropy of this with the prior. Applied to the problem of updating the prior of AB, namely

$$ab : p_1, a\neg b : p_2, \neg ab : p_3, \neg a\neg b : p_4$$

with the distribution for $A$ of

$$a : q_1, \neg a : q_2 = 1-q_1$$

we use

|  |  | $q_1$ | $q_2$ |  |  |
|--|--|-------|-------|--|--|
|  |  | $a$ | $\neg a$ | update | |
| prior | AB |  |  |  |  |
| $p_1$ | $ab$ | $p_1q_1 / N_1$ | 0 | $p_1q_1 / N_1$ | |
| $p_2$ | $a\neg b$ | $p_2q_1 / N_1$ | 0 | $p_2q_1 / N_1$ | |
| $p_3$ | $\neg ab$ | 0 | $p_3q_2 / N_2$ | $p_3q_2 / N_2$ | |
| $p_4$ | $\neg a\neg b$ | 0 | $p_4q_2 / N_2$ | $p_4q_2 / N_2$ | |

$$N_1 = p_1+p_2 \qquad N_2 = p_3+p_4$$

For the case in which we update with several variable evidences we perform an update, one at a time and repeat with the final update becoming the new prior until we obtain convergence. This is depicted diagrammatically in figure 3:

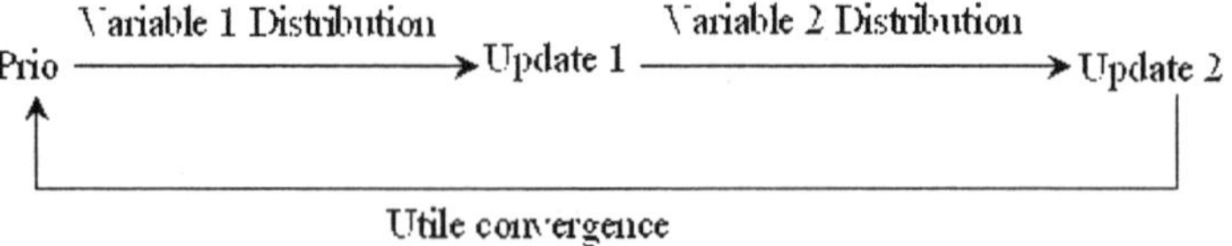


Figure 3.

The clique message-passing algorithm of Bayesian Nets is modified to effectively be equivalent to this modified updating. The modification does not add significant complexity to the algorithm.

## 4. A Little Mass Assignment Theory

We suppose the variable X can take values in $\{1, 2, 3, 4, 5, 6\}$. Let us suppose that we wish the values of X to be the mutually exclusive partition {even and odd}. If we were told that X is even then we would say that the distribution over the integer set is

$$\Pr(2|even) = \Pr(4|even) = \Pr(6|even) = \frac{1}{3}$$

This assumes an equally likely prior.

If we replace the partition with the fuzzy partition {small, medium, large} where

> **small** $= 1 / 1 + 2 / 0.7 + 3 / 0.3$
> **medium** $= 2 / 0.3 + 3 / 0.7 + 4 / 1 + 5 / 0.2$
> **large** $= 5 / 0.8 + 6 / 1$

and are told that X is **small** we should be able to derive a distribution over X assuming an equally likely prior. This we call the least prejudiced distribution.

The mass assignment for the fuzzy set **small** is

> MAsmall $= \{1\} : 0.3, \{1, 2\} : 0.4, \{1, 2, 3\} : 0.3$

Assuming the prior distribution of X is that each integer is equally likely then the least prejudiced distribution is obtained by allocating a mass associated with a set of values equally among those values. This gives

> LPDsmall $= 1 : 0.3 + 0.2 + 0.1 = 0.6, \ 2 : 0.2 + 0.1 = 0.3, \ 3 : 0.1$

If we had a different prior then this would be used to allocate the masses.

The mass assignment, [4], is a random set. We give it a different name to emphasize that the mass assignment theory is different to both random set and Dempster Shafer theories.

This definition of LPD can be extended to the case of continuous fuzzy sets and the expected value of the LPD is the mean $\mu$ of the fuzzy set that we used for defuzzification.

Suppose that we are told that $X$ is **about_2** where

> **about_2** $= 1 / 0.3 + 2 / 1 + 3 / 0.3$

then we are interested to determine Pr(**small** | **about_2**), Pr(**medium** | **about_2**) and Pr(**large** | **about_2**) giving the distribution over our partition. . The calculation of Pr(**f** | **g**) where **f** and **g** are fuzzy sets is called point value semantic unification.

The mass assignment for **about_2** is given by:

$$\text{MAabout\_2} = \{2\} : 0.7, \{1, 2, 3\} : 0.3$$

so that the LPD of **about_2** is

$$\text{LPDabout\_2} = 1 : 0.1, \ 2 : 0.7 + 0.1 = 0.8, \ 3 : 0.1$$
$$\text{LPDabout\_2}(x) = \Pr(x \mid \textbf{about\_2})$$

Thus

$$\Pr\big(\textbf{small}\big|\textbf{about}\_2\big) =$$
$$= 0.3 \Pr\big(\{1\}\big|\textbf{about}\_2\big) + 0.4 \Pr\big(\{1,2\}\big|\textbf{about}\_2\big) + 0.3 \Pr\big(\{1,2,3\}\big|\textbf{about}\_2\big) =$$
$$= 0.3 \cdot 0.1 + 0.4 \cdot (0.1 + 0.8) + 0.3 \cdot (0.1 + 0.8 + 0.1) = 0.69$$

Suppose we are given a database with attributes X, Y and Z that can take values in [0, 10], [5, 12] and {true, false} respectively. Suppose further that we partition X and Y into the fuzzy partitions {**fi**} and {**gi**} respectively.

A new range of values for attributes in a database can be a set of words where each word is defined by a fuzzy set. The original database attribute values are transformed into distributions over these words to form a reduced database. This reduced database can be used to obtain decision trees using $\text{ID}_3$, the required probabilities for Bayesian analysis and other applications. The entries in the original database can be precise, vague, fuzzy or unknown

Let one line in the database be X = *x*, Y = *y* and Z = *true*, then this can be rewritten as the lines

$$X = \textbf{fi}, \quad Y = \textbf{gj}, \quad Z = \textbf{true}$$

with probability $\Pr(\textbf{fi} \mid x) \, \Pr(\textbf{gj} \mid y)$ for all *i, j,* where x and y can be point values, intervals or fuzzy sets.

We can repeat this for all lines of the database and calculate the joint probability distribution for XYZ from which the Bayesian Net conditional probabilities can be calculated.

A similar approach can be used to determine decision trees from databases for classification and prediction.

## 5. Conclusions

The methods given here for the management of uncertainty, machine learning and modeling provides a unified and consistent approach to computing with words eliminating the ad hoc nature usually encountered. For example, one method of defuzzification suggests itself rather than the numerous ad hoc methods used in the literature.

In this paper we have described a Bayesian Net approach to user modeling and personalization. We extended the theory of Bayesian Nets to allow node variables to take fuzzy values. These word values were represented by fuzzy sets. The updating message algorithm was extended to allow probability distribution instantiations of the variables. This was necessary since instantiated values from messages were translated into distributions over words.

The approach was illustrated using a message traffic filtering application. A similar approach could be used for many other applications such as personalizing web pages to give users relevant information, providing TV programs, books and films of interest, intelligent filtering of emails. Advertisers could use the approach to select who should receive their adverts.

Fuzzy sets were used to provide better generalization and the use of continuous variables.

A similar approach has been used for data mining and other machine learning applications.

## References

[1] J. F. Baldwin, Knowledge from Data using Fril and Fuzzy Methods, In Fuzzy Logic in A.I., Baldwin, J. F. (ed.), John Wiley, pp. 33-76, 1996.

[2] J. F. Baldwin and T. P. Martin, Basic Concepts of a Fuzzy Logic Data Browser with Applications, In Software Agents and Soft Computing: Concepts and Applications, Nwana, H. S., Azarmi, N., eds., Springer (LNAI 1198), pp. 211-241, 1997.

[3] F. Jenson, An Introduction to Bayesian Networks, Springer - Verlag, 1996.

[4] J. F. Baldwin, T. P. Martin and B. W. Pilsworth, FRIL - Fuzzy and Evidential Reasoning in AI, Research Studies Press (John Wiley), 1995.

# Extending Fuzzy Temporal Profile Model for Dealing with Episode Quantification

Senén BARRO
Paulo FÉLIX
Purificación CARIÑENA
Abraham OTERO
*Departamento de Electrónica e Computación*
*Facultade de Física. Universidade de Santiago de Compostela*

**Abstract.** The aim of the Fuzzy Temporal Profile model is the representation and subsequent recognition of the evolution of a physical parameter. With this model it is possible to represent the behavior of a parameter over time on the basis of a linguistic description that assembles a set of signal events and episodes, the temporal support of which is the instant and the interval, respectively. Thus, by means of a signal event we represent the value of the parameter at a given time instant, and by means of a signal episode, we represent the persistence of a given property of the parameter over a time interval. In this article we extend the previous capabilities of the model, integrating the representation of certain linguistic sentences in which a degree of quantification is introduced into the description of signal episodes.

## 1. Introduction

The aim of the Fuzzy Temporal Profile model (FTP) is to contribute to the computational representation of knowledge by describing the behavior of a system, characterized by the evolution of one of its significant parameters, and to provide algorithms for implementing a signal pattern detection task.

The FTP model belongs to the set of proposals that are based on a structural perspective in pattern representation: methods which are closely linked with a fundamentally descriptive perspective in the representation of signal patterns [4, 8, 16, 18, 23, 25]. There are two fundamental ideas upon which the FTP model is based:

- Linguistic knowledge acquisition, on a register as close as possible to that used by human experts to convey their knowledge. We have developed an artificial language [13], which allows the description of the evolution of a physical parameter, its projection in the terms that define the FTP model, and its incorporation into a more general model of representation and reasoning over temporal facts [2].

- Modeling of the imprecision and uncertainty that characterize human knowledge, i.e., we attempt to capture, as far as possible, the richness of nuances contained in descriptions made by experts. In order to do so, the FTP model is based on constraint networks [21], [5] and the fuzzy set theory [29]. The former supplies a representational structure that facilitates the computational projection of a linguistic description. The latter deals with the handling of vagueness and uncertainty, which are characteristic of natural language terms.

In this work we go one step further towards the incorporation into the FTP model of the representation of linguistic descriptions containing some form of quantification, as is the case for sentences of the type, *"in most of the last half an hour the temperature has risen moderately"*. It is assumed that this quantification shares the imprecise nature of natural language, and thus, we will resolve the representation of the semantics that it induces on a sentence by means of fuzzy sets.

The structure of the paper is as follows: we start with a brief review of the fundamental concepts of the Fuzzy Temporal Profile model. We then go on to study the manner in which quantification is modeled using the FTP. Later we show how the FTP model can be used as a signal pattern recognition tool.

Lastly, we suggest some conclusions.

## 2. Initial Definitions

Given as discourse universe the set of real numbers **R**, we can extend by means of the Zadeh's extension principle [29] the concept of number to what we will call **fuzzy number** $C$ [9], represented by a possibility distribution $\pi_C$, which defines a mapping from $R$ to the real interval [0, 1]. Thus, given a precise number $v \in \mathbf{R}, \pi_C(v) \in [0,1]$ represents the possibility of $C$ being precisely $v$. The extreme values 1 and 0, respectively, represent the absolute and null possibility of $C$ being equal to $v$. By means of $\pi_C$ we can define a fuzzy subset $\tilde{c}$ of **R**, which contains the possible values of $C$, assuming that $\tilde{c}$ is a disjoint subset, in the sense that its elements represent mutually excluding alternatives for $C$.

Considering $\mu_C$ as the membership function that is associated to $\tilde{c}$, we have: $\forall v \in \mathbf{R}, \pi_C(v) = \mu_C(v)$. In general, and unless explicitly stated otherwise, we may use, indistinctly both membership functions and possibility distributions that are associated to the different concepts we define. We will always assume that $\pi_C$ is *normalized*, i. e, $\exists v \in \mathbf{R}, \pi_C(v) = 1$, and that it is *unimodal*, i.e.:

$$\forall v, v', v'' \in \mathbf{R}, v < v' < v'', \ \pi_C(v') \geq \min \left\{ \pi_C(v), \pi_C(v'') \right\}$$

These two properties are satisfied by representing $\pi_C$, for example, by means of a trapezoidal representation $C = (\alpha, \beta, \gamma, \delta), \alpha \leq \beta \leq \gamma \leq \delta$, where $\beta$ and $\gamma$ represent the beginning and the end of the *core*, $core(C) = \{v \in R | \pi_C(v) = 1\}$, and $\alpha$ and $\delta$ represent the beginning and the end of the *support*, $supp(C) = \{v \in R | \pi_C(v) > 0\}$. Although the validity of FTP model is not restricted to a specific representation of possibility distributions, in practice it will suffice to work with trapezoidal distributions, given that in the possibility theory, the important aspect is the order of the possibility degrees attached to the different values of the universe of discourse, rather than the precise assignment of possibility degrees [9].

We introduce the concept of **fuzzy quantity** in order to represent increments, extensions or, in general, the difference between two numbers. A fuzzy quantity $D$ is represented by means of a normalized and unimodal possibility distribution $\pi_D$ over **R** In this way, given an

$i \in \mathbf{R}$, $\pi_D(i) \in [0,1]$ represents the possibility of $D$ being precisely equal to $i$. Given an ordered pair of fuzzy numbers, we can talk of a fuzzy quantity given by the fuzzy subtraction $D = E \ominus A$, following the fuzzy arithmetic proposed in [17] .

We define **fuzzy interval** by means of its initial and final fuzzy numbers and its extension, which is a fuzzy quantity and represents the difference between them. $I_{(A,E,D)}$ denotes the interval delimited by the values $A$ and $E$, the difference between them being $D$. Even though the distributions of $A$ and $E$ may overlap, a positive extension of the interval will reject any assignment to $A$ of a value that is the same or greater than $E$.

We assume a discrete representation of the time domain $\tau = \{t_0, t_1, \ldots, t_i, \ldots\}$, which is isomorphic to the set of natural numbers $N$. Here $t_0$ represents the temporal origin, and given an $i$ belonging to the set of natural numbers $N$, $t_i$ represents a *precise instant*. We consider a total order relation between the precise instants $(t_0 < t_1 < \ldots < t_i < \ldots)$, and a uniform distance between them, in such a way that for every $i \in N$, $t_{i+1} - t_i = \Delta t$, $\Delta t$ being a constant. Thus $t_i$ represents a distance $i \times \Delta t$ to the time origin $t_0$. We will call $\Delta$ the *discretization factor*, and its selection will normally coincide with the sampling period of the signal on which we are working.

In the temporal domain, the concept of fuzzy number serves to represent that of *fuzzy instant* [2, 10]. In the physical parameter value domain, the evolution of which we will represent, the concept of fuzzy number serves to represent that of *fuzzy value*. The concept of fuzzy quantity serves to represent those of *fuzzy duration* or *fuzzy temporal extension* between the fuzzy instants, and that of *fuzzy increment* between fuzzy values. Lastly, the fuzzy interval serves to represent *fuzzy temporal intervals* and *fuzzy segments*.

## 3. The Fuzzy Temporal Profile Model

The Fuzzy Temporal Profile model (FTP) is inspired by the observations made by Attneave with regard to visual perception: the fundamental information on the shape of a curve is concentrated at certain significant points that have a high degree of curvature. The main aim is that of representing an evolution profile relative to a physical parameter $v(t)$, which takes real values in time. The FTP model [12] allows the expert to make a linguistic description of a profile of interest that will be projected onto a network of fuzzy constraints between a set of nodes performing the role of significant points.

Each significant point is defined as a pair of variables: one corresponding to the physical parameter and the other to time. The FTP's constraints limit the fuzzy duration, the fuzzy increase and the fuzzy slope values between each pair of significant points.

**Definition 1.** We define **significant point** $X_i$ associated with a variable $v(t)$ as the pair $X_i = <V_i, T_i>$, where $V_i$ is a variable that represents an unknown value of the physical parameter, and $T_i$ is a temporal variable representing an unknown time instant. In the absence of any constraints, the variables $V_i$ and $T_i$ may take any precise value $v_i$ and $t_i$ respectively.

The language introduced in [13] allows the description of two types of temporal facts: the *event*, which is represented by a significant point, and the *episode*, which is delimited by two significant points. An event is given by the occurrence of a particular value of the physical parameter in a temporal instant (*"the temperature is much higher a little later"*). An episode is given by the occurrence of a particular evolution during a temporal interval (*"the temperature rises moderately during the ensuing minutes"*). We now go on to introduce the constraints that enable us to represent the evolution of a parameter as a set of relations between signal events and episodes.

**Definition 2.** A binary constraint $L_{ij}$ on two temporal variables $T_i$ and $T_j$ is defined by means of a normalized and unimodal possibility distribution $\pi_{L_{ij}}$, whose discourse universe is $Z, \forall l \in Z : \pi_{L_{ij}}(l) \in [0,1]$. Given a precise value $l_{ij}, \pi_{L_{ij}}(l_{ij})$ represents the possibility that the temporal distance between $T_i$ and $T_j$ takes precisely the value $l_{ij}$.

The constraint $L_{ij}$ jointly restricts the possible value domains of the variables $T_i$ and $T_j$. In the absence of other constraints, the assignments $T_i = t_i$ and $T_j = t_j$ are possible if $\pi_{L_{ij}}(t_j - t_i) > 0$ holds. The possibility distribution $\pi_{L_{ij}}$ associated to a binary constraint, which we represent as $L_{ij}$, induces a fuzzy subset in the temporal distance domain. Formally, the distribution $\pi_{L_{ij}}$ corresponds to the possibility distribution of a fuzzy increase. Thus, we may interpret a binary constraint $L_{ij}$ as the assignment of a fuzzy increase, which we call *fuzzy duration*, to the distance between the variables $T_i$ and $T_j$.

We model those qualitative relations that appear in the bibliography. Thus, amongst the instants we represent are those of convex point algebra [27], and amongst the intervals we represent are those from Allen's interval algebra [1]. Furthermore, a fuzzy-set-based representation enables the model to capture the imprecision present in the quantitative relations between temporal events, and which can be found in expressions of the type: *"approximately 5 minutes after"*. Figure 1 shows an example of how one of the temporal relations between intervals is projected onto the network.

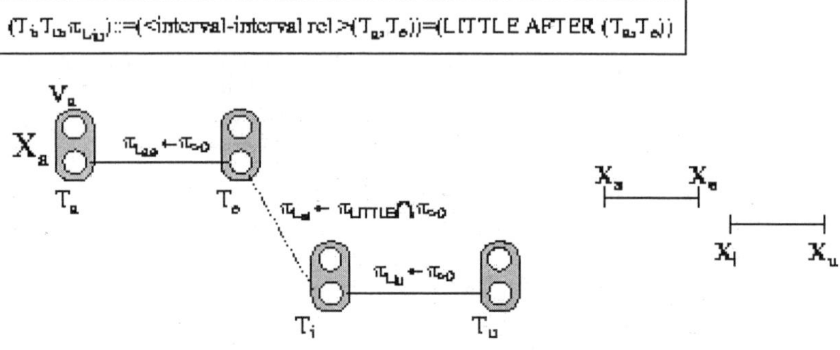


Figure 1. Projection of the relation between temporal intervals 'A LITTLE AFTER' on the constraints of the model. Above can be seen the rewriting rule [13] that makes it possible to specify an interval in relation to another, previously described one. This is projected onto the constraint network by introducing two significant points, $X_i$ and $X_j$, along with the corresponding constraints.

**Definition 3.** A binary constraint $D_{ij}$ on two variables of the domain $V_i$ and $V_j$ is defined in a similar way to the constraint $L_{ij}$, by means of a normalized and unimodal possibility distribution $\pi_{D_{ij}}$, whose discourse universe is $\mathbf{R}, \forall d \in \mathbf{R} : \pi_{D_{ij}}(d) \in [0,1]$.

We may interpret a binary constraint $D_{ij}$ as the assignment of a *fuzzy increase* to the distance between $V_i$ and $V_j$.

We add an additional significant point $X_0 =< L_0, D_0 >$ to the model, which represents a precise origin for the time and value axes. It is supposed that all the significant points are different and that they verify total temporal order, which impedes the assignment of two different values to the same instant. We only consider those significant points that are ordered by the relation $L_{ij} > 0$, since any constraint $L_{ij} < 0$ can be substituted by its symmetrical constraint $L_{ij}$, which is positive, and equivalent to the original one.

**Definition 4.** A quaternary constraint $M_{ij}$ on two significant points $X_i$ and $X_j$, is defined by means of a normalized and unimodal possibility distribution $\pi_{M_{ij}}$, whose discourse universe is $\mathbf{R}, \forall m \in \mathbf{R} : \pi_{M_{ij}}(m) \in [0,1]$. Given a precise value $m_{ij}, \pi_{M_{ij}}(m_{ij})$ represents the possibility that the slope of the line that joins $X_i$ and $X_j$ be precisely $m_{ij}$.

The constraint $M_{ij}$ jointly restricts the domains of $V_i, V_j, T_i$ and $T_j$. In the absence of other constraints, the assignments $V_i = v_i, V_j = v_j, T_i = t_i$ and $T_j = t_j$ are possible if $\pi_{M_{ij}}\left((v_j - v_i)/(t_j - t_i)\right) > 0$. We can interpret a constraint $M_{ij}$ as the assignment of a fuzzy value, which we call *fuzzy slope*, to the line which joins $X_i$ and $X_j$.

**Definition 5.** A fuzzy constraint $R_{ij}$ on two significant points $X_i$ and $X_j$ is a 3-tuple made up of a fuzzy duration $L_{ij}$, a fuzzy increment $D_{ij}$ and a fuzzy slope $M_{ij}$.

$$R_{ij} =< L_{ij}, D_{ij}, M_{ij} >$$

It is supposed that all constraints of the type $L_{ii}(D_{ii})$ are equal to the conventional set $\{0\}$. That is, it is verified that $\pi_{L_{ii}|D_{ii}}(0) = 1$ and $\pi_{L_{ii}|D_{ii}}(x) = 0 \, \forall x \neq 0$. Furthermore, it is supposed that the constraint $M_{ii}$ is equivalent to the universal constraint, i.e., $\pi_{M_{ii}} = \pi_U, (\pi_U(x) = 1, \forall x \in U)$.

**Definition 6.** We define **Fuzzy Temporal Profile** $N = \{X, R\}$ as a finite set of significant points $X = \{X_0, X_1, \ldots, X_n\}$, and a finite set of constraints $R = \{< L_{ij}, D_{ij}, M_{ij} >, 0 \leq i, j \leq n\}$ defined on the variables that constitute these points.

An FTP can be represented by means of a directed graph (figure 2) in which the nodes correspond to significant points, and the arcs correspond to constraints on the variables of the nodes that they join. Those constraints that are not described explicitly do not appear in the figure, and are defined by universal constraints.

The aim of any constraint satisfaction problem is the search for solutions, i.e., sets of assignments to the variables of nodes that are compatible with the network constraints. Thus, matching an FTP with a real evolution $E = \{(v_{[1]}, t_{[1]}), \ldots, (v_{[w]}, t_{[w]})\}$ (where $v_{[r]}$ is a precise value of the parameter $v(t)$ and $t[r]$ is a precise instant in which $v_{(t)}$ takes this value), will be carried out by way of searching for sets of assignments to the significant points of the profile; the aim is to realize the signal segmentation that best fits the definition of the profile.

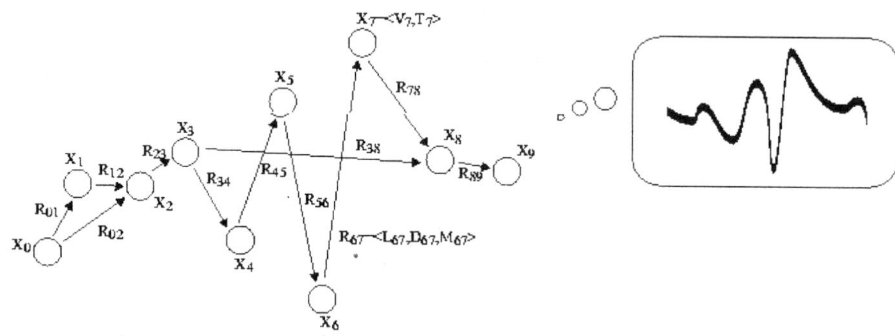

Figure 2. Example of a Fuzzy Temporal Profile.

**Definition 7.** We call $A_i$ the **assignment** of a sample of the evolution $\left(v_{[m]}, t_{[m]}\right) \in E$ to the significant point $X_i = <V_i, T_i>$. Thus, $A_i = <v_i, t_i> = <v_{[m]}, t_{[m]}>$ signifies that $V_i = v_{[m]}$ and $T_i = t_{[m]}$.

**Definition 8.** We say that a pair of assignments $\left(A_i, A_j\right)$ is valid under constraint $R_{ij}$ if and only if it satisfies the constraints that exist between the significant points $X_i$ and $X_j$ :

$$\pi_{D_{ij}}\left(v_j - v_i\right) > 0, \pi_{L_{ij}}\left(t_j - t_i\right) > 0, \pi_{M_{ij}}\left(\left(v_j - v_i\right)/\left(t_j - t_i\right)\right) > 0$$

**Definition 9.** *We say that an n-tuple of assignments* $A = \left(A_1, A_2, \ldots, A_n\right)$ *is a* **solution** *for the network* $N$ *corresponding to an FTP if and only if it fulfils all network constraints:*

$$\pi^N(A) = \min_{\substack{0 \le i, j \le n \\ t_0 = 0, v_0 = 0}} \pi_{R_{ij}}\left(A_i, A_j\right) = \min_{\substack{0 \le i, j \le n \\ t_0 = 0, v_0 = 0}} \left\{\pi_{D_{ij}}\left(v_j - v_i\right), \pi_{L_{ij}}\left(t_j - t_i\right), \pi_{M_{ij}}\left(\left(v_j - v_i\right)/\left(t_j - t_i\right)\right)\right\} > 0$$

where we carry out the conjunction of these conditions with the $T$-norm *minimum*. The possibility of an n-tuple $A$ being a solution to the network $N$ is given by the lowest degree of possibility with which the values of each one of the assignments satisfies each of the constraints of the network.

## 4. Semantic Considerations about Evolution

The FTP model that has been proposed up until now limits the representation of a profile to a set of signal events, i.e., their significant points, reducing the representation of an episode to the events that define the beginning and the end of the interval in which it takes place. Hence, in its current form, the model seems to adequately represent the semantics of expressions of the type ". . . *fifteen minutes later the temperature is somewhat lower*", where experts show their possible ignorance as to the evolution of the temperature during this fifteen-minute period, and in any case, their total lack of interest in what has occurred (figure 3). The semantics of these expressions does not restrict the evolution that takes place between each two significant points, so that, given two assignments $A_i$ and $A_j$ to the end significant points, any evolution is completely possible $\mu_{S_{ij}}\left(v_{[m]}, t_{[m]}\right) = 1, \forall t_{[m]} : t_i \leq t_{[m]} \leq t_j$.

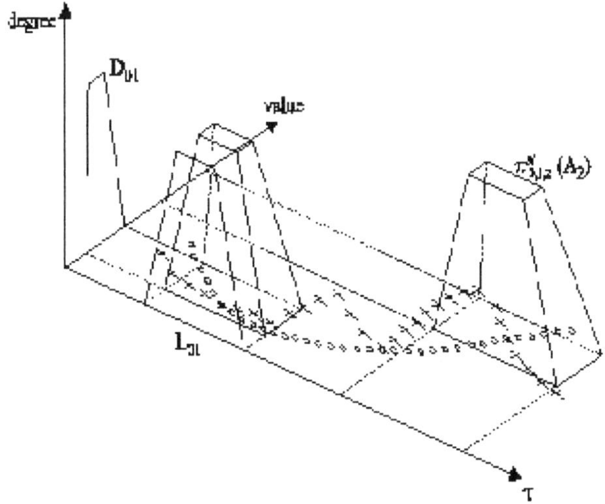


Figure 3. Graphic example of the representation of fuzzy constraints with which the real evolution is compared. Two different evolutions are shown, with the same degree of compatibility.

Nevertheless, natural language allows us to give a set of descriptions in which subtle distinctions are made on the way the evolution between two significant points takes place, as in the case of "... *throughout the next fifteen minutes the temperature rises moderately by 10 degrees*", whilst, in its present form, its projection onto the network constraints is the same as for the predicate "... *fifteen minutes later the temperature is ten degrees higher*", that is, $\pi_{D_{ij}} = \pi_{ten\_degrees}$ and $\pi_{L_{ij}} = \pi_{fifteen\_minutes}$.

With this idea in mind, the FTP model can be extended, to represent information concerning the evolution of a parameter between each two significant points. In order to do so, we have developed a representation, using fuzzy sets, that enables us to calculate the compatibility between the description of the section and a fragment of the real evolution of the physical parameter under study.

We introduce a fuzzy representation $S_{ij}$ of the semantics of a signal episode between two significant points. The objective is to model, by means of fuzzy sets, a set of prototypical evolutions found in the meaning of certain linguistic expressions, and to incorporate them into the FTP model. Thus, those linguistic sentences that are not restrictive with regard to the evolution of the parameter between the two significant points can be modeled by means of $\mu_{S_{ij}}\left(v_{[m]}, t_{[m]}\right) = 1, \forall t_{[m]} : t_i \leq t_{[m]} \leq t_j$ which is consistent with the use of FTP for modeling the semantics of events. We could try to model the semantics of an episode described by those sentences from the expert's language of the type "... *during the following fifteen minutes the temperature rises a little*", where it is understood that this rise is satisfied by all samples of the fragment of evolution with respect to the assignment carried out on the first significant point.

Thus, we aim to find a new membership function $\mu_{s_{ij}}$ for modeling the behavior of the physical parameter between each two significant points. In order to obtain this membership function we use the relative information that is associated to each section $(R_{ij})$ and the absolute information that is associated to the beginning and end points of the section $(R_{0i}$ and $R_{0j})$. By way of Zadeh's extension principle [29] on the expression of a straight line $y - y_1 = m \times (x - x_1)$, we obtain a fuzzy constraint on the points $(v,t)$ that belong to the temporal interval between the two significant points:

$$\mu_{ij}(v,t) = \max_u \left\{ \mu_{(v-D_{0i})\cap M_{ij}\otimes(t-L_{0i})}(u) \right\} \tag{1}$$

This linear equation can be understood as a set of two constraints, one on the spatial part, and the other on the temporal part of the same entity: a point $(v,t)$ in the space $R \times \tau$. The previous equation evaluates the degree of membership of a point $(v,t)$ to the fuzzy straight line that is given by the constraints of the section $R_{ij}$.

In our particular case, the membership function of the expression (1) is instantiated on the specific assignment $A_i = < v_i, t_i >$ for the first of the significant points of the section (the point for which it is said that "… *during the **following** fifteen minutes the temperature rises a little*" (see figure 4). Thus the expression that is applied for the comparison procedure is as follows:

$$\mu_{S_{ij}}\left(v_{[m]}, t_{[m]}\right) = \max_u \left\{ \mu_{(v_{[m]}-v_i)\cap M_{ij}\otimes(t_{[m]}-t_i)}(u) \right\} , \ t_i \le t_{[m]} \le t_j$$

The compatibility of the fragment of the evolution between the assignments to the end significant points is calculated by means of aggregation employing the operator minimum:

$$\mu_{S_{ij}}\left(A_i, A_j\right) = \min_{t_i \le t_{[m]} \le t_j} \max_u \left\{ \mu_{(v_{[m]}-v_i)\cap M_{ij}\otimes(t_{[m]}-t_i)}(u) \right\} \tag{2}$$

Other semantics can be modeled by following this idea, such as expressions of the type "*the temperature has risen moderately throughout the last few minutes*", "*throughout more or less half an hour the temperature is high*" amongst others [15]. We now go on to see a number of solutions for modeling these expressions, in which the use of quantifiers plays an important role.

## 5. Quantification

Quantification is used in a number of applications to represent a constraint on the number or percentage of objects that verify a given property. In natural language, the vagueness and imprecision that are adopted by quantification allow it to represent knowledge with a degree of flexibility that makes it a powerful tool in common reasoning. Examples of natural language quantifiers are *several, most, much, quite a few, large number, approximately five,* etc. Since the work of Zadeh [30], some authors have tried to define a mathematical model for the representation of this knowledge in the field of AI by using the theory of fuzzy sets [3, 7].

Two kinds of linguistic quantifiers are taken into consideration in the evaluation of quantified linguistic descriptions: these are called *absolute* and *relative* quantifiers.

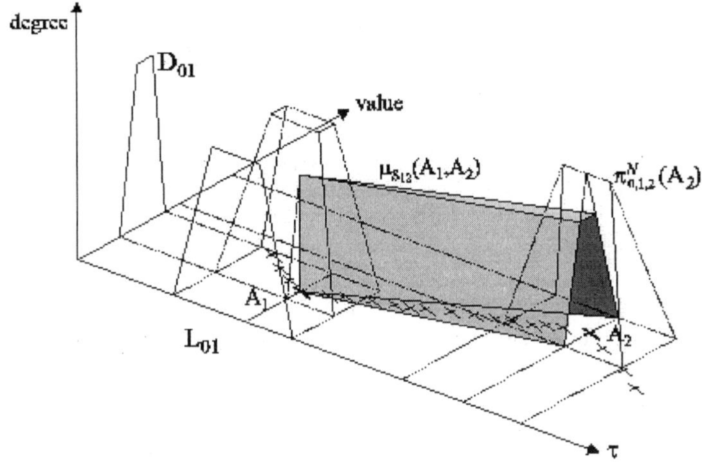


Figure 4. Example of FTP matching in a section represented by means of equation 2. The matching searches for the maximum compatibility between the values of the signal and the fuzzy set instantiated by assignments to significant points.

### 5. 1. Absolute Quantifiers

Absolute quantifiers are defined as possibility distributions over the non-negative integers, and represent fuzzy integer quantities or intervals. Examples of this type are *"approximately two"* and *"little"* in the sentence *"a little later the temperature is approximately two degrees higher"*. These quantifiers allow the definition of an absolute time extent and an absolute increase, represented by the following rewriting rules.

- For absolute time extent:

<absolute time extent>: : =([APPROXIMATELY]<temporal quantity>[< temporal unit>])
<temporal quantity>: : ={m|($\alpha$, $\beta$, $\gamma$, $\delta$)|LITTLE|MUCH |SOME|. . . }
<temporal unit>: : ={. . . |SECONDS|MINUTES|HOURS|. . . }

- For absolute increase:

<absolute increase>: : =([APPROXIMATELY] <quantity>[<unit$>])
<quantity>: : ={m|($\alpha$, $\beta$, $\gamma$, $\delta$)|LITTLE|MUCH|SOME |. . . }
<unit>: : ={. . . |BEATS/SEC. |DEGREES|. . . }
The semantics of these entities may be given by the expressions:

$$\pi_{absolute\_time\_extent} = \left\|[\pi_{\cong 0}] \oplus \pi_{quantity}\right| Unit$$

$$\pi_{absolute\_increase} = \left\|[\pi_{\cong 0}] \oplus \pi_{quantity}\right| Unit,$$

where bracketing an expression by means of '$\|Unit$' indicates its conversion into the corresponding units; quantities such as LITTLE, MUCH, etc. have a meaning which can be sensitive to the context of the specified reference; and the extension associated to APPROXIMATELY must depend on the distribution that it modulates.

The semantics of these expressions is projected onto network constraints that define the FTP. In the case of the aforementioned example, the sentence describes a temporal and a value distance which add a sign (+ or -) to the time extent and to the increment, giving rise to the constraints $L_{ij}$=*little later*, and $D_{ij}$= *approximately two degrees higher*.

### 5. 2. Relative Quantifiers

Relative quantifiers are defined as possibility distributions over the real interval, and represent fuzzy proportions. Examples of this type are "most", "few", "several", etc.

In the representation of temporal information, and thus with regard to the FTP model, quantification plays an especially relevant role by characterizing the relation between the verification of a given condition in the temporal interval and its verification on the time instants that belong to this interval, or rather, on the signal samples that coincide in the temporal interval. Here we study, in particular, relative quantifiers found in expressions such as *"the temperature rose a few degrees over the majority of the last half hour"*, which appear in the following rewriting rules:

<predication>: : =({EVENT|EPISODE}<assertion><temporal reference>)

<temporal reference>: : =([<quantifier>]<temporal interval>)

<quantifier>: : ={IN|THROUGHOUT|IN_A_GREAT_PART_OF|. . . }

where 'assertion' is understood as a description of the behavior of the value of the physical parameter.

Firstly, we will study the modeling of the *existential* and *universal* quantifiers, in order to go on to describe other ones that model intermediate situations.

### 5. 2. 1. Existential Quantifier $(\exists)$

Strictly speaking, this is not a relative quantifier, nevertheless, it can be conceived as the end-point of a continuous set of them. We identify it using the temporal adverb *"in"*. The corresponding temporal reference is projected onto a new time interval that is defined by the instants $T_i$ and $T_j$ in the following manner:

$$\left(T_i, T_j, \pi_{ij}^{L}\right) ::= \left(IN(T_a, T_e)\right) \equiv \pi_{ai}^{L} \leftarrow \pi_{\geq 0}; \pi_{ej}^{L} \leftarrow \pi_{\leq 0}; \pi_{ij}^{L} \leftarrow \pi_{>0}$$

With the aim of modeling the semantics of the section contained between $T_i$ and $T_j$ we distinguish between those assertions that describe the value of the parameter in this time interval ("*in the following minutes the temperature is very high*") and those that describe an evolution ("*in the following minutes the temperature rises moderately*").

- Value assertions. The existential quantifier in a value assertion requires that the latter should hold in at least one of the instants belonging to the interval between $T_i = t_i$ and $T_j = t_j$, where $t_i$ and $t_j$ correspond to the assignments carried out on the respective significant points. This responds to the intuitive meaning of expressions of the type "*in the following half an hour the temperature is high*". Thus the degree of verification of the value assertion for a signal fragment that is delimited between and is obtained by the following expression:

$$\mu_{S_{ij}}\left(\widetilde{A}_i, \widetilde{A}_j\right) = \bigvee_{t_i \leq t_{[m]} \leq t_j} \mu_{S_{ij}}\left(v_{[m]}, t_{[m]}\right) \wedge \mu_{I(T_i, T_j)}\left(t_{[m]}\right) \tag{3}$$

where $\wedge$ represents a t-norm operator, $\vee$ represents a t-conorm operator, $\mu_{S_{ij}}$ is the membership function that characterizes the value assertion and $I(T_i, T_j)$ if the fuzzy set formed by the instants that possibly are between $T_i$ and $T_j$. This interval is given by two temporal variables that correspond to the initial and final instants, $T_i$ and $T_j$, respectively, and a temporal constraint between these variables $L_{ij}$, corresponding to the duration of the interval. In general, the instants $T_i$ and $T_j$ are defined by the constraints $L_{mi}$ and $L_{ni}$, from the arbitrary nodes $T_m$ and $T_n$, or by absolute temporal references where $T_m = T_n = T_0$. The calculation is simplified on the basis of previous assignments to the significant points. The representation of the value assertion in the framework of the FTP model can be carried out by means of a change in the concept of assignment to the variables of a significant point: $\widetilde{A}_i$ is the assignment to the significant point $X_i =\, <V_i, T_i>$ given by a fuzzy value $\widetilde{v}_i$ assigned to the variable $V_i$, and a precise instant $t_i$ assigned to the variable $T_i$. Thus, $\widetilde{A}_i =\, <\widetilde{v}_i, t_i>$ means that $V_i = \widetilde{v}_i$ and $T_i = t_i$. In the specific problem that concerns us, we take as possible assignments to the significant points $X_i$ and $X_j$ of the section $R_{ij}$, different pairs $\widetilde{A}_i = \left(D_{0i}, t_i\right), \widetilde{A}_j = \left(D_{0j}, t_j\right)$, each one of them will be instantiated on the precise instant $t_{[m]}$ of a sample of the evolution $\left(v_{[m]}, t_{[m]}\right) \in E$. In this case it can be easily proved that

$$\mu_{R_{ij}}\left(\widetilde{A}_i, \widetilde{A}_j\right) = \pi_{L_{ij}}\left(t_j - t_i\right)$$

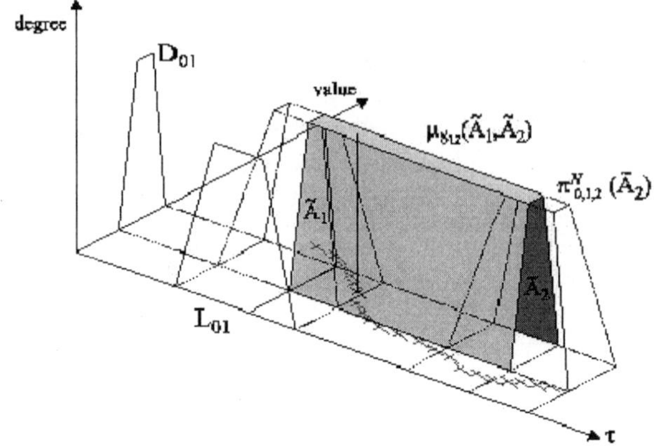


Figure 5. Example of FTP matching in a section described by the assertion *"the temperature is high in the following half an hour"* represented by means of equation 4. As can be seen, there is total compatibility with the signal fragment shown, since one of the samples of the signal fragment is totally compatible with the value assertion.

By including the constraint in the form of the network solutions, we are artificially subspecifying the problem, since it is understood that the increment and slope constraints, $D_{ij}$ and $M_{ij}$, respectively, are not defined on signal samples, rather over two fuzzy assignments in the indicated form. Let us take for example, that it is said that *"the temperature is high for 10 minutes"*; this supposes that the increment constraint $D_{ij}$ is defined as $\pi_{D_{ij}} = 0$, in the sense that the temperature is high, both at the start and at the end, as well as throughout the whole section, but it is not necessarily the same.

Once the assignment of the significant end points has been carried out, each one of the samples included between them is matched with the following membership function:

$$\mu_{S_{ij}}\left(v_{[m]}, t_{[m]}\right) = \pi_{D_{0i}}\left(v_{[m]}\right) \quad t_i \le t_{[m]} \le t_j \tag{4}$$

*Evolution assertions.* The existential quantifier in an evolution assertion requires that this should hold in at least one of the subintervals $I\left(T_i, T_j\right)$ that belong to the interval between $T_a$ and $T_e$, possibly excluding the two end points. Thus the degree of verification of the evolution assertion for a signal fragment that is delimited between the assignments that are temporally localized in $t_i$ and $t_j$ is obtained by means of the following expression:

$$\mu_{S_{ij}}\left(A_i, A_j\right) = \bigvee_{t_i \le t_{[m]} \le t_j} \mu_{S_{ij}}\left(v_{[m]}, t_{[m]}\right) \wedge \mu_{I(T_i, T_j)}\left(t_{[m]}\right) \tag{5}$$

The manner in which the assertion is verified depends wholly on the meaning it has, and given the richness of language, and the lack in many cases of a common agreement amongst speakers in a community when assigning a meaning, it would be highly complex to undertake a complete rationalization of the meanings of all the linguistic descriptions of evolution. Here we will demonstrate the FTP model's capability for formalizing the meaning of a number of them.

Let us take, for example, expressions of the type: *"the temperature falls sharply in the last half an hour"*. It would seem to be intuitive that the verification of this assertion is equivalent to the existence of at least one subinterval inside which verification is required for the universal quantifier. In discrete domains, such as the one in which the FTP model is applied, it will be sufficient for the evolution assertion to hold between two consecutive signal samples (see figure 6), which we can model by means of the following expression:

$$\mu_{S_{ij}}\left(v_{[m]}, t_{[m]}\right) = \max_{u}\left\{\mu_{\left(v_{[m]}-v_{[m-1]}\right)\wedge M_{ij}\otimes\left(t_{[m]}-t_{[m-1]}\right)}(u)\right\} \tag{6}$$

With regard to expressions of the type *"the temperature falls four degrees in the last half an hour"*, it would seem reasonable to think that their verification entails simply finding a section whose significant points will be $X_i$ and $X_j$, and such that $D_{ij}$ = *falls four degrees*, aside for the satisfaction of any type of persistence amongst these points (e.g., by means of the application of the expression 2) being considered relevant.

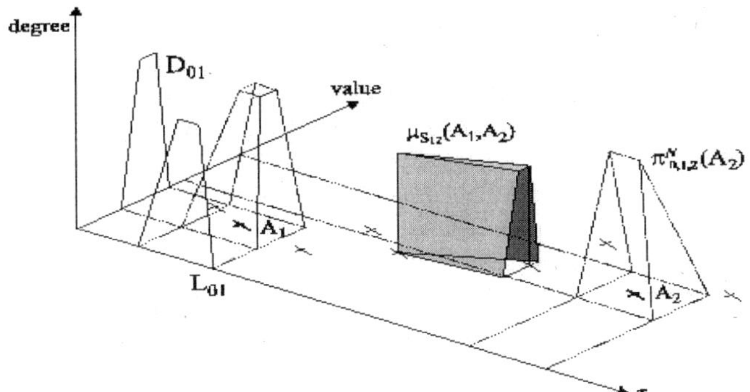


Figure 6. Example of FTP matching in a section described by the assertion *"the temperature rises moderately in the following fifteen minutes"* represented by means of equation 6. As can be seen, only two of the fragment samples included between $A_1$ and $A_2$ satisfy this description with a degree greater than zero, which is sufficient.

### 5. 2. 2. Universal Quantifier ( $\forall$ )

This we identify by means of the temporal adverb 'throughout'. The corresponding temporal reference is projected onto a new time interval that is defined by the instants $T_i$ and $T_j$, as follows:

$$\left(T_i, T_u, \pi^L_{iu}\right) ::= \left(THROUGHOUT\left(T_a, T_e\right)\right) \equiv \pi^L_{ai} \leftarrow \pi_{=0};$$
$$\pi^L_{eu} \leftarrow \pi_{=0}; \mu^L_{iu} \leftarrow \pi_{>0}$$

In a similar manner to the previous case, we distinguish between those assertions that describe the value of the parameter in a certain time interval (*"throughout the following minutes the temperature is high"*) from those that describe an evolution (*"throughout the following minutes the temperature rises moderately"*). On the other hand, in both cases the persistence of a particular property of the parameter is described *throughout* a given time interval.

*Value assertions.* In a value assertion the universal quantifier requires that this should hold for all of the instants contained in the interval included between $T_i = t_i$ and $T_j = t_j$, where $t_i$ and $t_j$ correspond to the assignments carried out on the respective significant points. This responds to the intuitive meaning of expressions of the type *"the temperature is a little lower throughout the following quarter of an hour"*. Thus the degree of verification of the value assertion for a signal fragment delimited between $t_i$ and $t_j$ is obtained by means of the following expression:

$$\mu_{S_{ij}}\left(\tilde{A}_i,\tilde{A}_j\right)= \bigwedge_{t_i \leq t_{[m]} \leq t_j} \mu_{S_{ij}}\left(v_{[m]},t_{[m]}\right) \vee \left(1-\mu_{I(T_i,T_j)}(t_{[m]})\right) \tag{7}$$

*Evolution assertions.* In an evolution assertion the universal quantifier requires that the description that it contains of the rhythm of variation of the parameter should hold for all the subintervals that belong to the interval between $T_i$ and $T_j$. Thus in expressions of the type *"the temperature rises moderately by approximately four degrees throughout the last half an hour"*, excluding the description of the global increase (*"approximately 4 degrees"*), the verification of which is carried out by the assignment to the significant end-points, it is expected that *"rises moderately throughout the last half an hour"* should hold for all the subintervals in the last half an hour. The degree of verification of the evolution assertion for a signal fragment delimited between and is obtained by means of the following expression:

$$\mu_{S_{ij}}\left(A_i,A_j\right)= \bigwedge_{t_i < t_{[m]} \leq t_j} \mu_{S_{ij}}\left(v_{[m]},t_{[m]}\right) \vee \left(1-\mu_{I(T_i,T_j)}(t_{[m]})\right) \tag{8}$$

where the description of the rhythm of variation of the parameter $\mu_{S_{ij}}$ can be modeled with an expression such as the following:

$$\mu_{S_{ij}}\left(v_{[m]},t_{[m]}\right)= \min_{t_i < t_{[k]} \leq t_{[m]}} \max_{u}\left\{\mu_{(v_{[k]}-v_{[k-1]}) \wedge M_{ij} \otimes (t_{[k]}-t_{[k-1]})}(u)\right\}, \quad t_i < t_{[m]} \leq t_j$$

which verifies the validity of the description of the rhythm of variation of the parameter in each of the two consecutive signal samples in the fragment delimited between $t_i$ and $t_j$, and therefore, for each one of the subintervals contained therein.

### 5. 2. 3. Intermediate Quantifiers

In order to increase the expressive capability of the linguistic description of the FTP, we now go on to consider modeling other relative quantifiers, such as *"in part of"*, *"in most of"*, *"in a small part of"* etc., the meaning of which as a fuzzy proportion is a halfway point between existential and universal quantification.

We will evaluate separately the modeling of quantification in value assertions and evolution assertions.

- *Value assertions.* We define a relative quantifier $Q$ by means of a possibility distribution $\pi_Q(p)$, which is a mapping from [0, 1] to [0, 1]. $Q$ describes in an imprecise manner the percentage (a fraction of 1) of the instants that are contained in a given interval for which the signal fulfils the value assertion. Figure 7 shows a possible representation of some quantifiers, where, for example, *"in most of"* requires the fulfillment of the value assertion in approximately 70% of the signal fragment samples. We assume that the possibility distribution $\pi_Q(p)$ is given by Zadeh's S function:

$$\pi_Q(p) = S_{Zadeh}(p; \alpha, \beta, \gamma) = \begin{cases} 0 & \text{if } p \leq \alpha \\ 2\left(\dfrac{p-\alpha}{\gamma-\alpha}\right)^2 & \text{if } \alpha < p \leq \beta \\ 1 - 2\left(\dfrac{p-\alpha}{\gamma-\alpha}\right)^2 & \text{if } \beta < p \leq \gamma \\ 1 & \text{if } p > \gamma \end{cases}$$

even when it can also be carried out by means of a linear description, or with any other one that is suited to the characteristics of the semantics we wish to model.

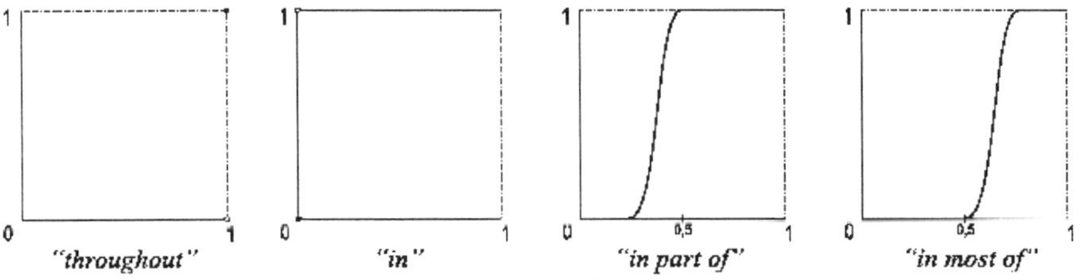



Figure 7. Definition of the possibility distributions of some temporal relative quantifiers $Q$.

Thus we have a set of quantifiers which enable us to define a fuzzy proportion that ranges from that of an existential meaning to a universal one. This graduation can be expressed by means of a compensatory aggregation operator $A$ [20, 31], which is dependent on a parameter $\lambda \in [0,1]$, which conditions the degree of interpolation between t-norm and t-conorm in each case. The compensatory operators are habitually defined as a linear or exponential combination, which enables us to define the verification of the value assertion as:

i) Linear combination:

$$\mu_{S_{ij}}(\tilde{A}_i, \tilde{A}_j) = (1-\lambda)(7) + \lambda(3) \tag{9}$$

where (7) and (3) are the respective expressions.
ii) Exponential combination:

$$\mu_{S_{ij}}(\widetilde{A}_i, \widetilde{A}_j) = (7)^{(1-\lambda)} \times (3)^{\lambda} \tag{10}$$

In this manner, in both cases it is verified that if $\lambda = 0$ the universal quantifier is obtained, and if $\lambda = 1$ the existential quantifier is obtained. There are other proposals for quantification outside the environment of the representation of temporal information. We have opted·for the one explained herein due to its computational efficiency $(O(n))$; due to the intuitive nature of the results that it supplies, compared with [26]; due to its monotone behavior, which is absent, for example, in [28]; and due to a degree of insensitivity to noise, as small modifications in the membership values do not give rise to drastic changes, in contrast to [6].

The value of $\lambda$ is directly related with the relative cardinality of the fuzzy set that is represented by the value assertion with regard to the temporal interval for a given signal fragment, and we model this on the basis of Zadeh's definition of relative cardinality, by means of the expression:

$$\lambda = Q \left( \frac{\sum\limits_{t_{[m]}} \mu_{S_{ij}} \wedge \mu_{I(T_i,T_j)}}{\sum\limits_{t_{[m]}} \mu_{I(T_i,T_j)}} \right)$$

It is possible to define another family of quantifiers, which we call $Q'$, and the meaning of which is the contrary or the negation of the aforementioned ones [3]. Examples such as "*at no time in*", "*in a small part of*" belong to this family. The representation of these quantifiers is also carried out by means of a possibility distribution $\pi_{Q'}(p)$, where $p$ represents the percentage of points within the temporal support for which the value assertion must be satisfied (see figure 8).

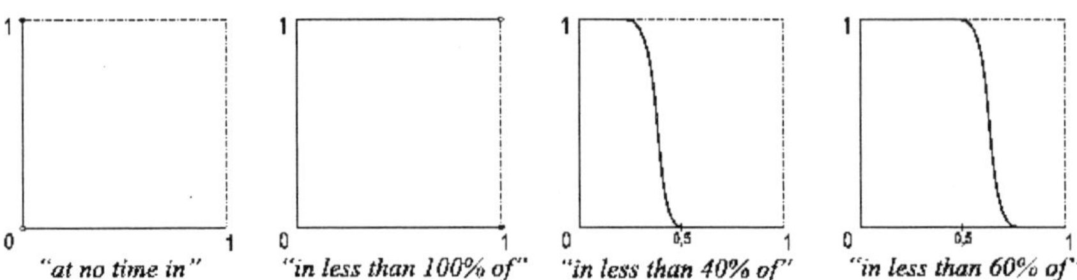


Figure 8. Definition of the possibility distributions of some temporal relative quantifiers $Q'$.

It should be said that the exponential compensatory operator is much more pessimistic, as, for the same set of data, it obtains a verification value that is lower than or equal to that obtained using the linear compensatory operator. Then again, the selection of the t-norms and t-conorms used in the calculation gives rise to quantitatively different results, according to the order of the operators [19] which also enables us to determine, to a certain extent, the degree of optimism or pessimism in the verification calculation [3].

- *Evolution assertions.* The relative quantifiers of the families $Q$ and $Q'$, introduced above, and represented by means of possibility distributions, model the proportion, (a fraction of 1), of the sum of the lengths of those disjoint subintervals in which the evolution assertion is verified with regard to the total length of the reference interval. In expressions of the type *"the temperature is roughly constant in most of the last half an hour"*, and in discrete domains, such as that of the samples signals upon which the FTP is applied, we will only need to apply a compensatory aggregation operator to the expressions 5 and 8:

i) Linear combination:

$$\mu_{S_{ij}}\left(\widetilde{A}_i, \widetilde{A}_j\right) = (1-\lambda)(8) + \lambda(5) \tag{11}$$

ii) Exponential combination:

$$\mu_{S_{ij}}\left(\widetilde{A}_i, \widetilde{A}_j\right) = (8)^{(1-\lambda)} \times (5)^{\lambda} \tag{12}$$

In the case of evolution assertions of the type *"the temperature rises 4 degrees in a small part of the following quarter of an hour"* it would seem reasonable to think that this describes an increase that takes place in a time interval that is proportionally related with that of the reference interval (*"the following fifteen minutes"*) which is given by the expression *"a small part of"*. The verification of this evolution assertion is thus given by the detection of section whose end significant points are $X_i$ and $X_j$, and such that $L_{ij} = Q \otimes L_{ae}$, where $Q = a$ *small part* $L_{ae} = $ *fifteen minutes*.

## 6. Recognition of Fuzzy Temporal Profiles

Once an FTP has been defined, its practical application requires the development of a set of algorithms that allow a profile recognition task to be implemented on the real evolution of the corresponding physical parameter. These algorithms attempt to operate in a similar manner to human experts when visually examining a fragment of the evolution of a physical parameter in search of a known morphology. At this point, we make a simplifying supposition: the real evolution upon which we work $E = \left\{\left(v_{[1]}, t_{[1]}\right), \ldots, \left(v_{[w]}, t_{[w]}\right)\right\}$ has a uniform and homogenous density, it sampling period being smaller than, or equal to the discretization factor with which the profile has been defined.

It should be pointed out that in spite of the fact that the fuzzy nature of the FTP model makes it particularly suitable for the handling of imprecision, in a number of real applications it may be necessary to employ signal processing techniques [22] (habitually digital filters, used in order to eliminate noise, elimination of outliers, enhancement of components to a given scale, etc. ) as a stage prior to the use of models such as the FTP.

The algorithms that make up the recognition task are based on the calculation of a measurement of compatibility between a fragment of the real evolution of a physical parameter, and the description that is made in the definition of an FTP. This matching is based on a signal segmentation procedure that carries out the assignment to the significant points of the profile. After each assignment we calculate the compatibility of the signal fragment between each two significant points with the corresponding section membership function. In this manner, matching inherits the NP-complete nature of the search for constraint network solutions, although its application to signal processing allows the development of certain heuristics that make the problem treatable.

As has already been seen, the assignment to a significant point shows a different image according to the semantics of each one of the sections to which this point belongs. Thus, for each significant point we carry out a *generalized assignment* $\tilde{A}_i = <\tilde{v}_i, t_i>$, whose compatibility with sections for which the assignment is given by $\tilde{v}_i = v_{[p]}, \tilde{v}_j = v_{[q]}, t_i = t_{[p]}$ and $t_j = t_{[q]}$ will take the form:

$$\pi_{R_{ij}}\left(\tilde{A}_i, \tilde{A}_j\right) = \min\left\{\pi_{D_{ij}}\left(\tilde{v}_j - \tilde{v}_i\right), \pi_{L_{ij}}\left(t_j - t_i\right), \pi_{M_{ij}}\left(\left(\tilde{v}_j - \tilde{v}_i\right)/\left(t_j - t_i\right)\right)\right\}$$

For those sections for which the assignation is given by $\tilde{v}_i = D_{0i}, \tilde{v}_j = D_{0j}, t_i = t_{[p]}$ and $t_j = t_{[q]}$ the compatibility of the assignment is calculated with the expression:

$$\pi_{R_{ij}}\left(\tilde{A}_i, \tilde{A}_j\right) = \pi_{L_{ij}}\left(t_j - t_i\right)$$

**Definition 10.** We call $\tilde{A} = \left(\tilde{A}_0, \tilde{A}_1, \tilde{A}_2, ..., \tilde{A}_n\right)$ a **valid segmentation** of the evolution $E$ induced by an FTP if and only if it satisfies:

$$\mu^N\left(\tilde{A}\right) = \min_{\substack{0 \le i, j \le n \\ \tilde{v}_0 = 0, t_0 = 0}} \left\{\pi_{R_{ij}}\left(\tilde{A}_i, \tilde{A}_j\right), \mu_{S_{ij}}\left(\tilde{A}_i, \tilde{A}_j\right)\right\} > 0 \tag{13}$$

where $S_{ij}$ is the semantics that corresponds to the section between each two significant points. $\mu^N\left(\tilde{A}\right)$ is the degree of compatibility of this segmentation with the definition of the profile.

In order to obtain each one of the segmentations of the signal we have designed a search tree-based procedure, so that, following an ordered method, a significant number of spurious assignments may be ruled out, thus reducing the computational cost of the process. In this way, each of the segmentations is built up incrementally. Thus, a definition of the validity of an incomplete segmentation is required.

**Definition 11.** We say that an ordered k-tuple of generalized assignments $\tilde{A}^{[k]} = \left(\tilde{A}_{h_1}, \tilde{A}_{h_2}, \cdots, \tilde{A}_{h_k}\right)$ to $k$ points of the FTP $(k < n)$ *is* **locally valid** if and only if it satisfies:

$$\mu^N_{h_1, ..., h_k}\left(\tilde{A}_{h_1}, ..., \tilde{A}_{h_k}\right) =$$
$$= \min_{\tilde{A}_{h_i}, \tilde{A}_{h_j} \in \left(\tilde{A}_0, \tilde{A}_{h_1}, ..., \tilde{A}_{h_k}\right)} \left\{\pi_{R_{h_i h_j}}\left(\tilde{A}_{h_i}, \tilde{A}_{h_j}\right), \mu_{S_{ij}}\left(\tilde{A}_{h_i}, \tilde{A}_{h_j}\right)\right\} > 0 \tag{14}$$

where $h = (h_1, h_2, \ldots, h_k)$ is any assignment order. $\mu^N_{h_1, \ldots, h_n}\left(\widetilde{A}_{h_1}, \ldots, \widetilde{A}_{h_k}\right)$ is the degree of compatibility of this incomplete segmentation with the definition of the profile.

From the definition itself it is directly derived how the extension of a local assignment to any new significant points does not improve the degree of compatibility [11]:

$$\left(\widetilde{A}_{h_1}, \ldots, \widetilde{A}_{h_k}\right) \subseteq \left(\widetilde{A}_{h_1}, \ldots, \widetilde{A}_{h_k}, \ldots, \widetilde{A}_{h_l}\right) \Rightarrow \mu^N_{h_1, \ldots, h_k}\left(A^{[k]}\right) \geq \mu^N_{h_1, \ldots, h_k, \ldots, h_l}\left(A^{[l]}\right)$$

The compatibility can be obtained incrementally, based on the calculation carried out on the previous assignment:

$$\mu^N_{h_1, \ldots, h_{k+1}}\left(\mathsf{A}^{[k+1]}_h\right) =$$

$$= \min_{\widetilde{A}_{h_i} \in A^{[k]}_h, \widetilde{A}_{h_{k+1}}} \left\{ \mu^N_{h_1, \ldots, h_k}\left(\mathsf{A}^{[k]}_h\right), \pi_{R_{h_i h_{k+1}}}\left(\widetilde{A}_{h_i}, \widetilde{A}_{h_{k+1}}\right), \mu_{S_{h_i h_{k+1}}}\left(\widetilde{A}_{h_i}, \widetilde{A}_{h_{k+1}}\right) \right\}$$

The search tree has as many levels as there are significant points, and it branches out on the possible assignments to each one of them. The first node of the tree represents the assignment to the significant point $X_0$ which we have set at $(0, 0)$, and the leaves represent complete assignments. Thus, we incrementally construct a segmentation on the signal, by means of successive assignments to the significant points of the profile, where the degree of compatibility with the real evolution of the variable under consideration is calculated in a partial manner. In order to do so, we use the typical *first-in-depth* search method [24]. With the aim of delimiting the search to sufficiently satisfactory solutions, we consider a level below $c_{\inf}$, which serves to prune all those branches from which a compatibility higher than this level is not obtained.

Given that the search is based on an *a priori* order in carrying out assignments, for the sake of simplicity in the resulting expression, we take the temporal order of the significant point themselves, although at a later point, it shall be seen that in certain cases it is more efficient to follow a different order.

At each step of obtaining the global compatibility between the evolution and the FTP we extend a k-tuple $\left(A_0, \widetilde{A}_1, \ldots, \widetilde{A}_k\right)$ of locally valid assignments to the significant point $X_{k+1}$. If there is an instant so that the local consistency hence we carry out the generalized assignment to the significant point $X_{k+1}$. If there is an instant $t_{[m]} : \left(v_{[m]}, t_{[m]}\right) \in E$ so that the local consistency $\mu^N_{0, 1, \ldots, k+1}\left(A_0, \widetilde{A}_1, \ldots, \widetilde{A}_{k+1}\right) > c_{\inf}$, hence we carry out the generalized assignment $\widetilde{A}_{k+1}$ to the significant point $X_{k+1}$. If no generalized assignment is found that satisfies the previous condition, we return to the assignment of the previous significant point, $X_k$. When a global assignment $\left(A_0, \widetilde{A}_1, \ldots, \widetilde{A}_n\right)$ with a compatibility greater than $c_{\inf}$ is found, this is considered as the best segmentation to date. As we are not interested in obtaining all possible segmentations, rather in calculating as rapidly as possible the compatibility of the best ones, we then update $c_{\inf} = \mu^N\left(A_0, \widetilde{A}_1, \ldots, \widetilde{A}_n\right)$ and we go back in search of a better assignment. The inclusion of a higher level $c_{\sup} > c_{\inf}$ can be considered for the selection of the best ones, which, together with $c_{\inf}$, narrows down the range of the search, and thus increases the efficiency of the matching process. One drawback of this is that narrowing down the range increases indeterminacy in the detection, and leads to a fall in the quality of the resulting information.

```
procedure MATCHING(i, min, max);
begin
 maxx=min;
 if (i=n) then
 return(max);
```

$$L_i \leftarrow \{ t_{[m]} : \mu^N_{0,1,\ldots,i}(A_0, \tilde{A}_1, \ldots, \tilde{A}_i) \geq \min \};$$

```
while (L_i ≠ 0) do
begin
to take and erase a t_[m] from L_i;
```

$$\text{maxx} = \mu^N_{0,1,\ldots,i}(A_0, \tilde{A}_1, \ldots, \tilde{A}_i);$$

$$\text{maxx} = \min(\max, MATCHING(i+1, \min, \text{maxx}));$$

```
end;
return(maxx);
end;
```

Figure 9. Matching procedure.

The optimum calculation for the degree of membership of each one of the samples $\left( v_{[m]}, t_{[m]} \right)$ to the FTP is given by the maximum of the compatibility degrees of each one of the segmentations to whose temporal interval it belongs:

$$\mu^N \left( v_{[m]}, t_{[m]} \right) = \max_{t_1 \leq t_{[m]} \leq t_n} \left\{ \mu^N \left( A_0, \tilde{A}_1, \ldots, \tilde{A}_n \right) \right\}$$

In figure 9 we present a high-level version of the matching algorithm. For each significant point a list $L_i$ is constructed in which those signal samples $\left( v_{[m]}, t_{[m]} \right)$ whose assignment is possible are stored. The algorithm is recursive, and in one single procedure resolves assignment and fault management by going back to the previous assignment.

Even though we will go on to deal with the efficiency of the matching process in greater depth, it is possible to now introduce certain very general heuristics in order to prune the assignment tree more efficiently.

Thus we should take into account that if a local assignment $\left( \tilde{A}_1, \ldots, \tilde{A}_k \right), k < n,$ maintains the same degree of compatibility when it is extended to a global assignment, all possible extensions that are based on the significant point $X_k$ can be omitted, as no better assignment will be obtained.

If to this we add an order in the possible assignments to each significant point, so that, in each local assignment we take the one with the highest degree of compatibility with the corresponding constraints, this enables us to reject a significant number of partial assignments that do not improve on the last solution found.

### 6. 1. Profile Localization

Due to the high computational cost of the matching procedure, based on a search algorithm, we employ a set of heuristics in order to increase its computational efficiency, which results in a more rapid localization of those signal fragments that show the highest compatibility with the definition of the profile. These heuristics refer fundamentally to the order of assignment to the significant points, searching for the one that restricts to the greatest degree the search area on the signal data.

We search for an order on the signal segmentation that firstly realizes the assignment on the significant points that are defined by means of those constraints which localize them most precisely on the signal.

Thus, those significant points that are described with a temporal reference to the origin, such as "*a little after admission into the Intensive Care Unit*" have a fairly precise localization on the signal, although they still depend on the intrinsic vagueness of terms such as '*a little after*', or to the moment of admission that is referred to; once significant points are introduced with expressions corresponding to temporal increases or durations, of the type "*approximately 10 minutes later, the temperature is considerably lower*", their localization becomes more imprecise, due to the propagation of these constraints being realized with a fuzzy addition operation, in which vagueness increases.

In the case of the FTP not being defined with an absolute temporal localization $\left(\pi_{L_{0i}} = \pi_u, \forall i \in \{1, ..., n\}\right)$, the matching process has to cover the entire signal. In order to make this process more efficient, we initially search for the salient features of the FTP on the signal, i.e., an event or an episode that is relevant in its definition.

As an example of the use of a relevant event, figure 10 shows the result of initially localizing one of the significant points on the signal, which serves as an initial filtering in order to highlight those areas of the signal in which to start to search for the profile with a higher probability of finding it.

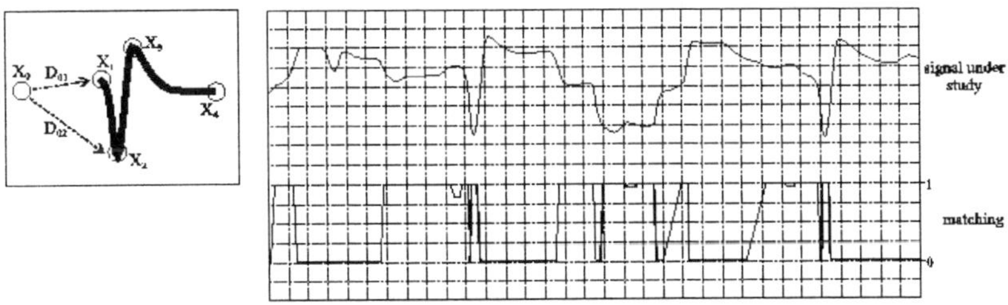


Figure 10. In this example we carry out the detection of the profile of the left of the figure. This shows the compatibility of each sample of the evolution with the fuzzy values corresponding to the constraints $D_{01}$ (thin line) and $D_{02}$ (thick line) corresponding to the significant points $X_1$ and $X_2$. It can be seen how the latter is the best starting point for the matching process.

### 6. 2. Non-optimal Matching Procedures

In the previous section we proposed a search tree-based segmentation procedure, with a theoretically high computational complexity, which grows with the number of significant points, for carrying out the optimal segmentation of the signals that are analyzed. Our aim has been to endow the segmentation procedure with the maximum degree of expressiveness allowed by the representation of an evolution by means of FTPs. Even so, the complexity of the matching procedure need not be a handicap in the use of FTP. This is possible by way of suitable design: even though the model allows the representation of vagueness such as "*later the temperature rose again*" where the temporal support is not delimited, experts usually describe profiles with a reduced number of sections in which for expediency vagueness is restricted. In any case, complying with the theoretical results, this complexity makes it unadvisable to apply these matching algorithms in systems working with hard real-time demands.

The proposed matching algorithms are a starting point, from which to lower the quality in the pattern recognition process in return for greater computational efficiency in the recognition task. The idea is to satisfy local criteria that determine segmentation, to the detriment of optimal global fulfillment. Thus, at the other extreme we can propose the typical solution, with linear computational complexity with regard to the number of samples processed, and which consists of carrying out the segmentation of a signal by maximizing its membership with respect to each of the two neighboring sections.

## 7. Conclusions

The Fuzzy Temporal Profile model is based on the linguistic acquisition of the information that describes a signal pattern of interest, and it is proposed as the study of a computational representation of the semantics contained in certain typical expressions which we use to describe the evolution of a physical parameter. It is precisely the capacity of FTPs to model vague or imprecise knowledge that makes their application possible in those domains where experts interpret the behavior of a system on the basis of identifying certain known profiles in the evolution of the parameters that characterize the system (patient monitoring, process control, etc.).

In this work we have extended the previous FTP model by modeling those linguistic descriptions that quantify the discourse universe upon which the verification of an assertion on the evolution of a parameter is required. Taking natural language as a starting point, we find that quantification is habitually described in imprecise terms, and in the majority of cases it is expressed more implicitly than explicitly [30]. It is for this reason that it was decided not to undertake a rationalization of quantification, rather to explain the manner in which the FTP model is capable of modeling a good number of the linguistic descriptions that are present in common discourse.

We represent the different quantifiers by means of the possibility distributions which describe in a fuzzy manner the percentage quantity of signal samples that verify a value or evolution assertion in a given time interval. For the calculation of the compatibility of a signal fragment with the linguistic description that is modeled in the FTP we employ compensatory operators, in which a numerical cardinality measure makes it possible to modify the behavior of the aggregation between a t-norm and a t-conorm. The selection of the specific t-norm and t-conorm operators determines the more or less optimistic nature of the calculation and is a design criterion, being chosen according to the best adaptation to the properties of the physical parameter and the subjectivity of whoever makes the linguistic description.

**Acknowledgments.** This work was funded by the Secretaría Xeral de I+D of the Xunta de Galicia and the Spanish Ministry of Education and Culture (CICYT) through research projects PGIDT00PXI20612 and TIC2000-0873-C02, respectively.

## References

[1]  J. Allen, Towards a general theory of action and time. *Artificial Intelligence* **23**, 1984, PP. 123-154.

[2]  S. Barro, R. Marín, J. Mira and A. Patón, A model and a language for the fuzzy representation and handling of time. *Fuzzy Sets and Systems*, **61**, 1994, pp. 153-175.

[3]  P. Cariñena, A. Bugarín and S. Barro, Un modelo para la cuantificación de la persistencia en proposiciones temporales borrosas. *Actas del VIII congreso español sobre Tecnologías y Lógica Fuzzy*, 1998, pp. 109-116. (in Spanish)

[4]  J. T. Y. Cheung and G. Stephanopoulos, Representation of process trends - Part I: A formal representation framework. *Computers Chemical Engineering*, **14** (4/5), 1990, pp. 495-510.

[5]  R. Dechter, I. Meiri and J. Pearl, Temporal constraint networks. *Artificial Intelligence*, **49**, 1991, pp. 61-95.

[6]  M. Delgado, D. Sánchez and M. A. Vila, Un método para la evaluación de sentencias con cuantificadores lingüísticos, in: *Proceedings of Estylf'98*, 1997 (in Spanish), pp. 193-198.

[7]  M. Delgado, D. Sánchez and M. A. Vila, Fuzzy cardinality based evaluation of quantified sentences. *International Journal of Approximate Reasoning*, **23**, 2000, pp. 23-66.

[8]  J. A. Drakopoulos and B. Hayes-Roth, tFPR: A fuzzy and structural pattern recognition system of multi—variate time--dependent pattern classes based on sigmoidal functions. *Fuzzy Sets and Systems*, **99**, 1998, pp. 57-72.

[9]  D. Dubois and H. Prade, Possibility theory: an approach to computerized processing of uncertainty. Plenum Press, New York, 1985.

[10] D. Dubois and H. Prade, Processing fuzzy temporal knowledge. *IEEE Transactions of Systems, Man and Cybernetics*, **19** (4), 1989, pp. 729-744.

[11] D. Dubois, H. Fargier and H. Prade, Possibility theory in constraint satisfaction problems: handling priority, preference and uncertainty. *Applied Intelligence*, **6**, 1996, pp. 287-309.

[12] P. Félix, S. Barro, R. Marín, M. J. Taboada and A. Engel, Patrones temporales borrosos en la supervisión de pacientes. Actas del V congreso español sobre Tecnologías y Lógica Fuzzy, 1995 (in Spanish), pp. 321-326.

[13] P. Félix, S. Fraga, R. Marín and S. Barro, Linguistic representation of fuzzy temporal profiles. *International Journal of Uncertainty, Fuzziness and Knowledge--Based Systems*, **7** (3), 1999, pp. 243-256.

[14] P. Félix, Perfiles Temporales Borrosos: Un modelo para la representación y reconocimiento de patrones sobre señal, Tesis Doctoral. Universidade de Santiago de Compostela, 1999 (in Spanish).

[15] P. Félix and S. Barro, A fuzzy model for the representation and recognition of linguistically described trends. Intelligent Data Analysis, in Press, 2001.

[16] I. J. Haimowitz, P. P. Le and I. S. Kohane, Clinical monitoring using regression-based trend templates. *Artificial Intelligence in Medicine*, **7**, 1995, pp. 473-496.

[17] A. Kaufmann and M. M. Gupta, Introduction to fuzzy arithmetic. Van Nostrand Reinhold, 1985.

[18] A. Lowe, M. J. Harrison and R. W. Jones, Diagnostic monitoring in anesthesia using fuzzy trend templates for matching temporal patterns. *Artificial Intelligence in Medicine*, **16**, 1999, pp. 183-199.

[19] M. Mizumoto, Pictorial Representations of Fuzzy Connectives, Part I: Cases of t-norms, t-conorms and averaging operators. *Fuzzy Sets and Systems*, **31**, 1989, pp. 217-242.

[20] M. Mizumoto, Pictorial Representations of Fuzzy Connectives, Part II: Cases of Compensatory Operators and Self-Dual Operators. *Fuzzy Sets and Systems*, **32**, 1989, pp. 45-79.

[21] U. Montanari, Networks of constraints: fundamental properties and applications to picture processing. *Information Science*, **7**, 1974, pp. 95-132.

[22] A. V. Oppenheim and R. W. Schafer, Discrete-time signal processing. Prentice Hall, 1989.

[23] D. Qian, Representation and use of imprecise temporal knowledge in dynamic systems. *Fuzzy Sets and Systems*, **50**, 1992, pp. 59-77.

[24] S. J. Russell and P. Norvig, Artificial Intelligence: a modern approach. Prentice Hall, 1996.

[25] F. Steimann, The interpretation of time--varying data with DIAMON-1. *Artificial Intelligence in Medicine*, **8**, 1996, pp. 343-357.

[26] M. A. Vila, J. C. Cubero, J. M. Medina and O. Pons, Using OWA operator in flexible query processing, in: Yager, R. R., Kacprzyk, J. (Eds.), The Ordered Weighted Averaging Operators: Theory, Methodology and Applications, 1997, pp. 258-274.

[27] M. Vilain and H. Kautz, Constraint propagation algorithms for temporal reasoning. *Proceedings of the AAAI-86*, 1986, pp. 377-382.

[28] R. R. Yager, On ordered weighted averaging aggregation operators in multi-criteria decision-making. *IEEE Transactions on Systems, Man and Cybernetics*, **18**, 1988, pp. 183-190.

[29] L. A. Zadeh, The concept of a linguistic variable and its application to approximate reasoning (Part 1). *Information Science*, **8**, 1975, pp.199-249.

[30] L. A. Zadeh, A computational approach to fuzzy quantifiers in natural languages, *Computing and Mathematics with Applications*, **9**(1), 1983, pp. 149-184.
[31] H. J. Zimmermann and P. Zysno, Latent connectives in human decision making. *Fuzzy Sets and Systems* **4**, 1980, pp. 37-51.

*Systematic Organisation of Information in Fuzzy Systems*
*P. Melo-Pinto et al. (Eds.)*
*IOS Press, 2003*



# Data Mining with Possibilistic Graphical Models

Christian BORGELT and Rudolf KRUSE
*Dept. of Knowledge Processing and Language Engineering*
*Otto-von-Guericke-University of Magdeburg*
*Universitätsplatz 2, D-39106 Magdeburg, Germany*
*E-mail: {borgelt,kruse}@iws.cs.uni-magdeburg.de*

**Abstract**: *Data Mining*, also called *Knowledge Discovery in Databases*, is a young research area, which has emerged in response to the flood of data we are faced with nowadays. It has taken up the challenge to develop techniques that can help humans discover useful patterns in data. One such technique, which certainly is among the most important, as it can be used for frequent data mining tasks like classifier construction and dependence analysis – is *graphical models* and especially learning such models from a dataset of sample cases. In this paper, we review the basic ideas of graphical modeling, with a focus on possibilistic networks, and we study the principles of learning graphical models from a dataset of sample cases.

## 1. Introduction

Today electronic information processing systems are used by almost every company, in departments like production, marketing, stockkeeping, or personnel. These systems were developed, to enable users finding certain information, for example the address of a customer, in a fast and reliable way. Today, however, due to increasingly powerful computers and advances in database and software technology, we may also think about using such collections of data not only for retrieving specific information that is needed at a given moment, but also to search for more general knowledge that is hidden in them. If, for instance, a supermarket analyses the receipts of its customers (which are easily collectible with scanner cashiers) and thus discovers that certain products are frequently bought together, turnover of these products may be increased by properly arranging them on the shelves of the market. Unfortunately, in order to find such hidden knowledge, the retrieval capacities of standard database systems and the methods of classical data analysis are rarely sufficient. These only allow us to retrieve individual data items as well as to compute simple aggregations like average regional sales. We may also test hypotheses like whether the day of the week has any influence on the production quality. More general patterns, structures, or regularities, however, go undetected. But often, knowing these patterns would make it possible, for example, to increase turnover or product quality. Consequently, in recent years we have seen the emergence of a new research area – often called "Knowledge Discovery in Databases" (KDD) or "Data Mining" (DM) – which focuses on automatically generating and testing hypotheses and models that describe the regularities in a given (large) dataset. Hypotheses and models found in this way are then used to make predictions and to justify decisions.

In this paper, we concentrate on a single data mining method, namely the automatic construction of graphical models from a dataset of sample cases. This method is very important, because it can be used to tackle such frequent data mining tasks as classifier construction and dependence analysis. We focus on possibilistic graphical models, which are introduced as fuzzyfications of relational graphical models.

## 2. Graphical Models: A Simple Example

The idea underlying graphical modeling is most easily explained with a simple example, which we study first in the relational and then in the possibilistic setting. The relational case has the advantage that we can neglect degrees of possibility, which may obscure the very simple structure. Possibilistic graphical models are then introduced as straightforward generalizations of relational models and are thus somewhat easier to understand than their probabilistic counterparts, although the basic structure is identical. Our example domain consists of a set of geometrical objects, as shown in Figure 1. These objects are characterized by three attributes: color (or hatching), shape, and size. As already indicated, we neglect degrees of possibility for the time being and consider only whether a certain state, i.e., a certain combination of attribute values, is possible or not. This enables us to represent the objects as a simple relation, which is shown in the table in Figure 1.

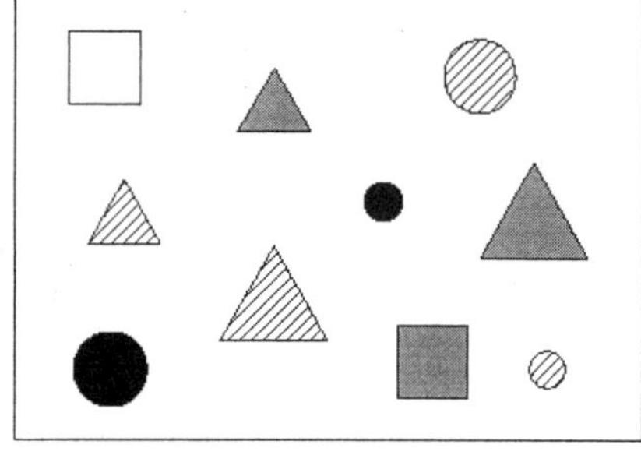

| color | shape | size |
|:-----:|:-----:|:----:|
| ■ | ○ | small |
| ■ | ○ | medium |
| ▨ | ○ | small |
| ▨ | ○ | medium |
| ▨ | △ | medium |
| ▨ | △ | large |
| □ | □ | medium |
| ▦ | □ | medium |
| ▦ | △ | medium |
| ▦ | △ | small |

Figure 1. A set of simple geometrical objects and the corresponding relation.

Suppose that an object of the set is chosen at random, but let us assume that not all attributes of the object can be observed. We may imagine, for example, that the object is drawn from a box at some distance, so that we can see the color, but cannot discern the shape or the size. We know, however, that there are only ten objects with certain values of the three attributes. How can we use this information to draw inferences about the unobserved properties?

Problems of this kind frequently occur in applications, for instance, in medical diagnosis: From textbooks and experience, a physician knows about the dependences between diseases and symptoms, perhaps in the context of other properties of the patient, like age or sex. But she/he can only observe or ask for symptoms as well as age, sex, the patient's history etc. Which disease or diseases are present, she/he has to infer with the help of her/his medical knowledge.

In our illustrative example the solution is, of course, trivial: Simply traverse the table and discard from it all objects with a different color than the one observed; then collect the possible shapes and sizes from the rest. However, this is possible only, because we have only a few objects and a few attributes. In medical diagnosis, this procedure is inapplicable, because the table we had to construct would be much too large to process. Therefore, we have to appropriately structure the medical knowledge of the physician; for example, by decomposing it into dependences between few attributes. Although our example is considerably simpler than the complex domains we have to handle in practice, it can be used to demonstrate how voluminous (tabular) knowledge can be decomposed, so that it becomes manageable. The table describing the geometrical objects can be decomposed, without loss, into two smaller tables, from which it can be reconstructed. We illustrate this by representing the domain as a three-dimensional space, each dimension of which we associate with an attribute. In this way, each possible combination of attribute values can be represented by a cube in this space, see Figure 2.

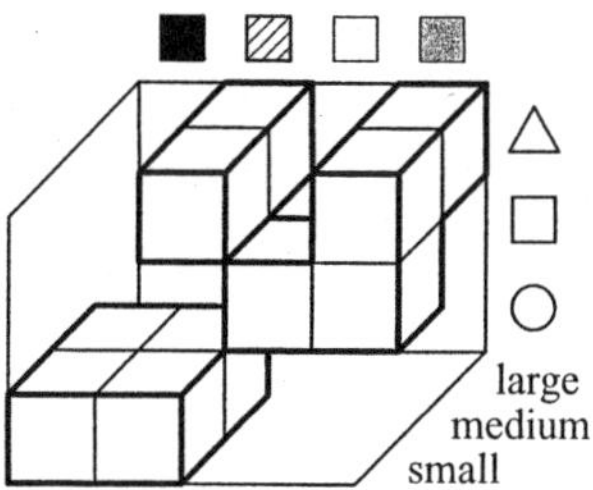

Figure 2. The reasoning space of the simple geometrical objects example shown in Figure 1. Each cube represents one tuple of the relation.

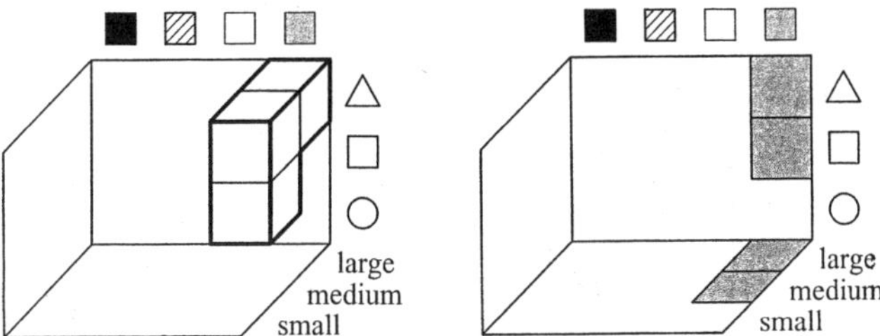

Figure 3. Reasoning in the space as a whole consists in restricting the relation to the "slice" that corresponds to the observation made.

Let us assume that the randomly chosen object is grey. In the representation just described, the naive way of reasoning consists in cutting out the "slice" that is associated with the color grey, as shown in Figure 3. In this way we infer that the object cannot be a circle, but must be square or a triangle, and that it cannot be small, but must be medium or large. However, this inference can also be drawn in a different way, since the knowledge about the objects can be decomposed into so-called projections to two-dimensional subspaces. All possible such projections are shown in Figure 4. They result as shadows thrown by the cubes if light sources are imagined (in sufficient distance) in front, to the right, and above the reasoning space shown in Figure 1. The relation can be decomposed into the projections to the back plane and to the left plane of the reasoning space, because it can be reconstructed from them. This is demonstrated in Figure 5: First we form the so-called cylindrical extensions of the two projections.

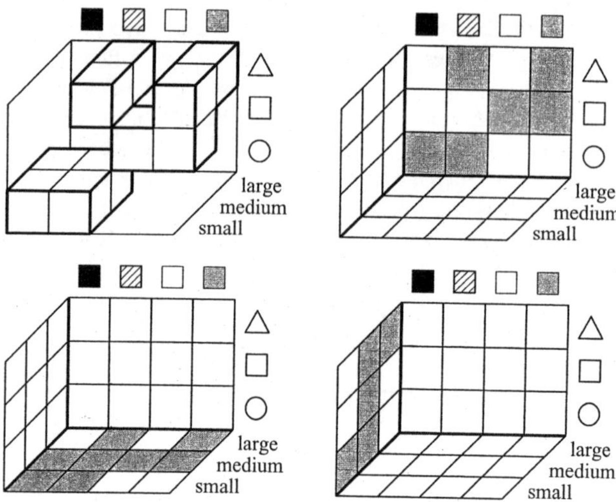

Figure 4. Projections of the relation showed in Figure 1 to the three possible two-dimensional sub-spaces. They are shadows thrown by the cubes if light sources are imagined in front, to the left, and above the reasoning space.

That is, we add all values of the missing dimensions. (The name "cylindrical extension" for this operation is derived from the common practice to sketch sets as circles: Adding a dimension to a circle yields a cylinder.) The resulting three-dimensional relations are intersected, i.e., only cubes contained in both are kept. The result is shown in Figure 5. Obviously it coincides with the original relation (cf. Figure 2).

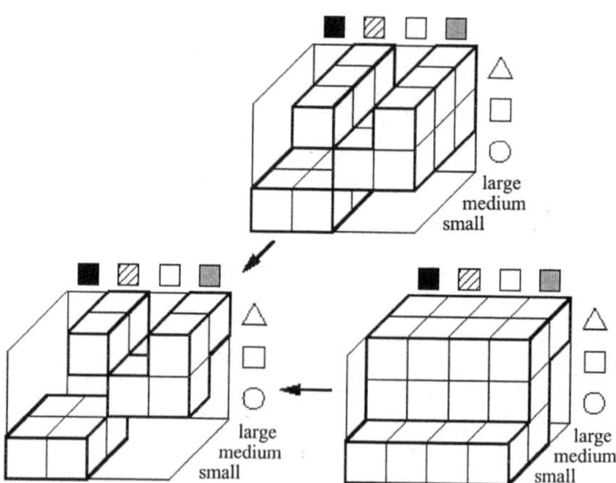

Figure 5. Cylindrical extensions of two projections of the relation depicted in Figure 1 and their intersection. This demonstrates that the relation can be decomposed into two two-dimensional projections.

The advantage of relational decomposition is that it can be exploited to draw inferences, without having to reconstruct the three-dimensional representation first. This is demonstrated in Figure 6.

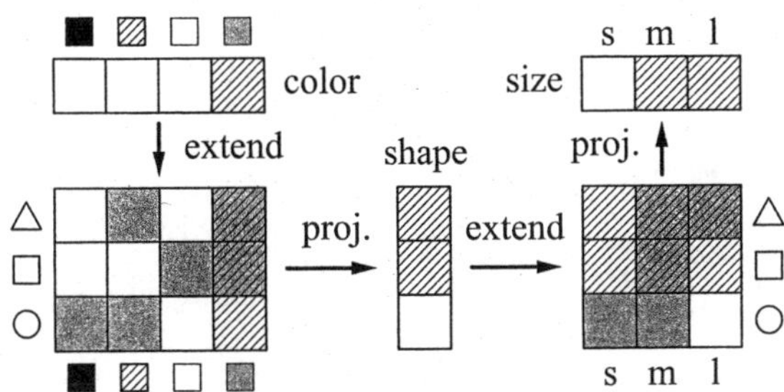


Figure 6. Propagating the evidence that the object is grey. It is not necessary to reconstruct the original relation:
We can work with the projections directly.

First, the observation that the object is grey is extended cylindrically to the subspace color×shape (hatched column) and intersected with the projection of the relation to this subspace (grey fields). The result is projected to the shape dimension. From this projection we read, just as we found out above, that the object cannot be a circle, but must be a square or a triangle.

Analogously, this result is extended cylindrically to the subspace shape×size (hatched row), intersected with the projection of the relation to this subspace (grey fields), and finally projected to the size dimension. This yields that the object cannot be small, but must be medium or large.

This reasoning procedure suggests representing the reasoning space as a graph or network, as shown in Figure 7. Each node of this network stands for an attribute and the edges indicate which projections are needed. It should be noted, though, that the subspaces are not always two-dimensional as in this very simple example. In applications the subspaces may have three, four, or more dimensions. Accordingly, the edges in the corresponding network then connect more than two nodes (thus forming the *hyperedges* of so-called *hypergraphs*).

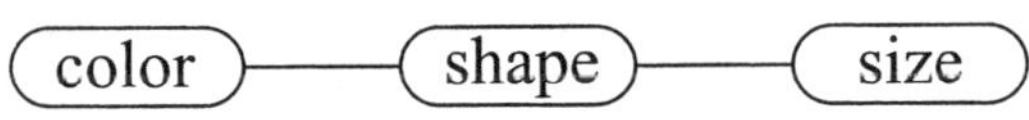


Figure 7. Network representation

Furthermore, it should be noted that the projections have to be chosen carefully:

Not just any two projections will do. This is demonstrated in Figure 8, where instead of the projection to the back plane we use the projection to the bottom plane. The intersection of the cylindrical extensions of these two projections, which is shown on the bottom left in Figure 8, differs considerably from the original relation, which is shown again on the top right. But not only have the projections to be chosen with care, sometimes it is not even possible to find a decomposition. To see this, consider Figure 9, in which two cubes are marked. Suppose first that the cube marked 1 is removed. It is easily verified that the resulting relation can no longer be decomposed into two projections to two subspaces. However, it is still possible to reconstruct the original relation by using all three possible two-dimensional projections: The intersection with the third projection (to the bottom plane) removes the additional cube 1. If, however, the cube marked 2 is removed, the relation cannot be decomposed any more. Removing this cube does not change any of the projections: In all three directions there is still another cube throwing the shadow.

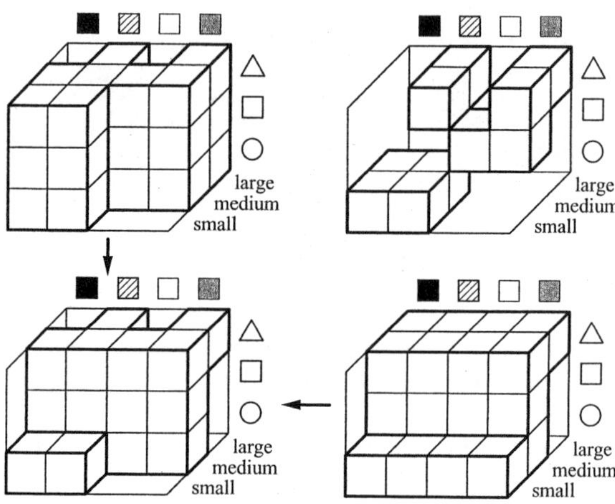

Figure 8. Not all choices of two projections are decompositions. If the wrong projections are selected, the intersection of their cylindrical extensions can contain many additional many additional tuples.

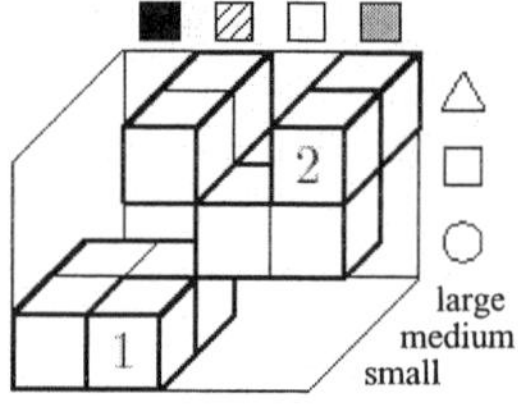

Figure 9. It is not always possible to decompose a relation. In this case approximations may have to be accepted.

Hence, the cube marked 2 is contained in all intersections of projections to subspaces. Such situations are common in practice. But since in applications it is usually impossible to manage the domains under consideration without decomposition, approximations have to be accepted. That is, if no (exact) decomposition is possible, it is tried to find a set of subspaces of limited size so that the intersection of the cylindrical extensions of the projections to these subspaces contains as few additional tuples as possible (obviously, an approximate decomposition can contain only additional tuples).

The idea underlying relational graphical models is easily generalized to probabilistic and possibilistic graphical models. Here we confine to the possibilistic case, though. (Details about probabilistic graphical models can be found, for instance, in [1][2][3][4]). In the possibilistic setting the (binary) information whether a combination of attribute values is possible or not is replaced by a *degree of possibility* [5], the semantics of which we consider in more detail below. As an example, consider the three-dimensional possibility distribution shown in Figure 10, which is defined on the same space as the relation considered above. The only difference is that tuples that were contained in the relation now have a high degree of possibility, while tuples that were missing have a low degree of possibility. The marginal distributions, which take the place of the shadow projections, are computed by taking the maximum over the dimension along which the projection is carried out.

Like the relation studied above, this possibility distribution can be decomposed into the marginal distributions on the two subspaces color× shape and shape× size.

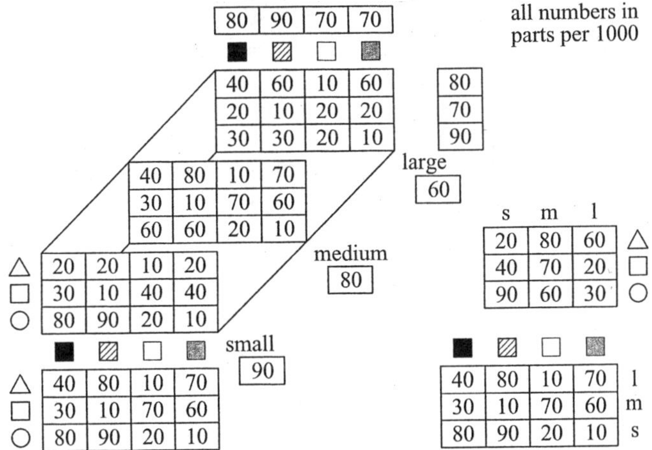

Figure 10. A simple possibility distribution that can be decomposed, just like the relation studied above, into marginal distributions on two subspaces.

From these marginal distributions, the original three-dimensional distribution can be reconstructed by computing the minimum of corresponding marginals. For instance, the value 20 for small black triangles is computed as the minimum of the value 40 for black triangles and the value 20 for small triangles. As in the relational case the possibility to decompose the distribution enables us to draw inferences using only the marginal distributions that form the decomposition without having to reconstruct the original three-dimensional distribution. This is demonstrated in Figure 11, assuming again that the randomly chosen object is observed to be green.

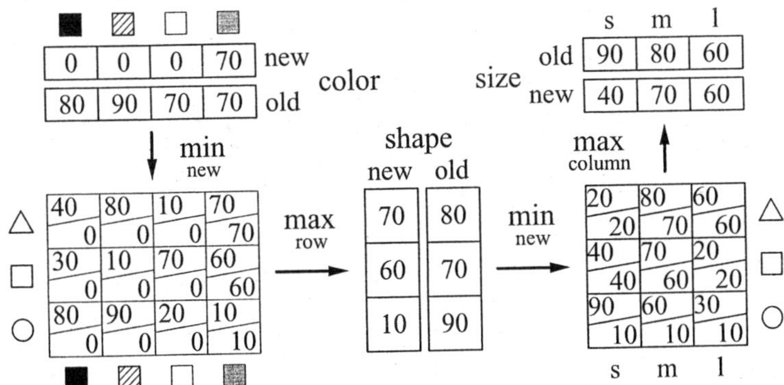

Figure 11. Propagation of the evidence that the object is grey, using only the two marginal distributions.

The reasoning procedure is exactly the same as in the relational case (cf. Figure 6): The evidence that the object is green is extended cylindrically to the subspace color×shape (setting all values in the same column to the marginal value) and intersected with the marginal distribution on this space (upper numbers) by taking the minimum. This yields the new distribution (lower numbers), which is projected to the shape dimension to obtain the degrees of possibility of the different shapes by taking the maximum over rows. The second step is analogous. The shape information is extended cylindrically to the subspace shape×size and intersected with the marginal distribution on this space (upper numbers) by taking the minimum. The resulting distribution (lower numbers) is projected to the size dimension by taking the maximum, thus yielding degrees of possibility for the different sizes.

## 3. Graphical Models: General Characterization

Based on the intuition conveyed with the simple examples of the preceding section, we now turn to a more formal characterization of graphical models.

### 3.1 Decomposition

The decomposition underlying relational graphical models is, of course, well known from the theory of relational databases [6] and actually relational database theory is strongly connected to the theory of graphical models. The connection is brought about by the notion of the *join-decomposability* of a relation, which in relational databases is exploited to store a high-dimensional relation with less redundancy and, obviously, using less storage space.

The idea underlying join-decomposability is that often a relation can be reconstructed from certain *projections* of it by forming their so-called natural join. Formally, this can be described as follows: Let $U = \{A_1, \ldots, A_n\}$ be a set of attributes and let $dom(A_i)$ be their respective domains. Furthermore, let $r_u$ be a relation over $U$. We represent this relation by its *indicator function*, which assigns a value of 1 to all tuples contained in the relation and a value of 0 to all tuples not contained in it. The tuples themselves are represented as conjunctions $\wedge_{A_i \in U} A_i = a_i$, which state a value for each of the attributes. Using an indicator function a projection $r_M$ of the relation $r_u$ to a subset $M$ of the attributes in $U$ can easily be defined by

$$r_M\left( \bigwedge_{A_i \in M} A_i = a_i \right) = \max_{\substack{\forall A_j \in U-M: \\ a_j \in dom(A_j)}} r_U\left( \bigwedge_{A_i \in M} A_i = a_i \right),$$

where the somewhat sloppy notation w.r.t. the maximum is meant to indicate that the maximum has to be taken over all values of all attributes in $U - M$.

With this notation a relation $r_U$ is called *join-decomposable* w.r.t. a family $\mathbf{M} = \{M_1, \ldots, M_m\}$ of subsets of $U$ iff

$$\forall a_1 \in dom(A_1): \ldots \forall a_n \in dom(A_n):$$

$$r_U\left( \bigwedge_{A_i \in U} A_i = a_i \right) = \min_{M \in \mathbf{M}} r_M\left( \bigwedge_{A_i \in M} A_i = a_i \right)$$

Note that the minimum operation used here is equivalent to the natural join of relational algebra. It is obvious that in such a situation it suffices to store the projections $r_M$ in order to capture all information contained in the relation $r_U$, because we can always reconstruct the original relation.

The decomposition scheme we just outlined for the relational case is easily transferred to the possibilistic case: We only have to extend the range of values of the indicator function to the real interval $[0; 1]$, i.e., we use *possibility distributions* instead, thus "fuzzifying" relational graphical models. In this way, a gradual possibility of a tuple is modeled. The decomposition formula is identical:

$$\forall a_1 \in dom(A_1): \ldots \forall a_n \in dom(A_n):$$

$$\pi_U\left(\bigwedge_{A_i \in U} A_i = a_i\right) = \min_{M \in \mathbf{M}} \pi_M\left(\bigwedge_{A_i \in M} A_i = a_i\right)$$

To define semantics of *degrees of possibility* we rely on the *context model* [7]: Suppose that for a description of the modeled domain we can distinguish between a set $C = \{c_1, \ldots, c_k\}$ of contexts. These contexts may be given, for example, by physical or observation-related frame conditions. Furthermore, suppose that we can describe the relative importance or frequency of occurrence of these contexts by assigning a probability $P(c)$ to each of them.

Finally, suppose that we can state for each context $c$ a set $\Gamma(c)$ of possible states – described by tuples – the modeled domain may be in under the physical or observation-related frame conditions that characterize the context. We assume each set $\Gamma(c)$ to be the *most specific correct set-valued specification* of the state $t_0$ of the modeled domain, which we can give for the context $c$. By "most specific set-valued specification" we mean that we can guarantee that $\Gamma(c)$ contains $t_0$, but that we cannot guarantee that a proper subset of $\Gamma(c)$ contains $t_0$. Given these ingredients, we define the *degree of possibility* that a tuple $t$ describes the actual state $t_0$ of the modeled section of the world as the weight (probability) of all contexts in which t is possible.

Formally, the above description results in a *random set* [8][9] (i.e., a set-valued random variable) $\Gamma : C \to 2^T$ as an imperfect (i.e., imprecise and uncertain) specification of the actual state $t_0$ of the modeled section of the world. From it we derive a possibility distribution by simply computing its *one-point coverage*

$$\pi_\Gamma : T \to [0,1], \quad \pi_\Gamma(t) = P(\{c \in C \mid t \in \Gamma(c)\}).$$

With this interpretation a possibility distribution represents uncertain and imprecise knowledge as can be seen by comparing it to a probability distribution and to a relation.

A probability distribution covers *uncertain*, but *precise* knowledge. This becomes obvious if one notices that a possibility distribution in the interpretation described above reduces to a probability distribution if

$$\forall c \in C : |\Gamma(c)| = 1,$$

i.e., if for all contexts the specification of $t_0$ is precise. On the other hand, a relation represents *imprecise*, but *certain* knowledge. Thus, not surprisingly, a relation can also be seen as a special case of a possibility distribution in the interpretation given above, namely if there is only one context. Hence the context-dependent specifications are responsible for the imprecision, the contexts for the uncertainty.

### 3.2 Graphical Representation

Graphs (in the sense of graph theory) are a very convenient tool to describe decompositions if we identify each attribute with a node. In the first place, graphs can be used to specify the sets $M$ of attributes underlying the decomposition. How this is done depends on whether the graph is directed or undirected. If it is undirected, the sets M are the maximal cliques of the graph, where a clique is a complete subgraph and it is maximal if it is not contained in another complete subgraph. If the graph is directed, we can be more explicit about the distributions in the decomposition: We can use conditional distributions, since we may use the direction of the edges to specify which is the conditioned attribute and which are the conditions. Note, however, that this does not make much of a difference in the relational and the possibilistic case, since here conditional distributions are simple identified with the corresponding marginal distributions, i.e.,

$$\pi\left( A_j = a_j \middle| \bigwedge_{A_i \in M} A_i = a_i \right) = \pi\left( A_j = a_j \wedge \bigwedge_{A_i \in M} A_i = a_i \right)$$

Secondly, graphs can be used to describe (conditional) dependence and independence relations between attributes via the concept of *separation* of nodes. What is to be understood by "separation" depends again on whether the graph is directed or undirected.

If it is undirected, separation is defined as follows: If X, Y, and Z are three disjoint subsets of nodes in an undirected graph, then Z separates X from Y iff after removing the nodes in Z and their associated edges from the graph there is no path, i.e., no sequence of consecutive edges, from a node in X to a node in Y. Or, in other words, Z separates X from Y iff all paths from a node in X to a node in Y contain a node in Z.

For directed graphs, which have to be acyclic, the so-called *d-separation criterion* is used [1][10]: If X, Y, and Z are three disjoint subsets of nodes, then Z is said to *d-separate* X from Y iff there is no path, i.e., no sequence of consecutive edges (of any directionality), from a node in X to a node in Y along which the following two conditions hold:

1. every node with converging edges either is in Z or has a descendant in Z,
2. every other node is not in Z.

These separation criteria are used to define *conditional independence graphs*: A graph is a conditional independence graph w.r.t. a given multi-dimensional distribution if it captures by node separation only correct conditional independences between sets of attributes. Conditional independence means (for three attributes A, B, and C with A independent of C given B) that

$$\pi(A = a, B = b, C = c) = \min\{\pi(A = a|B = b), \pi(C = c|B = b)\}$$

This formula indicates the close connection of conditional independence and decomposability. Formally, this connection between conditional independence graphs and graphs that describe decompositions is established by theorems that a distribution is decomposable w.r.t. a given graph if and only if this graph is a conditional independence graph of the distribution.

For the probabilistic setting, this theorem is usually attributed to [11], who proved it for the discrete case, although (according to [3]) this result seems to have been discovered in various forms by several authors. In the possibilistic setting similar theorems hold, although certain restrictions have to be introduced [12][13].

Finally, the graph underlying a graphical model is very useful to derive evidence propagation algorithms, since evidence propagation can be reduced to simple computations of node processors that communicate by passing messages along the edges of a properly adapted graph. A detailed account can be found, for instance, in [2][4].

## 4 Learning from Data: A Simple Example

Having reviewed the ideas underlying graphical models, we now turn to the problem how we can find a decomposition if there is one and how we can find a good approximation otherwise. If there is a human expert of the modeled application domain, we may ask him to specify an appropriate conditional independence graph together with the necessary distributions. However, we may also try to find a decomposition automatically by analyzing a dataset of example cases. In the following we study the basic ideas underlying such learning from data using again our simple geometrical objects example.

Suppose that we are given the table shown in Figure 1 and that we desire to find a (n approximate) relational decomposition, maybe satisfying certain complexity constraints. Of course, we could check all possible graphical models and count the additional tuples. However, such an approach, though feasible for our simple example, is prohibitive in practice due to the very high number of possible graph structures. It would be convenient if we could read from a projection (marginal distribution) whether we need it in a decomposition or not. Fortunately, there is indeed a simple heuristic criterion, which can be used for such local assessments and provides good chances to find an appropriate decomposition.

The basic idea is very simple: The intersection of the cylindrical extensions of the projections should contain as few additional tuples as possible, in order to be as close as possible to the relation to decompose. It is surely plausible that the intersection contains few combinations of attribute values if this holds already for the cylindrical extensions.

However, the number of combinations in the cylindrical extensions depends directly on the number of possible combinations of attribute values in the projections (forming the cylindrical extension only adds all values of the missing dimensions). Therefore we should select such projections, which contain as few combinations of attribute values as possible. In doing so we should take into account the size of the subspace projected to. The larger this subspace is, the larger the number of combinations will be, although this is not relevant for finding a good decomposition. Therefore we should consider not the absolute, but the relative number of value combinations. For our simple example the values of this criterion are shown in Table 1.

Table 1. Selection criteria for projections. Choosing the subspaces with the smallest (second column) or highest (third column) values yields the decomposition

| subspace | relative number of possible comb. | Hartley info. gain |
|---|---|---|
| color x shape | $\frac{6}{12} = \frac{1}{2} = 50\%$ | $\log_2 \frac{12}{6} = 1$ |
| color x size | $\frac{8}{12} = \frac{2}{3} \approx 67\%$ | $\log_2 \frac{12}{8} \approx 0.58$ |
| shape x size | $\frac{5}{9} = \frac{5}{9} \approx 56\%$ | $\log_2 \frac{9}{5} \approx 0.85$ |

It is easy to verify that choosing the subspaces with the smallest number of possible value combinations leads to the correct decomposition. The third column lists the binary logarithm of the reciprocal value of the relative numbers, which is also known as Hartley information gain. It is discussed in more detail below. Although this simple selection criterion works quite nicely in our simple example, it should be noted that it is not guaranteed to yield the best choice of projections. To see this consider Figure 12, which shows another three-dimensional relation together with the three possible projections to two-dimensional subspaces.

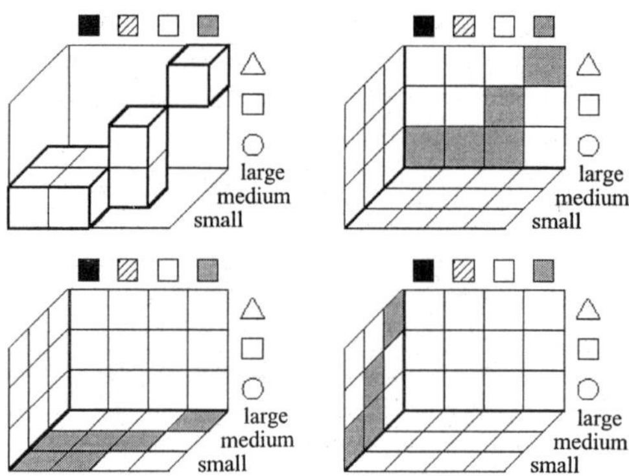

Figure 12. The selection criteria are only heuristics that not always find a possible decomposition: the relation in this figure is decomposable into the projections to the back plane and the bottom plane. However, the selection criteria do not yield this decomposition.

Although this relation can be decomposed into the projections to the back plane and to the bottom plane, the selection criterion just studied does not find this decomposition, but selects the back plane and the left plane.

## 5. Learning from Data: General Characterization

In general, there are three main approaches to learn a graphical model:
- Test whether a distribution is decomposable w.r.t. a given graph.
  This is the most direct approach. It is not bound to a graphical representation, but can also be carried out w.r.t. other representations of the subsets of attributes to be used to compute the (candidate) decomposition of the given distribution.

- Find a conditional independence graph by conditional independence tests.
  This approach exploits the theorems mentioned in the preceding section, which connect conditional independence graphs and graphs that describe decompositions. It has the advantage that by a single conditional independence test, if it fails, several candidate graphs can be excluded.
- Find a suitable graph by measuring the strength of dependences.
  This is a heuristic, but often highly successful approach, which is based on the frequently valid assumption that in a conditional independence graph an attribute is more strongly dependent on adjacent attributes than on attributes that are not directly connected to it.

Note that none of these methods is perfect. The first approach suffers from the usually huge number of candidate graphs. The second often needs the strong assumption that there is a perfect map (a conditional independence graph that captures all conditional independences by node separation). In addition, if it is not restricted to certain types of graphs (for example, polytrees), one has to test conditional independences of high order (i.e., with a large number of conditioning attributes), which tend to be unreliable unless the amount of data is enormous. The heuristic character of the third approach is obvious. A relational example in which it fails has been studied in the previous section (cf. Figure 12) and similar ones can be found for the possibilistic setting.

A (computationally feasible) analytical method to construct optimal graphical models from a database of sample cases has not been found yet. Therefore an algorithm for learning a graphical model from data usually consists of

1. an *evaluation measure* (to assess the quality of a given network) and
2. a *search method* (to traverse the space of possible networks).

It should be noted, though, that restrictions of the search space introduced by an algorithm and special properties of the evaluation measure sometimes disguise the fact that a search through the space of possible network structures is carried out. For example, by conditional independence tests all graphs missing certain edges can be excluded without inspecting these graphs explicitly. Greedy approaches try to find good edges or subnetworks and combine them in order to construct an overall model and thus may not appear to be searching. Nevertheless the above characterization is apt, since an algorithm that does not explicitly search the space of possible networks usually searches (heuristically) on a different level, guided by an evaluation measure. For example, some greedy approaches search for the best set of parents of an attribute by measuring the strength of dependence on candidate parents; conditional independence test approaches search the space of all possible conditional independence statements.

### 5.1 Computing Projections

A basic operation needed to learn a graphical model from a dataset of sample cases is a method to determine the marginal or conditional distributions of a candidate decomposition. Such an operation is necessary, because these distributions are needed to assess the quality of a given candidate graphical model. If the dataset is precise, i.e., if in all tuples there is exactly one value for each attribute, then computing a projection is trivial, since it consists in counting tuples and computing relative frequencies.

However, if the data is imprecise, i.e., contains missing values or set-valued information, things are slightly more complicated. Fortunately, with the context model interpretation of a degree of possibility, we have direct means to handle imprecise values. We simply interpret each imprecise tuple as a description of the set $\Gamma(c)$ of states of the world that are possible in some context $c$. We can do so, because an imprecise tuple can be rewritten as a set of tuples, namely the set of all precise tuples compatible with it.

Nevertheless, we face some problems, because we can no longer apply naive methods to determine the marginal distributions (a detailed explanation can be found in [13]). However, there is a simple preprocessing operation by which the database to learn from can be transformed, so that computing maximum projections becomes trivial. This operation is based on the notion of *closure under tuple intersection*. That is, we add (possibly imprecise) tuples to the database in order to achieve a situation, in which for any two tuples from the database their *intersection* (i.e., the intersection of the represented sets of precise tuples) is also contained in the database. Details can be found in [13][14].

### 5.2 Evaluation Measures

An *evaluation measure* serves to assess the quality of a given candidate graphical model w.r.t. a given database of sample cases, so that it can be determined which of a set of candidate graphical models best fits the given data. A desirable property of an evaluation measure is decomposability, i.e., the total network quality should be computable as an aggregate (e.g. sum or product) of local scores, for example a score for a maximal clique of the graph to be assessed or a score for a single edge.

Most such evaluation measures are based on measures of dependence, since for both the second and the third basic approach listed above it is necessary to measure the strength of dependence of two or more variables, either in order to test for conditional independence or in order to find the strongest dependence's. Here we confine ourselves to measures that assess the strength of dependence of two variables in the possibilistic case. The transfer to conditional tests (by computing a weighted sum of the results for the different instantiations of the conditions) and to more than two variables is straightforward.

Possibilistic evaluation measures can easily be derived by exploiting the close connection of possibilistic networks to relational networks (see above). The idea is to draw on the $\alpha$-*cut* view of a possibility distribution. This concept is transferred from the theory of fuzzy sets [15].

In the $\alpha$-*cut* view a possibility distribution is seen as a *set of relations* with one relation for each degree of possibility $\alpha$. The indicator function of such a relation is defined by simply assigning a value of 1 to all tuples for which the degree of possibility is no less than $\alpha$ and a value of 0 to all other tuples.

It is easy to see that a possibility distribution is decomposable if and only if each of the $\alpha$-*cut* relations is decomposable. Thus we may derive a measure for the strength of possibilistic dependence of two variables by integrating a measure for the strength of relational dependence over all degrees of possibility $\alpha$.

To make this clearer, we reconsider the simple example studied above. Figure 13 shows the projection to the back plane of our example reasoning space, i.e., to the subspace color $\times$ shape. We can measure the strength of dependence of color and shape by computing the *Hartley information gain* [16].

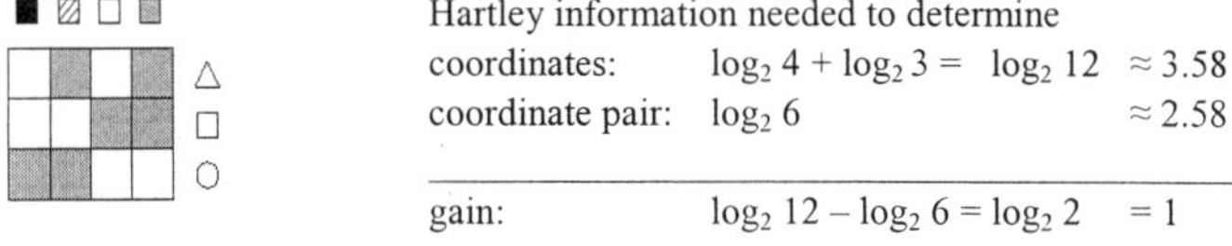



Hartley information needed to determine

coordinates: $\quad \log_2 4 + \log_2 3 = \log_2 12 \approx 3.58$

coordinate pair: $\quad \log_2 6 \qquad\qquad\qquad \approx 2.58$

gain: $\qquad\qquad \log_2 12 - \log_2 6 = \log_2 2 \quad = 1$

Figure 13. Computation of Hartley information gain.

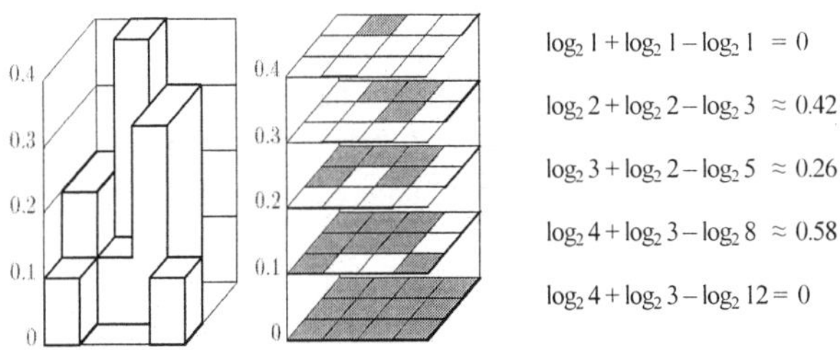

$\log_2 1 + \log_2 1 - \log_2 1 = 0$

$\log_2 2 + \log_2 2 - \log_2 3 \approx 0.42$

$\log_2 3 + \log_2 2 - \log_2 5 \approx 0.26$

$\log_2 4 + \log_2 3 - \log_2 8 \approx 0.58$

$\log_2 4 + \log_2 3 - \log_2 12 = 0$

Figure 14. Illustration of the idea of specificity gain.

$$I_{gain}^{(Hartley)}(C,S) = \log_2\left(\sum_{c \in dom(C)} r_c(C=c)\right) + \log_2\left(\sum_{s \in dom(S)} r_s(S=s)\right)$$

$$- \log_2\left(\sum_{c \in dom(C)} \sum_{s \in dom(S)} r_{cs}(C=c, S=s)\right)$$

$$= \log_2 \frac{\left(\sum_{c \in dom(C)} r_c(C=c)\right)\left(\sum_{s \in dom(S)} r_s(S=s)\right)}{\sum_{c \in dom(C)} \sum_{s \in dom(S)} r_{cs}(C=c, s=s)},$$

where $C$ stands for the color and $S$ for the size of an object.

The idea underlying this measure is as follows: Suppose we want to determine the actual values of the two attributes $C$ and $S$. Obviously, there are two possible ways to do this: In the first place, we could determine the value of each attribute separately, thus trying to find the "coordinates" of the value combination. Or we may exploit the fact that the value combination is restricted by the relation shown in Figure 13 and try to determine the value combination directly. In the former case we need the Hartley information of the set of values of $C$ plus the Hartley information of the set of values of S, i.e. $\log_2 4 + \log_2 3 \approx 3.58$ bits. In the latter case we need the Hartley information of the possible tuples, i.e. $\log_2 6 \approx 2.58$ bits, and thus gain one bit. Since it is plausible that we gain the more bits, the more strongly dependent the two attributes are (because in this case a value of one of the attributes leaves fewer choices for the value of the other), we may use the Hartley information gain as a direct indication of the strength of relational dependence of the two attributes.

The Hartley information gain is generalized to the *specificity gain* [13][17][18] as shown in Figure 14: It is integrated over all α-cuts of a given (two-dimensional) possibility distribution and thus measures the average strength of relational dependence on the different α-levels.

$$S_{gain}(A, B) = \int_0^{\sup \pi} \log_2 \left( \sum_{a \in dom(A)} [\pi]_\alpha (A = a) \right) + \log_2 \left( \sum_{b \in dom(B)} [\pi]_\alpha (B = b) \right)$$

$$- \log_2 \left( \sum_{a \in dom(A)} \sum_{b \in dom(B)} [\pi]_\alpha (A = a, B = b) \right) d\alpha$$

Surveys of other evaluation measures – which include probabilistic measures – can be found in [13][18].

### 5.3 Search Methods

As already indicated above, a search method determines which graphs are considered in order to find a good graphical model. Since an exhaustive search is impossible due to the huge number of graphs (there are $2^{\binom{n}{2}}$ possible undirected graphs over $n$ attributes), heuristic search methods have to be used.

Usually these heuristic methods introduce strong restrictions w.r.t. the graphs considered and exploit the value of the chosen evaluation measure to guide the search. In addition they are often greedy w.r.t. the model quality.

The simplest instance of such a search method is, of course, the Kruskal algorithm [19], which determines an optimum weight-spanning tree for given edge weights. This algorithm has been used very early in the probabilistic setting by [20], who used the *Shannon information gain* (also called *mutual information* or *cross entropy*) of the connected attributes as edge weights.

In the possibilistic setting, we may simply replace the Shannon information gain by the *specificity gain* in order to arrive at an analogous algorithm [13][17].

A natural extension of the Kruskal algorithm is a greedy parent selection for directed graphs, which is often carried out on a topological order of the attributes that is fixed in advance[1]: At the beginning the value of an evaluation measure is computed for a parentless child attribute. Then in turn each of the parent candidates (i.e. the attributes preceding the child in the topological order) is temporarily added and the evaluation measure is recomputed.

The parent candidate yielding the highest value of the evaluation measure is selected as a first parent and permanently added. In the third step each remaining parent candidate is added temporarily as a second parent and again the evaluation measure is recomputed. As before, the parent candidate that yields the highest value is permanently added.

The process stops if either no more parent candidates are available, a given maximal number of parents is reached, or none of the parent candidates, if added, yields a value of the evaluation measure exceeding the best value of the preceding step.

---

[1] A topological order is an order of the nodes of a graph such that all parent nodes of a given node precede it in the order. That is, there cannot be an edge from a node to a node, which precedes it in the topological order. By fixing a topological order in advance, the set of possible graphs is severely restricted and it is ensured that the resulting graph is acyclic.

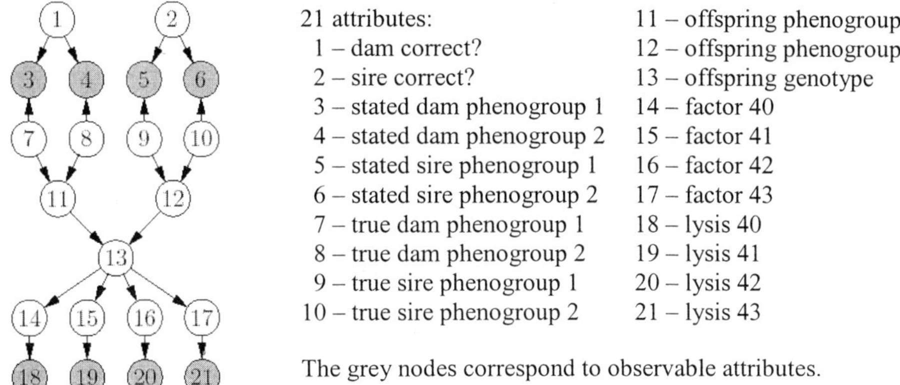

21 attributes:

| | |
|---|---|
| 1 – dam correct? | 11 – offspring phenogroup 1 |
| 2 – sire correct? | 12 – offspring phenogroup 2 |
| 3 – stated dam phenogroup 1 | 13 – offspring genotype |
| 4 – stated dam phenogroup 2 | 14 – factor 40 |
| 5 – stated sire phenogroup 1 | 15 – factor 41 |
| 6 – stated sire phenogroup 2 | 16 – factor 42 |
| 7 – true dam phenogroup 1 | 17 – factor 43 |
| 8 – true dam phenogroup 2 | 18 – lysis 40 |
| 9 – true sire phenogroup 1 | 19 – lysis 41 |
| 10 – true sire phenogroup 2 | 20 – lysis 42 |
| | 21 – lysis 43 |

The grey nodes correspond to observable attributes.

Figure 15. Domain expert designed network for the Danish Jersey cattle blood type determination example.

This search method has been used by [21] in the wellknown K2 algorithm. As an evaluation measure they used what has become known as the K2 metric. This measure has later been generalized by [22] to the *Bayesian-Dirichlet metric.*

Of course, in the possibilistic setting we may also apply this search method, again relying on the specificity gain as the evaluation measure. In order to handle multiple parent attributes with it, we simply combine all parents into one pseudo-attribute and compute the specificity gain for this pseudo-attribute and the child attribute.

A more extensive discussion of search methods for learning graphical models from data, which includes a simulated annealing approach, can be found, for example, in [13].

## 6. An Example Application

As an example of an application we consider the problem of blood group determination of Danish Jersey cattle in the F-blood group system [23]. For this problem there is a Bayesian network (a probabilistic graphical model based on a directed acyclic graph) designed by human domain experts, which serves the purpose to verify parentage for pedigree registration. The world section modeled in this example comprises 21 attributes, eight of which are observable. The size of the domains of these attributes ranges from two to eight values. The total reasoning space has $2^6 \cdot 3^{10} \cdot 6 \cdot 8^4 = 92\,876\,046\,336$ possible states. This number makes it obvious that the knowledge about this world section must be decomposed in order to make reasoning feasible, since it is clearly impossible to store a probability or a degree of possibility for each state. Figure 15 lists the attributes and shows the conditional independence graph of the Bayesian network.

Table 2. An example of conditional probability distribution that is associated with the conditional independence graph shown in Figure 15

| sire correct | true sire ph.gr. 1 | stated sire ph.gr. 1 | | |
|---|---|---|---|---|
| | | F1 | V1 | V2 |
| yes | F1 | 1 | 0 | 0 |
| yes | V1 | 0 | 1 | 0 |
| yes | V2 | 0 | 0 | 1 |
| no | F1 | 0.58 | 0.10 | 0.32 |
| no | V1 | 0.58 | 0.10 | 0.32 |
| no | V2 | 0.58 | 0.10 | 0.32 |

Table 3. An extract from the Danish Jersey cattle database

| | | | | | | | | | | | | | | | | | | | | |
|---|---|---|---|---|---|---|---|---|---|---|---|---|---|---|---|---|---|---|---|---|
| n | y | y | f1 | v2 | f1 | v2 | f1 | v2 | f1 | v2 | v2 | v2 | v2v2 | n | y | n | y | 0 | 6 | 0 | 6 |
| n | y | y | f1 | v2 | ** | ** | f1 | v2 | ** | ** | ** | ** | f1v2 | y | y | n | y | 7 | 6 | 0 | 7 |
| n | y | y | f1 | v2 | f1 | f1 | f1 | v2 | f1 | f1 | f1 | f1 | f1f1 | y | y | n | n | 7 | 7 | 0 | 0 |
| n | y | y | f1 | v2 | f1 | f1 | f1 | v2 | f1 | f1 | f1 | f1 | f1f1 | y | y | n | n | 7 | 7 | 0 | 0 |
| n | y | y | f1 | v2 | f1 | v1 | f1 | v2 | f1 | v1 | v2 | f1 | f1v2 | y | y | n | y | 7 | 7 | 0 | 7 |
| n | y | y | f1 | f1 | ** | ** | f1 | f1 | ** | ** | f1 | f1 | f1f1 | y | y | n | n | 6 | 6 | 0 | 0 |
| n | y | y | f1 | v1 | ** | ** | f1 | v1 | ** | ** | v1 | v2 | v1v2 | n | y | y | y | 0 | 5 | 4 | 5 |
| n | y | y | f1 | v2 | f1 | v1 | f1 | v2 | f1 | v1 | f1 | v1 | f1v1 | y | y | y | y | 7 | 7 | 6 | 7 |

As described above, a conditional independence graph enables us to decompose the joint distribution into a set of marginal or conditional distributions. In the Danish Jersey cattle example, this decomposition leads to a considerable simplification: Only 308 conditional probabilities have to be specified. An example of a conditional probability table, which is part of the decomposition, is shown in Table 2. It states the conditional probabilities of the phenogroup 1 of the stated sire of a given calf conditioned on the phenogroup 1 of the true sire and whether the sire was correctly identified. The numbers in this table are derived from statistical data and the experience of human domain experts.

Besides the domain expert designed reference structure there is a database of 500 real world sample cases (an extract of this database is shown in Table 3). This database can be used to test learning algorithms for graphical models, because the quality of the learning result can be determined by comparing the constructed graph to the reference structure. However, there is a problem connected with this database, namely that it contains a fairly large number of unknown values – only a little over half of the tuples are complete. (This can already be guessed from the extract shown in Table 3: the stars denote missing values). Missing values and set-valued information make it difficult to learn a Bayesian network, because an unknown value can be seen as representing imprecise information: It states that all values contained in the domain of the corresponding attribute are possible, without any known preferences between them. Nevertheless it is still feasible to learn a Bayesian network from the database in this case, since the dependencies are rather strong and thus the small number of complete tuples is still sufficient to recover the underlying structure. However, learning a possibilistic network from the same dataset is much easier, since possibility theory was especially designed to handle imprecise information (see above). Hence no discarding or special treatment of tuples with missing values or set-valued information is necessary.

In order to check this conjecture, we implemented the learning methods discussed above (together with their probabilistic counterparts) in a prototype program called INES (Induction of Network Structures) [13]). The networks induced with different evaluation measures are very similar to the domain expert designed reference structure, even though the reference structure is a Bayesian network, which may differ from the corresponding possibilistic network, since it employs a different notion of conditional independence. Evaluations of the learned networks show that their quality is comparable to that of learned probabilistic networks and the reference structure w.r.t. reasoning.

## 7. Conclusion

In this paper we reviewed possibilistic graphical models and discussed approaches to learn them from a database of sample cases. Based on the context model interpretation of a degree of possibility imprecise data are easily handled in such a possibilistic approach. W.r.t. learning algorithms a lot of work done in the probabilistic counterpart of this research area can be transferred: All search methods are directly usable, only the evaluation measures have to be adapted. Experiments carried out with an example application show that learning possibilistic networks from data is a noteworthy alternative to the established probabilistic methods if the data to learn from is imprecise.

**References**

[1] J. Pearl, Probabilistic Reasoning in Intelligent Systems: Networks of Plausible Inference, Morgan Kaufmann, San Mateo, CA, USA 1988. (2nd edition 1992)
[2] F. V. Jensen, An Introduction to Bayesian Networks, UCL Press, London, United Kingdom 1996.
[3] S. L. Lauritzen, Graphical Models, Oxford University Press, Oxford, United Kingdom 1996.
[4] E. Castillo, J. M. Gutierrez and A. S. Hadi, Expert Systems and Probabilistic Network Models, Springer, New York, NY, USA 1997.
[5] D. Dubois and H. Prade, Possibility Theory, Plenum Press, New York, NY, USA 1988.
[6] J. D. Ullman, *Principles of Database and Knowledge-Base Systems*, Vol. **1** & **2**, Computer Science Press, Rockville, MD, USA, 1988.
[7] J. Gebhardt and R. Kruse, The Context Model – An Integrating View of Vagueness and Uncertainty, *Int. Journal of Approximate Reasoning* **9**, pp. 283-314, North-Holland, Amsterdam, Netherlands 1993.
[8] H. T. Nguyen, On Random Sets and Belief Functions, *Journal of Mathematical Analysis and Applications* **65**, pp. 531 – 542, Academic Press, San Diego, CA, USA 1978.
[9] K. Ilcstir, H. T. Nguyen and G. S. Rogers, A Random Set Formalism for Evidential Reasoning, In: [24], pp. 209-344
[10] T. S. Verma and J. Pearl, Causal Networks: Semantics and Expressiveness, In: [25], pp. 69-76
[11] J. M. Hammersley and P. E. Clifford, Markov Fields on Finite Graphs and Lattices, Unpublished manuscript, 1971. Cited in: [26]
[12] J. Gebhardt, Learning from Data: Possibilistic Graphical Models, Habilitation Thesis, University of Braunschweig, Germany 1997.
[13] C. Borgelt and R. Kruse, Graphical Models – Methods for Data Analysis and Mining, J. Wiley & Sons, Chichester, United Kingdom 2002.
[14] C. Borgelt and R. Kruse, Efficient Maximum Projection of Database-Induced Multivariate Possibility Distributions, *Proc. 7th IEEE Int. Conf. on Fuzzy Systems (FUZZ-IEEE'98, Anchorage, Alaska, USA)*, CD-ROM, IEEE Press, Piscataway, NJ, USA 1998.
[15] R. Kruse, J. Gebhardt and F. Klawonn, Foundations of Fuzzy Systems, J. Wiley & Sons, Chichester, United Kingdom 1994.
[16] R. V. L. Hartley, Transmission of Information, *The Bell Systems Technical Journal* **7**, pp. 535-563, Bell Laboratories, USA 1928.

[17] J. Gebhardt and R. Kruse, Tightest Hypertree Decompositions of Multivariate Possibility Distributions, *Proc. 7th Int. Conf. on Information Processing and Management of Uncertainty in Knowledge-based Systems (IPMU'96, Granada, Spain)*, pp. 923-927, Universidad de Granada, Spain 1996.

[18] C. Borgelt and R. Kruse, Evaluation Measures for Learning Probabilistic and Possibilistic Networks, *Proc. 6th IEEE Int. Conf. on Fuzzy Systems (FUZZ-IEEE'97, Barcelona, Spain)*, Vol. **2**, pp. 1034-1038. IEEE Press, Piscataway, NJ, USA 1997.

[19] J. B. Kruskal, On the Shortest Spanning Subtree of a Graph and the Traveling Salesman Problem, *Proc. American Mathematical Society* **7**(1), pp. 48-50. American Mathematical Society, Providence, RI, USA 1956.

[20] C. K. Chow and C. N. Liu, Approximating Discrete Probability Distributions with Dependence Trees, *IEEE Trans. on Information Theory* **14**(3), pp. 462-467, IEEE Press, Piscataway, NJ, USA 1968.

[21] G. F. Cooper and E. Herskovits, A Bayesian Method for the Induction of Probabilistic Networks from Data, *Machine Learning* **9**, pp. 309-347, Kluwer, Dordrecht, Netherlands 1992.

[22] D. Heckerman, D. Geiger and D. M. Chickering, Learning Bayesian Networks: The Combination of Knowledge and Statistical Data, *Machine Learning* **20**, pp. 197-243, Kluwer, Dordrecht, Netherlands 1995.

[23] L. K. Rasmussen, *Blood Group Determination of Danish Jersey Cattle in the F-blood Group System (Dina Research Report* **8**), Dina Foulum, Tjele, Denmark 1992.

[24] I. R. Goodman, M. M. Gupta, H. T. Nguyen and G. S. Rogers, eds, Conditional Logic in Expert Systems. North-Holland, Amsterdam, Netherlands 1991.

[25] Shachter, R.D., Levitt, T.S., Kanal, L.N., Lemmer, J.F. eds, *Uncertainty in Artificial Intelligence* 4, North Holland, Amsterdam, Netherlands 1990.

[26] V. Isham, An Introduction to Spatial Point Processes and Markov Random Fields, *Int. Statistical Review* **49**, pp. 21-43, *Int. Statistical Institute*, Voorburg, Netherlands 1981.

# Automatic Conversation Driven by Uncertainty Reduction and Combination of Evidence for Recommendation Agents

Luis M. ROCHA

*Los Alamos National Laboratory, MS B256, Los Alamos, NM 87545, U.S.A.*
*e-mail: rocha@lanl.gov*
*WWW: http://www.c3.lanl.gov/~rocha*

**Abstract:** We present an adaptive recommendation system named *TalkMine*, which is based on the combination of Evidence from different sources. It establishes a mechanism for automated conversation between recommendation agents, in order to gather the interests of individual users of web sites and digital libraries. This conversation process is enabled by measuring the uncertainty content of knowledge structures used to store evidence from different sources. *TalkMine* also leads different databases or websites to learn new and adapt existing keywords to the categories recognized by its communities of users. *TalkMine* is currently being implemented for the research library of the Los Alamos National Laboratory under the *Active Recommendation Project* (http://arp.lanl.gov). The process of identification of the interests of users relies on the combination of several fuzzy sets into evidence sets, which models an ambiguous "and/or" linguistic expression. The interest of users is further fine-tuned by a human-machine conversation algorithm used for uncertainty reduction. Documents are retrieved according to the inferred user interests. Finally, the retrieval behavior of all users of the system is employed to adapt the knowledge bases of queried information resources. This adaptation allows information resources to respond well to the evolving expectations of users.

## 1.  The Active Recommendation Project

The *Active Recommendation Project* (ARP), part of the Library Without Walls Project, at the Research Library of the Los Alamos National Laboratory is engaged in research and development of recommendation systems for digital libraries. The *information resources* available to ARP are large databases with academic articles. These databases contain bibliographic, citation, and – sometimes – abstract information about academic articles.

Typical databases are *SciSearch*® and *Biosis*®; the first contains articles from scientific journals from several fields collected by ISI (Institute for Scientific Indexing), while the second contains more biologically oriented publications. We do not manipulate directly the records stored in these information resources; rather, we created a repository of XML (about 3 million) records, which point us to documents stored in these databases [1].

### *1.1 Characterizing the Knowledge stored in an Information Resource*

For each information resource we can identify several distinct relations among documents and between documents and semantic tags used to classify documents appropriately. For instance, documents in the WWW are related via a hyperlink network, while documents in bibliographic databases are related by citation and collaboration networks. Furthermore, documents are typically related to semantic tags such as keywords used to describe their content. Indeed, the relation between documents and keywords allows us to infer the semantic value of documents and the inter-associations between keywords. Such semantic relation is stored as a very sparse *Keyword-Document Matrix* **A**. Each entry $a_{i,j}$ in the matrix is boolean and indicates whether keyword $k_i$ indexes ($a_{i,j} = 1$) document $d_j$ or not ($a_{i,j} = 0$). From a matrix such as this, we can then compute other matrices defining the strength of association amongst keywords and amongst documents.

These matrices holding measures of closeness, formally, are proximity relations [2][3] because they are reflexive and symmetric fuzzy relations. Their transitive closures are known as similarity relations (Ibid). The collection of this relational information, all the proximity relations as well as $A$, is an expression of the particular knowledge an information resource conveys to its community of users. Notice that distinct information resources typically share a very large set of keywords and records. However, these are organized differently in each resource, leading to different collections of relational information. Indeed, each resource is tailored to a particular community of users, with a distinct history of utilization and deployment of information by its authors and users. For instance, the same keywords will be related differently for distinct resources. Therefore, we refer to the relational information of each information resource as a *Knowledge Context* (More details in [4]).

In [1] we have discussed how these proximity relations are used in ARP. However, the ARP recommendation system described in this article (*TalkMine*) requires only the Keyword Semantic Proximity (*KSP*) matrix, obtained from $A$ by the following formula:

$$KSP(k_i, k_j) = \frac{\sum_{k=1}^{m}(a_{i,k} \wedge a_{j,k})}{\sum_{k=1}^{m}(a_{i,k} \vee a_{j,k})} = \frac{N_\cap(k_i, k_j)}{N_\cup(k_i, k_j)} = \frac{N_\cap(k_i, k_j)}{N(k_i) + N(k_j) - N_\cap(k_i, k_j)} \quad (1)$$

The semantic proximity between two keywords, $k_i$ and $k_j$, depends on the sets of records indexed by either keyword, or the intersection of these sets. $N(k_i)$ is the number of records keyword $k_i$ indexes, and $N_\cap(k_i, k_j)$, the number of records both keywords index. This last quantity is the number of elements in the intersection of the sets of records that each keyword indexes. Thus, two keywords are near if they tend to index many of the same records. Table I presents the values of *KSP* for the 10 most common keywords in the ARP repository.

Table 1: Keyword Semantic Proximity for 10 most frequent keywords

| | cell | studi | system | express | protein | model | activ | human | rat | patient |
|---|---|---|---|---|---|---|---|---|---|---|
| cell | 1.000 | 0.022 | 0.019 | 0.158 | 0.084 | 0.017 | 0.085 | 0.114 | 0.068 | 0.032 |
| studi | 0.022 | 1.000 | 0.029 | 0.013 | 0.017 | 0.028 | 0.020 | 0.020 | 0.020 | 0.037 |
| system | 0.019 | 0.029 | 1.000 | 0.020 | 0.017 | 0.046 | 0.022 | 0.014 | 0.021 | 0.014 |
| express | 0.158 | 0.013 | 0.020 | 1.000 | 0.126 | 0.011 | 0.071 | 0.103 | 0.078 | 0.020 |
| protein | 0.084 | 0.017 | 0.017 | 0.126 | 1.000 | 0.013 | 0.070 | 0.061 | 0.041 | 0.014 |
| model | 0.017 | 0.028 | 0.046 | 0.011 | 0.013 | 1.000 | 0.016 | 0.016 | 0.026 | 0.005 |
| activ | 0.085 | 0.020 | 0.022 | 0.071 | 0.070 | 0.016 | 1.000 | 0.058 | 0.053 | 0.021 |
| human | 0.114 | 0.020 | 0.014 | 0.103 | 0.061 | 0.016 | 0.058 | 1.000 | 0.029 | 0.021 |
| rat | 0.068 | 0.020 | 0.021 | 0.078 | 0.041 | 0.026 | 0.053 | 0.029 | 1.000 | 0.008 |
| patient | 0.032 | 0.037 | 0.014 | 0.020 | 0.014 | 0.005 | 0.021 | 0.021 | 0.008 | 1.000 |

From the inverse of *KSP* we obtain a distance function between keywords:

$$d(k_i, k_j) = \frac{1}{KSP(k_i, k_j)} - 1 \qquad (2)$$

$d$ is a distance function because it is a nonnegative, symmetric real-valued function such that $d(k, k) = 0$. It is not an Euclidean metric because it may violate the triangle inequality: $d(k_1, k_2) \leq d(k_1, k_3) + d(k_3, k_2)$ for some keyword $k_3$. This means that the shortest distance between two keywords may not be the direct link but rather an indirect pathway. Such measures of distance are referred to as semi-metrics [5].

### 1.2. Characterizing Users

Users interact with information resources by retrieving records. We use their retrieval behavior to adapt the respective knowledge contexts of these resources (stored in the proximity relations). But before discussing this interaction, we need to characterize and define the capabilities of users: our agents. The following capabilities are implemented in enhanced "browsers" distributed to users.

***Present interests*** described by a set of keywords $\{k_1, \cdots, k_p\}$.

***History of Information Retrieval (IR)***. This history is also organized as a knowledge context as described in 2.1, containing pointers to the records the user has previously accessed, the keywords associated with them, as well as the structure of this set of records. This way, we treat users themselves as information resources with their own specific knowledge contexts defined by their own proximity information.

***Communication Protocol***. Users need a 2-way means to communicate with other information resources in order to retrieve relevant information, and to send signals leading to adaptation in all parties involved in the exchange.

Regarding point 2, the history of IR, notice that the same user may query information resources with very distinct sets of interests. For example, one day a user may search databases as a biologist looking for scientific articles, and the next as a sports fan looking for game scores. Therefore, each enhanced browser allows users to define different *"personalities"*, each one with its distinct history of IR defined by independent knowledge contexts with distinct proximity data (see Figure 1).

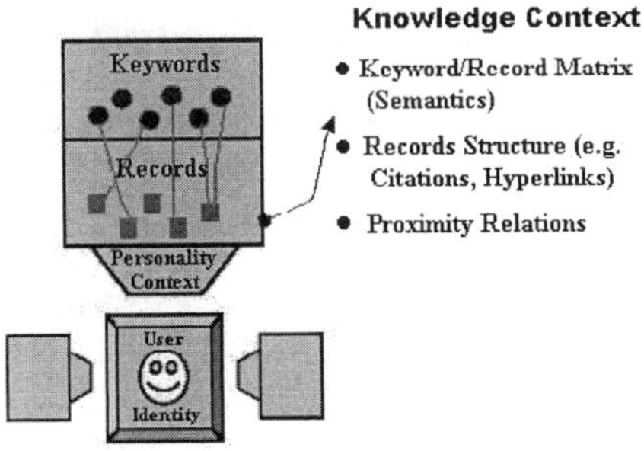


Figure 1. Each user can store different personalities in enhanced browsers. Each personality is stored as a knowledge context created from previous history of IR. The actual identity of the user can remain private.

Because the user history of IR is stored in personal browsers, information resources do not store user profiles. Furthermore, all the collective behavior algorithms used in ARP do not require the identity of users. When users communicate (3) with information resources, what need to be exchanged are they're present interests or query (1), and the relevant proximity data from their own knowledge context (2). In other words, users make a query, and then share the relevant knowledge they have accumulated about their query, their "world-view" or context, from a particular personality, without trading their identity. Next, the recommendation algorithms integrate the user's knowledge context with those of the queried information resources (possibly other users), resulting in appropriate recommendations. Indeed, the algorithms we use define a communication protocol between knowledge contexts, which can be very large databases, web sites, or other users. Thus, the overall architecture of the recommendation systems we use in ARP is highly distributed between information resources and all the users and they're browsing personalities (see Figure 2).

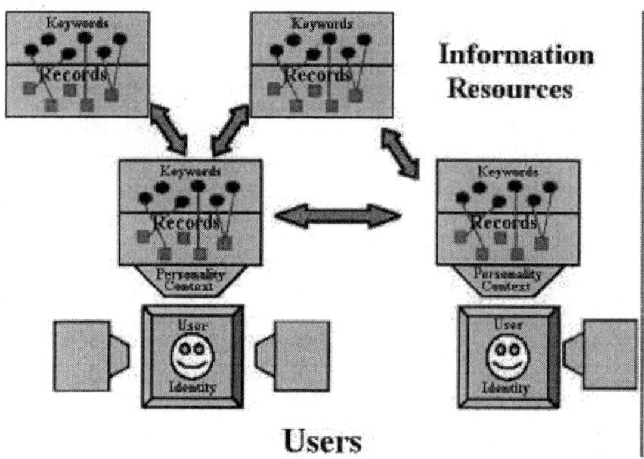


Figure 2. The algorithms we use in ARP define a distributed architecture based on communication between knowledge contexts from information resources and users alike

The collective behavior of all users is also aggregated to adapt the knowledge contexts of all intervening information resources and users alike. This open-ended learning process [6] is enabled by the *TalkMine* recommendation system described below.

## 2. Categories and Distributed Memory

### 2.1. *A Model of Categorization from Distributed Artificial Intelligence*

*TalkMine* is a recommendation system based on a model of linguistic categories [7], which are created from conversation between users and information resources and used to recombine knowledge as well as adapt it to users. The model of categorization used by *TalkMine* is described in detail in [6][7][8]. Basically, as also suggested by [9], categories are seen as representations of highly transient, context-dependent knowledge arrangements, and not as model of information storage in the brain. In this sense, in human cognition, categories are seen as linguistic constructs used to store temporary associations built up from the integration of knowledge from several neural sub-networks. The categorization process, driven by language and conversation, serves to bridge together several distributed neural networks, associating tokens of knowledge that would not otherwise be associated in the individual networks. Thus, categorization is the chief mechanism to achieve knowledge recombination in distributed networks leading to the production of new knowledge [6][7].

*TalkMine* applies such a model of categorization of distributed neural networks driven by language and conversation to recommendation systems. Instead of neural networks, knowledge is stored in information resources, from which we construct the knowledge contexts with respective proximity relations described in section 1. *TalkMine* is used as a conversation protocol to categorize the interests of users according to the knowledge stored in information resources, thus producing appropriate recommendations and adaptation signals.

### 2.2. *Distributed Memory is Stored in Knowledge Contexts*

A knowledge context of an information resource (section 1.1) is not a connectionist structure in a strong sense since keywords and records are not distributed as they can be identified in specific nodes of the network [10]. However, the same keyword indexes many records, the same record is indexed by many keywords, and the same record is typically engaged in a citation (or hyperlink) relation with many other records. Losing or adding a few records or keywords does not significantly change the derived semantic and structural proximity relations (section 1) of a large network. In this sense, the knowledge conveyed by such proximity relations is distributed over the entire network of records and keywords in a highly redundant manner, as required of sparse distributed memory models [11]. Furthermore, Clark [9] proposed that connectionist memory devices work by producing metrics that relate the knowledge they store. As discussed in section 1, the distance functions obtained from proximity relations are semi-metrics, which follow all of Clark's requirements [6]. Therefore, we can regard a knowledge context effectively as a distributed memory bank. Below we discuss how such distributed knowledge adapts to communities of users (the environment) with Hebbian type learning.

In the *TalkMine* system we use the KSP relation (formula (1)) from knowledge contexts. It conveys the knowledge stored in an information resource in terms of a measure of proximity among keywords. This proximity relation is unique to each information resource, reflecting the semantic relationships of the records stored in the latter, which in turn echo the knowledge of its community of users and authors. *TalkMine* is a content-based recommendation system because it uses a keyword proximity relation. Next we describe how it is also collaborative by integrating the behavior of users. A related structural algorithm, also being developed in ARP, is described in [1].

### 3. Evidence Sets: Capturing the Linguistic "And/Or" in Queries

### 3.1. *Evidence Sets Model Categories*

*TalkMine* uses a set structure named *evidence set* [7][8][12][13], an extension of a fuzzy set [14], to model of linguistic categories. The extension of fuzzy sets is based on the Dempster-Shafer Theory of Evidence (DST) [15], which is defined in terms of a set function $m : \Pi(X) \to [0, 1]$, referred to as a *basic probability assignment*, such that $m(\phi) = 0$ and $\sum_{A \subseteq X} m(A) = 1$. $\Pi(x)$ denotes the power set of $X$ and $A$ any subset of $X$. The value $m(A)$ denotes the proportion of all available evidence which supports the claim that $A \in \Pi(x)$ contains the actual value of a variable $x$. DST is based on a pair of nonadditive measures: *belief* (Bel) and *plausibility* (Pl) uniquely obtained from m.

Given a basic probability assignment $m$, Bel and Pl are determined for all $A \in \prod(X)$ by the equations:

$$Bel(A) = \sum_{B \subseteq A} m(B), \quad Pl(A) = \sum_{B \cap A \neq \phi} m(B)$$

the expressions above imply that belief and plausibility are dual measures related by: $Pl(A) = 1 - Bel(A^c,)$ for all $A \in \prod(X)$, where $A^c$ represents the complement of A in X. It is also true that $Bel(A) \leq Pl(A)$ for all $A \in \prod(x)$. Notice that "$m(A)$ measures the belief one commits exactly to $A$, not the total belief that one commits to $A$." [15, page 38] $Bel(A)$, the total belief committed to A, is instead given by the sum of all the values of $m$ for all subsets of $A$.

Any set $A \in \prod(x)$ with $m(A) > 0$ is called a focal element. A body of evidence is defined by the pair $(\phi, m)$, where $\phi$ represents the set of all focal elements in $X$, and $m$ the associated basic probability assignment. The set of all bodies of evidence is denoted by $B(x)$.

An evidence set $A$ of $X$, is defined for all $x \in X$, by a membership function of the form:

$$A(x) \rightarrow (\phi^x, m^x) \in B[0,1]$$

where $B[0,1]$ is the set of all possible bodies of evidence $(\phi^x, m^x)$ on $I$, the set of all subintervals of $[0,1]$. Such bodies of evidence are defined by a basic probability assignment $m^x$ on $I$, for every $x$ in $X$. Thus, evidence sets are set structures which provide interval degrees of membership, weighted by the probability constraint of DST. They are defined by two complementary dimensions: membership and belief. The first represents an interval (type-2) fuzzy degree of membership, and the second a subjective degree of belief on that membership(see Figure 3).

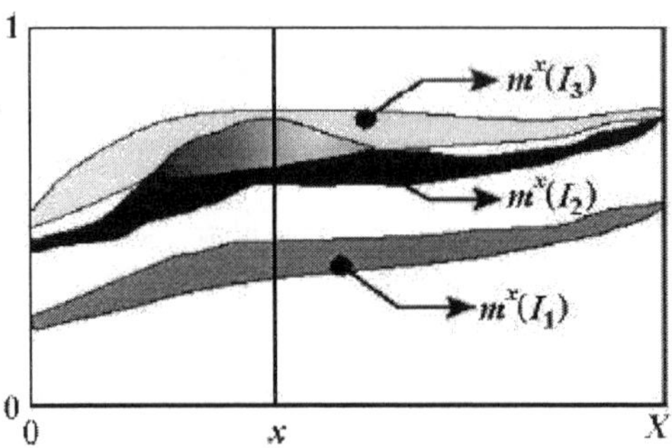


Figure 3. Evidence-Set with 3 focal elements for each $x$

Each interval of membership $I_j^x$, with its correspondent evidential weight $m^x(I_j^x)$, represents the degree of importance of a particular element $x$ of $X$ in category $A$ *according* to a particular *perspective*. Thus, the membership of each element $x$ of a evidence set $A$ is defined by distinct intervals representing different, possibly conflicting, perspectives. This way, categories are modeled not only as sets of elements with a membership degree (or prototypicality [7]), but as sets of elements which may possess different interval membership degrees for different contexts or perspectives on the category.

The basic set operations of complementation, intersection, and union have been defined and establish a belief-constrained approximate reasoning theory of which fuzzy approximate reasoning and traditional set operations are special cases [7][8]. Intersection (Union) is based on the minimum (maximum) operator for the limits of each of the intervals of membership of an evidence set. For the purposes of this article, the details of these operations are not required, please consult [7] for more details.

### 3.2. The Uncertainty Content of Evidence Sets

Evidence sets are set structures, which provide interval degrees of membership, weighted by the probability constraint of DST. Interval Valued Fuzzy Sets (IVFS), fuzzy sets, and crisp sets are all special cases of evidence sets. The membership of an element $x$ in a crisp set is perfectly certain: the element is either a member of the set or not. The membership of an element $x$ in fuzzy set is defined as degree value in the unit interval; this means that the membership is *fuzzy* because the element is a member of the set with degree $A(x)$ and <u>simultaneously</u>, is also not a member with complementary degree $1-A(x)$. The membership of an element $x$ in an IVFS is defined as an interval $I$ contained in the unit interval; this means that the membership is both fuzzy and *nonspecific* [12][16], because the element is a member of the set with a nonspecific degree that can vary in the interval $I$. Finally, membership of an element $x$ in an evidence set is defined as a set of intervals constrained by a probability restriction; this means that the membership is fuzzy, nonspecific, and *conflicting,* since the element is a member of the set with several degrees that vary in each interval with some probability.

To capture the uncertainty content of evidence sets, the uncertainty measures of [17] were extended from finite to infinite domains [13]. The total uncertainty, $U$, of an evidence set $A$ was defined by: $U(A) = (IF(A), IN(A), IS(A))$. The three indices of uncertainty, which vary between 1 and 0, IF (*fuzziness*), IN (*nonspecificity*), and IS (*conflict*) were introduced in [8][13], where it was also proven that IN and IS possess good axiomatic properties wanted of information measures. IF is based on [18][19] and [2] measure of fuzziness. IN is based on the Hartley measure [13], and IS on the Shannon entropy as extended by [17] into the DST framework. For the purposes of this article, all we need to know is that these measures vary in the unit interval, for full details see [13].

### 3.3. Obtaining an Evidence Set from Fuzzy Sets: The Linguistic "And/Or"

Fundamental to the *TalkMine* algorithm is the integration of information from different sources into an evidence set, representing the category of topics (described by keywords) a user is interested at a particular time. In particular, as described below, these sources of information contribute information as fuzzy sets. This way, we need a procedure for integrating several fuzzy sets into an evidence set.

Turksen [20] proposed a means to integrate fuzzy sets into IVFS (or type-2 fuzzy sets). He later proposed that every time two fuzzy sets are combined, the uncertainty content of the resulting structure should be of a higher order, namely, the fuzziness of two fuzzy sets should be combined into the fuzziness and nonspecificity of an IVFS [16]. Turksen's Fuzzy Set combination is based on the separation of the disjunctive and conjunctive normal forms of logic compositions in fuzzy logic.

A disjunctive normal form (DNF) is formed with the disjunction of some of the four primary conjunctions, and the conjunctive normal form (CNF) is formed with the conjunction of some of the four primary disjunctions, respectively: $A \cap B, A \cap \overline{B}, \overline{A} \cap B, \overline{A} \cap \overline{B}$ and $A \cup B, A \cup \overline{B}, \overline{A} \cup B, \overline{A} \cup \overline{B}$. In two-valued logic the CNF and DNF of a logic composition are equivalent: CNF = DNF.

Turksen [20] observed that in fuzzy logic, for certain families of conjugate pairs of conjunctions and disjunctions, we have instead DNF $\subseteq$ CNF for some of the fuzzy logic connectives. He proposed that fuzzy logic compositions could be represented by IVFS's given by the interval [DNF, CNF] of the fuzzy set connective chosen [20].

Using Turksen's approach, the union and intersection of two fuzzy sets $F_1$ and $F_2$ result in the two following IVFS, respectively:

$$IV^{\cup}(x) = \left[ F_1(x) \underset{DNF}{\cup} F_2(x), \ F_1(x) \underset{CNF}{\cup} F_2(x) \right]$$

$$IV^{\cap}(x) = \left[ F_1(x) \underset{DNF}{\cap} F_2(x), \ F_1(x) \underset{CNF}{\cap} F_2(x) \right]$$

(3)

where,

$$A \underset{CNF}{\cup} B = A \cup B, \quad A \underset{DNF}{\cup} B = (A \cap B) \cup (A \cap \overline{B}) \cup (\overline{A} \cup B),$$

$$A \underset{CNF}{\cap} B = (A \cup B) \cap (A \cup \overline{B}) \cap (\overline{A} \cup B), \ and \ A \underset{DNF}{\cap} B = A \cap B,$$

for any two fuzzy sets $A$ and $B$, with union and intersection operations chosen from the families of t-norms and t-conorms following the appropriate axiomatic requirements [2].

In *TalkMine* only the traditional maximum and minimum operators for union and intersection, respectively, are used. Clearly, all other t-norms and t-conorms would also work. The intervals of membership obtained from the combination of two fuzzy sets can be interpreted as capturing the intrinsic nonspecificity of the combination of fuzzy sets with fuzzy set operators.

Due to the introduction of fuzziness, the DNF and CNF do not always coincide. This lack of coincidence reflects precisely the nonspecificity inherent in fuzzy set theory: because we can arrive at different results depending on which normal form we choose, the combination of fuzzy sets is ambiguous.

Turksen [16] suggested that this ambiguity should be treated as nonspecificity and captured by intervals of membership. In this sense, fuzziness "breeds" nonspecificity. Figure 4 depicts the construction of two IVFS from two fuzzy sets $F_1$ and $F_2$ according to the procedure described by formulae (3). Formulae (3) constitute a procedure for calculating the union and intersection IVFS from two fuzzy sets, which in logic terms refer to the "Or" and "And" operators. Thus, $IV^{\cup}$ describes the linguistic expression "$F_1$ or $F_2$", while $IV^{\cap}$ describes "$F_1$ and $F_2$", – capturing both fuzziness and nonspecificity of the particular fuzzy logic operators employed.

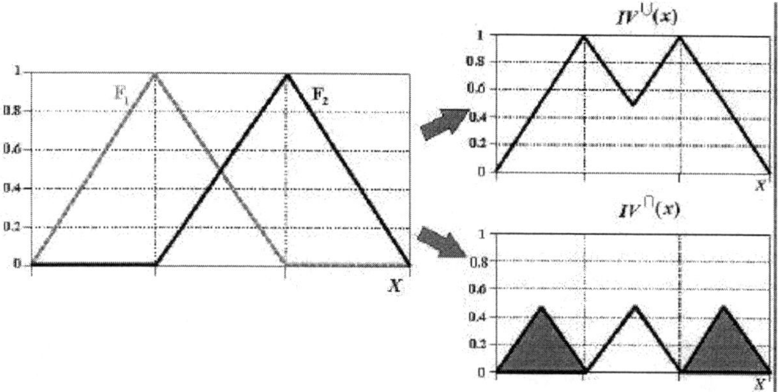



Figure 4. Combination of two fuzzy sets $F_1$ and $F_2$ into two IVFS according to formulae 3. The union IVFS, $IV^\cup$ is a fuzzy set since DNF and CNF coincide, which does not happen for $IV^\cap$.

However, in common language, often "and" is used as an unspecified "and/or". In other words, what we mean by the statement "I am interested in $x$ and $y$", can actually be seen as an unspecified combination of "$x$ and $y$" with "$x$ or $y$". This is particularly relevant for recommendation systems where it is precisely this kind of statement from users that we wish to respond to.

One use of evidence sets is as representations of the integration of both $IV^\cup$ and $IV^\cap$ into a linguistic category that expresses this ambiguous "and/or". To make this combination more general, assume that we possess an evidential weight $m_1$ and $m_2$ associated with each $F_1$ and $F_2$ respectively. These are probabilistic weights $(m_1 + m_2 = 1)$, which represent the strength we associate with each fuzzy set being combined. The linguistic expression at stake now becomes "I am interested in x and y, but I value x more/less than y". To combine all this information into an evidence set we use the following procedure:

$$ES(x) = \left\{ \; \left\langle IV^\cup(x), min(m_1, m_2) \right\rangle, \left\langle IV^\cap(x), max(m_1, m_2) \right\rangle \; \right\} \tag{4}$$

Because $IV^\cup$ is the less restrictive combination, obtained by applying the maximum operator, or suitable t-norm to the original fuzzy sets $F_1$ and $F_2$, its evidential weight is acquired via the minimum operator of the evidential weights associated with $F_1$ and $F_2$. The reverse is true for $IV^\cap$. Thus, the evidence set obtained from (4) contains $IV^\cup$ with the lowest evidence, and $IV^\cap$ with the highest. Linguistically, it describes the ambiguity of the "and/or" by giving the strongest belief weight to "and" and the weakest to "or". It expresses: "I am interested in $x$ and $y$ to a higher degree, but I am also interested in x or y to a lower degree". This introduces the third kind of uncertainty: conflict. Indeed, the ambiguity of "and/or" rests on the conflict between the interest in "and" and the interest in "or". This evidence set captures the three forms of uncertainty discussed in Section 2.3: fuzziness of the original fuzzy sets $F_1$ and $F_2$ nonspecificity of $IV^\cup$ and $IV^\cap$, and conflict between these two as they are included in the same evidence set with distinct evidential weights. Figure 5 depicts an example of the evidence set obtained from $F_1$ and $F_2$, as well as its uncertainty content (fuzziness. Nonspecificity, and conflict).

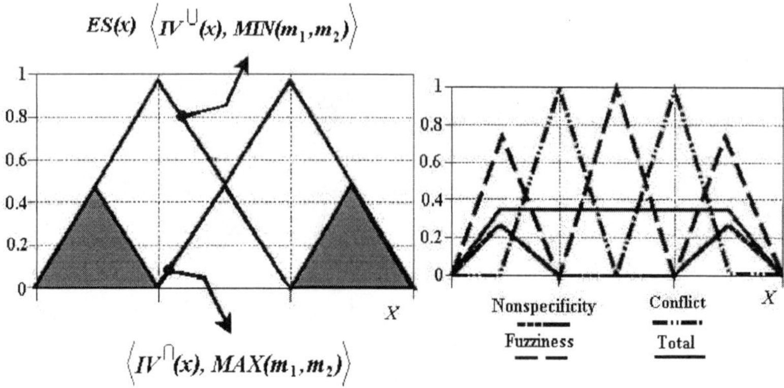


Figure 5. Evidence Set obtained from $F_1$ and $F_2$ and respective uncertainty content

Finally, formula (4) can be easily generalized for a combination of $n$ fuzzy sets $F_i$ with probability constrained weights $m_i$ :

$$ES(x) = \left\{ \left\langle IV^{\cup}_{F_i/F_j}(x), \frac{min(m_i, m_j)}{n-1} \right\rangle, \left\langle IV^{\cap}_{F_i/F_j}(x), \frac{max(m_i, m_j)}{n-1} \right\rangle \right\} \tag{5}$$

This procedure can be used to combine evidence in the form of fuzzy sets from $n$ weighted sources. It produces intervals obtained from the combination of each pair of fuzzy sets with a union and an intersection operator. Intersection is given the highest weight. The evidence set obtained is the ambiguous, common language, "and/or" for $n$ items.

## 4. TalkMine: Integrating Several Sources of Knowledge via Conversation

### 4.1. Inferring User Interest

The act of recommending appropriate documents to a particular user needs to be based on the integration of information from the user (with her history of retrieval) and from the several information resources being queried. With *TalkMine* in particular, we want to retrieve relevant documents from several information resources with different keyword indexing. Thus, the keywords the user employs in her search, need to be "decoded" into appropriate keywords for each information resource. Indeed, the goal of *TalkMine* is to project the user interests into the distinct knowledge contexts of each information resource, creating a representation of these interests that can capture the perspective of each one of these contexts.

Evidence Sets were precisely defined to model categories (knowledge representations) which can capture different perspectives. As described in Section 1.2, the present interests of each user are described by a set of keywords $\{k_1, \cdots, k_p\}$. Using these keywords and the keyword distance function (2) of the several knowledge contexts involved (one from the user and one from each information resource being queried), the interests of the user, "seen" from the perspectives of the several information resources, can be inferred as an evidence category using (5).

Let us assume that $r$ information resources $R_t$ are involved in addition to the user herself.

The set of keywords contained in all the participating information resources is denoted by $K$. As described in Section 1, each information resource is characterized as a knowledge context containing a KSP relation (1) among keywords from which a distance function $d$ is obtained (cfr. (2)). $d_0$ is the distance function of the knowledge context of the user, while $d_1, \cdots, d_r$ are the distance functions from the knowledge contexts of each of the information resources.

### 4.1.1 Spreading Interest Fuzzy Sets

For each information resource $R_t$ and each keyword $k_u$ in the users present interests $\{k_1, \cdots, k_p\}$, a *spreading interest fuzzy set* $F_{t,u}$ is calculated using $d_t$:

$$F_{t,u}(k) = max\left[e^{\left(-\alpha . d_t(k,k_u)^2\right)}, \varepsilon\right], \qquad \forall k \in R_t, t = 1 \ldots r, u = 1 \ldots p \qquad (6)$$

This fuzzy set contains the keywords of $R_t$ which are closer than $\varepsilon$ to $k_u$, according to an exponential function of $d_t$. $F_{t,u}$ spreads the interest of the user in $k_u$ to keywords of $R_t$ that are near according to $d_t$. The parameter $\alpha$ controls the spread of the exponential function. $F_{t,u}$ represents the set of keywords of which are near or very related to keyword $k_u$. Because the knowledge context of each $R_t$ contains a different $d_t$, each $F_{t,u}$ will also be a different fuzzy set for the same $k_u$, possibly even containing keywords that do not exist in other information resources. There exist a total of $n = r.p$ spreading interest fuzzy sets $F_{t,u}$. Figure 6 depicts a generic $F_{t,u}$.

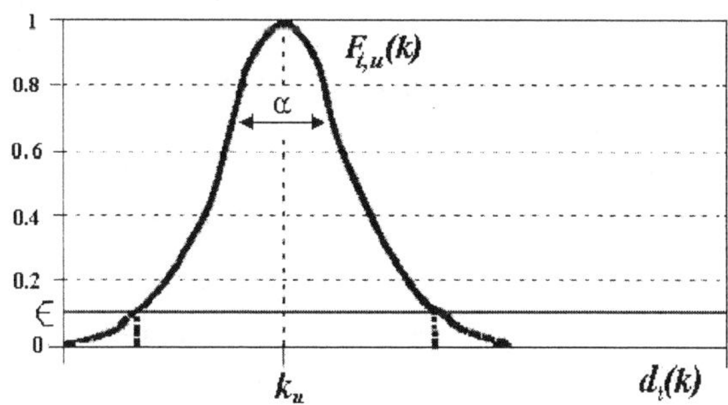


Figure 6. The exponential membership function of $F_{t,u}(k)$ spreads the interest of a user on keyword $k_u$ to close keywords according to distance function $d_t(k)$ for each information resource $R_t$.

### 4.1.2. Combining the Perspectives of Different Knowledge Contexts on the User Interest

Assume now that the present interests of the user $\{k_1, \cdots, k_p\}$ are probabilistically constrained, that is, there is a probability weight associated with each keyword: $\mu_1, \cdots, \mu_p$, such that $\mu_1 + \ldots + \mu_p = 1$. Assume further that the intervening r information resources $R_t$ are also probabilistically constrained with weights: $\upsilon_1, \cdots, \upsilon_r$, such that $\upsilon_1 + \ldots + \upsilon_r = 1$. Thus, the probabilistic weight of each spreading interest fuzzy set $F_i = F_{t,u}$, where $i = (t-1)p + u$, is $m_i = \upsilon_t . \mu_u$.

To combine the $n$ fuzzy sets $F_i$ and respective probabilistic weights $m_i$, formula (5) is employed. This results in an evidence set ES(k) defined on $K$, which represents the interests of the user inferred from spreading the initial interest set of keywords in the knowledge contexts of the intervening information resources. The inferring process combines each $F_{t,u}$ with the "and/or" linguistic expression entailed by formula (5). Each contains the keywords related to keyword $k_u$ in the knowledge context of information $F_{t,u}$ resource $R_t$, that is, the perspective of $R_t$ on $k_u$. Thus, ES(k) contains the "and/or" combination of all the perspectives on each keyword $k_u = \{k_1, \cdots, k_p\}$ from each knowledge context associated with all information resources $R_t$.

As an example, without loss of generality, consider that the initial interests of an user contain one single keyword $k_1$, and that the user is querying two distinct information resources $R_1$ and $R_2$. Two spreading interest fuzzy sets, $F_1$ and $F_2$, are generated using $d_1$ and $d_2$ respectively, with probabilistic weights $m_1 = \upsilon_1$ and $m_2 = \upsilon_2$. ES(k) is easily obtained straight from formula (4). This evidence set contains the keywords related to $k_1$ in $R_1$ "and/or" the keywords related to $k_1$ in $R_2$, taking into account the probabilistic weights attributed to $R_1$ and $R_2$. $F_1$ is the perspective of $R_1$ on $k_1$ and $F_2$ the perspective of $R_2$ on $k_1$.

### 4.2. Reducing the Uncertainty of User Interests via Conversation

The evidence set obtained in Section 4.1 with formulas (5) and (6) is a first cut at detecting the interests of a user in a set of information resources. But we can compute a more accurate interest set of keywords using an interactive conversation process between the user and the information resources being queried. Such conversation is an uncertainty reducing process based on Nakamura and Iwai's [21] IR system, and extended to Evidence Sets by Rocha [6][7].

In addition to the evidence set ES(k) constructed in Section 4.1, a fuzzy set $F_0(k)$ is constructed to contain the keywords of the knowledge context $R_0$ of the user which are close to the initial interest set $\{k_1, \cdots, k_p\}$ according to distance function $d_0$. As discussed in Section 1, the user's history of IR is itself characterized as a knowledge context $R_0$ with its own KSP relation and derived distance function $d_0$. $F_0(k)$ is given by:

$$F_0(k) = \bigcup_{u=1}^{p} F_{0,u}(k) \tag{7}$$

where $F_{0,u}(k)$ is calculated using formula (6). $F_0(k)$ represents the perspective of the user, from her history of retrieval, on all keywords $\{k_1, \cdots, k_p\}$. Given ES(k) and $F_0(k)$, for a default value of $\alpha = \alpha_0$, the algorithm for *TalkMine* is as follows:

1. Calculate the uncertainty of ES(k) in its forms of fuzziness, nonspecificity, and conflict (see Section 3.2). If total uncertainty is below a pre-defined small value the process stops, otherwise continue to 2.

2. The most uncertain keyword $k_j \in ES(k)$ is selected.

3. If $k_j \in R_0$, then goto 4 (AUTOMATIC), else goto 6 (ASK).

4. If $F_0(k_j) > 0.5 + \delta$, then goto 7 (YES).

5. If $F_0(k_j) \leq 0.5 - \delta$, then goto 8 (NO), else goto 6 (ASK).

6. ASK user if she is interested in keyword $k_j$. If answer is yes goto 7 (YES), else goto 8 (NO).

7. An evidence set YES(k) is calculated using the procedure of section 4.1 for a single keyword $k_j$ and all $r$ information resources $R_t$. The spread of the exponential functions is controlled with parameter $\alpha$ so that answers to previous keywords $k_j$ are preserved. *ES(k)* is then recalculated as the evidence set union of YES(k) and *ES(k)* itself.

8. An evidence set NO(k) is calculated as the complement of YES(k) used in 7. *ES(k)* is then recalculated as the evidence set intersection of NO(k) and *ES(k)* itself.

9. Goto 1.

The parameter $\delta$ controls how much participation is required from the user in this interactive process, and how much is automatically deduced from her own knowledge context used to produce $F_0(k)$. $\delta \in [0, 0.5]$; for $\delta = 0$, all interaction between user and information resources is mostly automatic, as answers are obtained from $F_0(k)$, except when $k_i \notin R_0$; for $\delta = 0.5$, all interaction between user and information resources requires explicit answers from the user. If the user chooses not to reply to a question, the answer is taken as NO. Thus, $\delta$ allows the user to choose how automatic the question-answering process of *TalkMine* is.

Regarding the change of spread employed in steps 7 and 8 for the construction of the YES(k) and NO(k) evidence sets. A list of the keywords the user (or $F_0(k)$ automatically) has responded YES or NO to is kept. The membership value of these keywords in the final ES(k) produced must be 1 or 0, respectively. Thus, the union and intersections of ES(k) with YES(k) and NO(k) in 7 and 8, must be defined in a such a way as to preserve these values. If the spread obtained with $\alpha_0$ would alter the desired values, then a new $\alpha$ is employed in formula (6) so that the original values are preserved $\pm \epsilon$. Because of this change of spreading inference of the YES(k) and NO(k) evidence sets, the sequence of keywords selected by the question-answering process in step 2 affects the final *ES(k)*. That is, the selection of a different keyword may result in a different *ES(k)*.

The final *ES(k)* obtained with this algorithm is a much less uncertain representation of user interests as projected on the knowledge contexts of the information resources queried, than the initial evidence set obtained in Section 4.1. The conversation algorithm lets the user reduce the uncertainty from the all the perspectives initially available.

The initial evidence set produced in Section 4.1 includes all associated keywords in several information resources. The conversation algorithm allows the user and her knowledge context to select only the relevant ones. Thus, the final *ES(k)* can be seen as a low-uncertainty linguistic category containing those perspectives on the user's initial interest (obtained from the participating information resources) which are relevant to the user and her knowledge context [6][7].

Notice that this category is not stored in any location in the intervening knowledge contexts. It is temporarily constructed by integration of knowledge from several information resources and the interests of the user expressed in the interactive conversational process. Such a category is therefore a temporary container of knowledge integrated from and relevant for the user and the collection of information resources.

Thus, this algorithm implements many of the, temporary, "on the hoof" [9] category constructions as discussed in [6].

### *4.3. Recommending Documents*

After construction of the final ES(k), *TalkMine* must return to the user documents relevant to this category. Notice that every document $n_i$ defines a crisp subset whose elements are all the keywords $k \in K$, which index $n_i$ in all the constituent information resources. The similarity between this crisp subset and ES(k) is a measure of the relevance of the document to the interests of the user as described by ES(k). This similarity is defined by different ways of calculating the subsethood [22] of one set in the other. Details of the actual operations used are presented in [7]. High values of these similarity measures will result on the system recommending only those documents highly related to the learned category ES(k).

### *4.4. Adapting Knowledge Contexts*

From the many *ES(k)* obtained from the set of users of information resources, we collect information used to adapt the KSP and semantic distance of the respective knowledge contexts. The scheme used to implement this adaptation is very simple: the more certain keywords are associated with each other, by often being simultaneously included with a high degree of membership in the final *ES(k)*, the more the semantic distance between them is reduced. Conversely, if certain keywords are not frequently associated with one another, the distance between them is increased. An easy way to achieve this is to have the values of $N(k_i)$, $N(k_j)$ and $N_\cap(k_i, k_j)$ as defined in formula (1), adaptively altered for each of the constituent $r$ information resources $R_t$. After *ES(k)* is constructed and approximated by a fuzzy set $A(x)$, these values are changed according to:

$$N^t(k_i) = N^t(k_i) + w.A(k_i), t = 1 \ldots r, k_i \in R_0 \cup R_1 \cup \ldots \cup R_r \tag{8}$$

and

$$N_\cap^t(k_i, k_j) = N_\cap^t(k_i, k_j) + w.\min[A(k_i), A(k_j)], t = 1 \ldots r, k_i, k_j \in R_0 \cup R_1 \cup \ldots \cup R_r \tag{9}$$

where $w$ is the weight ascribed to the individual contribution of each user. The adaptation entailed by (8) and (9) leads the semantic distance of the knowledge contexts involved, to increasingly match the expectations of the community of users with whom they interact. Furthermore, when keywords with high membership in $ES(k)$ are not present in one of the information resources queried, they are added to it with document counts given by formulas (8) and (9). If the simultaneous association of the same keywords keeps occurring, then an information resource that did not previously contain a certain keyword, will have its presence progressively strengthened, even though such keyword does not index any documents stored in this information resource.

## 5. Collective Evolution of Knowledge with Soft Computing

*TalkMine* models the construction of linguistic categories. Such "on the hoof" construction of categories triggered by interaction with users, allows several unrelated information resources to be searched simultaneously, temporarily generating categories that are not really stored in any location. The short-term categories bridge together a number of possibly highly unrelated contexts, which in turn creates new associations in the individual information resources that would never occur within their own limited context.

Consider the following example. Two distinct information resources (databases) are searched using *TalkMine*. One database contains the documents (books, articles, etc) of an institution devoted to the study of computational complex adaptive systems (e.g. the library of the *Santa Fe Institute*), and the other the documents of a Philosophy of Biology department. I am interested in the keywords GENETICS and NATURAL SELECTION. If I were to conduct this search a number of times, due to my own interests, the learned category obtained would certainly contain other keywords such as ADAPTIVE COMPUTATION, GENETIC ALGORITHMS, etc. Let me assume that the keyword GENETIC ALGORITHMS does not initially exist in the Philosophy of Biology library.

After I conduct this search a number of times, the keyword GENETIC ALGORITHMS is created in this library, even though it does not contain any documents about this topic. However, with my continuing to perform this search over and over again, the keyword GENETIC ALGORITHMS becomes highly associated with GENETICS and NATURAL SELECTION, introducing a new perspective of these keywords. From this point on, users of the Philosophy of Biology library, by entering the keyword GENETIC ALGORITHMS would have their own data retrieval system point them to other information resources such as the library of the *Santa Fe Institute* or/and output documents ranging from "The Origin of Species" to treatises on Neo-Darwinism – at which point they would probably bar me from using their networked database!

Given a large number of interacting knowledge contexts from information resources and users (see Figure 2), *TalkMine* is able to create new categories that are not stored in any one location, changing and adapting such knowledge contexts in an open-ended fashion. Openendedness does not mean that *TalkMine* is able to discern all knowledge negotiated by its user environment, but that it is able to permutate all the semantic information (KSP and $d$ described in Section 1) of the intervening knowledge contexts in an essentially open-ended manner.

The categories constructed by *TalkMine* function as a system of collective linguistic recombination of distributed memory banks, capable of transferring knowledge across different contexts and thus creating new knowledge. In this way, *TalkMine* can adapt to an evolving environment and generate new knowledge given a sufficiently diverse set of information resources and users. Readers are encouraged to track the development of this system at http://arp.lanl.gov.

*TalkMine* is a collective recommendation algorithm because it uses the behavior of its users to adapt the knowledge stored in information resources. Each time a user queries several information resources, the category constructed by *TalkMine* is used to adapt those (cfr. Section 4). In this sense, the knowledge contexts (cfr. Section 1) of the intervening information resources becomes itself a representation of the knowledge of the user community. A discussion of this process is left for future work.

*TalkMine* is a soft computing approach to recommendation systems as it uses Fuzzy Set and Evidence Theories, as well as ideas from Distributed Artificial Intelligence to characterize information resources and model linguistic categories. It establishes a different kind of humanmachine interaction in IR, as the machine side rather than passively expecting the user to pull information, effectively pushes relevant knowledge.

This pushing is done in the conversation algorithm of *TalkMine*, where the user, or her browser automatically, selects the most relevant subsets of this knowledge. Because the knowledge of communities is represented in adapting information resources, and the interests of individuals are integrated through conversation leading to the construction of linguistic categories and adaptation, *TalkMine* achieves a more natural, biological-like, knowledge management of distributed information systems, capable of coping with the evolving knowledge of user communities.

## References


[1] Rocha, L. M., Bollen, J., Biologically motivated distributed designs for adaptive knowledge management, In: *Design Principles for the Immune System and Other Distributed Autonomous Systems*, Cohen, I., Segel, L., (Eds.), Santa Fe Institute Series in the Sciences of Complexity, Oxford University Press, pp. 305-334, 2001.

[2] Klir, G. J., Yuan, B., *Fuzzy Sets and Fuzzy Logic: Theory and Applications*, Prentice Hall, 1995.

[3] Miyamoto, S., *Fuzzy Sets in Information Retrieval and Cluster Analysis*, Kluwer, 1990.

[4] Rocha, L. M., Combination of evidence in recommendation systems characterized by distance functions, Proceedings of the FUZZ-IEEE 2002, In Press, 2002.

[5] Galvin, F., Shore, S.D., Distance functions and topologies, *The American Mathematical Monthly*, vol. **98**, no. 7, pp. 620-623, 1991.

[6] Rocha, L. M., Adaptive Recommendation and Open-Ended Semiosis, *Kybernetes*, vol. **30**, no. 5-6, pp. 821-851, 2001.

[7] Rocha, L. M., Evidence sets: modeling subjective categories, *International Journal of General Systems*, vol. **27**, pp. 457-494, 1999.

[8] Rocha, L. M., *Evidence Sets and Contextual Genetic Algorithms: Exploring Uncertainty, Context and Embodiment in Cognitive and biological Systems*, PhD. Dissertation, State University of New York at Binghamton, UMI Microform 9734528, 1997.

[9] Clark, A., *Associative Engines: Connectionism, Concepts, and Representational Change*, MIT Press, 1993.

[10] Van Gelder, T., What is the 'D' in 'PDP': a survey of the concept of distribution, In: *Philosophy and Connectionist Theory*, Ramsey, W., Lawrence Erlbaum, 1991.

[11] Kanerva, P., *Sparse Distributed Memory*, MIT Press, 1988.

[12] Rocha, L. M., Cognitive categorization revisited: extending interval valued fuzzy sets as simulation tools concept combination, *Proc. of the 1994 Int. Conference of NAFIPS/IFIS/NASA*, IEEE Press, pp. 400-404, 1994.

[13] Rocha, L. M., Relative uncertainty and evidence sets: a constructivist framework, *International Journal of General Systems*, vol. **26**, no. 1-2, pp. 35-61, 1997.

[14] Zadeh, L. A., Fuzzy Sets, *Information and Control*, vol. **8**, pp. 338-353, 1965.

[15] Shafer, G., *A Mathematical Theory of Evidence*, Princeton University Press, 1976.

[16] Turksen, I. B., Non-specificity and interval-valued fuzzy sets, *Fuzzy Sets and Systems*, vol. **80**, pp. 87-100, 1996.

[17] Klir, G. J., Developments in uncertainty-based information, In: *Advances in Computers*, Yovits, M., (Ed.), vol. **36**, pp. 255-332, 1993.

[18] Yager, R. R., On the measure of fuzziness and negation, Part I: membership in the unit interval, *Int. J. of General Systems*, vol. **5**, pp. 221-229, 1979.

[19] Yager, R. R., On the measure of fuzziness and negation, Part II: lattices, *Information and Control*, vol. **44**, pp. 236-260, 1980.

[20] Turksen, B., Interval valued fuzzy sets based on normal forms, *Fuzzy Sets and Systems*, vol. **20**, pp. 191- 210, 1986.

[21] Nakamura, K., Iwai, S., A representation of analogical inference by fuzzy sets and its application to information retrieval systems, In: *Fuzzy Information and Decision Processes*, Gupta, M. M., Sanchez, E., (Eds.), North-Holland, pp. 373-386, 1982.

[22] Kosko, B., *Neural Networks and Fuzzy Systems: A Dynamical Systems Approach to Machine Intelligence*, Prentice-Hall, 1992.

# Part IV

# Applications

*Systematic Organisation of Information in Fuzzy Systems*
*P. Melo-Pinto et al. (Eds.)*
*IOS Press, 2003*



# Solving Knapsack Problems
# using a Fuzzy Sets-based Heuristic

Armando Blanco, David A. Pelta, José-L. Verdegay

*Depto. de Ciencias de la Computación e Inteligencia Artificial*
*Universidad de Granada*
*Periodista Daniel Saucedo Aranda, s/n*
*18071 Granada, Spain*
*E-Mail: {armando,dpelta,verdegay}@ugr.es*

**Abstract** The Fuzzy Adaptive Neighborhood Search (*FANS*) method was previously presented and succesfully compared against a genetic algorithm and simulated annealing doing minimization of functions in $\mathcal{R}^n$. Here we propose to evaluate this fuzzy sets-based heuristic on instances of knapsack problems with multiple restrictions, doing comparison against the cited algorithms in order to gain knowledge about the behaviour of our method.

## 1  Introduction

Within the framework of Decision Support Systems, simple general purpose optimization tools are key elements to decision makers because they enable them to obtain initial solutions with minimal knowledge of the problem being solved. In this way, such initial solutions may serve as a guide for further improvements. Among those simple approaches are neighborhood or local search based heuristic techniques.

In [2, 4, 5], we presented the main ideas of a Fuzzy Adaptive Neighborhood Search (*FANS*) heuristic. Its main motivation is to provide a general purpose optimization tool, suitable to be embedded in a decision support system, which is easy to tailor to specific problems by means of appropriate definition of its components.

*FANS* is termed Fuzzy, because the solutions are qualified in terms of *fuzzy valuations*, and *adaptive*, because its behaviour is adapted as a response to the state of the search.

By means of a fuzzy valuation, represented in *FANS* by fuzzy sets, a fuzzy measure of the generated solutions is obtained. Fuzzy valuations may represent concepts like *Acceptability*, thus a degree of acceptability is calculated for each solution and such degrees are used by *FANS* at the decision stages. The use of schedulers enable *FANS* to change its behaviour as a function of the state of the search. In particular, the operator used to generate solutions is changed when the search seems trapped, leading to an intermediate escaping mechanism. When this mechanism fails, a classical restart operator is applied.

Being a heuristic, *FANS* needs some parameters and components to be defined. Suitable component definitions and parameters, lead *FANS* to reflect the qualitative behaviour of traditional techniques like HillClimbing, Random Walks, etc.

In [4, 5], we showed how *FANS* outperformed a genetic algorithm (*GA*) and simulated annealing (*SA*) over a set of real function minimization problems when the three algorithms were given a fixed amount of cost function evaluations. In those problems the only restriction was the range available for each variable, so it was easy to ensure that all the generated solutions were feasible.

Now, we propose to test *FANS* over Knapsack Problems with multiple restrictions in order to confirm the potential of *FANS* as a general purpose optimization technique. In this case, infeasible solutions will exist and the algorithm must deal with them in some way. For the sake of the limited number of pages, we will not describe here the basics of fuzzy sets; the interested reader is referred to [7].

The paper is organized as follows: in Section 2 test problems are presented. Then in Section 3 the main components of *FANS* are defined, and the characteristics of the *GA* and *SA* implemented are shown. The experiments and results obtained are in Section 4. Finally, Section 5 is devoted to conclusions and further work.

## 2   Test Instances

Knapsack Problem with Multiple Restrictions is the more general class of knapsack problems. It is an integer linear problem with the following formulation:

$$\text{Max} \sum_{i=1}^{n} p_i \, x_i \tag{1}$$

$$\text{s.t} \sum_{i=1}^{n} w_{ij} \, x_i \leq C_j, \ \ j = 1..m$$

where $n$ is the number of items, $m$ the number of restrictions, $x_i \in \{0, 1\}$ indicates if the $i$ item is included or not in the knapsack, $p_i$ is the profit associated with the item $i$, and $w_{ij} \in [0, .., r]$ is the weight of item $i$ with respect to constraint $j$.

Also, the following conditions are considered $w_{ij} < C_j \ \forall\, i, j$ (every item alone fits in the knapsack) and $\sum_{i=1}^{n} w_{ij} > C_j \ \forall\, j$ (the whole set of items don't fit in the knapsack).

For our experiments we select 9 instances from a standard set of 55 problems available from [1]. The instances are named *pb5*, *pb7*, *Weing1*, *Weing3*, *Weing7*, *Weish10*, *Weish14*, *Weish18*, *Weish27*. The number of variables range from 20 to 105 with 2 to 30 restrictions.

## 3   Algorithms

In this section we present the main components of *FANS*, together with their interaction. Also, some design decisions for the *SA* and *GA* implementations are later shown.

The knapsack solutions are represented by binary vectors $X$, where position $i$ represents the variable $x_i$, and this representation is used in *FANS, SA* and *GA*. The initial solutions for each method will contain just a unique position in 1, reflecting that $w_{ij} < C_j \ \forall\, i, j$.

In our experiments, infeasible solutions will not be taken into account and they will be discarded. No reparation procedure will be used.

## 3.1   FANS *Components*

*FANS* operation relies on four main components: an operator, to construct new solutions; a fuzzy valuation, to qualify them; an operator scheduler, to adapt the operator's behaviour; and a neighborhood scheduler, to generate and select a new solution.

In order to apply *FANS* to a particular problem, these components must be defined. As usual, when more problem-dependant definitions are used, the better the performance. We decide to use simple definitions in order to test the quality of the search strategy induced by *FANS*. Below, we describe the definition proposed for each component.

### Modification Operator $k$-BitFlip

This operator chooses $k$ positions randomly and flips the associated bit value. *Back mutation* is not allowed.

### Fuzzy Valuation: Acceptable

The generated solutions will be qualified in terms of "Acceptability", a concept reflecting the following idea: with a solution at hand, those generated solutions improving the current cost, will have a higher degree of acceptability than those with lower cost. Solutions diminishing the cost a little, will be considered "acceptable" but with lower degree. Finally those solutions demeliorating too much the current cost will not be considered as acceptable.

So, given the objective function $f$, the current solution $s$, $q$ a neighbor solution, and $\beta$ the limit for what is considered as acceptable, the following definition comprise those ideas:

$$\mu(q, s) = \begin{cases} 0.0 & \text{if } f(q) < \beta \\ (f(q) - \beta)/(f(s) - \beta) & \text{if } \beta \le f(q) \le f(s) \\ 1.0 & \text{if } f(q) > f(s) \end{cases} \tag{2}$$

For our experiments we used $\beta = f(s)(1 - \gamma)$, with $\gamma = 0.05$ .

### Operator Scheduler

The $k$-BitFlip operator will be adapted trough changes on the $k$ parameter. The used scheme is rather simple: being $k_t$ the actual value, then $k_{t+1}$ will be a random integer value in $[1, 2k_t]$. Also, if $k_{t+1} > top - n/10$, then $k_{t+1} = top$.

### Neighborhood Scheduler

Given the cost function $f$, the operator $\mathcal{O}$, the fuzzy valuation $\mu$ and the current solution $s$, we define the operational neighborhood of $s$ as

$$N(s) = \{\hat{x} \mid \hat{x} = \mathcal{O}(s)\} \tag{3}$$

and the "semantic" neighborhood of $s$ as

$$\hat{N}(s) = \{\hat{x} \mid \mu(\hat{x}) > \lambda; \ \hat{x} \in N(s)\} \tag{4}$$

Taking into account both definitions, we provide two neigborhood schedulers:

**a) Quality Based Grouping Scheme** $R|S|T$**:** this scheduler tries to generate $R$ "Acceptable" neighborhood solutions in $maxTrials$ trials, then those solutions are grouped into $S$ sets on the basis of their acceptability degree, and finally $T$ solutions are returned [5]. We used $maxTrials = 12$ and a $R = 5|S = 3|T = 1$ scheme.

The $S = 3$ sets or clusters are represented by overlapped triangular membership functions with boundaries adjusted to fit the range $[\lambda, 1.0]$, being $\lambda = 0.99$ the minimun level of acceptability required. The sets represents the terms *Low, Medium, High* for the linguistic variable "Quality". At the end of the process, $T = 1$ solution must be returned. Here, we choose to return any solution of the highest quality available.

**b) First Found Scheme:** at most $maxTrials = 12$ trials are available to obtain a solution $\hat{x} \in \hat{N}(x)$. The first one found is returned.

**Components Interaction**

Figure 1 shows *FANS* pseudo code. The iterations end when some external condition holds. The neigborhood scheduler $NS$ is called at the beginning of each iteration, with parameters: current solution $S_{cur}$; fuzzy valuation $\mu()$ and modification operator $\mathcal{O}$. Two situations may ocurr: an "acceptable" neigborhood solution $S_{new}$ was found or not.

In the first case $S_{new}$ is taken as the current solution and $\mu()$ parameters are adapted. In this way, we are varying our notion of "Acceptability" as a function of the context. If $NS$ could not return any acceptable solution, an exception condition is raised. No solutions were acceptable in the neighborhood induced by the operator. In this case, the operator scheduler $OS$ is executed, returning a modified version of $\mathcal{O}$. The next time $NS$ is executed, it will have a modified operator to search for solutions.

The $trappedCondition()$ exception is raised when $Top$ iterations were done without improvements in the best solution found. In such a case, the $doRestart()$ procedure is executed applying a perturbation over the current solution: $\frac{1}{3}$ of randomly chosen variables in one are set to zero. Then, the cost of the current solution is reevaluated, $\mu()$ is adapted, and finally the process is restarted.

*3.2    Genetic Algorithm and Simulated Annealing*

The *GA* used may be regarded as a "traditional" one. Mutation is applied to all individuals with certain probability. As mutation operator, the $k$-BitFlip operator is used with $k = 2$. Because no repair procedure is used, we applied the following procedure: if the solution obtained after mutation is infeasible, it is discarded and at most four more trials are done to obtain a feasible one. If no feasible solution was obtained, the original one is kept.

As crossover operator, two classical ones are implemented: 1 point and uniform crossover. Thus, we obtain 2 algorithms: *GAop* y *GAux* respectively. Elitism is also used in both versions and we used the following settings for other parameters: crossover and mutation probabilities $P_{xover} = 0.8$, $P_{mut} = 0.2$, population size $PopSize = 100$; and Tournament Selection with tournament size $q = 2$ within a $(\mu = 50 + \lambda = 75)$ scheme.

Our implementation of *SA* follows the guidelines presented in [3]. The $k$-BitFlip operator is also used with $k = 2$. The initial temperature was set to $T_0 = 5$ and proportional cooling is used with $T_{k+1} = 0.9\,T_k$. Temperature is adapted when 15 neighbors were accepted, or when $2 \times n$ neighbors were generated, being $n$ the size of the problem.

```
Procedure FANS:
Begin
   While ( not-finalization ) Do
     /* the neighborhood scheduler is called */
     NS->Run(O,μ(),S_cur,S_new,ok) ;
     If (S_new is good enough in terms of μ()) Then
        S_cur = S_new;
        adaptFuzzyVal(μ(),S_cur) ;
     Else
        /* No good enough solution was found  */
        /* with the current operator */
        /* We change it with the operator scheduler */
        OS->Run(O) ;
     endIf
     If (trappedCondition()) Then
        doRestart() ;
     endIf
   endDo
End.
```

Figure 1: *FANS* Pseudocode

To the best of our knowledge, no practical methodology exists to help in the determination of the parameters' values. So we used the standard approach of running a reduced set of experiments in order to detect a set of values which provide satisfactory results.

## 4  Experiments and Results

In order to analyze the performance of *FANS*, we conduct a set of experiments and made comparisons among the following five algorithms: $F_{rst}$, is *FANS* with scheduler $R|S|T$; $F_{ff}$, is *FANS* with scheduler $FirstFound$; *GAux*, is the *GA* with uniform crossover; *GAop*, is the *GA* with one-point crossover; and *SA*, the Simulated Annealing implementation.

We compare the performance of the algorithms under minimal knowledge of the problem (reflected by the use of simple operators), and when they are given a fixed number of resources (i.e cost function evaluations and number of generated solutions). In this way, we can assume the results are consequence of the search strategies and not of additional knowledge.

For each problem and algorithm 30 runs were made; each one ending when $maxEvals = 15000$ cost function evaluations were done or when $4 \times maxEvals$ solutions were generated. This limit is needed because only feasible solutions are evaluated.

The results are analyzed in terms of the error for each problem and globally over the whole test set. The first results are presented on Table 1, where part (a) shows the mean of the errors over 30 runs for each algorithm on each problem. The error is calculated as:

$$error = 100 \times (Optimum - Obtained\,Value)/(Optimum) \qquad (5)$$

The Table 1 shows that both versions of *FANS* achieved the lower values, except for problems Weing3, Weing7 y Weish27. *SA* achieved the better value on Weing7 and *GAux* did it on the other two problems.

(a)

|          | $F_{rst}$ | $F_{ff}$ | SA    | GAop | GAux |
|----------|-----------|----------|-------|------|------|
| pb5      | 1.04      | 0.92     | 6.52  | 3.37 | 3.07 |
| pb7      | 0.94      | 1.19     | 4.13  | 3.83 | 4.23 |
| weing1   | 0.20      | 0.19     | 8.07  | 0.92 | 1.37 |
| weing3   | 1.54      | 1.32     | 22.04 | 1.85 | 0.91 |
| weing7   | 0.50      | 0.51     | 0.48  | 1.13 | 0.93 |
| weish10  | 0.27      | 0.14     | 1.22  | 1.34 | 1.18 |
| weish14  | 0.85      | 0.78     | 1.93  | 1.85 | 0.91 |
| weish18  | 0.73      | 0.71     | 1.39  | 0.95 | 0.89 |
| weish27  | 3.02      | 2.89     | 2.85  | 3.21 | 1.18 |

(b)

|          | $F_{rst}$ | $F_{ff}$ | SA | GAop | GAux |
|----------|-----------|----------|----|------|------|
| pb5      | x         | x        |    |      |      |
| pb7      | x         | x        |    |      |      |
| weing1   | x         | x        |    | x    | x    |
| weing3   |           |          |    | x    | x    |
| weing7   |           |          |    |      |      |
| weish10  | x         | x        | x  | x    | x    |
| weish14  | x         | x        |    | x    | x    |
| weish18  | x         | x        |    | x    | x    |
| weish27  |           |          |    |      | x    |
| # Opt.   | 39        | 39       | 9  | 17   | 31   |

Table 1: Results of the comparison. On (a), the mean error over 30 runs for each problem and algorithm. On (b), an $x$ means the algorithm of the column reached the optimum of the problem on the row on any run.

On part (b) of Table 1, an $X$ appears if the algorithm on the column reached the optimum of the problem on the row in any of the 30 runs. It's seen that $F_{rst}$, $F_{ff}$ and *GAux* obtained the optimum on 6 of 9 problems, while *GAop* on 5 of 9 and *SA* just in 1 of 9.

It is hard to determine why *FANS* failed to achieve the optimums for those problems, because it is not clear what makes an instance easy or hard. One aspect is the correlation between profits and weights, which is well stablished for classical knapsack problems but not for multiple restrictions knapsack problems. Other elements are needed but this discussion is out of the aim of this work.

The last row (# Opt) in Table 1 (b), indicates the number of executions ending at the optimum for a total of $9 \times 30 = 270$ runs. $F_{rst}$ y $F_{ff}$ achieves the higher values followed by *GAux*. *SA* and *GAop* are quite ineffective from this point of view.

To conclude, taking the mean and variance of the errors over the whole set of problems (results not shown) it was clear that both version of *FANS* achieved the lowest values, followed by *GAux*. The mean error in *GAux* y *GAop* was almost twice of that in $F_{rst}$ y $F_{ff}$; and *SA* values were 5 times higher. To confirm if the differences in the mean of errors had statistical significance, t-tests were done with a confidence level of 95%. The results shown in Table 2 enabled us to confirm that both versions of *FANS* outperformed *GAux*, *GAop* y *SA*. Both *GA* outperformed *SA*, and no significant differences in the means were found among them, in spite of the apparent superiority of *GAux*.

|        | $F_{rst}$ | $F_{ff}$ | $GAux$ | $GAop$ | $SA$ |
|--------|-----------|----------|--------|--------|------|
| $F_{rst}$ |        | =        | +      | +      | +    |
| $F_{ff}$  | =      |          | +      | +      | +    |
| $GAux$    | -      | -        |        | =      | +    |
| $GAop$    | -      | -        | =      |        | +    |
| $SA$      | -      | -        | -      | -      |      |

Table 2: T-test results. A + sign at $(i, j)$ means algorithm $i$ was better than $j$ with confidence level of 95%. A - sign stands for $i$ worst than $j$ and = stands for no significative differences.

## 5   Conclusions and Future Work

In this work we gained evidence about the suitability of our method, *FANS*, as a general purpose optimization tool. Although our intention was not to develop a specific tool to deal with knapsack problems, but to use them to evaluate *FANS* behaviour, the results show the good performance of *FANS* in spite of the simple definitions used for the components.

It seems clear that using problem oriented elements, like a repairing scheme for infeasible solutions, or a greedy heuristic to construct initial solutions, the current results may be improved, making *FANS* a more useful tool to deal with multiple restriction knapsack problems.

Now, *FANS* is being applied to problems arising in bioinformatics, where the user's knowledge is very important to assess the quality of the results. In particular, we are applying *FANS* to simple models of the protein folding problem (see [6] for recent results), and we have encouraging results for the molecular matching problem.

## Acknowledgments

Research supported in part by Projects PB98-1305 and TIC 99-0563. David Pelta is a grant holder from Consejo Nac. de Invest. Científicas y Técnicas (Argentina).

## References

[1] J. Beasley. The or-library: a collection of test data sets. Technical report, Management School, Imperial College, London SW7 2AZ., 1997. http://mscmga.ms.ic.ac.uk/info.html.

[2] A. Blanco, D. Pelta, and J. L. Verdegay. Learning about fans behaviour: Knapsack problems as a test case. In *Proceedings of the International Conference in Fuzzy Logic and Technology, Eusflat 2001*, pages 17–21, 2001.

[3] A. Diaz, F. Glover, H. Ghaziri, J. Gonzalez, M. Laguna, P. Moscato, and F. Tseng. *Heuristic Optimization and Neural Nets*. Ed. Paraninfo, 1996. in Spanish.

[4] D. Pelta, A. Blanco, and J. L. Verdegay. A fuzzy adaptive neighborhood search for function optimization. In *Fourth International Conference on Knowledge-Based Intelligent Engineering Systems & Allied Technologies, KES 2000*, volume 2, pages 594–597, 2000.

[5] D. Pelta, A. Blanco, and J. L. Verdegay. Introducing fans: a fuzzy adaptive neighborhood search. In *Eigth International Conference on Information Processing and Management of Uncertainty in Knowledge-based Systems, IPMU 2000*, volume 3, pages 1349–1355, 2000.

[6] D. Pelta, N. Krasnogor, A. Blanco, and J. L. Verdegay. F.a.n.s. for the protein folding problem: Comparing encodings and search modes. In *Proceedings of the 4th Metaheuristics International Conference MIC2001*, volume 1, pages 327–331, 2001.

[7] H. J. Zimmermann. *Fuzzy Sets Theory and Its Applications*. Kluwer Academic, 1996.

# Evolution of Analog Circuits for Fuzzy Systems

Adrian STOICA
*Jet Propulsion Laboratory*
*California Institute of Technology*
*4800 Oak Grove Drive*
*Pasadena, CA 91109*
*818-354-2190*

**Abstract**. The paper presents several aspects related to the use of evolutionary algorithms (EA) for analog circuit design, and to the automatic circuit reconfiguration on programmable analog devices. It demonstrates the power of EA in the synthesis of circuits implementing circuits for specific fuzzy operators and for deriving very compact implementations of complete fuzzy systems. Uniquely, circuits with a totally new type of superimposed multi-functionality can be synthesized. It discusses the power of using programmable devices to do fuzzy configurable computing. To obtain circuits that work both in simulations and on real hardware a technique called mixtrinsic evolution has been developed, which expended becomes a model identification technique with larger applicability. Finally, it is shown how evolution can be accelerated by operating with "fuzzy" switches, with partial opening, and the concept of "fuzzy topologies" for electronic circuits is suggested as a model to characterize behaviors in those experiments.

## 1. Introduction

Recent research in evolutionary synthesis of electronic circuits and evolvable hardware [1][2] has generated a set of ideas and demonstrated a set of concepts that have direct application to the computational circuits used for information processing in fuzzy systems.

The idea behind evolutionary circuit synthesis / design and Evolvable Hardware (EHW) is to employ a genetic search / optimization algorithm that operates in the space of all possible circuits and determines solution circuits that satisfy imposed specifications, including target functional response, size, speed, power, etc.

The genetic search in EHW is tightly coupled with a coded representation that associates each circuit to a "genetic code" or chromosome. The simplest representation of a chromosome is a binary string, a succession of 0s and 1s that encode a circuit. However, circuits can also be represented using integer or real strings, or parsing trees [3].

First, a population of chromosomes is randomly generated. The chromosomes are converted into circuit models and their responses are compared against specifications, and individuals are ranked based on how close they come to satisfying them. In preparation for a new iteration, a new population of individuals is generated from the pool of best individuals in the previous generation. This is subject to a probabilistic selection of individuals from a best individual pool, followed by two operations: random swapping of parts of their chromosomes, the *crossover* operation, and random flipping of chromosome bits, the *mutation* operation.

The process is repeated for several generations, resulting in increasingly better individuals. Randomness helps to avoid getting trapped in local optima. Monotonic convergence (in a loose Pareto sense) can be forced by unaltered transference to the next generation of the best individual from the previous generation. There is no theoretical guarantee that the global optimum will be reached in a useful amount of time; however, the evolutionary / genetic search is considered by many to be the best choice for very large, highly unknown search spaces. The search process is usually stopped after a number of generations, or when closeness to the target response has reached a sufficient degree. One or several solutions may be found among the individuals of the last generation.

This paper also reviews the concept of a Field Programmable Transistor Array (FPTA), a fine-grained re-configurable architecture targeted for Evolvable Hardware experiments. This transistor array combines the features of fine granularity, being programmable at the transistor level, with the architectural versatility for the implementation of standard building blocks for analog and digital designs, being also efficient for the implementation of fuzzy computational functions [3].

This paper overview some of these concepts. Section 2 illustrates evolutionary techniques for automatic synthesis of electronic circuits implementing fuzzy operators and functions. Section 3 discusses re-configurable devices for fuzzy computing. Section 4 presents a very compact implementation (seven transistors) of a complete fuzzy system. Section 5 presents a new technique, called mixtrinsic evolution, which is used for automatic modeling / identification of correlated fuzzy models of different granularity / resolution / flavor. Section 6 shows how circuit evolution can be accelerated (and modeling in general) through gradual morphing through fuzzy topologies. Section 7 presents a new class of specific circuits, polymorphic electronics, as circuits with superimposed multiple functionality, in particular fuzzy functionality. Section 8 concludes the work.

## 2. Evolutionary design of electronic circuits implementing fuzzy operators

Evolutionary algorithms (EA) have proven to be powerful search mechanisms able to automatically find novel circuits / topologies that satisfy desired functional and implementation / efficiency-related requirements. In particular EA can be applied to determine circuits that implement operators such as conjunctions and disjunctions modeled by triangular norms, which is a central element for fuzzy systems.

While circuits implementing simple $t$-norms / co-norms, such as min, max, product and probabilistic sum, have been explored quite early, more complex $t$-norms, such as parametric $t$-norms (e.g. Frank's $t$-norms explored in [4]), while shown valuable both in theoretical studies as well as in practical applications (such as in the fuzzy controllers by Gupta $et\ al.$ in the early 90s, see [5][6]) are still lacking circuit solutions.

The reasons are that such circuits (referring here to analog circuits – computing the functions digitally or storing tables in memory being computationally expensive solutions in terms of power / time / area of silicon) are hard to design. Each such circuit would practically mean a new patent, and one may need many of them, one for each value of the parameter of the $t$-norm (unless of course one designs a parametric circuit, which would be an even harder task).

Evolutionary techniques have proven efficiency (and proficiency) in creating such new designs for implementation of parametric $t$-norm circuits. Several such circuits are presented in [7] and will be shown for illustration. See Figures 1, 2 and 3 for exemplification (Figure 1 shows the FPTA cell discussed next). Figures 4 and 5 compare evolved- and target- response for a $t$-norm and a $s$-norm respectively.

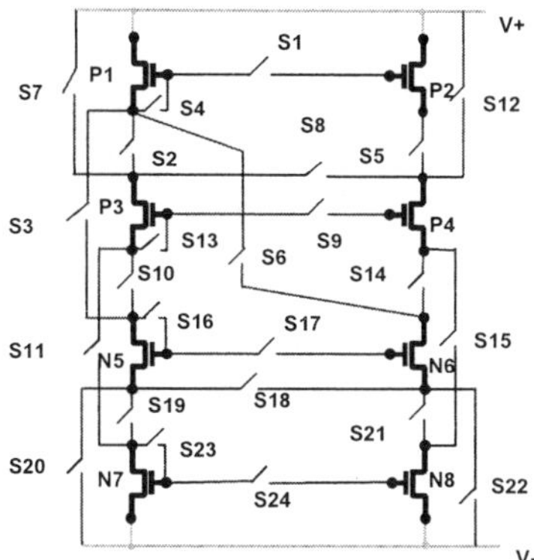

Figure 1. Schematic of an FPTA cell consisting of 8 transistors and 24 switches.

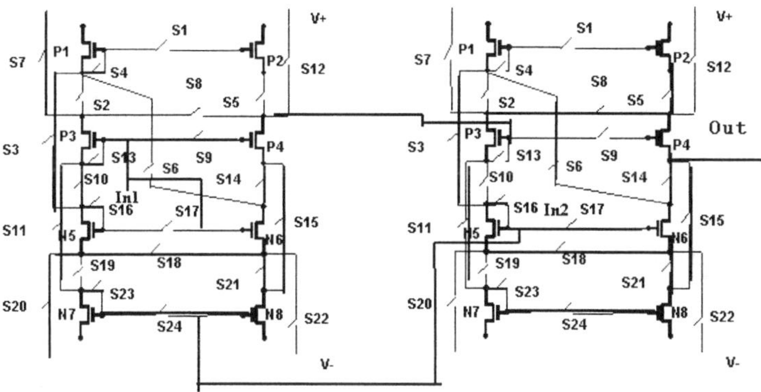

Figure 2. Evolved circuit mapped onto the FPTA for *t*-norm (*s*=1). Inputs *In*1 and *In*2, output at the right (*Out*).

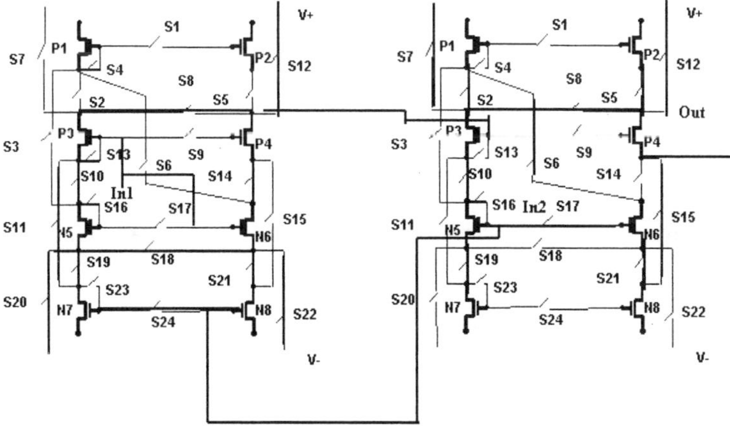

Figure 3. Evolved circuit mapped onto the FPTA for *t*-norm (*s*=100). Inputs *In*1 and *In*2, output at the right (*Out*).

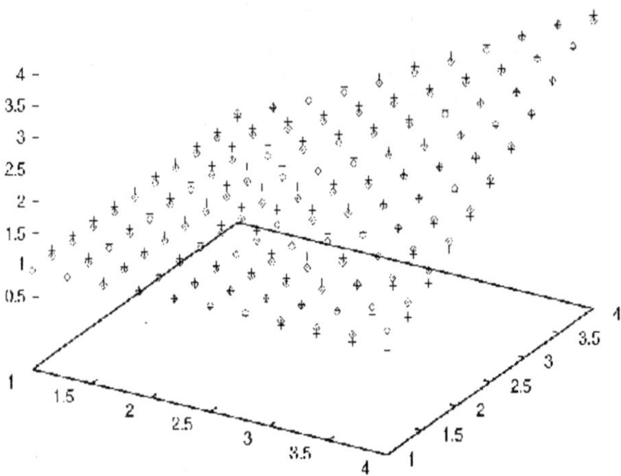

Figure 4. Response of a circuit implementing the fundamental *t*-norm for s=100 (◊). Target characteristic shown with (+).

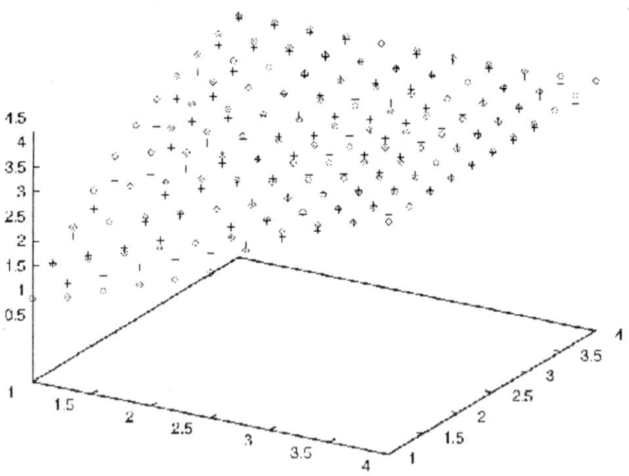

Figure 5. Response of a circuit implementing the fundamental *s*-norm for s=100 (◊). Target characteristic shown with (+).

## 3. Re-configurable / evolvable devices for fuzzy configurable computing

The Field Programmable Transistor Arrays (FPTA) [7][8][9][10] are devices on which one can program circuits at the lowest granularity and achieve the highest control of the topology, i.e. by reconfiguration at the transistor level. These devices were conceived as platforms for rapid experiments in evolutionary synthesis; instead of simulating circuits (and in fact very many of them, as required by EA), one would just instantiate them on the configurable chips and evaluate the measured response. (Evolutionary design could benefit from various models of different granularity, sub-circuits of several transistors can be "frozen" and used as components of higher level of granularity. For various levels of modeling, evolvability and granularity levels see [9]).

First, from the perspective of fuzzy computational platforms, these devices are versatile / malleable architectures on which one can rapidly instantiate the desired fuzzy circuitry. Second, these are adaptable and can be used to implement systems that change the operators from one processing stage to the other; i.e. one can start with MIN-MAX logic and change later to another as needed. Third, the FPTA are platforms for fuzzy configurable computing in the same way FPGAs are for conventional / digital computing. In an ASIC version the whole processing described by an algorithm is mapped at once in a fixed circuitry, with building blocks corresponding to various stages of the algorithm.

In an FPGA / FPTA implementation, only a part of the algorithm is mapped first in an instance of the processing machine, after which the chip reconfigures itself to become the blueprint for the next step, and so on. The fourth aspect is related to the use of FPTA to evolve fuzzy circuits directly in hardware. This not only is rapid, but also real, i.e. valid without doubts, unlike the case of evolution in simulations which may lead to results that may differ when tested in reality due to modeling approximation and fabrication effects. In other words evolution on real chips takes in consideration and adapts to account for effects of fabrication, temperature of operation, etc.

## 4. Fuzzy controller implementations with less than ten transistors

The results mentioned above, related to analog circuit of implementations for fuzzy operators are particularly relevant from the point of view of building fuzzy computation engines, especially those of programmable nature, meaning that some software code would change the function of the system. Most of the traditional fuzzy systems in use however, are quite simple in nature and the computation can be expressed in terms of a simple surface.

An example is the control surface of a two-input fuzzy controller. A fuzzy circuit could be synthesized to approximate this surface. If the circuit is synthesized on a programmable cell structure, such as the FPTA cell, it can in fact be "programmed" to change function by a new encoding bit-string found through the evolutionary search. In this case the search replaces a conventional compiler taking inputs from specifications and mapping to a configuration based on the very clear architecture of the device. If a fixed solution suffices, then a very compact circuit that could be implemented in an ASIC, integration with other silicon components is also possible. Since the previous examples illustrated evolution on a programmable architecture, the following example illustrates evolution of a circuit with fixed topology.

The example chosen is that of a fuzzy controller provided as a demo for the popular Matlab software [11]. The "ball juggler" is one of the demos of the Matlab Fuzzy Logic Toolbox. The fuzzy controller for the ball juggler has two inputs and one control output. A screen capture illustrating the membership functions is shown in Figure 6. The controller is a simple Sugeno-type with 9 rules. A screen capture of the control surface is shown in Figure 7.

A circuit approximating the control surface was evolved and is presented in Figure 8. The response, presented together with the target surface for comparison, is shown in Figure 9. The average error achieved was of 1. 93%, and the maximum error to the target surface was 6. 7%.

The circuit is rather robust, and was tested at variations in transistor sizes, supply voltage and temperature. The testing results are: decreasing the transistor sizes by a factor of 10 did not change the circuit response and the deviation from the target; average error of 1. 98% and maximum error of 6. 96% when decreasing the power supply voltage to 4. 75V; average error of 1. 94% and maximum error of 6. 65% when increasing the power supply voltage to 5. 25V; average error of 1. 89% and maximum error of 6. 3% when decreasing the temperature to 0°C; average error of 1. 98% and maximum error of 7. 2% when increasing the temperature to 55°C.

*A. Stoica / Evolution of Analog Circuits for Fuzzy Systems*

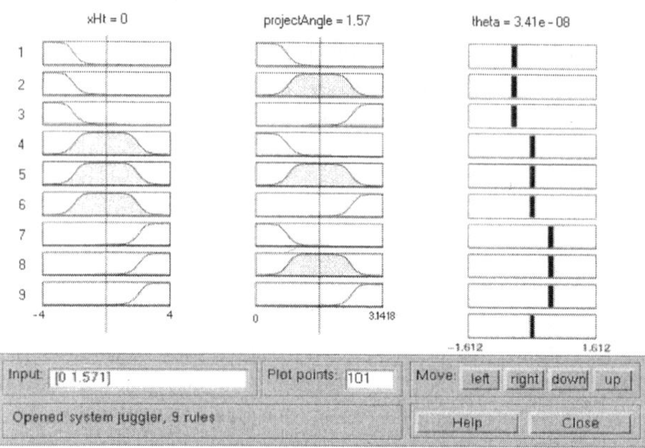

Figure 6. Membership functions for the ball juggler fuzzy controller.

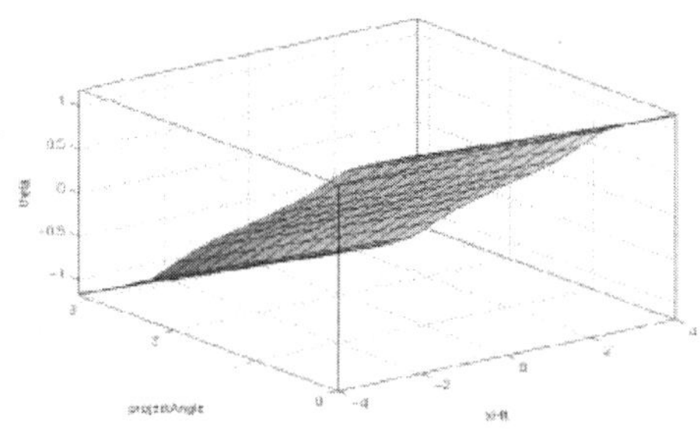

Figure 7. Surface of the ball juggler fuzzy controller.

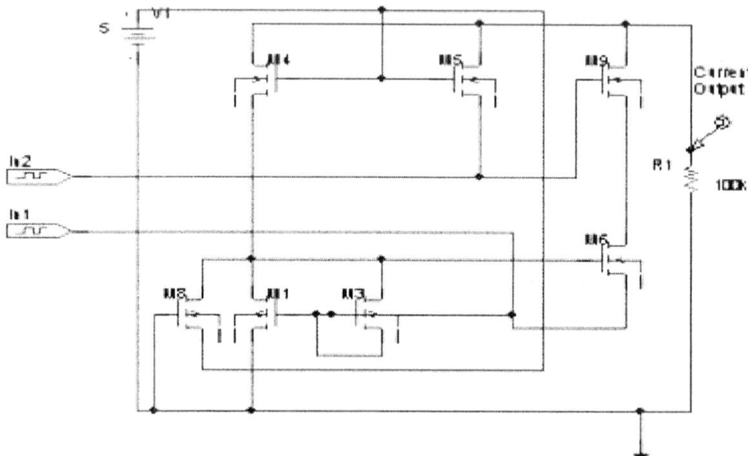

Figure 8. Evolved circuit realizing the ball juggler fuzzy controller. Transistors substrate connections at 5V for PMOS and ground for NMOS.

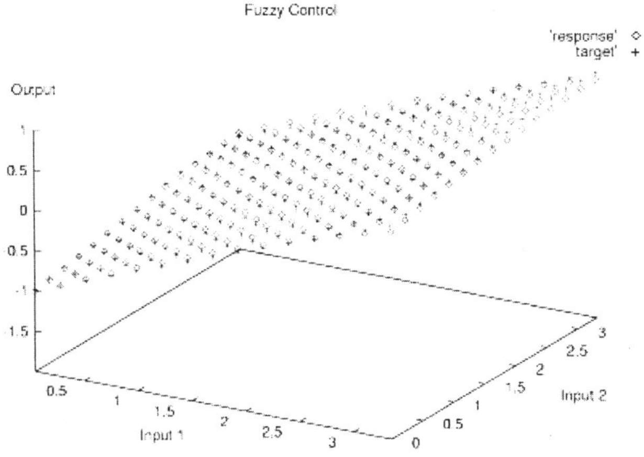


Figure 9.

Comparison between response and target for the evolved fuzzy controller.

Finally, a different model, (specific for a HP 0.5 MOS fabrication process) led to qualitatively the same result, with slight increase in the error. That error became small again when evolution targeted a circuit in that specific process.

## 5. Mixtrinsic evolution

Several fuzzy researchers, foremost Prof. Zadeh, emphasize the importance of concepts such as information granularity and processing of information at different levels of granularity. In complex electronic systems, to cope with the complexity induced by the explosion in number of components reflected by Moore's Law, and also recently the extra challenges related to systems-on-a-chip – the idea of using high-level approaches (models, languages) imposes as an indisputable necessity. One needs however to preserve the connection and representation power for the structures – at finest levels of granularity / resolution.

Thus, models of different complexity are needed for given entity. Mixtrinsic evolution (introduced in [12]) is a method of modeling (model construction and identification) which allows simultaneous development of coherent / corresponding models. Traditional modeling uses one model type only. In evolutionary modeling a population of same-type models is evaluated. For example, in evolving electronic circuits on SPICE models of the FPTA – in simulations – one can use a resistive model for the interconnecting switches. Alternatively one can use the more realistic model for the switch, considering two back-to-back transistors.

The two circuits simulate at different speeds, the resistive one much faster, while less accurate. The problem is that evolving solutions using only one representation may result in (and this has been shown in practice in various contexts) "solutions" (characterized in first instance by a genetic code that from which the circuit is derived) that do not work in the context of the other representation. For example, a. derived model obtained using resistive switches may not work when tested on the transistor-based representation, and vice versa.

The idea of mixtrinsic evolution is to have a mixed population in which candidate solutions may be at different levels of resolution (or perhaps even of very different nature). One approach is to assign the solutions (which in evolutionary design model are coded in chromosomes) to alternative instantiations of different resolution changing from a generation to another. Another approach is to have a combined fitness function that characterizes the modeling ensemble (i.e. each candidate solution in all its instantiations at different levels of resolution).

Both approaches were demonstrated to work, illustrated on problems on which not applying the method results in solutions that are not equivalent at different levels of resolution. (In some other experiments we used SW models and HW programmed circuits and converged to solutions with similar behaviors).

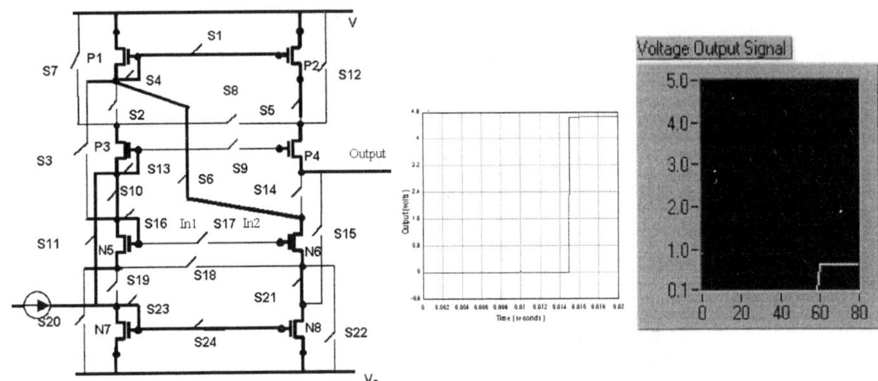


Figure 10. Portability problem: AND gate evolved in SW that does not work in HW. Response in SW at the left, HW response at the right.

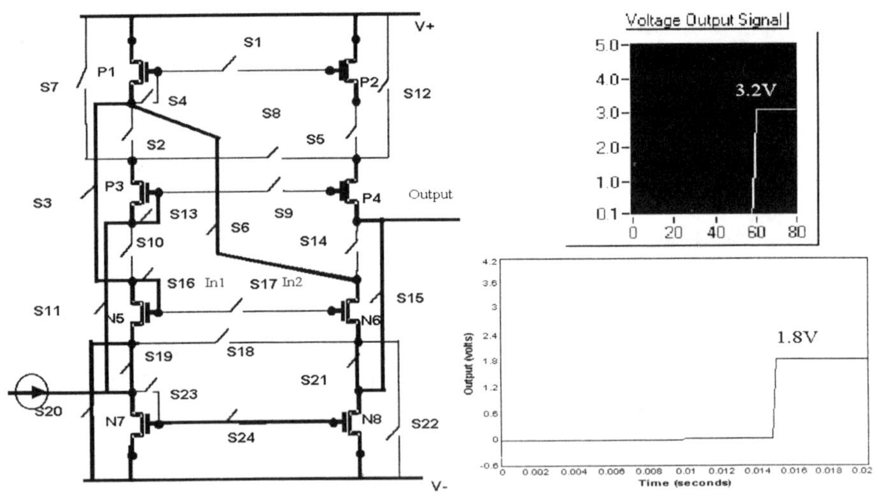


Figure 11. AND gate evolved in HW that does not work in SW. Response in HW at the left, SW response at the right.

Figures 10 and 11 depict the portability problem for an AND gate, showing that solutions evolved in SW do not work in HW and vice-versa. Figure 12 depicts a solution where mixtrinsic evolution was applied, and it can be seen that the logic gate works both in SW and in HW. Currently the concept of mixtrinsic evolution is being tested for analog circuits such as Gaussian neurons.

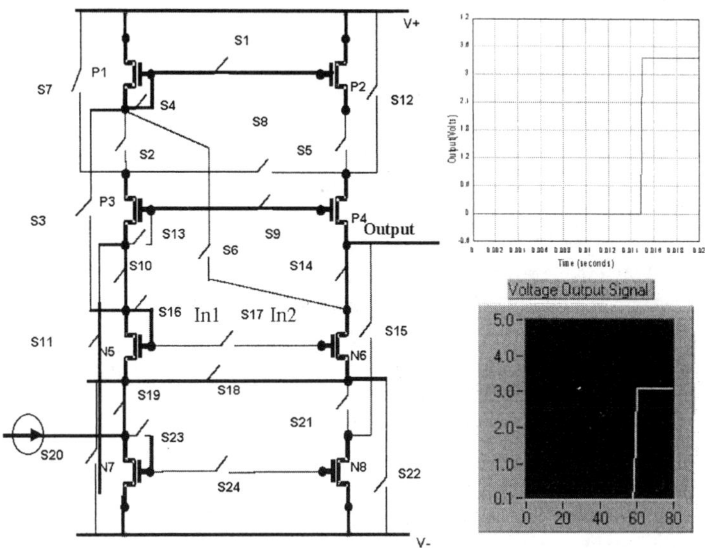


Figure 12. Mixtrinsic evolution of an AND gate, working both in HW and in SW. SW response at the top, HW response at the bottom.

## 6. Gradual morphing through fuzzy topologies

The idea comes from early experiments with the FPTA. Switches interconnecting transistors are not ON / OFF but can be gradually controlled with analog voltages on the gates of transistors that make the switch. In current implementation analog Low and High signals are the same for all switches on the chip. This allows us to map the 1 / 0-based genetic code / program of the switch to a Low / High resistance, not only to ON / OFF (i.e. very low, and very high), but to a great number of in-between values. Thus, the switches can be partly open, allowing the signals to more freely propagate / spread over the topology. The basic idea behind this technique, which accelerate evolution compared to the case when the search proceeds only with ON / OFF status, is inspired from annealing.

A temperature parameter is high at the beginning of evolution and decreasing to null over generations. When it is "hot" there is a lot of "thermal movement" and the Low / High values for the resistance of the switch, are close to each other, somewhere in the middle range (say 2k / 10k). As the temperature decreases, the two separate, low becoming lower, high becoming higher (e.g (1k / 30k), (500 / 100k) ending up for example at (5 / 100M).

This annealing was demonstrated to accelerate evolution, in some experiments over 10 times [13]. When we operate on a physical FPTA chip we could stop evolution early, since solutions usually show up in the first few generations and are fully operational. If the interest is to evolve a design that would go on an ASIC (or a circuit to be patented, where either there are, or there are no connections between components) then evolution has to maintain the acquired solution while "cooling" to ON / OFF connectivity.

The gradual opening seems to permit signal propagation to points where OFF switches in a traditional ON / OFF Boolean search may prevent it, making it more a "needle in-the-hay" search problem. Another way of looking at the circuits connected by gradual switches is to see not one, but several superimposed (conventional / Boolean) topologies at once – a fuzzy topology. These superposition / fuzziness allows for a more parallel search.

## 7. Polymorphic electronics

Polymorphic electronics (polytronics) is a new category of electronic circuits with superimposed built-in functionality. A function change does not require switches / reconfiguration as in traditional approaches. Instead, the change comes from modifications in the characteristics of devices involved in the circuit, in response to controls such as temperature, power supply voltage (VDD), control signals, light, etc.

For example, a temperature-controlled polytronic AND / OR gate behaves as AND at 27°C and as OR at 125°C. Polytronic circuits in which the control is by temperature, morphing signals, and VDD respectively are demonstrated in [14]. Polymorphic circuits were synthesized by evolutionary design/evolvable hardware techniques. These techniques are ideal for the polytronics design, a new area that lacks design guidelines/know-how, yet the requirements/objectives are easy to specify and test. The circuits are evolved / synthesized in two different modes. The first mode explores an unstructured space, in which transistors can be interconnected freely in any arrangement (in simulations only). The second mode uses the FPTA model, and the circuit topology is sought as a mapping onto a programmable architecture (these experiments were performed both in simulations and on FPTA chips). The experiments demonstrate the polytronics concept and the synthesis of polytronic circuits by evolution. The capacity of storing / hiding "extra" functions provides for watermark / invisible functionality, thus polytronics may find uses in intelligence / security applications. Built-in environment-reactive behavior (e.g. changing function with temperature) may also find uses in a variety of space and military applications. From the perspective of fuzzy systems, one can think of polymorphic fuzzy circuits, i.e. circuits with superimposed fuzzy functions.

As an example, Figure 13 depicts a polymorphic circuit controlled by an external signal $V_{morph}$ and evolved in an unconstrained search space, i.e., not using the FPTA model. The circuit behaves as an OR for $V_{morph} = 0$, as an XOR for $V_{morph} = 1.5V$ and as an AND for $V_{morph} = 3.3V$. The schematic in this figure provides the information on transistors width and length, parameters that have been evolved together with the circuit topology. Also shown in this figure the response of the evolved polymorphic gate.

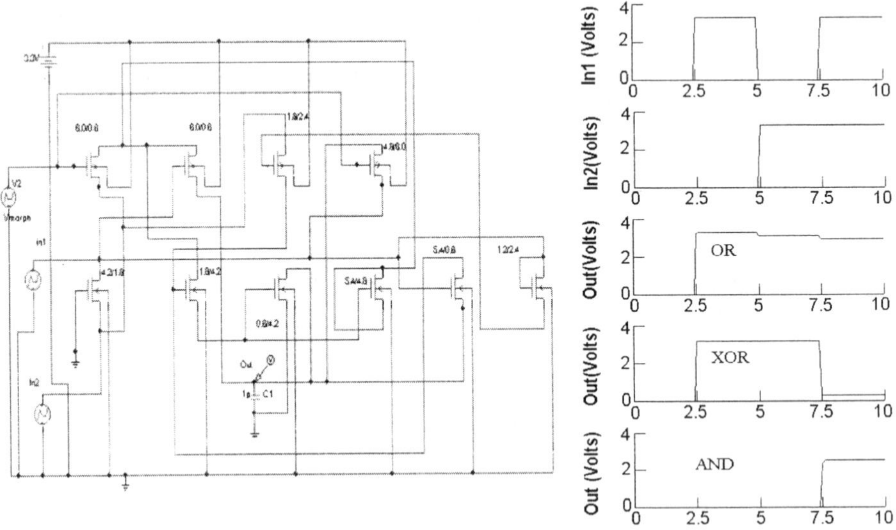

Figure 13. Evolved Polymorphic circuit exhibiting three different functionalities, OR, XOR and AND.

## 8. Conclusion

This paper illustrated aspects of using evolvable hardware for the design of unconventional circuits such as combinatorial circuits for fuzzy logics and polymorphic circuits. Additionally, it described mixtrinsic evolution, a method to improve the portability of evolved circuits when using different models and how evolution can leverage on gradual morphing by fully exploring the operational range of electronic switches.

Another result presented here has shown that it is possible to evolve very small circuits realizing a complete solution for fuzzy controllers. One can look at circuits with transistors as functional approximators – how many transistors are needed is only a question of how accurate a function needs to be. Two main conclusions are suggested by the experiments in this paper: that very compact solutions for complete fuzzy systems can be custom designed by evolutionary algorithms, and that specific programmable analog devices could be used as a general purpose platform to rapidly prototype and deploy any fuzzy system. Thus, the programmable analog device solution comes between the high-performance (in speed and power) but expensive and inflexible full ASIC solution and the less-performance but cheaper and flexible microprocessor solution.

### References

[1]   A. Stoica, A. Fukunaga, K. Hayworth and C. Salazar-Lazaro, Evolvable Hardware for Space Applications. In Sipper, M., Mange, D., Perez-Uribe, A. (Eds.), Evolvable Systems: From Biology to Hardware. Lecture Notes in Computer Science, Springer, Berlin, 1998, pp. 166-173.

[2]   A. Stoica, D. Keymeulen, V. Duong and C. Lazaro, Automatic Synthesis and Fault-Tolerant Experiments on an Evolvable Hardware Platform. In Profet, R. et al. (Eds.), *Proc. of IEEE Aerospace Conference*, March 18-25, 2000, Big Sky, MO, IEEE Press (CD, paper # 7_1102).

[3]   J. R. Koza, F. H. Bennett III, D. Andre and M. A. Keane, Genetic Programming III – Darwinian Invention and Problem Solving, Morgan Kaufman, San Francisco, 1999.

[4]   D. Butnariu and E. P. Klement, Triangular norm-based measures and games with fuzzy coalitions, Kluwer Academics, 1993.

[5]   M. M. Gupta and J. Qi, Design of fuzzy logic controllers based on generalized *t*-operators, *Fuzzy Sets and Systems*, vol. **40**, 1991, pp. 473-389.

[6]   M. M. Gupta and J. Qi, Theory of *t*-norms and fuzzy inference methods, *Fuzzy Sets and Systems*, vol. **40**, 1991, pp. 431-450.

[7]   A. Stoica, R. Zebulum, D. Keymeulen, Y. Jin and T. Daud, Evolutionary Design of Smart Analog Circuits. In *Proceedings of ANNIE2000 (Smart Engineering System Design)*, St. Louis, MO, November 5-8, 2000, AMSE press, pp. 245-250.

[8]   A. Stoica, R. Zebulum, D. Keymeulen, R. Tawel, T. Daud and A. Thakoor, Reconfigurable VLSI Architectures for Evolvable Hardware: from Experimental Field Programmable Transistor Arrays to Evolution-Oriented Chips, *IEEE Transactions on VLSI Systems*, vol. **9**, no. 1, February 2001, pp. 227-232.

[9]   A. Stoica, Toward evolvable hardware chips: experiments with a programmable transistor array. *Proceedings of 7th International Conference on Microelectronics for Neural, Fuzzy and Bio-Inspired Systems*, Granada, Spain, April 7-9, IEEE Comp. Sci. Press, 1999, 156-162.

[10] A. Stoica, On hardware evolvability and levels of granularity, *Proc. of the International Conference "Intelligent Systems and Semiotics 97: A Learning Perspective*, NIST, Gaithersburg, MD, Sept. 22-25, 1997.

[11] http: //www.mathworks.com

[12] A. Stoica, R. Zebulum and D. Keymeulen, Mixtrinsic Evolution. In Fogarty, T., Miller, J., Thompson, A., Thompson, P. (Eds.), *Proceedings of the Third International Conference on Evolvable systems: From Biology to Hardware (ICES2000)*, April 17-19, 2000, Edinburgh, UK. New York, USA, Springer Verlag, pp. 208-217.

[13] A. Stoica, D. Keymeulen, R. Tawel, C. Lazaro and Wei-te Li, Evolutionary experiments with a fine-grained reconfigurable architecture for analog and digital CMOS circuits. In Stoica, A., Keymeulen, D. Lohn, J. (Eds.) *Proc. of the First NASA / DoD Workshop on Evolvable Hardware*, July 19-21, 1999, Pasadena, CA IEEE Computer Society Press, pp. 77-84.

[14] A. Stoica, R. Zebulum and D. Keymeulen, Polymorphic Electronics. *International Conference on Evolvable Systems*, October 2001, Tokyo, Japan, pp. 291-302.

*Systematic Organisation of Information in Fuzzy Systems*
*P. Melo-Pinto et al. (Eds.)*
*IOS Press, 2003*

289

# A Methodology for Incorporating Human Factors in Fuzzy-Probabilistic Modeling and Risk Analysis of Industrial Systems

Miroslaw KWIESIELEWICZ and Kazimierz T. KOSMOWSKI
*Technical University of Gdansk, Faculty of Electrical and Control Engineering*
*Narutowicza 11/12, 80-952 Gdansk, Poland*

**Abstract.** The paper addresses some methodological aspects of fuzzy-probabilistic modeling of complex hazardous industrial systems. Due to several sources of uncertainty involved in risk evaluations and necessity to incorporate human and organ factors, a fuzzy-probabilistic modeling framework is proposed for handling uncertainty. A methodology has been developed for systematic incorporating these factors into the risk model. Proposed methodology is a synthesis of techniques for performing human reliability analysis (HRA) in the context of probabilistic safety analysis (PSA) based on available quantitative and qualitative information. Human and organizational factors are nested in a hierarchical structure called influence diagram in a similar way as in analytic hierarchy process (AHP) methodology. Several levels of hierarchy are distinguished: direct, organizational, regulatory and policy. Systematic procedure for weighting and scaling of these factors in this hierarchical structure has been proposed.

The study emphasizes that PSA and related HRA are performed in practice for representative events: initiating events, equipment/ human failures, and accident sequences (scenarios) that represent categories of relevant events. Transitions of the plant states in accident situations can not be assumed always as random. Taking into account limitations of the Bayesian framework it is proposed to apply for the uncertainty modeling a fuzzy-probability method and when justified the fuzzy interval framework related to the possibility theory. The paper also outlines a method of cost-benefit analysis of risk-control options (RCOs) in comparison with a basis option (B) for representative values of the fuzzy intervals. RCOs considered are raked based on defined effectiveness measure. At the end of the paper some research challenges in the domain are shortly discussed.

## 1. Introduction

Research concerning causes of industrial accidents indicate that organizational deficiencies or inadequacies resulting in human errors (broadly understood) are determining factors in 70-80% of cases [1]. Because several defenses against foreseen faults and accidents are usually used to protect the hazardous plants, multiple faults and errors have contributed to most of accidents. A closer look at major accidents indicates that the observed coincidence of multiple errors cannot be explained by a stochastic coincidence of independent events. Therefore, it justifies the conclusion that in industrial systems - operating in aggressive, competitive environment - the erroneous relations between actors within an organization may result in non-appropriate or faulty individual / team behavior, who in certain circumstances are not able to prevent an accident and cope with.

This paper deals with a complex distributed industrial system in which various sources of hazards can be identified [2]. Many types of accidents are possible in such systems that can result in more or less severe consequences. The paper proposes a methodology for incorporating human factors in probabilistic modeling and risk analysis. Taking into account the requirement given in the standard [3], that the risk analysis should include the evaluation of risk measures as well as uncertainty assessment, some methods are proposed for interval risk analysis and assessment of the risk reduction strategies. It is an extension of the methodology proposed by Rosness [2]. As the result the risk control options (RCOs) are elaborated [4] for rational safety management within specific industrial system.

In the approach proposed the uncertainties are represented as fuzzy intervals and evaluated according to axioms of the possibility theory [5]. Such approach is especially useful for the predictive risk analysis in complex industrial systems when aggregated risk indices for various accidents should be evaluated but there is scarcity of plant specific reliability data and expert opinions are extensively used. Some additional information of qualitative nature based on the assessment of important factors in given socio-technical system should be also included. Applying traditional approach in such situation for handling uncertainties, represented by probabilistic distributions (a Bayesian approach), would be questionable and too troublesome / costly in practical implementation. However, if for some events in the risk model the uncertainty is represented by defined probability density functions, these functions are transformed into possibility distribution (equivalent to a fuzzy interval) before final aggregation of the risk related information.

The paper emphasizes the role of predictive risk analysis in the safety management and risk control. The integrated attributive risk analysis (IARA) in socio-technical systems is outlined. Various factors potentially influencing accidents are represented in hierarchical structures called the influence diagrams (IDs), similarly as in the FSA methodology [4]. It was proved in the paper that this method is a generalization of a failure oriented SLIM method [6][7] called here SI-FOM (Success Index - Failure Oriented Method). For treating of the organizational factors a concept in the hierarchical influence diagrams (HIDs) has been developed. The similarities and differences of HIDs compared with (1) structures used in a AHP (*Analytic Hierarchy Process*) method developed by Saaty [8] and (2) the influence diagrams (IDs) used in the FSA (*Formal Safety Assessment*) methodology developed by IMO [4] are examined.

Some illustrations of the risk evaluation under uncertainties are outlined. It was also indicated how to make the risk assessments based on the mean modal values of fuzzy intervals for the frequency and consequences of consecutive accident sequences. The losses due to an accident can be defined as multidimensional. Some methods have been developed for the sensitivity analysis and comparing the risk indices of given risk control option (RCO) with a basis option (B). The results of predictive risk analysis are useful for comparative assessments of several RCOs and multi-attribute decision-making to select a RCO reducing most effectively (or rationally) the risk.

## 2. Safety management and predictive risk analysis

Maintaining an acceptable level of safety or risk in an industrial installation can be viewed as a control problem. As it is known two general control modes of a system can be distinguished [1][9]: (1) *feedback control* and (2) *feedforward control*. In feedback control the control actions are derived based on a deviation of the actual system state from a reference state. Such control is usually preferred when improving the quality of the process or the system state, subject to control is of interest. In some cases, however, feedforward control is necessary to achieve some goals. In feedforward control, the control actions are preplanned, and no direct corrections depending on the actual system state are possible, when the time horizon is too short for observing changes. This is also the case when the actual target state is not directly accessible for measurement within the time frame required.

Similarly, these two conceptually different approaches are applied in practice when managing the safety of given industrial system and controlling the risk associated with its operation. Whether direct feedback control of safety level is possible or not, depends on the time constant of technological change, and the time needed for empirical identification of the safety level achieved [1]. The risk is not directly measured and for its assessment, for given time horizon, a model is required to enable rational feedforward risk control. This model is based on probabilistic and deterministic methods applied for *predictive risk analysis* in specific industrial systems. The control strategy for achieving safety and risk goals in an industrial installation is implemented in practice as a combination of a feedforward control (knowing the predictive risk assessment), and feedback control with regard to qualitative / quantitative information. This information forms preconditions for the risk analysis and its quality [10]. The violation of these preconditions in given socio-system, caused e.g. by inadequate management, may significantly increase the likelihood of an accident and therefore should be included in the risk analysis [11].

The general idea of the safety and risk control in given industrial plant, treated as a socio-technical system, is illustrated in Fig. 1. 'A' denotes the feedback control loop. The feedback control strategy is elaborated on the safety related qualitative / quantitative information with regard to factors related to performance, quality assurance, safety culture and organizational/ policy influences etc. This information is directly useful for elaborating corrective actions in the case of deficiencies. The qualitative information forms preconditions for the predictive risk analysis in open loop 'B' (Fig. 1).

Fig. 1 illustrates also that the risk analysis is immanently associated with classifying of objects (generally understood) and failure events. Only seldom the plant specific reliability databases (equipment / hardware, software, human) are available and the analysts have to use general databases or some heuristic rules adapting data to new conditions. It makes substantial difficulties and problems in the risk estimation, because objects in reality and instances in models can differ (their attributes are usually not identical). Moreover, for physical objects (modules of equipment with specified boundaries) the operating conditions can differ significantly. These aspects are seldom included in databases but should be considered in the risk models.

An important issue to consider is that in socio-technical systems the human/ organization aspects are of prime importance. The assumption that failures/ errors in such systems are random, can be questioned [11][12]. Even if it was assumed that failures of the equipment items occur randomly, the related failure processes might not be stationary in real industrial systems. It is associated with operation conditions due to different safety culture, quality assurance procedures and preventive maintenance strategies, and makes substantial difficulties in assessing parameters of the risk model for specific industrial system.

Thus, two kinds of the risk models can be distinguished:

General Risk Model (GRM) developed on the basis of general data / knowledge and models available applied to given, but treated as generally defined system, and

Specific Risk Model (SRM) developed on the basis of data / knowledge and models adapted to given situation and applied to specific system (equipment, software, functionality, human factors, organization, environment, and infrastructure).

The rationale is that the risk analysis should appropriately support the safety management in given socio-technical system.

The requirement for specific risk evaluation is associated with the issue of representing and treating uncertainties in the risk model [9]. Because plant specific data are sparse or not available, analysts should somehow modify the general data / knowledge and models, preferably with regard to expert opinions. To overcome theoretical limitations and practical difficulties associated with risk evaluation under uncertainties for specific socio-technical system the following approach is proposed:

A. As long as in the risk modeling process the assumption is justified that failure events / errors are random (relevant data / knowledge are available) the traditional reliability / probabilistic modeling is applied; the uncertainty of the probability / frequency estimates is modeled using probability density function (a Bayesian probabilistic approach).

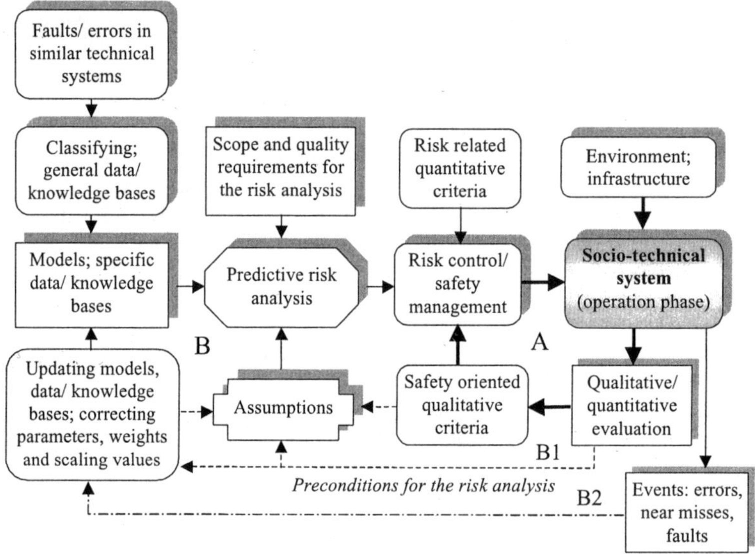


Figure 1. Control of plant safety (A - feedback control based on qualitative evaluation, B - feedforward control by predictive risk analysis).

B. When there is substantial problem in classifying objects / failure events or there is lack of evidence that failures / errors are random, the uncertainty of the probability / frequency estimates is represented using possibility distribution (a possibilistic approach).

C. Aggregations of estimates require the same theoretical framework. The possibility theory and the fuzzy interval calculus are used for this purpose. Therefore, uncertainties represented as in case A are transformed into the possibility distributions (fuzzy intervals).

D. Due to difficulties in precise defining objects, failure events, accident sequences and consequences (losses) of accidents, the results of probabilistic modeling and risk evaluations are represented by fuzzy intervals.

E. Such approach is effective in performing the predictive risk analysis (aggregation of risk from various hazards) in a complex industrial system [9]. Theoretical aspects of probability-possibility transformation and interval evaluations are described in chapter 4.

## 3. Incorporation human factors into probabilistic models and risk analysis

### *3. 1. Probabilistic modeling of complex industrial systems and factors influencing risk*

Several accident categories, such as: fire, explosion, releases of toxic substances etc., can be distinguished in probabilistic modeling a complex hazardous system [4]. Fig. 2 illustrates a general concept of probabilistic modeling in predictive risk analysis. Probabilistic models are developed step by step in the processes of deductive and inductive reasoning. As shown in Fig. 1 the risk modeling procedure should include some basic quality requirements [10]. Important roles in predictive risk analysis play the influencing factors [11][4]. Three categories of factors can be distinguished (Fig. 2) which influence: (a) initiation of abnormal event / accident and its frequency, (b) course of accident sequences and their conditional probabilities, and (c) level of consequences (losses). These factors should be identified in a specific socio-technical system to enable correct predictive risk analysis. They are included in the influence diagrams: IDs (Fig. 3) constructed for distinguished accident categories [4].

The approach enables a systematic functional / structural decomposition of the plant. For this purpose three basic methods are used: event trees (ETs) and fault trees (FTs) and influence diagrams (IDs). Events of human errors are represented in FTs and / or, but indirect human factors are incorporated in hierarchical structures of HIDs. Human error events can be latent, due to the design deficiencies or organizational inadequacies, and active committed by human operators in the course of accident. A method proposed for evaluation of human component reliability is a generalization of SLIM technique for a hierarchy of factors.

The systemic approach should be applied in the risk analysis in which the accident sequences are influenced to some extent by: hardware failures, human errors and external events (adverse environment). Several levels of factors can be distinguished in the risk model [11][4]. It is convenient to show relations between factors placed on neighboring levels of the influence diagram [4]. An example of the influence diagram is shown in Fig. 3. These factors are evaluated for given socio-technical system in the audit process using a number scale from 0 to 10 [9].

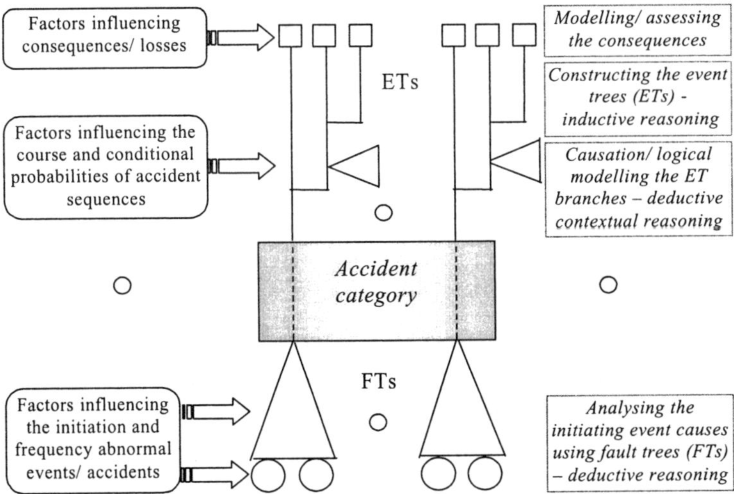


Figure 2. Influencing factors in predictive risk analysis of complex industrial system.

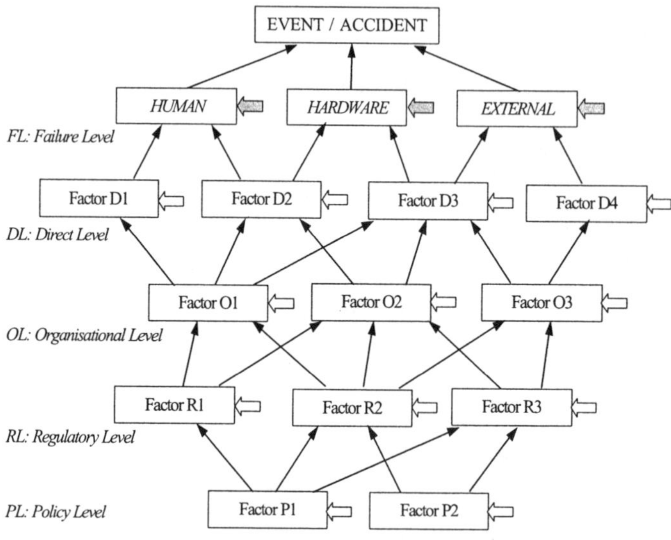


Figure 3. Influence diagram consisting of hierarchy of factors.

## 3. 2. *Generalization of SLI method for influence diagrams*

Several methods have been developed for the human reliability analysis (HRA) within probabilistic safety studies [13]. It is important to perform a situation specific HRA to include appropriate influence factors. An appreciated method for doing that is SLIM [6], which is oriented on success probabilities to accomplish specified tasks. Probabilistic modeling in risk analysis is failure oriented and it is more convenient to apply a modification of SLIM method named here SI-FOM (*Success Index - Failure Oriented Method*).

Equations including the failure probabilities $q_j$ and the success indices $SI_j$ for $j$-th task are

$$lg\, q_j = c \cdot SI_j + d \tag{1}$$

$$SI_j = \sum_i w_i r_{ij} \tag{2}$$

where $w_i$ - normalized weight coefficient assigned to $i$-th influence factor $\left( \sum_i w_i = 1 \right)$, $r_{ij}$ - scaling value of $i$-th factor in $j$-th task (normalized scaling value is: $0 \le r_{ij} \le 1$).

If for cases considered the success indices $SI_j$ were assessed and two probabilities $q_j$ are known then coefficients $c$ and $d$ can be evaluated and the failure probabilities calculated from the formula

$$q_j = 10^{c\, SI_j + d} \tag{3}$$

Unknown probabilities for other tasks can be calculated on a relative basis from the formula

$$q_j / q_k = 10^{c\,SI_j + d} / 10^{c\,SI_k + d} = 10^{c(SI_j - SI_k)} \tag{4}$$

The SI-FOM approach can be generalized for the influence diagrams (IDs). As illustrates Fig. 4 for each factor/ node three *scaling values* are distinguished:

$r$ – aggregated for given node,

$r^w$ – calculated for related nodes on higher hierarchy level (situated lower in Fig. 3 and 4) with regard to relevant weights $w$,

$r^*$ - assessed factor in specific system included on given hierarchy level in ID.

Examples of factors $r^*$ and scales of their assessment are given in [9].

For $j$-th node on the level $h$-th ($h$ = 1, 2, ... to last level considered in the model minus 1) the weights sum to 1: $\sum_n w_{h,j/n} = 1$. The scaling values are calculating from the formulas

$$r^w_{h,j} = \sum_n w_{h,j/n}\, r_{h+1,n} \tag{5}$$

$$r_{h,j} = g(r^w_{h,j} + r^*_{h,j}) \tag{6}$$

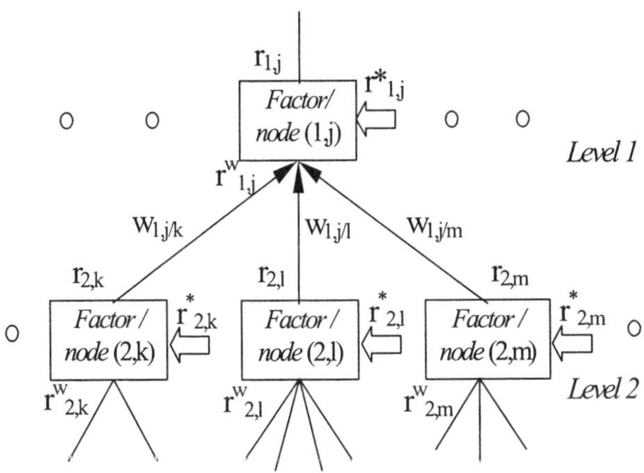


Figure 4. Weights and scaling values for factors in hierarchical influence diagram.

where: $r^*_{h,j}$ - the normalized scaling value assigned by expert to $j$-th factor on $h$-level; $r_{h+1,n}$ - the normalized aggregated scaling value for $n$-th node on level $h+1$ connected with $j$-th node on level $h$ ($h$ = 1 for the node on level 1 – Fig. 4); $g(\cdot)$ is a function (can be linear or non-linear) holding aggregation principles in IDs - for worst case: $g(0)=0$ and for best case: $g(1+1)=1$. The formula (6) for a linear relation is (13)

$$r_{h,j} = (r^w_{h,j} + r^*_{h,j}) / 2 \tag{7}$$

For the level 0 of the influence diagram the overall aggregated scaling value is $SI = r_{0,1}^w = r_0$. Resulting probability of a failure event or accident for the situation $x$ in given socio-technical system and accident category (known scaling values for factors distinguished in ID) can be calculated on the basis of $SI_x$ from the formula

$$q_x = 10^{c\,SI_x + d} \tag{8}$$

In the formula (8) coefficients $c$ and $d$ are unknown and must be somehow evaluated. Because in practice a relative comparison of given risk control option (RCO) to the basis state (B) is usually of interest, it can be written

$$q_{RCO} / q_B = 10^{c\,SI_{RCO} + d} / 10^{c\,SI_B + d} = 10^{c(SI_{RCO} - SI_B)} \tag{9}$$

Thus, only one coefficient $c$ should be known ($SI_{RCO}$ and $SI_B$ are calculated for given influence diagram with changing appropriate scaling values, respectively for RCO and B). For new RCO there will be reduction of the accident probability if $SI_{RCO} > SI_B$ ($c < 0$). It can be shown that similar formula is valid for the accident frequency

$$f_{RCO} / f_B = 10^{c(SI_{RCO} - SI_B)} = 10^{-e(SI_{RCO} - SI_B)} \tag{10}$$

Similar formula might be also written for the risk $R$ (if given RCO would not change the levels of consequences). The coefficient $e = -c$ ($e > 0$) is to be evaluated with regard to occupational risk assessments across the industry [4]. In the FSA methodology developed by IMO, it was assumed that the change from the worst practice ($SI=0$) to the best practice ($SI=1$) would change the event frequency by three orders of magnitude ($e=3$). It has been selected on the basis that individual risks span approximately $10^3$ from the border of tolerability to the level where society currently places no demand for further risk reduction however low the cost. This approach should be verified with regard to the range of risk from the most to least dangerous occupational risk across a range of industries [4].

### 3. 3. Aggregation of factors ranked in the sense of AHP and its generalization

The decision process concerning a risk control option (RCO) ranking is in general the multi-attribute one with a hierarchical structure. Hence the AHP approach [8] is proposed to handle it. Since the aggregation process used in AHP method is similar to that used in IMO one, a modified AHP method having features of both methods is introduced.

Multi-attribute ranking of variants, in our case risk control options (RCOs), consists of three main phases: (i) definition of decision problem structure, (ii) ranking variants with regard to attributes and sub-attributes (iii) hierarchical aggregation in order to obtain a final ranking. The study in this paper focuses attention on the aggregation. For simplicity let us assume a two level decision structure. Thus for m attributes and $k$ variants (options) the final ranking can be calculated

$$r_k^0 = \sum_{i=1}^{m} w_i^1 r_{ik}^0 \tag{11}$$

where $r_{ik}^1$ is a value of ranking of $k$-th variant under $i$-th attribute, calculated at a lower level, and $w_i^1$ are weights associated with attributes. Rankings for the level 1 are calculated similarly

$$r_{ik}^1 = \sum_{j=1}^{m_i} w_{ij}^1 r_{ijk}^2 \, , \tag{12}$$

where $w_{ij}^1$ is a weight associated with $j$-th sub-attribute at level 2 referred to $i$-th attribute from level 1 (sub-attribute $i, j$), and $r_{ijk}^2$ is a ranking value of $k$-th variant under the sub-attribute $(i, j)$.

Exemplary aggregation calculations scheme for two variants, two attributes and two sub-attributes is shown in Fig. 5. Thus, the ranking vector for the level one is

$$r_{11}^1 = w_{11}^1 r_{111}^2 + w_{12}^1 r_{121}^2 \, , \tag{13}$$

$$r_{12}^1 = w_{11}^1 r_{112}^2 + w_{12}^1 r_{122}^2 \, , \tag{14}$$

$$r_{21}^1 = w_{21}^1 r_{211}^2 + w_{22}^1 r_{221}^2 \, , \tag{15}$$

$$r_{22}^1 = w_{21}^1 r_{212}^2 + w_{22}^1 r_{222}^2 \, . \tag{16}$$

For the zero level we get

$$r_1^0 = w_1^0 r_{11}^1 + w_2^0 r_{21}^1 = w_1^0 \left( w_{11}^1 r_{111}^2 + w_{12}^1 r_{121}^2 \right) + w_2^0 \left( w_{21}^1 r_{211}^2 + w_{22}^1 r_{221}^2 \right) \tag{17}$$

and

$$r_2^0 = w_1^0 r_{12}^1 + w_2^0 r_{22}^1 = w_1^1 \left( w_{11}^1 r_{112}^2 + w_{12}^1 r_{122}^2 \right) + w_2^1 \left( w_{21}^1 r_{212}^2 + w_{22}^1 r_{222}^2 \right) \tag{18}$$

The aggregation method presented can be easily extended for any finite numbers of levels. Note that calculations for ID of IMO and AHP diagram have similar features. Recall that there are three scaling factors in the ID approach: (i) factor aggregated for a node, (ii) factor calculated for a given node at an upper level and (iii) factor estimated for a given system. Calculations in ID for a given level consist of two phases (i) aggregation using the weighted sum (5) (ii) correction made according to (6). The function $g(\cdot)$ is most often the arithmetic mean (3).

If in ID method we assume that the factors $r^*$ in Fig. 4 are equal to 0 and the function $g(\cdot)$ is linear then the calculation procedure leads to AHP approach with exactly one variant (Fig. 7). Moreover using the notation for the IDs and assuming that for a given level a node in ID refers to an attribute in AHP the aggregation according to (5) leads to the aggregation in sense of AHP (11)- (12). Assuming linear form of the scaling function and zero value of estimated factors in a system considered the AHP approach can be used for calculations in IDs. More general calculations need however a modification of the former one. In such case when assuming in AHP (Fig. 6) additionally scaling factors $r^*$ as in ID (Fig. 4) we get a kind of generalization of the

AHP approach with scaling. Using the linear scaling (7) instead of the non-linear one (6) suggests a modification of the former to a form of a weighted sum

$$r_j^h = r_j^{\wedge h} v_j^{\wedge h} + r_j^{*h} v_j^{*h}$$

$$(19)$$

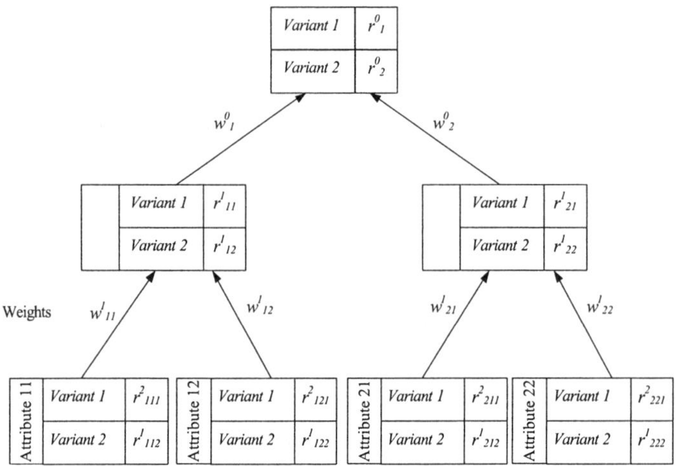


Figure 5. Ranking scheme for two variants, two attributes and two sub-attributes.

where $v_j^{\wedge h}, v_j^{*h}$ are weights of the factors $r_j^{\wedge h}$ and $r_j^{*h}$ respectively. Thus equation (14) from the point of view of AHP can be treated as an influence of a super expert or exterior decision-maker on the decision process considered. Moreover, the introduced here modified AHP algorithm can be also used for IDs calculations. Diagrams with such generalized aggregation we call the hierarchical influence diagrams (HIDs)

As it was shown above the AHP approach to multi-atribute decision making can be used to rank risk control decision options with both qualitative and quantitative factors. In the numerical examples presented numbers from a given scale were associated with qualitative variables. It is seems however more natural to use fuzzy variables to describe qualitative factors. Therefore a fuzzy extension of the generalized would be very useful for practical calculations.

A fuzzy extension of the aggregation procedure (5) - (6) can be defined as

$$\tilde{r}_j^{\wedge h} = \sum_n \tilde{w}_n^h \, \tilde{r}_n^{h+1}$$

$$(20)$$

$$\tilde{r}_j^h = g(\tilde{r}_j^{\wedge h} + \tilde{r}_j^{*h})$$

$$(21)$$

where: $\tilde{r}_j^{*h}$ is fuzzy normalized rating assigned by expert to $j$-th factor on $h$-level; $\tilde{r}_n^{h+1}$ is normalized aggregated rating for $n$-th node on level $h+1$ connected with $j$-th node on level $h$, and $g(\cdot)$ is a crisp function of fuzzy variables, $\tilde{w}_n^h$ are fuzzy weights.

Instead of formula (21) a fuzzy extension of the linear one (19) can be used

$$\tilde{r}_j^h = \tilde{r}_j^{\wedge h}\tilde{v}_j^{\wedge h} + \tilde{r}_j^{*h}\tilde{v}_j^{*h}$$

(22)

For calculations fuzzy number arithmetic introduced by Dubois and Prade [5] is proposed. Alternatively evaluations on $\alpha$-cuts can be used. It is necessary to stress that crisp numbers are special cases of fuzzy numbers do the approach proposed can be used both for qualitative and quantitative data.

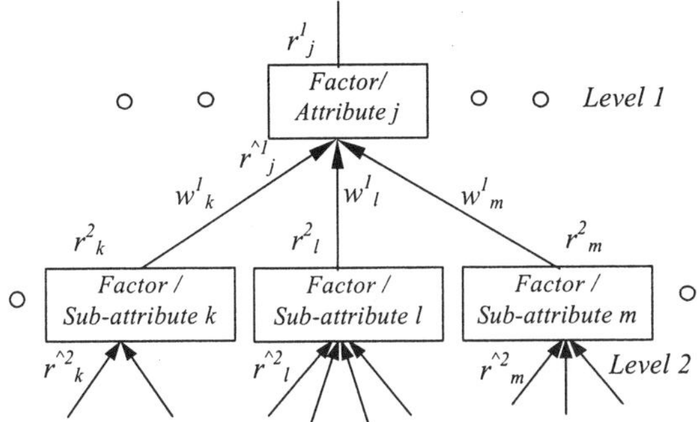


Figure 6. Weights and ratings in the AHP method in comparison with the ID method

## 4. Representing and treating uncertainties

### 4. 1 Transformation probability – possibility

*Dubois and Prade approach*

The probability-possibility transformation introduced by Dubois and Prade [14] is based on the possibility theory consistent with the Dempster-Shafer theory of evidence [14][15]. The main assumption made [14] is that a statistical experiment can be biased and then in such situation frequencies associated with elementary events are not necessarily equal. Therefore two measures of uncertainty are introduced instead of probability measure, namely necessity and possibility measures. The later can be viewed as a possibility measure in sense of Zadeh [16] The both measures can be completely specified through a possibility distribution.

Assume a set of elementary events: $X = \{x_i; i = 1,2...,n\}$ and an ordered probability distribution $\mathbf{p} = (p_1, p_2,...,p_n)$ $(p_1 \geq p_2 \geq ... \geq p_n)$, where $p_i = Pro(\{x_i\})$, $\sum_{i=1}^{n} p_i = 1$ and *Pro* is a probability measure. Moreover assume a set of nested events $A_i = \{x_1, x_2,...,x_i\}$ with $A_0 = \varnothing$. Thus, the possibility distribution $\mathbf{\delta} = (\pi_1, \pi_2,...,\pi_n)$ can be calculated [14] as

$$\forall i \quad \pi_i = \sum_{j=1}^{n} min(p_i, p_j)$$

(23)

whereas the inverse transformation as

$$\forall i = 1,\ldots,n \quad p_i = \sum_{j=i}^{n}\frac{1}{j}\left(\pi_i - \pi_{j+1}\right); \quad \pi_{n+1} = 0. \tag{24}$$

*Klir approach*

Assume that we have to ordered possibility and probability distributions $\mathbf{p} = (p_1, p_2, \ldots, p_n)$ and $\pi = (\pi_1, \pi_2, \ldots, \pi_n)$ respectively. Klir [17] proposes the following probability possibility-transformation

$$\pi_i = \left(\frac{p_i}{p_1}\right)^{\beta}, \quad i = 1,2,\ldots,n \tag{25}$$

where the coefficient $\beta$ should be evaluated according to the following uncertainty balance equation [17]

$$H(p) = N(\pi) + S(\pi), \quad 0 < \beta < 1 \tag{26}$$

where $H(p)$ is the Shanon [18] entropy, $N(\pi)$ U-uncertainty [19] and $S(\pi)$ strife function [17]. The sum $N(\pi) + S(\pi)$ expresses the total uncertainty associated with the possibility distribution $\pi$ and can be viewed as the generalization of the Shanon entropy for the possibilistic case. The inverse transformation is then calculated according to

$$p_i = \frac{\pi_i^{1/\beta}}{\sum_{k=1}^{n}\pi_k^{1/\beta}}, \quad i = 1,2,\ldots,n, \tag{27}$$

Given the probability distribution $\mathbf{p}$ evaluation of $\beta$ leads to solving the following equation

$$f(\beta) = \sum_{i=1}^{n} p_i \, log_2 \, p_i + \sum_{i=2}^{n}\left(\left(\frac{p_i}{p_1}\right)^{\beta} - \left(\frac{p_{i+1}}{p_1}\right)^{\beta}\right) log_2 \frac{i^2}{\sum_{j=1}^{i}\left(\frac{p_j}{p_1}\right)^{\beta}} = 0 \tag{28}$$

The above equation can be solved using any numerical method because the function $f(\beta)$ is monotonic. See [17] for more details concerning the transformation.

### 4. 2 Fuzzy probability interval calculation

The following form for fuzzy probability interval is assumed

$$\tilde{a} = \left(l, \underline{a}, \overline{a}, u\right) \tag{29}$$

where $l$ and $u$ are lower and upper values respectively, $\underline{a}$ and $\overline{a}$, lower and upper modal values, respectively of the fuzzy interval $\tilde{a}$.

In particular cases we get - (1) fuzzy probability if $\underline{a} = \overline{a}$, (2) interval probability if $l = \underline{a}$ and $\overline{a} = u$, and (3) probability singleton if $l = \underline{a} = \overline{a} = u$. In order to make calculations on fuzzy intervals one can perform calculations on a chosen finite number of á-cuts (in our case real number intervals) next use the interval analysis [20] and finally recompose a solution fuzzy number. When performing more complex fuzzy computations a given variable often occurs more then once in a given formulae to evaluate. It leads to an unreasonable extension of fuzziness of a calculation result. In such situations it seems very attractive is to apply so called constrained fuzzy interval arithmetic [21], which prevents uncertainty increase. From practical point of view it could be useful to perform calculations on mean modal values

$$a = \frac{\underline{a} + \overline{a}}{2} \tag{30}$$

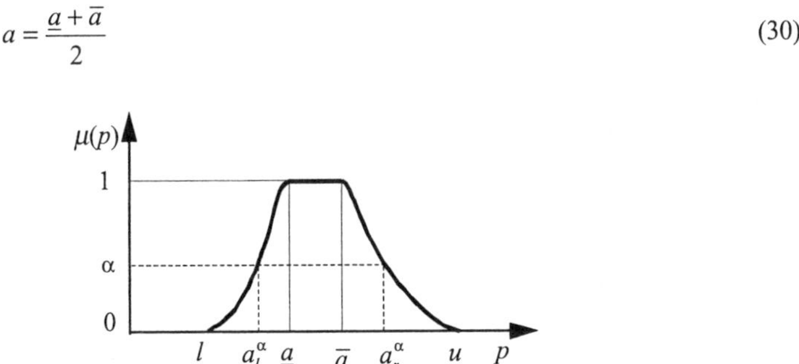


Figure 7. Fuzzy probability interval.

However, it is necessary to stress that for multiplication and division obtained results for modal values of fuzzy intervals and the modal value of resulting fuzzy interval are not equal. The difference depends on operand support and modal interval lengths [9]. For real cases met in the risk assessment the difference is usually in the range of several percents.

## 5. Illustrations of the interval and modal evaluations in risk analysis

### 5. 1. Fuzzy interval probability of events

Let assume that three human tasks are analyzed for those relevant success indices $SI_i$ have been assessed and two mean modal values of human failure probabilities $q_1$ and $q_3$ are known from a data source, e.g. TRC [22] or THERP technique [23]. It is known that in the situation considered the human operators perform a difficult diagnosis task. For calculating the failure probability $q_2$ the SI-FOM method will be employed. From two equations (1) knowing relevant indices $SI_j$ (2) the coefficients $c$ and $d$ are calculated. Taking into account (3) the following formula can be written

$$q_2 / q_3 = 10^{c\,SI_2 + d} / 10^{c\,SI_3 + d} = 10^{c(SI_2 - SI_3)} \tag{31}$$

If for the third task the probability expressed by the fuzzy interval $\tilde{q}_3$ were known, then the fuzzy interval probability $\tilde{q}_2$ could be calculated from the formula

$$\tilde{q}_2 = 10^{c(SI_2 - SI_3)}\,\tilde{q}_3 \tag{32}$$

The formula (32) should be used with caution to obtain consistency between $\tilde{q}_1$, $\tilde{q}_2$ and $\tilde{q}_3$. Such approach might be used to assess the quality of emergency procedures, knowing e.g. that given procedure is event oriented and the effect of introducing a symptom oriented procedure is considered.

### 5. 2. Correcting fuzzy interval frequency for modified influence factors

Let assume that the frequency of $k$-th accident sequence was assessed using traditional approach with assumption that processes of failures are stochastic. It was identified that this sequence is sensitive to a human error performing complex task including diagnosis and some organizational factors are important including the quality of procedures. The conditional probability of this error was estimated from a general data source using THERP technique [23]. Due uncertainty involved the accident frequency $f_k$ is expressed by the probability density function $p_k(f)$. For re-assessing the frequency of this sequence for new organizational conditions, described by related factors (not included in the THERP method), the influence diagram was constructed for given plant and accident type.

The first step is to make the probability - possibility transformation from the probability distribution $p_k(f)$ to fuzzy interval $\tilde{f}_k$ as described in chapter 4. Then denoting the current practice as B and improved organization treated as new risk control option (RCO) it can be written with regard to the formula (10)

$$\tilde{f}_k^{RCO} = 10^{-e(SI_k^{RCO} - SI_k^{B})}\,\tilde{f}_k^{B} = \varepsilon_k \tilde{f}_k^{B} \tag{33}$$

Thus, the frequency distribution (fuzzy interval) for new RCO will be moved left on the axis of frequency to a new position. It can be noted that a particular RCO which only influences the likelihood of an accident will not have any effect on the distribution of the risk level shown via the F-N diagram - the curve for mean modal values of $F$ and $N$ will simply be "lowered" according to factor $\varepsilon_k$ ($\varepsilon_k < 1$). However, any RCO that also influences the severity of an accident and / or conditional probabilities of accident sequences can affect significantly the distribution of the risk. It requires respectively re-quantification of consequences and / or conditional probabilities of branches in given event tree (see Fig. 2).

### 5. 3. Fuzzy interval risk evaluation

To reduce the modeling complexity an approach for the fuzzy interval risk assessment was developed in which a certain number of categories of initiating events and consequences are distinguished [24]. The consequences can be of different types (injuries, deaths, and economic losses). For such aggregated risk analysis using of the fuzzy interval approach is especially justified [12][25][26]. Below the aggregation of overall risk for defined accident sequences is described. Consecutive $k$-th accident ($k \in [1, K]$) in the risk model is described by fuzzy intervals, respectively the frequency $\tilde{f}_k$ [$a^{-1}$] and the consequence $\tilde{n}_k^l$ [units of $l$-th type loss] are known. The risk of $l$-th type associated with this sequence is

$$\tilde{R}_k^l = \tilde{f}_k \tilde{n}_k^l . \tag{34}$$

Evaluated risk $\tilde{R}_k^l$ is useful when comparing risks associated with different sequences or with cut-off criteria. For comparing two fuzzy intervals the approach described in [9] was adapted. The measure of uncertainty associated with fuzzy interval expressed by left deviation $\delta_1(\tilde{R}_k^l)$ and right deviation $\delta_2(\tilde{R}_k^l)$ can be also calculated to focus on more important sequences with higher uncertainty. Based of a set of pairs $\{ \tilde{f}_k, \tilde{n}_k^l \}$ a F-N curve can be constructed for ordered mean modal values $l$-type losses $n_k^l$ and associated frequency $f_k$. The aggregated risk associated with losses of $l$-th type will be

$$\tilde{R}^l = \sum_k \tilde{R}_k^l = \sum_k \tilde{f}_k \tilde{n}_k^l \tag{35}$$

Aggregated risk represented by a fuzzy interval is very useful in comparing risk with a criterion expressed by a real number or fuzzy interval [9]. The individual risk in given geographical location $x$, $y$ (the location can be consider inside and outside of the plant) is calculated from the formula

$$\tilde{R}_{x,y}^l = \sum_i \tilde{R}_{x,y;i}^l = \sum_i \tilde{f}_i \tilde{p}_{x,y;i} \tag{36}$$

where: $\tilde{R}_{x,y;i}^l$ - the individual risk from $i$-th accident outcome that potentially can cause a fatality at location $x$, $y$; $\tilde{f}_i$ - the frequency that $i$-th accident; $\tilde{p}_{x,y;i}$ - conditional probability that $i$-th accident outcome will result in a fatality at location $x$, $y$.

### 5. 4. The assessment of risk control option based on the mean modal values

The aggregated risk expressed by the mean modal values is calculated from the formula

$$R^l = \sum_k f_k n_k^l \tag{37}$$

Knowing the risk profile is useful for the safety related decision making, in particular for selecting a risk control option (RCO). The importance of $k$-th accident sequence is defined as

$$I_k^l = \frac{R_k^l}{R^l} = \frac{f_k n_k^l}{R^l} \tag{38}$$

On the basis of these ranks a set most important sequences in the risk profile is determined. Based on current values of influence factors associated with these sequences a set of risk control measures (RCM) is specified, grouped according to some rules into risk control options (RCOs). Several rules to define RCO are proposed in the FSA methodology [4]. The implementation effect of given RCO comparing with a basis option (B) can be evaluated on the incremental or incremental relative basis as follows:

$$\Delta R^l = R^{l;\, RCO} - R^{l;\, B} = \sum_k f_k^{RCO} n_k^{l;\, RCO} - \sum_k f_k^B n_k^{l;\, B} \tag{39}$$

$$\delta f_k = \frac{f_k^{RCO} - f_k^B}{f_k^B} = \frac{f_k^{RCO}}{f_k^B} - 1 \tag{40}$$

$$\delta n_k^l = \frac{n_k^{l;\, RCO} - n_k^{l;\, B}}{n_k^{l;\, B}} = \frac{n_k^{l;\, RCO}}{n_k^{l;\, B}} - 1 \tag{41}$$

$$\Delta R^l = \sum_k f_k^B (1 + \partial f_k)\, n_k^{l;\, B} (1 + \partial n_k^l) - \sum_k f_k^B n_k^{l;\, B} =$$
$$= \sum_k f_k^B n_k^{l;\, B} (\delta f_k + \delta n_k^l + \delta f_k\, \delta n_k^l) \tag{42}$$

The relative risk change for given RCO compared with a basis option B can be calculated as follows

$$\delta R^l = \frac{\Delta R^l}{R^{l;\, B}} = \frac{R^{l;\, RCO} - R^{l;\, B}}{R^{l;\, B}} = \frac{\sum_k R_k^{l;\, B} (\delta f_k + \delta n_k^l + \delta f_k\, \delta n_k^l)}{\sum_k R_k^{l;\, B}} \tag{43}$$

Given RCO can contribute to the relative change of the frequency ($\delta f_k$) and / or consequences ($\delta n_k^l$) for only some accident sequences, so the calculation according to (42) is usually not complicated. If several RCOs are proposed, the calculations are performed for each RCO. As it was mentioned the calculations for the mean values of fuzzy intervals should performed with cautions. If the membership function of a fuzzy interval is very asymmetric, the calculations can be also made for the representative value of the fuzzy interval.

Initial comparative calculations for a risk model under uncertainty have shown that the risk indices obtained for expected values (before transformation probability-possibility) and for representative values of fuzzy intervals are very closed. The cost of calculations for the second case is usually significantly reduced.

## 6. Conclusions

In the paper some issues associated with the risk analysis of complex distributed industrial plants with various sources of hazards identified, which can potentially cause many accidents sequences (scenarios) resulting in more or less severe consequences (losses), are discussed. Such plants are treated as socio-technical systems, as the human and organizational factors are of a prime importance in their reliable and safe operation.

A methodology for incorporating human factors in probabilistic modeling and risk analysis in such systems using the influence diagrams (IDs) was outlined. It was proved that the influence diagram approach is a generalization of failure oriented SLIM methods for a hierarchy of factors (HIDs). The similarities and differences of HIDs with diagrams used in the AHP method and the FSA (*Formal Safety Assessment*) methodology of IMO are described in details.

It is proved that aggregation algorithms proposed in the HID method is generalization of the ID method (FSA) and an extension of the AHP method thanks to introducing additional information (from observation or audit of given system, super expert opinion, or exterior decision-maker) on the decision process.

The risk analysis includes the evaluation of risk measures as well as the uncertainty assessment. A methodology is proposed for fuzzy interval risk analysis and assessment of risk reduction strategies through implementation of a risk control option (RCO). For uncertainty representation in predictive risk analysis the possibilistic framework seems to be more justified than traditional probabilistic Bayesian approach. Fuzzy interval representation of uncertainty and relevant calculus offer many indices to characterize the risk profile. The risk indices evaluated for mean modal (or representative) values are very close to those calculated on the basis of expected values in traditional approach.

However, the interval risk approach is much better suited when the aggregation of final risk indices is performed under uncertainties as well as for multi-criteria decision making to reduce rationally assessed overall risk. To limit uncertainty range in the risk evaluation the fuzzy interval calculations with constraints is proposed.

**Acknowledgments.** The research work outlined in this paper has been carried out within the Strategic Government Program (SPR-1) "Safety and health protection in the work environment" supported by the Committee of Scientific Research in Warsaw, Poland in the period 1998-2000; Main Co-ordinator – Central Institute of Labour Protection (CIOP) in Warsaw.

## References

[1] J. Rasmussen, Approaches to the control of the effects of human error in chemical plant safety. Denmark, Riso National Laboratory (Riso-M-2638), 1987.

[2] R. Rosness, Risk Influence Analysis, A methodology for identification and assessment of risk reduction strategies, *Reliability Engineering and System Safety* **60** (1998), pp. 153-164.

[3] IEC 300-3-9, International Standard, Risk analysis of technological systems, 1995

[4] IMO, Formal Safety Assessment: A methodology for formal safety assessment of shipping, MSC 66, London, International Maritime Organization (IMO), 1996

[5] D. Dubois and H. Prade, Possibility Theory, An Approach to Computerized Processing of Uncertainty, Plenum Press, New York, USA, 1988.

[6] D. E. Embrey, *et al.*, SLIM-MAUD, An approach to assessing human error probabilities using structured expert judgement. NUREG/CR-3518, 1984.

[7] K. T. Kosmowski, *et al.*, Development of Advanced Methods and Related Software for Human Reliability Evaluation within Probabilistic Safety Analyses, Berichte des Forschunszentrum Jülich; 2928, Germany, 1994.

[8] T. L. Saaty, The Analytic Hierarchy Process, McGraw-Hill, 1980.

[9] K. T. Kosmowski, *et al.*, Methods and tools for risk analysis in complex industrial systems. The research work carried out in Strategic Government Program (SPR-1), task 02. 6. 11, Technical University of Gdansk, 1999.

[10] J. Suokas and V. Rouhiainen, (Eds), Quality Management of Safety and Risk Analysis. Amsterdam, Elsevier, 1993.

[11] D. E. Embrey, Incorporating management and organizational factors into probabilistic safety assessment, *Reliability Engineering and System Safety*, **38** (1992), pp. 199-208.

[12] K. Y. Cai, Introduction to Fuzzy Reliability. Kluwer Academic Publishers. Boston, USA, 1996.

[13] P. Humphreys, (Ed.), Human Reliability Assessor Guide. RTS 88/95Q, Safety and Reliability Directorate, UK, 1988.

[14] D. Dubois and H. Prade, Unfair coins and necessity measures: towards a possibilistic interpretation of histograms, *Fuzzy Sets and Systems* **10** (1983) pp. 15-20.

[15] G. Shafer, A Mathematical Theory of Evidence. Princeton University Press, Princeton, New Jersey, 1976.

[16] L. A. Zadeh, Fuzzy sets as a basis for a theory of possibility, *Fuzzy Sets and Systems* **1** (1) (1978), pp. 3-28.

[17] G. J. Klir, Developments in Uncertainty-Based Information, In *Advances in Computers*, Yovits, M. C. (Ed.), Academic Press: Harcourt Brace Jovanovich, New York, 1993.

[18] C. E. Shannon, The mathematical theory of Communication, *The Bell System Technical Journal* **27** (1948) pp. 379-423, 623-656.

[19] M. Higashi and G. J. Klir, On the notion of distance representing information closeness: Possibility and probability distributions, *Intern. J. General Systems*, **9** (2), 1983, pp. 103-115.

[20] R. Moore, *Interval Analysis*. Englewood Cliffs: Prentice Hall, 1966.

[21] G. J. Klir, Fuzzy arithmetic with requisite constraints, *Fuzzy Sets and Systems* **91** (1997), pp. 165-175.

[22] E. M. Dougherty and J. R. Fragola, Human Reliability Analysis: A Systems Engineering Approach with Nuclear Power Plant Applications, A Wiley-Inter-science Publication, John Wiley & Sons Inc., New York, USA, 1988.

[23] A. D. Swain and H. E. Guttmann, Handbook of Human Reliability Analysis with Emphasis on Nuclear Power Plant Application. NUREG/CR-1278, USA, 1983.

[24] K. Duzinkiewicz, *et al.*, Fuzzy probabilistic modeling of complex hazardous plants and risk assessment under uncertainties, *IFAC Symposium on Fault Detection, Supervision and Safety for Technical Processes*, Patton, R. J., Chen, J. (Eds.), The University of Hull, U. K., August 26-28 (1997) pp. 753-758.

[25] M. H. Chung and K. I. Ahn, Assessment of the potential applicability of fuzzy set theory to accident progression event trees with phenomenological uncertainties, *Reliability Engineering and System Safety* **37** (1992) pp. 237-252.

[26] T. Onisawa, System reliability from the viewpoint of evaluation and fuzzy set theory approach, In: Reliability and Safety Analysis under Fuzziness (T. Onisawa and J. Kacprzyk Eds.) Physica-Verlag, Heidelberg, Germany, 1995.

# Systematic Approach to Nonlinear Modelling Using Fuzzy Techniques

Drago MATKO
*Faculty of Electrical Engineering University of Ljubljana*
*drago.matko@fe.uni-lj.si*

**Abstract.** The paper deals with the systematic approach of using Fuzzy Models as universal approximators. Four types of models suitable for identification are presented: The Nonlinear Output Error, The Nonlinear Input Error, The Nonlinear Generalized Output Error and The Nonlinear Generalized Input Error Model. The convergence properties of all four models in the presence of disturbing noise are reviewed and it is shown that the condition for an unbiased identification is that the disturbing noise is white and that it enters the nonlinear model in specific point depending on the type of the model. The application of the proposed modelling approach is illustrated with a fuzzy model based control of a laboratory scale heat exchanger.

## 1. Introduction

Fuzzy logic provides a mathematical tool to formulate mental models in a compact mathematical form. The intuitive and heuristic nature of the human mind, which is actually imprecise, can be incorporated in formal models which can essentially support the planning and decision making processes. Fuzzy logic has extended the classical mathematical models in the form of differential and difference equations to a broad class of models, which are easily understandable. An overview and classification of identification models can be found in [1]. Fuzzy logic originates from so called "intelligent" control and can be treated as a universal approximator (denoted by UA in the sequel) which can approximate continuous functions to an arbitrary precision [2], [3], [4]. In general UA may have several inputs and outputs. Without loss of generality, only one output will be treated; the approximators with more than one output can be treated as several approximators in parallel.

The paper is organized as follows: Different types of fuzzy models will be presented as UA and their application to dynamic models will be introduced in the next Section. In Section 3 four basic forms suitable for the identification of nonlinear dynamic models will be given and their convergence properties regarding the disturbing noise will be examined in Section 4. Finaly the application of the proposed modelling approach will be illustrated in the Section 5 as a case study of the fuzzy model based control application to a laboratory scale heat exchanger.

## 2. Rule Based Nonlinear Dynamic Models

According to [5], three types of "rule based" fuzzy models are known: the linguistic, the fuzzy relational and the Takagi - Sugeno fuzzy models [6]. A typical Takagi-Sugeno type rule can be written

$$R^j: \text{if } x_1 \text{ is } A_1^j \text{ and } \dots \text{ and } x_N \text{ is } A_N^j \text{ then } y = g^j(x_1 \dots x_N) \tag{1}$$

where $x_1 \dots x_N$ are inputs, $A_1^j$ a subset of the input space, $y$ output and $g^j$ a function (in general nonlinear, usually linear). In this case the output of the UA with $n$ inputs can be written in the following form

$$y(k) = \frac{\sum_{i1}, \sum_{i2}, \dots, \sum_{in} s_{i1,i2,\dots,in}(x) r_{i1,i2,\dots,in}}{\sum_{i1}, \sum_{i2}, \dots, \sum_{in} s_{i1,i2,\dots,in}} \tag{2}$$

where $s_{i_1,i_2,\dots,i_n}$ is the element of the multidimensional structure (tensor)

$$S = \mu_1 \otimes \mu_2 \otimes \dots \otimes \mu_n \tag{3}$$

which is obtained by the composition $\otimes$ (usually a product or an other T norm) of fulfilment grade vectors (of dimensions $mi$) of membership functions on the universe of discourse

$$\mu_i = [\mu_i^1, \mu_i^2, \dots, \mu_i^{mi}]^T \tag{4}$$

where $mi$ is the number of membership functions. $r_{i1,i2,\dots,in}$ are consequences of the Takagi-Sugeno type model according to Eq.(1). In the simplest case they are constants (Takagi - Sugeno models with crisp constant consequences).

Piece-wise linear interpolation between the rule consequents is obtained if the following conditions hold [7]

1.  The antecedent membership functions are triangular and form a partition, i. e.

$$\sum_{j=1}^{mi} \mu_i^j = 1, \forall i.$$

2. The product T norm is used to represent the logical "and", and the rule base is complete (i.e. rules are defined for all combinations of the antecedent terms).
3. The fuzzy-mean defuzzyfication is applied.

Under condition 1, the denominator of Eq. (2) becomes 1.
However even with the Mamdani type rules where the consequence is symbolic (fuzzy set):

$$R^j: \text{if } x_1 \text{ is } A_1^j \text{ and } \dots \text{ and } x_N \text{ is } A_N^j \text{ then } y = y \text{ is } B^j \tag{5}$$

the fuzzy system (fuzzyfication - inference - defuzzyfication) can be treated as nonlinear mapping between inputs and outputs. So in general we may write:

$$y(k) = f(x(k))\tag{6}$$

The input-output relations of industrial processes and controllers are dynamic. A common approach to dynamic models is to use time-shifted signals, which result in discrete-time models. The usage of derivatives, integrals or other transfer function would result in continuous-time models. Also, mixed-continuous-discrete time models are possible. In this paper the discrete time models will be treated and tapped-lines will be used to generate the time-shifted signals. If only the (tapped) input signal is used as input of the UA (elements of x), the resulting model is nonrecursive and has finite impulse response. Such models are a class of Nonlinear Finite Impulse Response models and are denoted by NFIR. If the (tapped) input and (tapped) output signals are used as input of the UA, the resulting model is recursive and may have an infinite impulse esponse. Four types of such models will be given in the next Section.

## 3. Systematic Approach to the Identification Models

Experimental modelling or identification is based on observation of input-output data. In the sequel it will be supposed that the unknown plant which has produced the input-output data can be described in the following mathematical form

$$\begin{aligned}
y(k) &= F\big(u(k),\ldots,u(k-N_p),y(k-1),\ldots,y(k-N_p),v(k)\big) = \\
&= f\big(u(k),\ldots,u(k-N_p),y(k-1),\ldots,y(k-N_p),+n(k)\big)
\end{aligned}\tag{7}$$

where $F(\ )$ and $f(\ )$ are unknown nonlinear functions, $N_p$ the order of the model, $v(k)$ the noise disturbing the nonlinear plant and $n(k)$ the nonlinear noise at the plant output. The resulting model is obtained by the best fit of the model response to the identified process response if the same input signal is applied to both of them. So the identification problem is formulated as an optimization problem utilising a criterion in the form of a functional e.g. the sum of squared errors

$$E(y, y_M) = \sum_{k=0}^{N} \epsilon^2(k)\tag{8}$$

where $y$ is the observed signal, $y_M$ is the model output and $\epsilon$ the error. The dentification procedure involves the structure identification of the plant and the estimation of unknown parameters. In practice, the structure is usually chosen ad hoc (engineering feeling) and then improved by an optimization procedure. Optimization is also used for the determination of parameters. Four cases of model description will be given next.

1.The case

$$\epsilon(k) = y(k) - y_M(k)\tag{9}$$

where

$$y_M(k) = \hat{f}(u(k),\ldots, u(k-N), y_M(k-1),\ldots, y_M(k-N), \tag{10}$$

is the output of a recursive model with input $u(k)$, is referred to as **Nonlinear Output Error Model**. The nonlinear function $f(\ )$ is an estimate of the nonlinear function $f(\ )$ and in ideal case, where the plant is identified perfectly, both non-linear functions become the same: $(\hat{f}(\ ) = f(\ ))$.

2. The case

$$\in (k) = u(k) - u_M(k) \tag{11}$$

where

$$u_M(k) = \hat{fi}(y(k),\ldots, y(k-N), u_M(k-1),\ldots, u_M(k-N)) \tag{12}$$

is the output of a recursive model with input $y(k)$, is known as **Nonlinear Input Error Model**. The nonlinear function $f_i(\ )$ represents the estimation of the nonlinear function $\hat{f}_i(\ )$ which is the inverse of the function $f(\ )$ in the sense that if the signal $y(k)$ is produced by Eq. 7 and no noise is present n(k)=0), then the signal produced by the inverse function is identical to the signal $u(k)$

$$u(k) = f_i\big(y(k),\ldots, y(k-N_p), u(k-1),\ldots, u(k-N_p)\big) \tag{13}$$

The inverse function $f_i$ exists if the function $f(\ )$ is injective. Also only the dynamic part of the plant without time delay is inverted. In ideal case where the plant is identified perfectly both inverse nonlinear functions become the same $(\hat{fi}(\ ) = fi(\ ))$.

3. If in the right hand side of Eq. (10) the output of the model $y_M(k)$ is replaced by the measured output of the process, i.e.

$$y_M(k) = \hat{f}_i u\,((k),\ldots, u(k-N), y(k-1),\ldots, y(k-N)) \tag{14}$$

the error defined by Eq. (9) becomes a **Nonlinear Generalized Output Error Model**. It has two inputs, namely the plant input signal $u(k)$ and the plant output signal $y(k)$ and one output signal $y_M(k)$. The nonlinear function $\hat{f}(\ )$ is an estimate of the nonlinear function $f(\ )$ and in ideal case where the plant is identified perfectly both nonlinear functions become the same $(\hat{f}(\ ) = f(\ ))$.

4. If in the right hand side of Eq. (12) $u_M(k)$ is replaced by the input of the process, i.e.

$$u_M(k) = \hat{f}_i(y(k),\ldots, y(k - N), u(k - 1),\ldots, u(k - N))\qquad(15)$$

the error defined by Eq. (11) becomes a **Nonlinear Generalized Input Error Model**. It has two inputs, namely the plant output signal $y(k)$ and the plant input signal $u(k)$ and one output signal $u_M(k)$. The nonlinear function $\hat{f}i(\ )$ represents the estimation of the nonlinear function $f_i(\ )$ and in ideal case where the plant is identified perfectly both inverse nonlinear functions become the same ($\hat{f}i(\ ) = fi(\ )$).

Fig. 1 represents all four cases, the Nonlinear Output, Input and both Generalized Error Identification Models respectively.

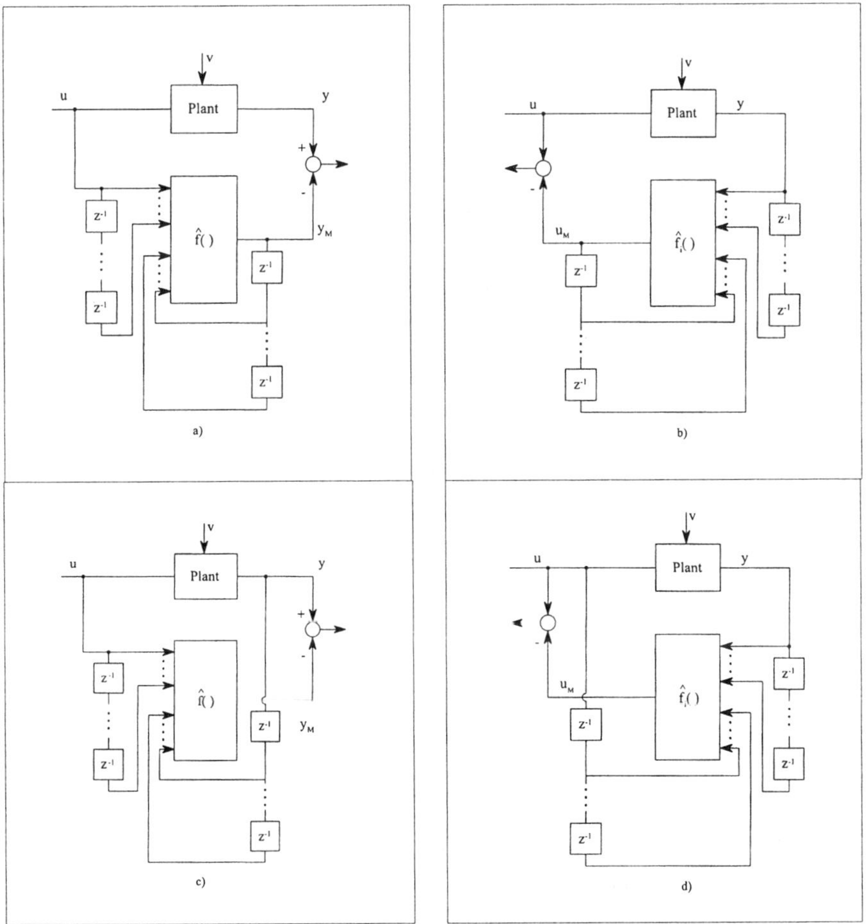


Figure 1. The output error (a), the input error (b), the generalized output error (c), and the generalized input error (d) identification models.

All four forms of the error models are closely related. The input error and the Generalized input error models are inverse to the output error and the Generalized output error models respectively in the sense described above. The output and input models are complementary to the Generalized output and Generalized input models respectively in the sense that Generalized models are suitable for identification since with known or supposed structure of the nonlinearity (which remain fixed during the optimization) the estimation of unknown parameters becomes a linear problem and the least squares technique can be used. This is the case of Takagi-Sugeno type models with Center of Singletons defuzzification and predetermined fuzzy sets of the premise space (input partition) or the case of the radial base function networks with preselected locations of the regressors. The output/input models are applicable in prediction. So they are called also prediction or simulation models.

The four models described by Eqns. (10, 12, 14, 15) can be treated as universal dynamic approximators (UDA) where the nonlinear function is realised by fuzzy logic universal approximators (UA). According to the topology of the models in Fig. 1 the input, output and both Generalized models can be called also parallel, series and series-parallel models respectively. It should be noted that there are two kinds of the Generalized error (series-parallel) models, origination in the output and input error models respectively.

### 4. Convergence Properties

An important point of identification is the convergence in the presence of the noise. Since the white noise has minimal variance all optimization procedures seek the minimum of the criterion function (8) in the sense that the residuals $\in (k)$ become white noise. So an unbiased estimation is possible only in cases where the noise has a special character which will be discussed next for all four models.

#### *4.1. Nonlinear Output Error Model*

In this case the noise $v(k)$ is introduced additively to the output of the undisturbed plant $y_0(k)$:

$$y_0(k) = f\big(u(k),\ldots,u(k - N_p), y_0(k-1),\ldots,y_0(k - N_p)\big)$$
$$y(k) = y_0(k) + v(k) \tag{16}$$

The residual $\in (k)$ becomes now

$$\in (k) = y(k) - y_M(k) =$$
$$= f(u(k),\ldots, u(k - N_p), y_o(k - 1),\ldots, y_o(k - N_p) + v(k) - \tag{17}$$
$$- f(u(k),\ldots, u(k - N), y_M(k - 1),\ldots, y_M(k - N)).$$

If the structure of the identification model is the same as the structure of the nonlinear plant, the optimization procedure for the minimization of (8) tries to make the residual $\in (k)$ white, so if $v(k)$ is white noise, the minimum is $\in (k) = v(k)$ and consequently $(\hat{f}( ) = f( ))$ and

$y_M(k) = y_0(k)$. The condition for an unbiased estimation is that the noise is additively added to the plant output and that it is white as illustrated in Fig. 2-a.

### 4.2. Nonlinear Input Error Model

In this case the noise $v(k)$ is introduced additively to the input of the undisturbed plant $u_0(k)$

$$u_0(k) = f_i\big(y(k),\ldots,y(k-N_p),u_0(k-1),\ldots,u_0(k-N_p)\big)$$
$$u(k) = u_0(k) + v(k) \tag{18}$$

The residual $\in(k)$ becomes now

$$(k) = u(k) - u_M(k) =$$
$$= f_i(y(k),\ldots,y(k-N_p),\,u_o(k-1),\ldots,u_o(k-N_p)) + v(k) - \tag{19}$$
$$-\hat{f_i}(y(k),\ldots,y(k-N),\,u_M(k-1),\ldots,u_M(k-N)).$$

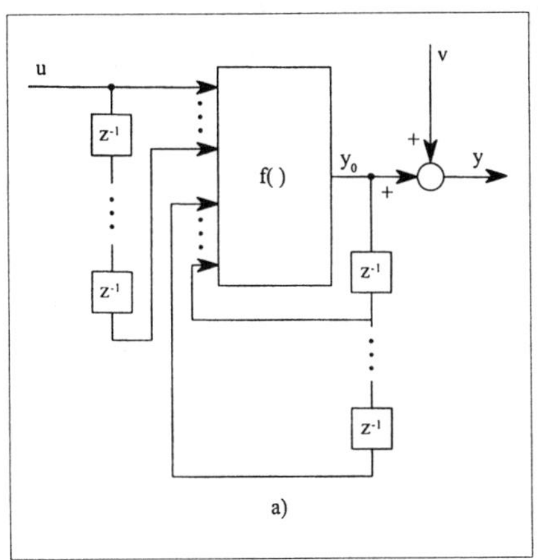


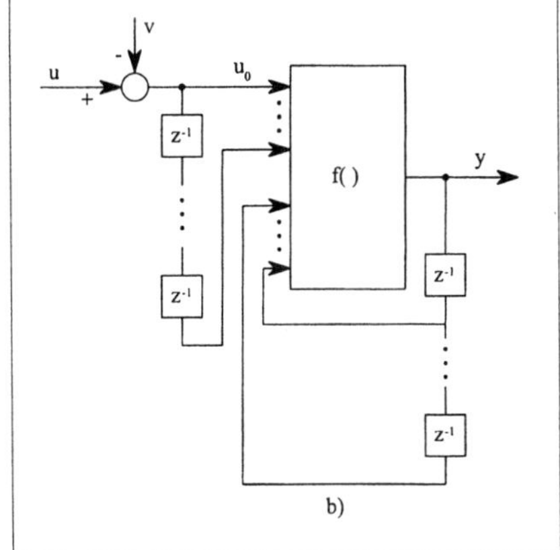

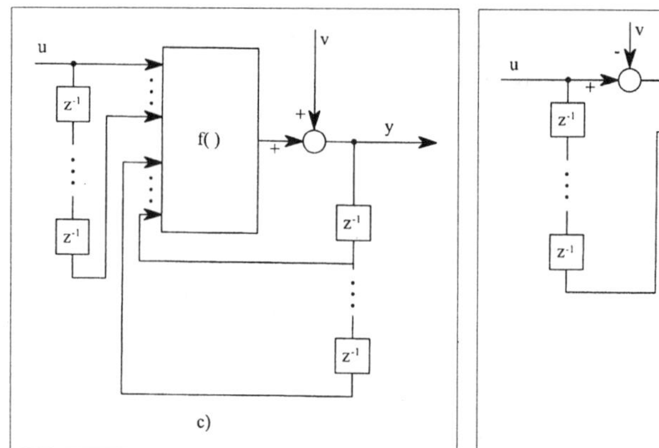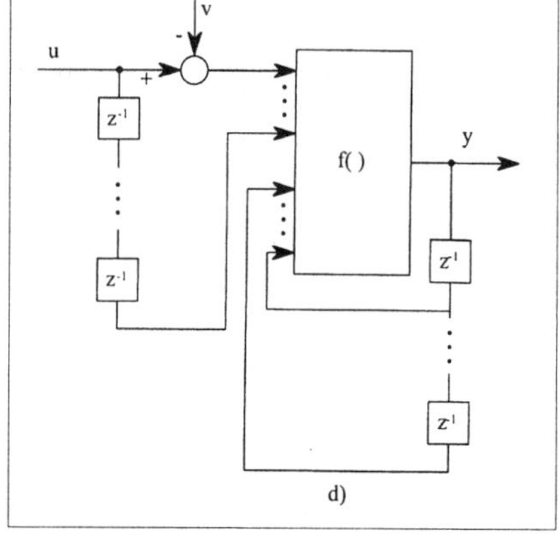

Figure 2. The representation of noise, which results in an unbiased estimation for the output error model.

If the structure of the identification model is the same as the structure of the nonlinear plant, the optimization procedure for the minimization of (8) tries to make the residual $\in (k)$ white, so if $v(k)$ is white noise, the minimum is $\in (k) = v(k)$ and consequently $(\hat{fi}(\ ) = fi(\ ))$ and $u_M(k) = u_0(k)$. The condition for an unbiased estimation is that the noise is additively added to the plant input and that it is white as illustrated in Fig. 2-b.

### 4.3. Nonlinear Generalized Output Error Model

In this case the noise $v(k)$ is introduced inside the plant as illustrated in Fig. 2-c. The plant equation becomes now:

$$y(k) = \hat{f}(u(k),\ldots, u(k - N_p), y(k - 1),\ldots, y(k - N_p) + v(k)) \qquad (20)$$

and the residual

$$\begin{aligned} \in (k) &= y(k) - y_M(k) = \\ &= f(u(k),\ldots, u(k - N_p), y(k - 1),\ldots, y(k - N_p)) + v(k) - \\ &\quad -\hat{f}(u(k),\ldots, u(k - N), y(k - 1),\ldots, y(k - N)). \end{aligned} \qquad (21)$$

If the structure of the identification model is the same as the structure of the nonlinear plant, the optimization procedure for the minimization of (8) tries to make the residual $\in (k)$ white, so if $v(k)$ is white noise, the minimum is $\in (k) = v(k)$ and consequently $(\hat{f}(\ ) = f(\ ))$.

The condition for an unbiased estimation is that the noise additively added to the plant as illustrated in Fig. 2-c is white.

### *4.4. Nonlinear Generalized Input Error Model*

In this case the noise $v(k)$ is introduced inside the plant as illustrated in Fig. 2-d. The plant equation becomes now

$$u(k) = f_i\big(y(k),\ldots,y(k-N_p),u(k-1),\ldots,u(k-N_p)\big) + v(k) \tag{22}$$

and the residual

$$
\begin{aligned}
\in(k) &= u(k) - u_M(k) = \\
&= f_i(y(k),\ldots,y(k-N),\,u(k-1),\ldots,\,u(k-N) + v(k) - \\
&\quad -\hat{f}_i(y(k),\ldots,y(k-N),\,u(k-1),\ldots,\,u(k-N)\,.
\end{aligned}
\tag{23}
$$

If the structure of the identification model is the same as the structure of the nonlinear plant, the optimization procedure for the minimization of (8) tries to make the residual $\in(k)$ white, so if $v(k)$ is white noise, the minimum is $\in(k) = v(k)$ and consequently $(\hat{f}i(\ ) = fi(\ ))$. The condition for an unbiased estimation is that the noise additively added to the plant as illustrated in Fig. 2-d is white.

The estimation is unbiased if the structure of the model is the same as the structure of the plant and the white noise enters into the plant in a special location depending on the type of a model used in identification. With the output error model this is the output of the UDA; the plant output is in this case the sum of uncorrupted signal of UDA and the white noise. With the input error model this is the input of the plant; the input signal of the UDA is the sum of the uncorrupted input signal and the white noise. With Generalized output error model the white noise enters into the plant at the output of the UA; the corrupted output signal is fed back and represents the input of the UA. With Generalized input error model the white noise enters into the plant only at the first input of the UA.

## 5. Application to a Laboratory Scale Heat Exchanger

The identification models were applied to a fuzzy model based control of a laboratory scale heat exchanger. The accessory used, depicted schematically in Fig. 3, consists of a plate heat exchanger, through which hot water from an electrically heated reservoir is ontinuously circulated in counter-current flow to the cold process fluid (cold water). Thermocouples are located in the inlet and outlet streams of the exchanger, thus the primary and secondary flow rates can be visually monitored. Power to the heater is controlled by an external control loop. The flow of the heating fluid can be controlled by a proportional motor driven valve. The control variable is the control current of the valve (4-20 mA),

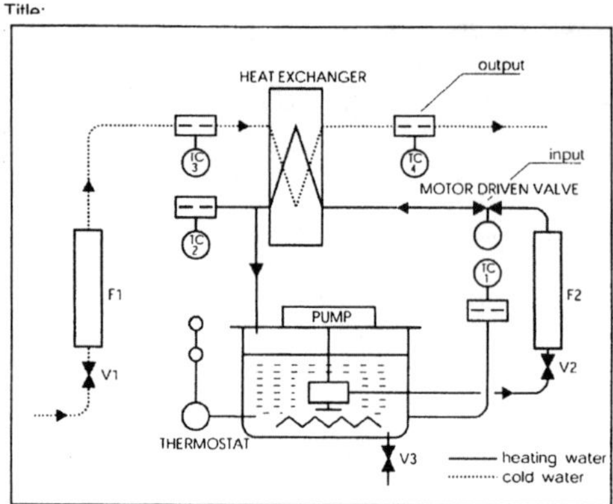


Figure 3. The scheme of the laboratory scale heat exchanger.

while the controlled variable is the temperature of the water in the secondary circuit at the heat exchanger outlet.

The plant was identified off-line on the basis of signals which assure a throughout excitation. The sampling time used was 4 s. As the dynamics of the plant exhibits approximately first order dynamics with a small time delay, the three step delayed control variable and one step delayed controlled variable were used as inputs to the Nonlinear Generalized Output Error model

$$y_M(k) = f(u(k-3), y(k-1)) \tag{24}$$

If, on the right hand side of Eq. (24), $u(k-1)$ were used, the resulting nonlinear dynamic system would be of the first order without any delay. By delaying the control variable by two additional steps, the dead time of 8 s was achieved. Two Generalized Output Error models were compared. With the first one, the universe of discourse of the control and the controlled variables were divided into five triangular, equally spaced membership functions respectively using the Takagi-Sugeno model with crisp constant consequences $\left(f^j = c^j\right)$. The central values of the membership functions for control and controlled variables were [4, 8, 12, 16, 20] mA and [13.00, 22.75, 32.50, 42.25, 52.00] °C, respectively. The product was used as the composition operator $\otimes$, and the rule base was complete.

In the second model, three triangular equally-spaced membership functions were used, together with the Takagi-Sugeno model with linear consequences $\left(f^j = c_1^j u(k-3) + c_2^j y(k-1) + c_3^j\right)$. Both models have approximately the same number of parameters to be determined (5 x 5 = 25 for the first, and 3 x 3 x 3 = 27 for the second model, respectively). The Takagi-Sugeno model was chosen to fulfil the three conditions of Section 2. The unknown parameters were estimated by the least squares method. In general, [8] first order consequences result in better tracking, however our result of the comparison was that the first model (five membership functions and crisp constant consequences) has a slightly better tracking capability. This somewhat surprising result was probably due to the high nonlinearity of the plant with which the three membership functions are hardly able to cope. For simplicity, the Takagi-

Sugeno model with crisp constant consequences was used in the design and real time implementation of the control algorithm.

It should be noted that the Nonlinear Generalized Output Error model was used in the identification procedure, while the Nonlinear Output Error model was used for long range prediction, as required by the predictive control algorithm. This is a simple practical solution; theoretically, the Nonlinear Output Error model should be also used for the identification, however in this case the optimization problem would be nonlinear, with all the known effects.

The treated plant was first identified by a linear first order Generalized Output Error model with three steps time delay. The results of the validation are shown in Fig. 4.

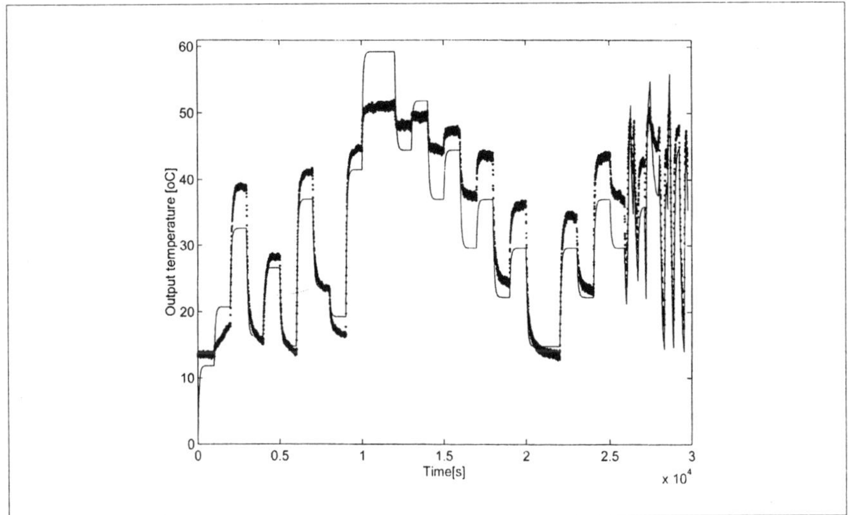

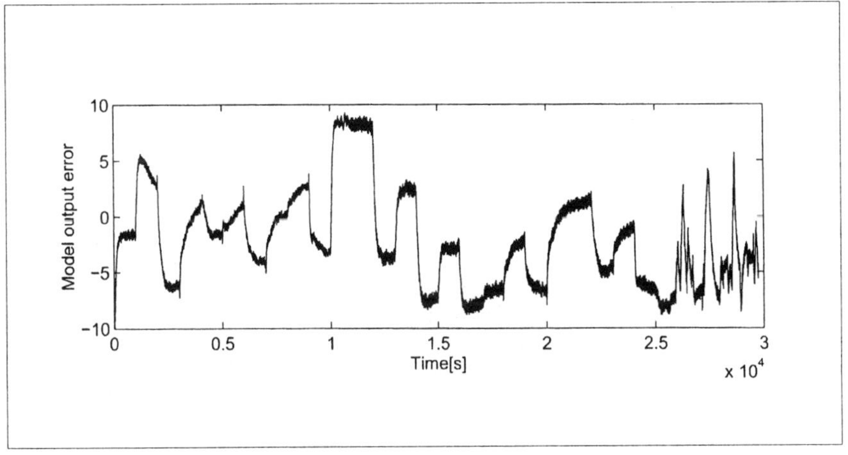

Figure 4. Comparison of the responses of the linear nominal model and the measured plant.

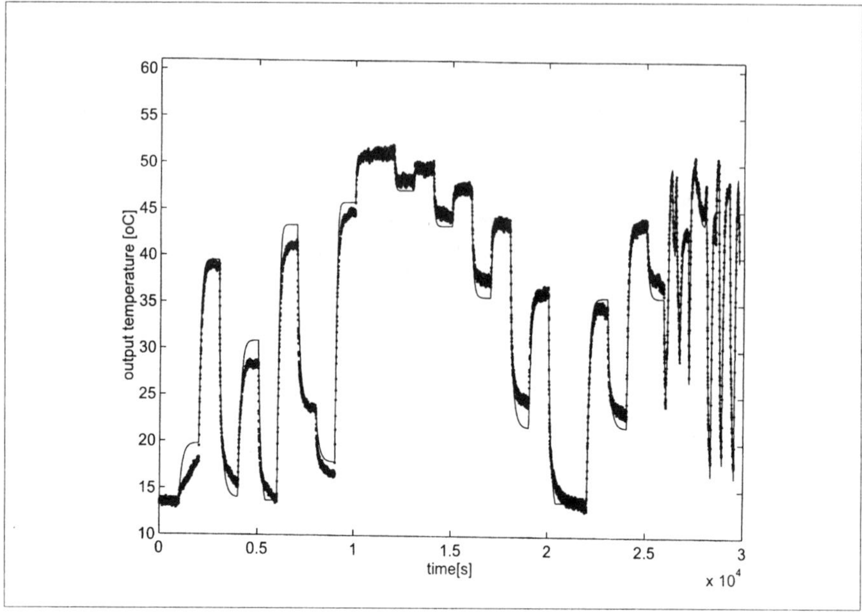

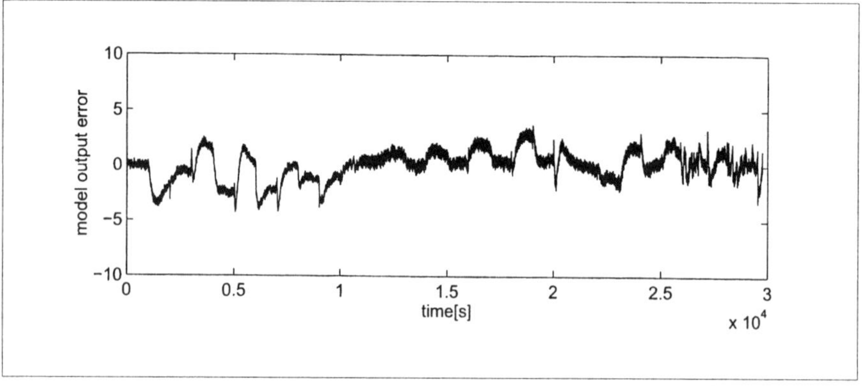

Figure 5. The validation of the Nonlinear Generalized Output Error model.

Next, the Nonlinear Generalized Output Error model was applied. Fig. 5 shows the validation of the applied model. It can be seen that the nonlinear model has much lower output error, which means much lower model uncertainty. Lower model uncertainty enables a choice of controller parameters, which give better tracking quality.

The design parameters of the model based predictive control (the control $N_1$ and the prediction $N_2$ horizon, respectively, and the weight factor $r$) were $N_1 = 2$, $N_2 = 30$, $N_u = 3$ and $r = 0.002$.

Fig. 6 depicts the closed loop response of the Nonlinear Output Error model-based predictive control. It can be seen that the closed loop response remains approximately the

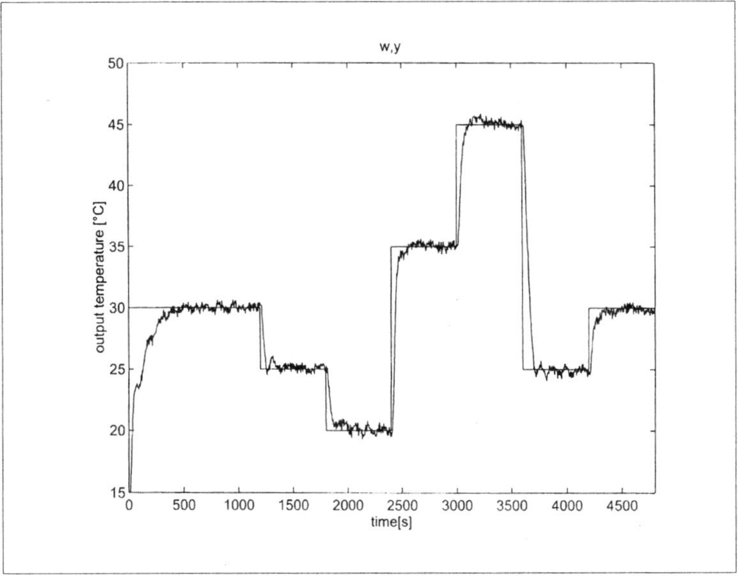


Figure 6. The closed loop response of the Nonlinear Output Error model based predictive control.

same throughout the entire range of the controlled variable, and that the response is stable and adequate.

## 6. Conclusions

Four forms of Fuzzy Identification Models are presented in the paper. Their convergence properties in the presence of noise is reviewed and it is shown that unbiased estimation is achieved if the disturbing noise is white and enters the plant at specific points: at the plant output for the Nonlinear Output Error Model, at the plant input for the Nonlinear Input Error Model, and inside the plant at the output and first input of the nonlinear block for Nonlinear Generalized Output Error and Nonlinear Generalized Input Error Models respectively. In the case of fixed (known or preselected) structure the estimation becomes linear in unknown parameters. An application of the proposed modelling technique to the laboratory scale heat exchanger illustrates the practicability of the proposed design technique.

### References

[1] J. Sjöberg, Q. Zhang, L. Ljung, A. Benveniste, B. Delyon, P-Y. Glorennec, H. Hjalmarsson and A. Juditsky, Nonlinear Black-box Modeling in System identification: a unified Overview, *Automatica* Vol. 31, No. 12, pp 1691-1724, 1995.

[2] F. Girosi and T. Poggio, Networks and the Best Approximation Property, C.B.I.P.Paper, No. 45, MIT 1994.

[3] J. Castro, Fuzzy Logic Controllers are Universal Approximators, IEEE Transaction on Systems, Man and Cybernetics, 1995, pp. 629 - 635.

[4] X. J. Zeng and G. H. Singh, Approximation theory of fuzzy systems - MIMO case, *IEEE Tr. on Fuzzy Systems*, Vol. 3, No. 2, pp.219 - 235, 1995.

[5]  R. Babuška and H. B.  Verbruggen, An overview of Fuzzy Modeling for Control, *Contr. Eng. Practice*, Vol. **4**, 1996, pp. 1593 - 1606.
[6]  T. Takagi and M. Sugeno, Fuzzy identification of systems and its application to modelling and control, *IEEE Tr. on Systems, Man and Cybernetics*, Vol. **15**, No.1, pp. 116 - 132, 1985.
[7]  R. Jager, Fuzzy Logic in Control, PhD. Thesis, Delft University of Technology, Delft, 1995
[8]  T. A. Johansen, Fuzzy model Based Control.

*Systematic Organisation of Information in Fuzzy Systems*
*P. Melo-Pinto et al. (Eds.)*
*IOS Press, 2003*



# Neural Networks Classifiers based on Membership Function ARTMAP

Peter SINCÁK[1*], Marcel HRIC[**], Ján VAŠCÁK[2***]

[*] *Siemens AG, Vienna, PSE Department, Vienna, Austria*
[**] *Center for Intelligent Technologies, Faculty of EE and Informatics,*
*Technical University, Letná 9, 04001Košice, Slovak Republic*
[***] *Center for Intelligent Technologies, Faculty of EE and Informatics,*
*Technical University, Letná 9, 04001Košice, Slovak Republic*
*sincak@tuke.sk, hric@neuron-ai.tuke.sk, Vascak@tuke.sk,*

**Abstract:** The project deals with the application of the computational intelligence tools for multispectral image classification. Pattern Recognition scheme is a global approach where a classification part is playing an important role to achieve the highest classification accuracy. Multispectral images are data mainly used in remote sensing and this kind of classification is very difficult to make an accuracy assessment of classification results. There is a feedback problem to adjust the parts of pattern recognition scheme. Precise classification accuracy assessment is almost not possible to obtain, because it is a very laborious procedure. In the paper, simple neural networks for multispectral image classification, ARTMAP like neural networks as more sophisticated tools for classification; modular approach to achieve the highest classification accuracy on multispectral images is presented. There is a strong link to progress in computer technology which gives much better conditions for modeling more sophisticated classifiers for multispectral images.

**Keywords:** pattern recognition principles, classifier design, classification accuracy assessment, contingency tables, Back-propagation neural networks, fuzzy BP neural networks, ART and ARTMAP neural networks, Modular neural networks

## 1. Introduction

Image processing is an important part of modern technology and its application possibilities include environmental monitoring, something that is very important for the future of humankind. Image processing can be the most important part of this system and determines the quality accuracy of information retrieval from the images. Generally we have to take into consideration the basic pattern recognition principles where classification plays an important role. These principles are presented in figure 1. As it is clear from the figure, the determination of feature space is playing a key role in overall pattern recognition principles. The feature selection and classification are based on this information and the level of ability to approximate the nonlinear discrimination function determinates the success of classification procedure. A very important aspect of this procedure is accuracy assessment of classification results. The contingency analysis

---

[1] P. Sincák is on temporary leave from Center for Intelligent Technologies at TU Kosice Slovakia and his stay as guest professor at Siemens, Vienna is supported by Maria-Curie fellowship from EU 5.FP activity
[2] On leave at Hirota Laboratory, Tokyo Institute of Technology Japan, vascak@hrt.dis.titech.ac.jp

is a proper way of accuracy assessment and it reflects the multi-class results of classification procedure. The important aspects of the pattern recognition principle are the adaptability of the overall procedure and the equal importance of all steps in the procedure including feature space determination and feature selection procedure. This pattern recognition principle reveals such questions as follows:

Is it more useful to "invest" more time to the best feature selection procedure?

Or is it more important to "invest" more time to develop the most sophisticated classification procedure for any type of features?

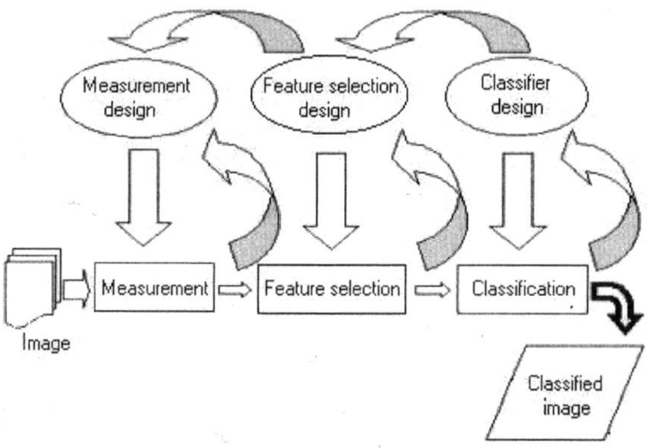


Figure 1. Basic principles of pattern recognition.

These two questions are playing the key-role in the future of research in the classification domain. Certainly these two problems are closely connected and usually feature selection techniques assume a specific type of classifier. We can find a number of techniques dealing with the first approach and a number of approaches dealing with the second problem. Basically the first approach is based on the assumption that a simple and fast classification procedure is used while the second problem is related to the assumption of a simple and almost raw data or data with minor preprocessing procedures. Certainly the combination of both approaches could be fruitful and could bring some good results for real-world application.

## 2. Motivation of the Project

Classification is mapping from feature space into space of classes. Considering supervised mode a determination of training sets is extremely important. Many times is a situation that the training sites produce non-homogeneous training data, which are represented in the feature space as union of clusters in different locations in the feature space. From this perspective the classification approach which uses labeled clustering is the most appropriate way of handling classification task.

In figure 2 is the basic concept of ARTMAP neural network, which is approaching a problem as labeled clustering task. The role of the Mapfield neural layer is to associate the various clusters into the desired class. The coefficient $P$ is adjusted in the training phase and reflects the plasticity of the system. Number of nodes in the recognition layer reflects number of identified clusters in the feature space. It very often happens that it is very difficult to decide if a certain point in feature space belongs to a certain class. Therefore an approach based on fuzzy sets has many advantages to reduce misclassification results. Sometimes it is more convenient to have results in the form of transparent information concerning relations of the observed point in the feature space to all classes of interest. Instead of the crisp classifier output we can be more satisfied with outputs based on fuzzy sets and namely on values of membership functions of observed input to fuzzy cluster and fuzzy classes. The notions of fuzzy clusters and fuzzy class are described in the next part of this paper.

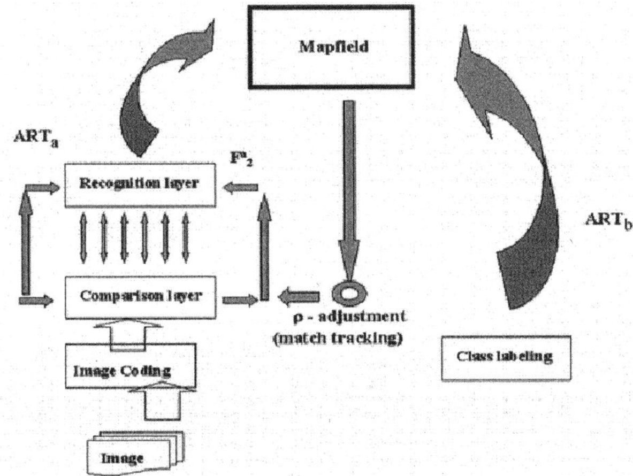


Figure 2. The basic concept of ARTMAP neural network.

The motivation is to provide for the end-user a smaller number of misclassifications and higher readability of the classification results. The output of these classification results is a vector of values describing relation of the input to each class of interest.

The desire is to have a highly parallel tool with incremental learning ability similar to ARTMAP family neural network.

## 3. Description of the Method

The project is based on the assumption that data in feature space are organized in fuzzy cluster. Fuzzy cluster $A$ is considered as a fuzzy set $A$ in multidimensional feature space representing a set of the ordered couples e.g.

$$A \in \left\{ [x_1, \mu(x_1)], \ldots, [x_n, \mu(x_n)] \right\} \tag{1}$$

where $A$ is a fuzzy set and $\{x, \mu_A(x)\}$ are ordered couples, where $x$ is a point in multidimensional feature space and $\mu_A(x)$ is a value of the membership function of $x$ to fuzzy cluster (set) $A$. There are many fuzzy clusters in feature space and a certain set of fuzzy clusters create a fuzzy class. Fuzzy class is the union of fuzzy clusters belonging to a considered class defined by training set e.g.

$$CL = \left\{ \bigcup_{i=1}^{n} A_i \right\} \tag{2}$$

Generally we can consider fuzzy class as a set of fuzzy sets $A_i$ representing the variety of the numerical representation of the class. Relation between $\mu_{CL}(x)$ and $\mu_A(x)$ must be as follows:

$$\mu_{CL}(X) = max_{i=1,n}\left(\mu_{Ai}(X)\right) \tag{3}$$

where $A_i$ is a fuzzy cluster, which belongs to class $CL$ and $n$ is a number of fuzzy clusters creating class $CL$. The membership function is considered as:

$$\mu_A(X) = \frac{1}{\left[1 + e^{\frac{X-mc}{\sigma}}\right]^F} \tag{4}$$

where $F$ is a positive value, $mc$ is mean the value and $\sigma$ is variance of gaussian type of membership function. The MF-ARTMAP is intended to be such a tool to calculate values of membership functions of $X$ to each class of interest in feature space. In figure 2 we can find an example of 2-dimensional membership function with various $F$ in single-dimensional directions in the space. The space is $n+1$ dimensional where $n$ is number of features and $(n+1)$-st dimension is membership function value.

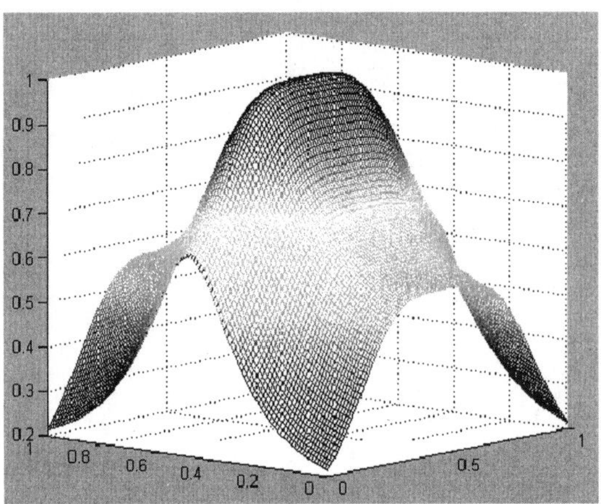

Figure 3. Two dimensional membership value with various F.

### 3.1. Description of the Neural Network Topology

Topology of MF ARTMAP is based on a similar architecture to ARTMAP. In figure 4 can be seen two MF-ARTMAP neural networks with 4 neural layers. The input layer is mapping the input into the comparison layer where the comparison between input pattern and mean values of existing clusters are compared. In Figure 4a is represented a starting situation, where only one cluster is identified, while in the Figure 4b, 3 clusters are already revealed so the input pattern is tested if it does not belong to one of the clusters. In this case the second layer is dynamically changing according to the number of clusters in the 3-rd layer. So, the 2-nd and 3-rd layers are extending according to the number of clusters found in the feature space. The recurrent connection between the 2-nd and 3-rd layer is the encoding of mean value and mean square deviation associated with a particular membership function based on a Gaussian type of representation. The 4-th layer is representing the mapfield like part of a neural network whose role is to integrate the clusters into a resulting class. The input to the mapfield is from the 3-rd layer and also from outside of the neural network as associated output into the overall MF-ARTMAP neural networks.

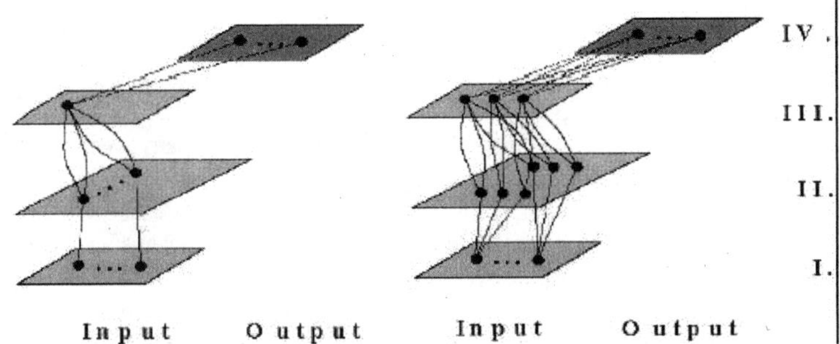


4a - MF-ARTMAP with one cluster                 4b - MF-ARTMAP with three clusters
Figure 4. Basic topology for MF-ARTMAP neural networks.

More detailed topology is in figure 5, where a situation with more clusters is represented. Basically it can be listed the number of neural layers in the MF-ARTMAP can be listed as follows:

Layer #1 – input mapping layer, number of neurons equal to "$n$", where $n$ is dimensionality of the feature space,

Layer #2 – comparison layer, number of neurons equal to "$n \times nc$", where $nc$ is the number of clusters identified in the recognition layer,

Layer #3 – recognition layer, number of neurons equal to "$nc$", where $nc$ is the number of clusters,

Layer #4 – mapfield layer, number of neurons equal to "$M$", where $M$ is the number of classes for classification procedure.

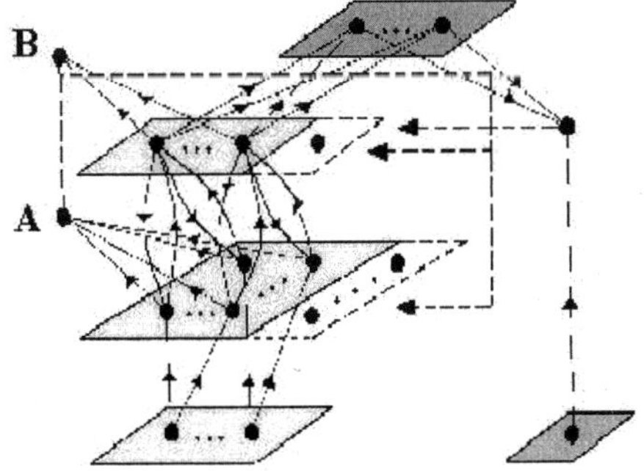


Figure 5. General MF-ARTMAP topology with a dynamic number of neurons in the *2nd* and 3-rd neural network layer.

More detailed description of the MF-ARTMAP is presented in the figure 5. The important role is done by neurons A and B, which reset and freeze some of procedures during training. These neurons are fed from the comparison layer with values that represent a difference between input patterns and mean values of the inputs feature values. If the maximum of these differences is greater than a threshold, then a new cluster is setup immediately. If not, then there is a chance that input pattern belongs to one of the clusters identified in the past procedures. Then a training and updating of the recurrent synapses is underway to adapt the shape of the membership function. The adaptation procedure is described in the following section. More info can be found in [1].

### 3.2. Parallel MF-ARTMAP

The notion of modular neural network has been known for many years. It is a very promising idea of solving complex problems by its distribution into more sub-problems that are easy to solve. Basically the "divide and conquer" principles are usually used in modular neural networks. There are some difficult questions about the separability problems and discrimination hyper-planes determination. The key problem is in making an answer to the following problem:

"Is it easier to separate one particular class from the feature space or to identify more classes among each other?"

So in fact the question is about difficulties of dichotomous classification comparing the multi-class approach. The first impression could be that dichotomous classification is always easier than the multi-class approach, but it is very difficult to conclude in general. For investigating these ideas, a parallel ARTMAP approach was designed and tested. The basic philosophy of this approach is illustrated in the following figure 6. The MF-ARTMAP is suitable for solving the conflict in this approach because the values of membership functions to fuzzy clusters are good indicators of conflict solution among more experts as it is indicated in the figure 6.

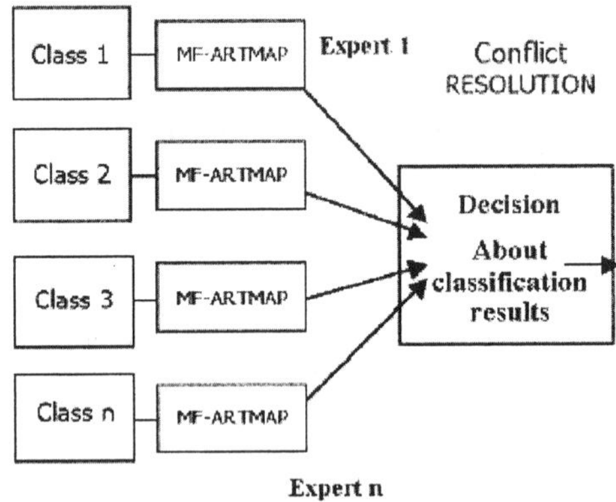


Figure 6. The Basic philosophy of Parallel ARTMAP.

The basic advantages of Parallel MF-ARTMAP are as follows:

*1. Ultra-fast learning abilities on highly parallel systems e.g. PC-farms. This feature can be very useful in cases of large databases with easier handling of larger amount of data.*
*2. Easy and comfortable extension of classes of interest by adding a new expert network and train on the new class training data. This is very interesting in case of frequent adding the new classes to list of classes of interest.*
*3. Easy identification of class "unknown" by measuring the membership function value of the unknown input to the fuzzy classes. If the value is lower than the given value – then it is rejected and proclaimed for class "unknown". This feature is very important when large data is considered with many classes.*
*4. Easy readability of fuzzy classes as unions of fuzzy clusters and identification of their basic parameters.*

On the other side the basic disadvantage of Parallel MF-ARTMAP is as follows:
Necessity of more parameters determination considering starting values of vigilance parameters for each expert network separately. This can be avoided by designing the meta-controller setting up al the parameters for the overall complex of experts e.g. fuzzy system.

## 4. Experimental Results

Experiments using benchmark and real-world data were done during this project. The aim was comparative analysis of known CI systems with those modified or developed during the research namely MF-ARTMAP, Parallel ARTMAP and MF-ARTMAP with adaptation F parameter. Basically the same real-word data were used and so the comparison can be done assuming the same training and testing data, which means the same amount of knowledge was used for classification procedure.

### *4.1 Accuracy Assessment*

Accuracy assessment was evaluated using contingency table analysis. A contingency table was used in a basic comparison study between all methods, which were investigated and developed. Some details about contingency table analysis for accuracy assessment of classification results can be found in [2].

### *4.2. Experiments on Benchmark Data*

There were 2 benchmark data used for testing classification results for comparative purposes. Circle in the rectangle and double spiral problem was used for dichotomous classification testing. These two benchmark data are used the most for estimating the classifier level of sophistication. If results on these benchmark data are good there, is a good assumption that the classifier will be successful in main applications. On the following tables are the results from the selected benchmark classification procedure. Results are on testing data.

Table 1. Results on the "circle in the square" dichotomial classification.

| Predicted class | Actual class | | | | | |
|---|---|---|---|---|---|---|
| | MF Artmap | | Parallel MF Artmap | | MF Artmap with F adaptation | |
| | A | B | A | B | A | B |
| A' | 97,62 | 2,92 | 98,14 | 1,46 | 98,32 | 1,82 |
| B' | 2,38 | 97,08 | 1,86 | 98,54 | 1,68 | 98,18 |

Table 2. Results on the "double spirals" dichotomous classification.

| Predicted class | Actual class | | | | | |
|---|---|---|---|---|---|---|
| | MF Artmap | | Parallel MF Artmap | | MF Artmap with F adaptation | |
| | A | B | A | B | A | B |
| A' | 87,54 | 10,76 | 87,96 | 7,86 | 88,72 | 9,95 |
| B' | 12,46 | 89,24 | 12,04 | 92,14 | 11,28 | 90,05 |

Table 3. Contingency table of Landsat TM classification on the test sites for MF-ARTMAP

| Predicted | Actual | | | | | | |
|---|---|---|---|---|---|---|---|
| | A | B | C | D | E | F | G |
| A' | 94,87 | 0,83 | 0,00 | 0,00 | 4,68 | 0,00 | 5,41 |
| B' | 0,00 | 84,30 | 0,00 | 3,77 | 6,43 | 0,00 | 0,00 |
| C' | 0,00 | 0,00 | 99,03 | 0,00 | 0,00 | 0,00 | 0,00 |
| D' | 0,64 | 9,71 | 0,00 | 96,01 | 1,17 | 0,00 | 1,35 |
| E' | 4,49 | 4,96 | 0,97 | 0,22 | 87,72 | 0,17 | 4,05 |
| F' | 0,00 | 0,00 | 0,00 | 0,00 | 0,00 | 99,74 | 0,00 |
| G' | 0,00 | 0,20 | 0,00 | 0,00 | 0,00 | 0,09 | 89,19 |

Table 4. Contingency table of Landsat TM classification on the test sites for Parallel MF-ARTMAP

| Predicted | Actual | | | | | | |
|---|---|---|---|---|---|---|---|
| | A | B | C | D | E | F | G |
| A' | 95,51 | 0,21 | 0,00 | 0,00 | 2,76 | 0,00 | 4,23 |
| B' | 0,00 | 83,16 | 0,00 | 3,01 | 7,18 | 0,00 | 0,00 |
| C' | 0,00 | 0,00 | 100 | 0,00 | 1,66 | 0,00 | 0,00 |
| D' | 0,00 | 11,91 | 0,00 | 96,66 | 1,10 | 0,00 | 0,00 |
| E' | 3,85 | 4,72 | 0,00 | 0,22 | 87,29 | 0,17 | 7,04 |
| F' | 0,00 | 0,00 | 0,00 | 0,00 | 0,00 | 99,49 | 5,36 |
| G' | 0,64 | 0,00 | 0,00 | 0,11 | 0,00 | 0,34 | 83,10 |

Table 5. Contingency table of Landsat TM classification on the test sites for MF-ARTMAP
with F adaptation

| Predicted | Actual | | | | | | |
|---|---|---|---|---|---|---|---|
| | A | B | C | D | E | F | G |
| A' | 94,92 | 0,23 | 0,00 | 0,00 | 3,55 | 0,00 | 3,41 |
| B' | 0,00 | 87,50 | 0,00 | 4,12 | 5,26 | 0,00 | 0,00 |
| C' | 0,00 | 0,00 | 99,03 | 0,00 | 0,00 | 0,00 | 0,00 |
| D' | 0,64 | 7,71 | 0,00 | 95,51 | 2,75 | 0,48 | 0,9 |
| E' | 4,44 | 2,96 | 0,97 | 0,37 | 88,44 | 0,11 | 2,5 |
| F' | 0,00 | 0,00 | 0,00 | 0,00 | 0,00 | 99,32 | 0,85 |
| G' | 0,00 | 1,60 | 0,00 | 0,00 | 0,00 | 0,09 | 92,34 |

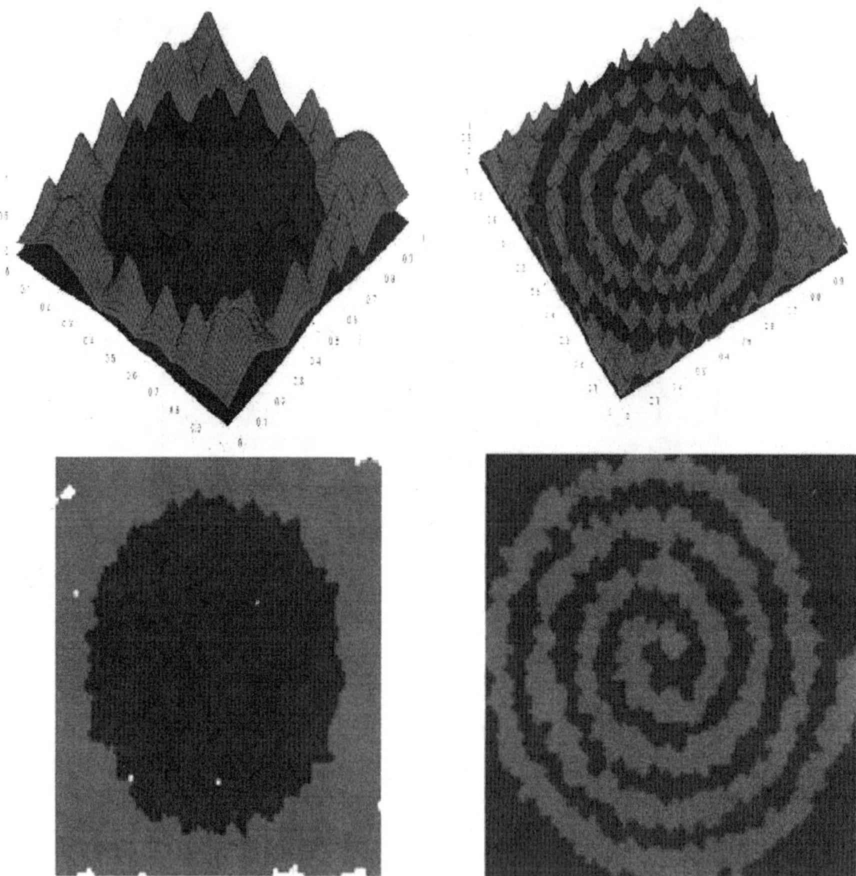

Figure 7. Classification of circle in the square and double spiral.

### 4.3. Experiments on Real-world Data

Experiments were done on benchmark and real-world data. Basically the behaviors of the methods were observed on multispectral image data with the aim to obtain the best classification accuracy on the test data subset.

The Košice data consists of a training set of 3164 points in the feature space and of a test set of 3167 points of the feature space. A point in the feature space has 7 real-valued coordinates of the feature space normalized into the interval (0,1) and 7 binary output values. The class of a fact is determined by the output which has a value of one; the other six output values are zero. The data represents 7 attributes of the color spectrum sensed from Landsat satellite. The representation set was determined by a geographer and was supported by the ground verification procedure.

The main goal was landuse identification using the most precise classification procedure for achieving accurate results. The image was taken over the eastern Slovakia region particularly from the City of Kosice region. There were seven classes of interest picked up for classification procedure as it can be seen in figure 8.

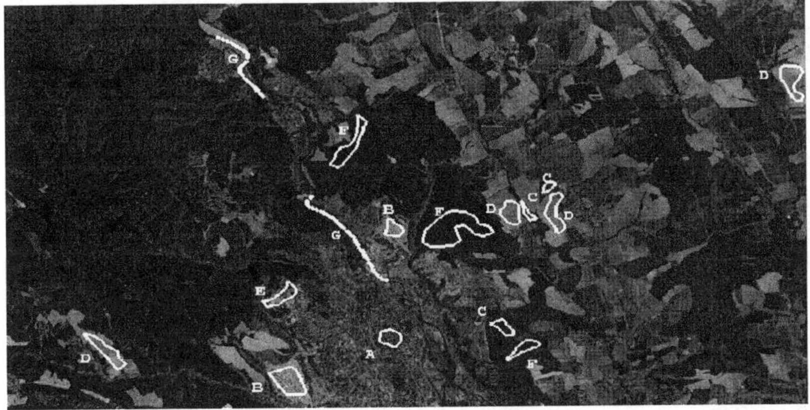

Figure 8. Original image. Highlighted areas were classified by expert (A – urban area, B – barren fields, C – bushes, D – agricultural fields, E – meadows, F – forests, G – water).

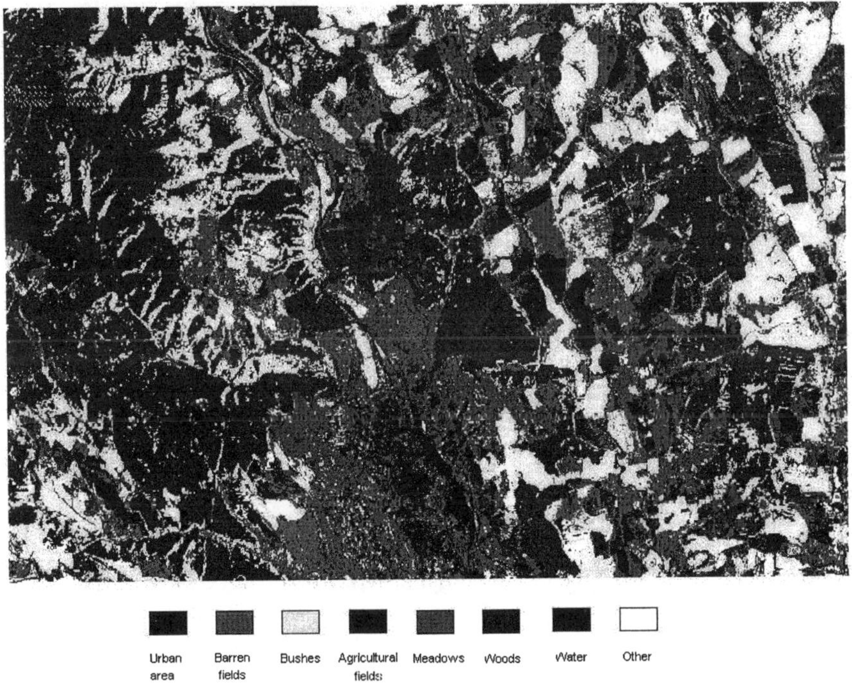


Figure 9. Classification results on Landsat TM data using Parallel MF ARTMAP approach

## 5. Conclusions

The paper presents further research on the Parallel MF-ARTMAP approach to the classification procedure. The advantage of this approach is higher readability of the neural network in providing the values of membership functions of input to fuzzy clusters or fuzzy classes identified by this approach. Results of these neural networks are comparable with MF ARTMAP [1] on benchmark and real-world data and in addition, provide more useful information about the measure of membership into all fuzzy classes and fuzzy clusters discovered in the feature space. Adaptation of F parameter seems to be a useful tool to investigate in the future. This seems to be an interesting advantage of this approach. Some further work is underway to apply the Parallel MF-ARTMAP in some financial data for classification purposes.

### References

[1]  P. Sincák, N. Kopèo, M. Hric and H. Veregin, MF-ARTMAP to identify fuzzy clusters in feature space, accepted to IIZUKA 2000, *6-th International Conference on Computational Intelligence*, IIZUKA, Japan, October 1-4, 2000 (1).

[2]  P. Sincák, H. Veregin and N. Kopèo, Conflation techniques in multispectral image processing, *Geocarto Int*, March, 2000, pp. 11-19.

### Further Reading

P. M. Atkinson and A. R. L. Tatnall, Neural networks in remote sensing, *Int. J. Remote Sensing,* vol. **18**, 4, 1997, pp. 711-725.

J. Richards, *Remote sensing digital image analysis: An introduction.* Springer - Verlag, Berlin, 1993.

B. G. Lees and K. Ritman, Decision-tree and rule-induction approach to integration of remotely sensed and GIS data in mapping vegetation in disturbed or hilly environments, *Environmental Management*, vol. **15**, 1991, pp. 823-831.

C. H. Chen, Trends on information processing for remote sensing, In *Proceedings of the International Geoscience and Remote Sensing Symposium (IGARSS)*, vol. **3**, Aug. 3-8 1997, pp. 1190-1192.

P. Werbos, Beyond Regression: New Tools for Prediction and Analysis in the Behavioral Sciences, Ph.D. thesis, Harvard University, 1974.

G. A. Carpenter, M. N. Gjaja, S. Gopal and C. E. Woodcock, ART neural networks for remote sensing Vegetation classification from Landsat TM and terrain data, *IEEE Transactions on Geoscience and Remote Sensing*, vol. **35**, no. 2, 1997, pp. 308-325.

D. Rutkowska and R. Nowicki, R., Implication-based neuro-fuzzy architectures, In *International journal of applied mathematics and computer science*, vol. **10**, no. 4, 2000, pp. 675-701.

S. Grossberg, Adaptive pattern classification and universal recording, I: Feedback, expectation, olfaction, and illusions, *Biological Cybernetics,* vol. **23**, 1976, pp. 187-202.

G. A. Carpenter, B. L. Milenova and B. W. Noeske, Distributed ARTMAP: a neural network for fast distributed supervised learning, *Neural Networks,* vol. **11**, no. 5, Jul. 1998, pp. 793-813.

J. R. Williamson, Gaussian ARTMAP: A neural network for fast incremental learning of noisy multidimensional maps, *Neural Networks,* vol. **9**, 1996, pp. 881-897.

R. K. Cunningham, R. K., *Learning and recognizing patterns of visual motion, color, and form*, Unpublished Ph.D. thesis, Boston University, Boston, MA, 1998.

R. Duda and P. Hart, Pattern Classification and Scene Analysis, Wiley, New York, 1973.

E. Ocelikova and T. Nguyen Hong, Diferential Equations for maximum entropy Image Restoration, In *Proc. of the 3rd International Conference "Informatics and Algorithms 99" Prešv* , September 9-10, 1999, pp. 214-218, ISBN 80-88941-05-9.

E. Ocelikova and D. Klimesova, Clustering by Boundary Detection, In *Proc. of the 4th International Scientific Technical Conference "Process control – ŘP 2000* , Pardubice, June 2000, pp. 108, ISBN 80-7194-271-5.

D. Klimesova and E. Ocelikova, GIS and Spatial Data Network, In: *Proc. of International Conference Agrarian Perspectives X-Sources of Sutainable Economic Growth in the Third Millenium. Globalization versus Regionalism,* Sept. 18-19, 2001, Prague, Czech republic, ISBN 80-213-0799-4.

P. Daponte, D. Grimaldi and L. Michaeli, Analogue neural network design for the preprocessing of measurement signals MEASUREMENT, vol. **26**, no. 1, (1999), pp.1-17, ISSN 0263-2241.

P. Daponte, D. Grimaldi and L. Michaeli, Design method of analog neural networks for pre-processing in measurement sensors MEASUREMENT, vol. **24**, no. 2, (1998), pp. 109-118, ISSN 0263-2241.

*Systematic Organisation of Information in Fuzzy Systems*
*P. Melo-Pinto et al. (Eds.)*
*IOS Press, 2003*



# Environmental Data Interpretation: Intelligent Systems for Modeling and Prediction of Urban Air Pollution Data

Francesco Carlo MORABITO, Mario VERSACI
*University "Mediterranea" of Reggio Calabria*
*Faculty of Engineering, DIMET*
*Via Graziella, Feo di Vito, Reggio Calabria*

**Abstract**. Environmental data processing is based on modeling and prediction of time series whose dynamic evolution is the result of the concurrence of many variables. The goal of this paper is to show how some recent advances in data driven approaches (like Artificial Neural Networks, ANN, and Fuzzy Inference Systems, FIS) can be of help to environmental problems solution. These kind of intelligent systems can be useful in environmental data analysis and interpretation from various perspectives: to perform knowledge discovery in large environmental databases ("environmental data mining"), to make prediction, to explain and interpret data and non-linear correlation among predicting variables. The output of the intelligent processing systems can also facilitate decision-making. Environmental data show some characteristics features and peculiarities (noise, non linearity, non-stationarity, missing data, ...) that largely justifies the use of data oriented models. Here, we propose some specific open-ended issues in environmental monitoring (in particular, in air pollution monitoring and control), which require a modern approach for their assessment: identification and diagnosis of a given situation based on the processing of time and spatially varying data; forecasting of regular events (short time); forecasting of rare end extreme events (mid and long time); evaluation of a solution; inverse modeling. The present paper illustrates practical applications in which intelligent systems have been deliberately introduced in the processing chains to solve problems that appears to be "unsolvable" by making use of more traditional statistical and model-based approaches. The paper will hopefully stimulate a wide interest on environmental data analysis and monitoring within the framework of supervised and unsupervised learning.

**Keywords**: Air Pollution, Fuzzy Ellipsoidal Systems, Neuro-Fuzzy Systems, Complexity, non-linear Systems.

## 1. Introduction

Waiting for the great bridge, to reach Sicily from Europe by cars and trucks there is the only possibility to take the ferry boat service from Villa San Giovanni to Messina and vice versa. These two cities are located on the two sides of the homonymous strait. As a direct consequence, the cities are interested by a huge vehicular traffic that is mainly responsible for the poor quality of the air because of urban pollution, particularly in the absence of wind, fortunately not uncommon in the area. In addition, the fact that the crossing traffic is, for the most part, heavy traffic and, consequently, discharges in the air organic and inorganic compounds (sometimes not completely burnt) can strongly affect the health of citizens living along the roads interested by the traffic. Moreover, also ferries are responsible for a periodic injection of pollutants in selected areas of Villa san Giovanni. Typically, it can be noted a seasonal dependency of the pollution level: indeed, very high values of the pollutants are achieved very in the summertime when the strong raising of the temperature encourages the outbreak of photochemical phenomena to compromise the quality of the air. In that case, also the ozone content has a synergistically dangerous impact on health.

Environmental data are typically very complex to model due to the underlying correlation among several variables of different types, which yield an intricate mesh of relationships. It is ordinary to have multivariate dependency with non-linear behavior. Furthermore, typical variables do not meet the very common Gaussian assumption. To simplify statistical complexity, many attempts to model these interactions fix some of the micro-climatic conditions or ignore their effects, thereby making any results dependant on a tight set of conditions [1]. These methods always involve expert knowledge and are not suitable for automatic systems. Standard statistical techniques fail to adequately model complex non-linear phenomena. In contrast, NN's are widely used because they showed ability to model non-linear data and their non-reliance on previously assumed equations. Predictions based on NN's depend on extensive collections of data, which are used to learn the model. Unfortunately, the knowledge that a NN gain about a problem is encoded in the connection weights whose strength is fixed after learning. For political reasons, it is impossible to make decisions based on black box models, particularly for taking unpopular actions, like stopping car traffic in a town. In addition, in the case of a transit-town, a local decision may have a strong impact on a national level. It is thus essential to derive a set of rules that can support the decisions. On the other hand, the prediction of a pollution episode may benefit of knowledge on the typical meteorological, seasonal and traffic conditions of a particular area. Also, extracting meaning from a trained NN may offer a novel insight on not observed relationships [2].

To investigate on the effect of traffic on air quality, various experimental campaigns of measurements have been carried out. The structure of the available environmental data shows that both spatial and time dependence are of interest; in addition, the data are very complex because of non-linearity, non-stationarity, measurement noise, missing data and the need of periodically test the instrumentation of the monitoring station. There is, thus, a need to use smart signal processing of the observed data, possibly incorporating a priori knowledge on the model. The present work reports a study of estimation and short time prediction of atmospheric pollutants carried out by making large use of fuzzy systems and neural networks tools. In particular, estimation and prediction of hydrocarbons in the air are carried out, because they are very dangerous for the health.

The methodologies of soft computing (mainly mixing fuzzy inference systems and neural networks) are naturally suited to help in forecasting the local behaviour of complex systems starting from the available set of measurements. The knowledge expressed by the database is invariably uncertain and inaccurate. The possibility of solving the problem is related to the hope that the events affecting the time series are correlated in some way to be determined. For example, in the case of car queues, an upward trend of the pollution parameter time series is observed, while the night-day rhythm introduces a cyclic component in the same time series. Neural networks can be trained on the observed data with the aim of building a simulated model of the system, at the least locally.

Fuzzy inference systems can be used to generate banks of linguistic rules, whose IF...THEN structure could be exploited to support local authorities decisions about constraining the vehicular traffic or redirect it. In the past, Neural Networks (NNs) have been already exploited in order to estimate and predict pollutant levels in the airs, and to forecast environmental time-series [1][3][4], in terms of multilayer perceptron models where the input vector is composed of past samples of the time series and the training algorithm is either the static backpropagation or the backpropagation through time [5].

This paper is organized as follows: the exploited experimental database is firstly presented in the next Section. Then, a qualitative analysis of the database and a fuzzy neural approach to estimate and predict HC levels in the urban air are proposed. Finally, the best results achieved are presented and some conclusions are drawn.

Some of the work here reported has been carried out in the framework of the SMAURN Project (Monitoring System of Urban Environment through Neural Networks), an European Community funded project on the environmental monitoring of Villa San Giovanni. The goal of the SMAURN project was to design a network of mobile terminals for monitoring some parameters and predicting in real time their evolution in a urban contest. The gathered information can be used to the system manager to advise local authorities in adopting action plans and administering environmental protection.

## 2. The experimental database

Atmospheric pollution arises from the adverse effects on the environment of a variety of substances contaminants emitted into the atmosphere by natural and man-made processes [6]. The physical and chemical processes undertaken during an air pollution episode are, to a large extent, understood. The effects of individual environmental factors and man-made pollutant on human life are also known, but this is mainly through single factor. What is largely unknown is the effect of differing levels of multiple factors, such as humidity, temperature, global radiation, wind speed and directions, light, ground-level ozone-concentration [3].

The temporal evolution of each pollutant depends on external variables in a complex fashion, and the kind of dependency can be rather different also in seemingly near spatial locations. This fact implies the need of designing and/or optimizing the location of the measuring station(s) in order to derive significant conclusions about the air quality. The problem above introduced is too complex to be solved by using analytical models. According to previous works on the subject [3], we propose to build a model of the interactions on a local basis of the kind NNs are able to extract from input-output data experimentally observed. This can be done if any information is available from databases obtained from experimental campaigns.

In this paper, we use an environmental database that refers to the Southern Italy area of the Messina Strait. Figure 1 shows a remotely sensed image of the area. The analyzed database has been obtained through an experimental campaign carried out by using a mobile monitoring station equipped with both sensors for measuring chemical quantities and atmospheric parameters. The mobile station was moved in different measuring points, always along the only access road to the ferry boats: each time period lasts about four weeks.

The time series used includes observation in the period from 27th July to 25th August 1998, along the road to ferries, in Villa San Giovanni. This period has been selected because of the intense vehicular traffic. The traffic is mainly formed by trucks. The database of measurements also includes vehicular traffic data: each variable is hourly sampled. For the period analyzed the database is divided in two parts, respectively used for training and testing. The training database includes 408 time samples, corresponding to 17 days of readings: the time period spanned was from August 2nd to August 19th, 1998.

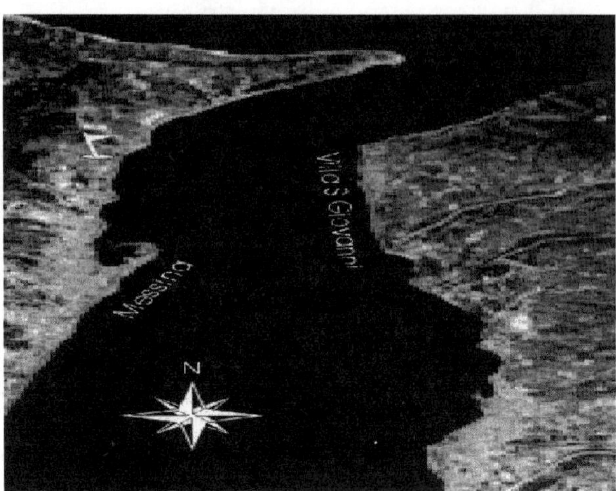


Figure 1. Villa San Giovanni on the Messina Strait: the area under investigation.

The database has the following structure: the data are organized in a matrix of 408 rows and 14 columns. The 14 variables are the time of observation, the following pollutant parameters: $SO_2$, CO, $O_3$, NO, $NO_2$, HC, PTS, PM10 (the last two variables represent, respectively, the total suspended solid particulate and the $<10\mu m$ diameter particulate, mostly known as strongly affecting public health), some the atmospheric quantities, wind speed, wind direction, temperature, atmospheric pressure and the traffic data, i.e., the number of vehicles per hour in the two opposite directions, from and to Messina. The following Table 1 reports the statistical parameters of the database (chemical pollutants are measured in $\mu g/m^3$). High-order statistics are also indicated but not exploited in the present analysis, while standard deviation can be used as a reference value for prediction performance.

Table 1 – Statistical quantities computed on the database

| | Max | Min | Average | Standard deviation | Skewness | Kurtosis |
|---|---|---|---|---|---|---|
| $SO_2$ | 109.407 | 0 | 7.4884 | 9.1087 | 4.3404 | 36.3145 |
| CO | 10.546 | 0 | 1.0067 | 0.913 | 2.6315 | 21.2147 |
| $O_3$ | 135.692 | 0 | 20.9619 | 24.041 | 1.3117 | 4.4806 |
| NO | 279.181 | 0 | 15.5604 | 25.7522 | 4.809 | 36.6351 |
| $NO_2$ | 168.718 | 1.722 | 18.2401 | 22.0091 | 2.8278 | 12.1922 |
| HC | 960.256 | 0 | 308.3483 | 159.6636 | 0.9048 | 4.039 |
| PTS | 396 | 0 | 59.7724 | 32.9931 | 3.538 | 30.3285 |
| PM10 | 497.558 | 0 | 51.2416 | 61.462 | 3.4567 | 18.6915 |
| Wind speed [m/sec] | 6.197 | 0.061 | 1.406 | 0.9008 | 1.3581 | 5.5984 |
| Wind direction [degree] | 354.945 | 91.86 | 221.2189 | 75.0959 | 0.1032 | 1.1855 |
| Temperature [°C] | 37.949 | 23.16 | 28.1623 | 2.1188 | 0.670 | 3.8443 |
| Pressure[mbar] | 1015.8 | 1004 | 1010.9 | 2.1322 | -0.6687 | 3.4464 |

## 3. A qualitative analysis of the database

Each column of the measurement database corresponds to a recorded time series of the pollutants: the entries correspond to hourly measurements, carried out in different ways by the various types of chemical/meteorological sensors. We have thus 24 samples for each day. We deal with two evident problems: missing data and outliers.

The outliers are mainly due to the incorrect working of the instruments or to incorrect methodology for collection and analysis. Generally, the maximum and minimum values can be considered as outliers and they must be examined with care, because they can cause deformation in the calibration of the predict model. Many methodologies can be used to individuate the presence of outliers; for example, control cards can be exploited for this purpose (Shewhard's card, Cusum's card) which are based on the check of the measurements consistency between past and present [7].

The characteristics of data about the air quality (correlation, regularity, non-constant standard deviation, non-normal distribution) invalidate the exploitation of control cards; then a lot of modification are necessary. The presence of an outlier can also constitute an indication of unusual and unexpected events. If we are able to give possible explications, we can run two ways: we can eliminate the suspect datum making a missing datum or we keep the outlier and exploit more effective statistical methods.

In our case, we have decided to keep the outliers because the tools we use may become robust with suitable implementation.

Missing data are mainly due to failures of the measurement instruments. The presence of missing data can invalidate the statistical analysis introducing a systematic component of errors about the estimation of parameters in the prediction model. In addition, if we estimate the parameters of the model by exploiting the observed data without taking into account the presence of missing data, the obtained estimations could be unreliable because many information concerning the missing data would disappear.

The in force regulation do not foresee any indication for this purpose, consequently, we have chosen to replace missing data by the averages of the measurements at the same time in the previous days. By using the correlations matrix (see Table 2), we can fit the correlation between the distribution of couples of pollutant variables and between traffic data, atmospheric quantities and pollution.

The relevant linear correlations between pairs of variables are rare. Figure 3 shows a typical correlation between CO and HC. In this study, we exploit the correlations between parameters in order to calibrate the model for pollutant estimations: the performance of the model typically improves if correlated variables are exploited.

Our attention is mainly addressed to HC: this is because, in the time period under study, it shows above threshold of attention values in 61% of the hourly measurements and above threshold of alarm in about the 10% of the same data (Figure 2). Also, it is known that HC diffusion is strongly related to the amount of vehicles along the road. Very similar conditions we experience for NO, $NO_2$ and CO.

Table 2 – Correlation matrix of the database

| | $SO_2$ | CO | $O_3$ | NO | $NO_2$ | HC | PTS | PM10 | Wind speed | Wind direct. | Temp. | Press. | Traffic Me-Villa | Traffic Villa-Me |
|---|---|---|---|---|---|---|---|---|---|---|---|---|---|---|
| $SO_2$ | 1.0000 | | | | | | | | | | | | | |
| CO | 0.2889 | 1.000 | | | | | | | | | | | | |
| $O_3$ | 0.1560 | -0.1509 | 1.000 | | | | | | | | | | | |
| NO | 0.1302 | 0.3050 | -0.0899 | 1.000 | | | | | | | | | | |
| NN $NO_2$ | 0.1282 | 0.3391 | -0.1167 | 0.9181 | 1.000 | | | | | | | | | |
| HC | 0.2062 | 0.7241 | -0.1071 | 0.2949 | 0.3260 | 1.000 | | | | | | | | |
| PTS | -0.0082 | -0.0018 | -0.0653 | 0.0848 | 0.0585 | -0.0531 | 1.000 | | | | | | | |
| PM10 | 0.0486 | 0.0848 | -0.0468 | 0.2321 | 0.1906 | 0.1463 | 0.0818 | 1.000 | | | | | | |
| Wind speed | -0.0763 | -0.2533 | 0.1383 | -0.1795 | -0.1459 | -0.3195 | -0.0759 | -0.0586 | 1.000 | | | | | |
| Wind direction | 0.1309 | 0.2467 | 0.1272 | 0.1660 | 0.2018 | 0.1767 | 0.0626 | 0.1124 | 0.0749 | 1.000 | | | | |
| Temper. | 0.1400 | 0.2570 | 0.0345 | 0.2832 | 0.2254 | 0.3122 | 0.1099 | 0.3830 | -0.1947 | 0.2003 | 1.000 | | | |
| Pressure | -0.0743 | 0.1653 | -0.1128 | 0.1910 | 0.1737 | 0.0082 | -0.1601 | 0.0703 | 0.0415 | 0.0055 | 0.2940 | 1.000 | | |
| Traffic Me-Villa | -0.0041 | 0.3522 | 0.0223 | 0.2032 | 0.2116 | 0.4811 | -0.0478 | 0.2883 | -0.05278 | 0.2197 | 0.4861 | 0.0776 | 1.000 | |
| Traffic Villa-Me | 0.0434 | 0.4662 | 0.0785 | 0.2361 | 0.2252 | 0.6264 | 0.2252 | 0.1902 | -0.0848 | 0.1146 | 0.3860 | 0.0535 | 0.8220 | 1.000 |

Table 3 – Attention and alarm threshold levels for different pollutant parameters, times of above threshold observations in the database

| | $SO_2$ | CO | $O_3$ | NO | $NO_2$ | HC | PTS | PM10 |
|---|---|---|---|---|---|---|---|---|
| Attention threshold | 125 | 20 | 200 | 1000 | 150 | 200 | 150 | 150 |
| Alarm threshold | 250 | 30 | 400 | 1000 | 250 | 500 | 200 | 200 |
| #attention threshold | 0 | 0 | 0 | 0 | 0 | 286 | 7 | 25 |
| #alarm threshold | 0 | 0 | 0 | 0 | 0 | 49 | 5 | 13 |
| % above attention threshold | 0 | 0 | 0 | 0 | 0 | 70.1% | 1.72% | 6.13% |
| % above alarm threshold | 0 | 0 | 0 | 0 | 0 | 12.01% | 1.23% | 3.19% |

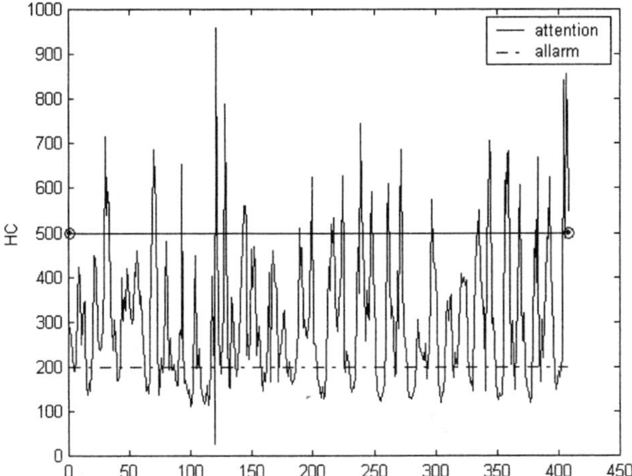


Figure 2. The time series of the HC: the continuous line indicates the alarm threshold level, and the dashed line indicates the attention threshold level.

The preliminary analysis carried out on the available database study clearly showed the characteristics of regularity (cyclicality) of the atmospheric parameters: as rather expected, for example, the temperature attains the maximum values during mid-day, while during night and early morning it reaches its minimum values (Figure 8). The same can be said for the atmospheric pressure, whose regularity is strictly related to the period of the year the observations have been taken (summertime). Since the weather also affects the presence and stabilization of the pollutants in the air, it is expected that the derived model is not able to generalize in other periods of the year. That means, a set of non-linear models should be designed to be used in different background weather conditions.

A strong reduction of the pressure is to be related to summer thunderstorms (see Figure 9).

Although the knowledge of atmospheric temperature and pressure is not sufficient to directly derive the pollutant level, the regularity of the observations is of help to decompose the modeling task in sub-models. This can be done by using frequency (Fourier) analysis and/or by deriving some appropriate fuzzy rules. The presence of Ozone ($O_3$) in the air is also strongly related to the weather.

By looking at the histograms of the main pollutants, it seems that a log-normal distribution well describe their distribution (Figure 4). As far as the HC is concerned, the distribution tends to show most high frequencies corresponding to very high pollution levels.

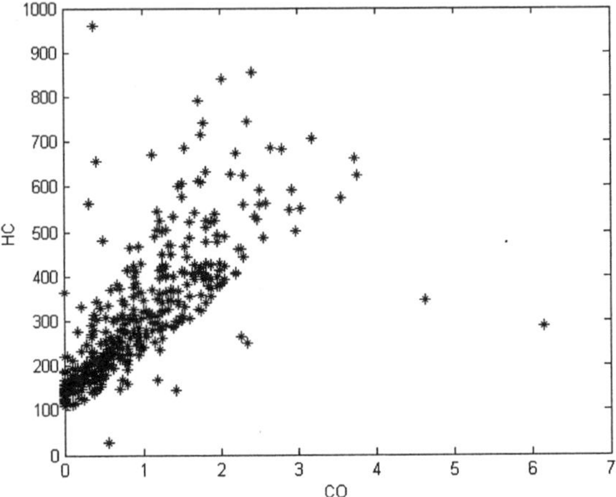

Figure 3. Scatter plot showing the correlation between two typical pollution variables (HC and CO). The kind of correlation implies the possibility of predicting one parameter from the knowledge of the other one.

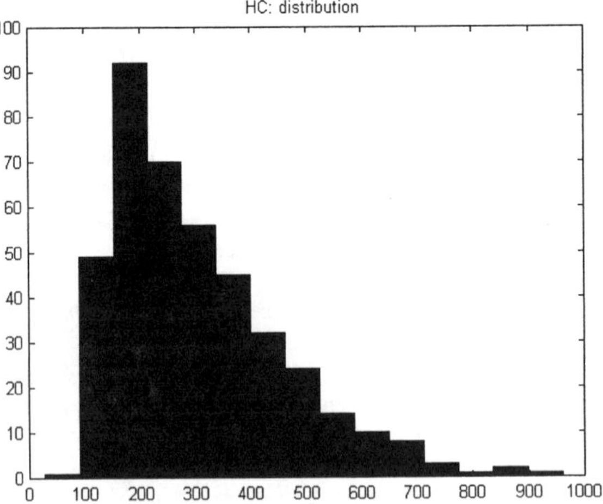

Figure 4. Histogram of the HC variable in the database.

According to the experimental evidence, the conditions of high level of pollution occur when the vehicular traffic is high, the wind is medium, and the temperature falls into the range 26-32°C (Figure 6). As a reference value, the hourly number of cars/trucks crossing the detection section in both directions is varying from about 200 to 1000 in a typical acquisition day. The trucks correspond to roughly the 20% of the total number, also varying from a night peak of about 35% to a minimum of 8%. The typical averaged velocity is around 30 km/h.

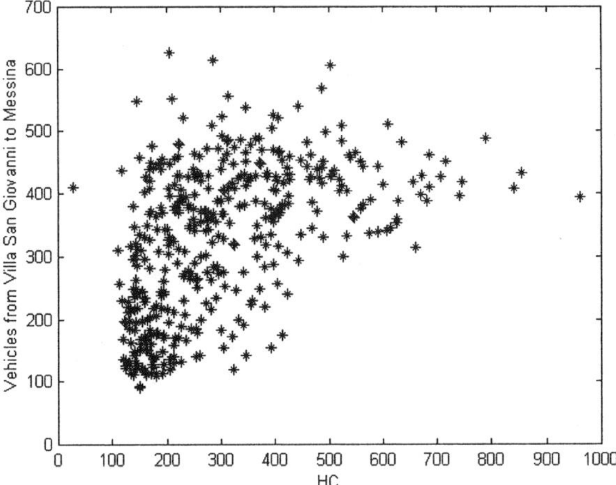

Figure 5. Scatter plot of the traffic data versus the HC level in the database.

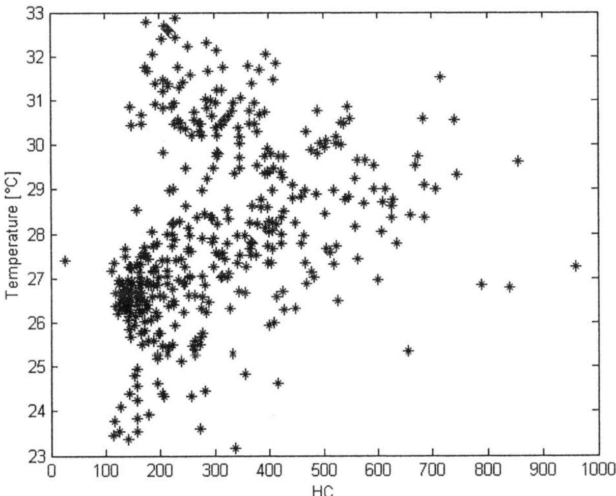

Figure 6. Scatter plot of T (°C) versus HC ($\mu g/m^3$).

Another relevant process to the phenomena of atmospheric pollution and of the dispersion of pollutants in the air results is the extension of thermal inversion. However, from the mere analysis of the database and the general information obtained by experts, we are not able to consider the impact of this aspect on the dynamic evolution of pollution. However, we can exploit other well-known concept to simplify in some way the investigation.

For example, in the case of the ozone, $O_3$, we know that it represents a secondary pollutant related to the photochemical reaction between NO, $NO_2$ and HC: the products are referred to as "photochemical pollution", since their presence is favored by sunlight. The presence of wind can also deeply affect their concentrations. In Figure 7, reporting the correlation values between the hourly concentrations of $O_3$ and $NO_2$, it is shown that there are two different possible mechanisms governing the $NO_2$ and $O_3$ formation-removal process, depending on the meteorological condition and on the location of the sensor [8]. This kind of information is used in the fuzzy approach that is proposed in the paper.

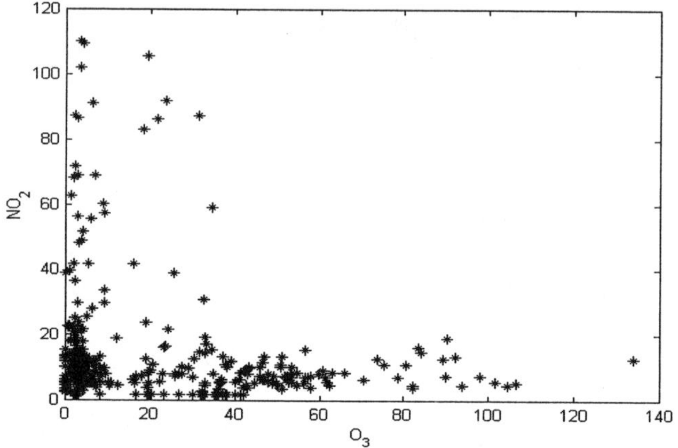

Figure 7. $NO_2 - O_3$ dependency in the database.

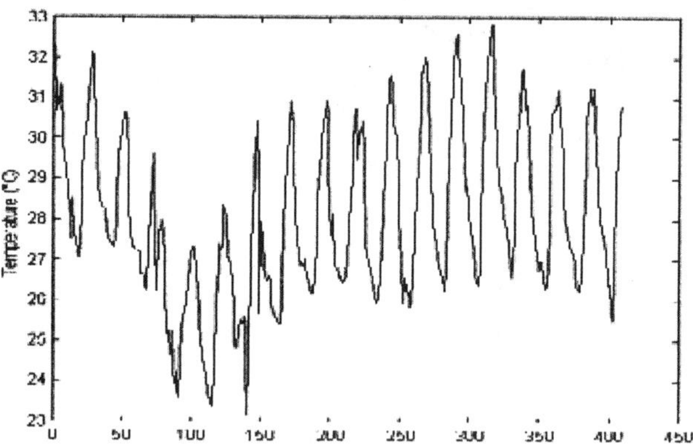

Figure 8. Temperature in the period under study: the regularity of data allows us to predict the pollutant levels (on the abscissa axis, the progressive number of sample is reported).

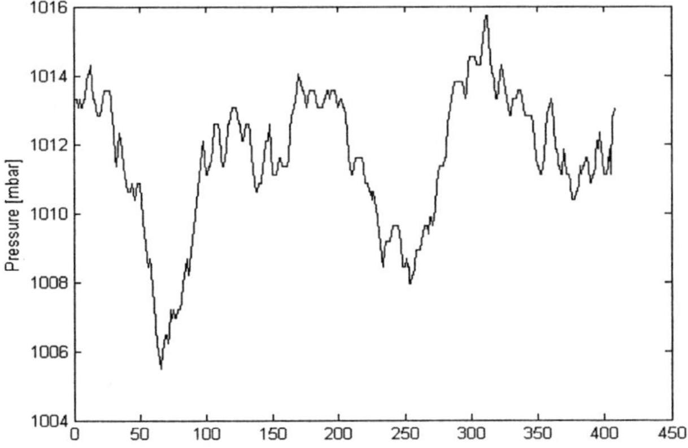

Figure 9. Atmospheric pressure (*mbar*) versus the number of sample (each sample corresponds to an hourly averaged measurement).

## 4. The Neural Network (NN) forecasting approach

The problem to estimate the polluting level can be formulated as the search of a suitable mapping between the set of available measurements and the selected set of pollutant parameters [9]. Typically, we know the present level of each pollutant variable and we have access to the previous hourly measurements of the variable. We should thus forecast the temporal evolution of the pollutant concentration by means of a time scheme of NNs. Most NN approaches to the problem of forecasting use a multilayer perceptron (MLP) trained using some variants of the backpropagation algorithm. Given the time series $x(1), ..., x(t)$, it is required to predict the value $x(t+k)$: with $k=1$, we have the one-sample ahead prediction [10].

The inputs to the MLP are commonly chosen as the previous $h$ values $x(t-h+1), ..., x(t)$ and the output will be the forecasted value. The NN is trained and tested on sufficiently large training and testing datasets, that are extracted from the time history. In addition to previous time series values, one can utilize as inputs the values or forecasts of other time series (external variables) that have a correlated relationship with the series to be forecasted. For our HC prediction problem, the atmospheric parameters, the traffic data and the CO concentration can conveniently be used. We have also studied the case in which such external variables are not available or not reliable: it seems that this kind of problem is not solvable, at least in terms of the requested accuracy. For this reason, we shall use external variables as additional inputs to the model, also in the case of fuzzy neural approaches. It will also be discussed in the next Section how the insight yielded by an off-line statistical analysis carried out on the database can be exploited to improve basic NN prediction performance.

## 5. Incorporating knowledge in the model

The qualitative observations carried out in Section 3 can be exploited in order to build an "expert" model of the forecasting problem. This can be done in principle by designing a hybrid network scheme in which the available a priori knowledge could be included in the form of syntactic rules. Actually, the available knowledge cannot be implemented in terms of crisp rules because of the uncertainty underlying the complex model (for example, the different dependency between pollutant variables in various wind direction conditions). The most suitable approach seems the adaptive network schemes based on fuzzy inference, that allows us to incorporate knowledge in terms of fuzzy rules. This kind of models are able to transform the prediction problem in a function approximation one, like NNs.

### *5.1 Pollution forecasting by means of Fuzzy Inference Systems (FIS)*

Precise quantitative analysis is suitable when high degree of precision is required. However, the task of deciding something about a pollution event by processing a set of measurements is by its very nature an uncertain one. The measurements are corrupted by noise, and, more importantly, the interaction among variables is difficult to interpret within a quantitatively perfect model. The same origin of the phenomena under study gives automatically rise to uncertainties or ambiguities about its evolution and the importance that play different conditions (e.g., meteorological parameters, vehicle traffic, …) on the estimation. Although the idealization of causes-effects relationships is useful aiming to the analytical formulation of the problems, which also allows to benchmark the algorithms, is far from being a realistic hypothesis. On the other hand, the ability to summarize information and to approximately describe a process plays an essential role in the characterization of complex phenomena.

Fuzzy Inference Systems (FISs) appears to be very good tools as they hold the nonlinear universal approximation property, and they are able to handle experimental data as well as a priori knowledge on the unknown solution, which is expressed by inferential linguistic rules in the form IF-THEN whose antecedents and consequents utilize fuzzy sets instead of crisp numbers. The inputs of the procedure are interpreted as fuzzy variables. Each fuzzy value carried out by a fuzzy variable is characterized by a fuzzy membership function (FMF). In turn, each FMF is expressing a membership measure to each of the linguistic properties. FMF are usually scaled between zero and unity, and they overlap. To improve the flexibility of our model, we used Gaussian FMFs throughout this work.

By summarizing, a FIS is designed according to the following procedure:
  1) fuzzification of the input-output variables;
  2) fuzzy inferencing through the bank of fuzzy rules;
  3) defuzzification of the fuzzy output variables to yield numerical values.

There are two important problems to be solved for using the FIS approach, namely, the model reduction, i.e., the choice of the relevant inputs among the available ones, and the improvement of model accuracy of the "naive" model that can be derived by a direct rule extraction from the database and by a fine tuning of the model parameters carried out by a learning approach.

### *5.2 Fuzzy curves and surfaces*

The FIS described in the previous section yields a model that can be useful as a first guess model of the systems. Such a FIS makes use of the available input-output pairs without exploiting the concept of learning: as a result, the estimation accuracy is invariably not quite good. The estimation performance can easily be improved by using an algorithm of automatic extraction of FIS from numerical data [11]. The MATLAB® GENFIS Clustering System [12] is here used: this can give us the possibility to build a Sugeno-type first-order model largely improvable by directly introducing learning by means of the MATLAB® ANFIS [11][13] code.

A network FIS scheme facilitates the computation of the gradient vector for computing the parameter corrections.

Once the gradient vector is obtained, a number of optimization routines can be applied to reduce the error.

A typical fuzzy rule of the FIS can be expressed as follows:

*IF the traffic is high AND the temperature is high AND the wind speed is small*
*THEN the pollutant concentration is high.*

The bank of rules could be able to heuristically describe the local model of interactions and, consequently, to give some hints about the possibility of predicting the pollutant concentration evolution.

This kind of representation of knowledge could also help to explain unpopular decisions that should be taken in order to avoid the increase of pollutants above alarm threshold limits.

As far as the modeling problem is concerned, the construction of simple bank of fuzzy rules is related to the possibility to cover the regions of the input-output space where data samples are present by means of overlapping patches: each patch represents a fuzzy rules. Unfortunately, the need of completely covering by patches the samples implies the growth of the fuzzy rules numbers beyond any practically interesting level. Thus, patches could not cover many samples. Since in a multidimensional space it is impossible to decide which regions are mostly important from the information content viewpoint, we decided to operate by exploiting a sort of weighted-average technique based on the concept of fuzzy curves and surfaces [14]. The same technique can be used to face the other problem of the FIS approach, namely, the ranking and selection of the best inputs for achieving model reduction.

In order to show how a fuzzy curve works, let us consider a Multiple-Input Single-Output (MISO) system. We assume that m training data are available, thus $x_{ik}$ (k=1,..., $m$) are the i[th] co-ordinate of each of the $m$ training patterns. The fuzzy curve is defined as follows:

$$c_i(x_i) = \Sigma k\, F_{ik}(x_i)\, y_k\, /\, \Sigma k\, F_{ik}(x_i)\,, \quad k=1, ..., m,$$

where $F_{ik}(x_i) = \exp[-((x_{ik} - x_i)/s)^2]$ is a Gaussian function (other different local functions could advantageously be introduced). Each training pattern is basically used as a fuzzy rule (with single antecedent) of the type:

$$IF\ x_i\ is\ F_{ik}(xi)\ THEN\ y\ is\ y_k.$$

The importance of the input in affecting the estimation of the output is determined on the basis of a figure of merit defined as the range of the fuzzy curve, $\rho = (c_{i\ max} - c_{i\ min})$. The determination of the $\rho$'s allows us to rank the input variable in order of importance and, thus, to properly reduce the complexity of the final system. Figure 10 shows a typical fuzzy curve describing the "weighted" relationship between two variables of the database under study: the fuzzy rules of main interest to describe the relationship are typically located where the system changes its behavior. In particular, we place the fuzzy rules on zero derivative points of the curve. A natural extension of the fuzzy curves is the fuzzy surfaces whose structure is as follows:

$$c_i(x_i, y_j) = \Sigma k\, F_{ik}(x_i) \cdot F_{ik}(y_j) \cdot y_k\, /\, \Sigma k\, F_{ik}(x_i) \cdot F_{ik}(y_j)\,, \quad k=1, ..., m,$$

The corresponding fuzzy rules are of the type with double antecedents whose connective is "and". In this way, the transformation occurs into a space where the application of fuzzy patches (and thus the determination of fuzzy rules) becomes most easy. A typical fuzzy surface is reported in Figure 11 where each sharp blob represents a fuzzy rule. Of course the fuzzy curves can be derived by suitable projections of the fuzzy surfaces reported in Figure 11. To derive fuzzy rules with double antecedent from fuzzy surfaces reduces the risk of combinatorial explosion of the rules' number. On the other hand, the rules extracted by fuzzy surfaces include the rules extracted by fuzzy curves.

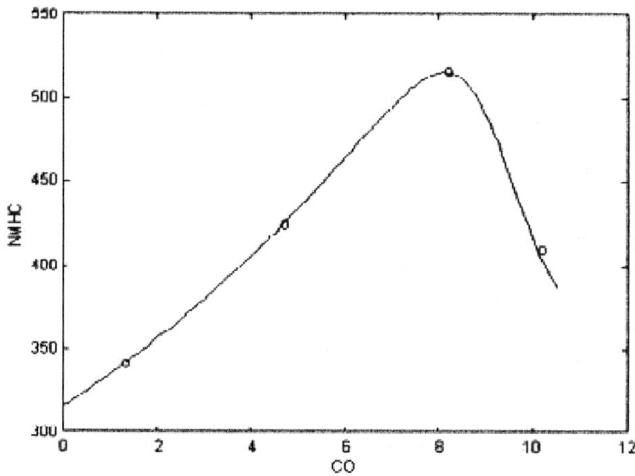

Figure 10. Correlation between two pollutant parameters extracted by fuzzy curves. By using this approach, the extraction of simple fuzzy rules is reduced to the research of maximum and minimum points.

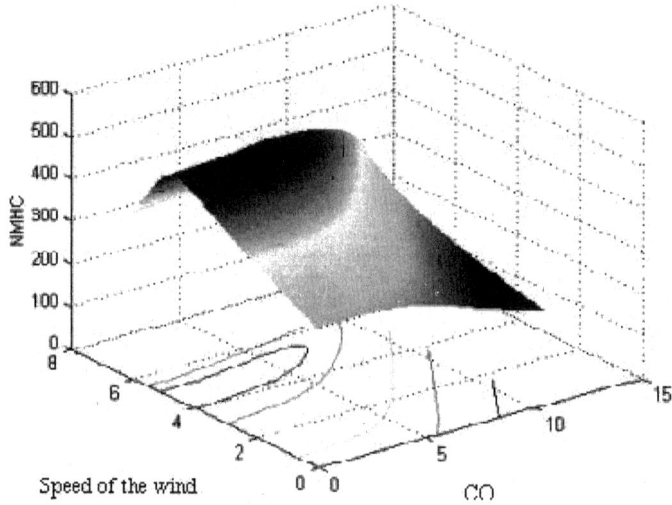


Figure 11. Fuzzy surface extracted from the database: the transformed space of three variables.

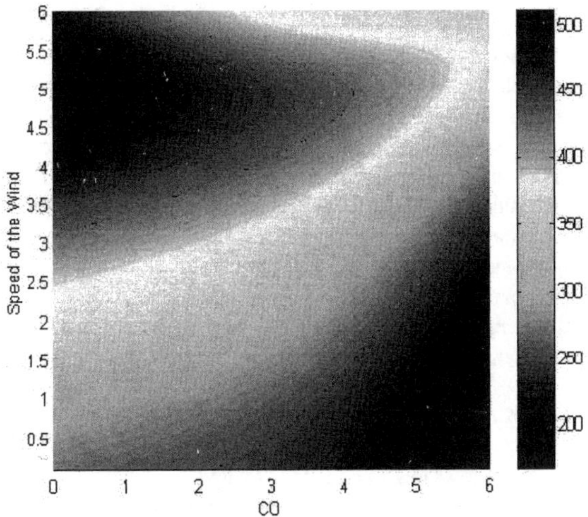

Figure 12. A different 2D view of the surface plotted in the Figure 11: by visual inspection, it is possible to extract a bank of fuzzy rules (traditional FIS).

A different perspective of the fuzzy surface reported in Figure 11 is shown in Figure 12: in this case, the extraction of fuzzy rules becomes straightforward. The fitness of the rules to the underlying model can be improved by means of fuzzy ellipsoidal patches, as will be shown in the next Sections.

### 5.3 The fuzzy modeling system

The following scheme reports the most important characteristics of the fuzzy neural system we propose to solve the modeling problem:

Output: HC (pollutant to be monitored)
#Rules: 66
Type of Rules: Modus Ponens
Typical Rule:
*If CO is VERY LOW and SPEED OF THE WIND is VERY HIGH*
*Then HC is VERY HIGH*

Fuzzy Partition of the Input-Output Spaces:

| | |
|---|---|
| Intense Red | = VERY HIGH |
| Red | = HIGH |
| Green | =MEDIUM |
| Intense Green | = OVER MEDIUM |
| Blue | = LOW |
| Intense Blue | = VERY LOW |

-Rules with double antecedents

- The whole of the rules describes the behaviour of the system

The designed system is only able to give us the order of magnitude of the pollutant. In fact, in spite of the remarkable number of rules (#44 fuzzy rules), each rule allows the use of only two variables as antecedents. In addition, we cannot use any patches in order to cover the transition of colors: this leaves large regions of the input-output space uncovered. To reduce the cardinality of the system by compacting the rules, we have considered the following steps:

<u>Step 1</u>: we have considered the rules, which present the same label of the output;

<u>Step 2</u>: if an input is included in a rule with the same label, this label is considered one time;

<u>Step 3</u>: if an input is included in a rule with different labels, the labels are all considered by making use of the "OR" connective.

Nevertheless, the expert's knowledge allows us to write ad hoc rules thus increasing the performance of the systems.

## 6. Estimation of HC level in the air: the fuzzy neural approach (best results)

By making use of the tools described in the previous sections, we build a FIS that is able to yield a significant prediction of the pollutant concentration. The accuracy of the estimation can be considered good, at least in order to correctly interpret the trend of the time series. In this section, we shall report the results achieved by making use of the Matlab® GENFIS and ANFIS toolboxes. The determination of the relevant subset of inputs has been carried out by using the fuzzy curve technique. Based on the computation of the $\rho$ parameter, in the case of the HC estimation, the following parameters have been selected:

1) CO
2) Number of Vehicles from Messina to Villa San Giovanni
3) $NO_2$
4) NO
5) Number of Vehicles from Villa San Giovanni to Messina
6) Wind speed
7) $SO_2$
8) Temperature
9) PM10
10) Wind direction
11) PTS
12) $O_3$
13) Atmospheric Pressure.

As rather expected, the best inputs for predicting HC are the CO (see also Figure 3) and the number of vehicles. Indeed, the traffic is mainly responsible of the pollution. The wind velocity is taken into account because it has an impact on the diffusion of the pollution variable. The following Table 4 reports the best results achieved in the estimation of the HC variable by using the relevant inputs selected with the fuzzy curve the technique of the automatic extraction of the bank of rules FIS's from the available database.

In the reported results, we present the testing error, since a relevant correlation between the training error and the testing error has been noted. In other words, choosing a set of inputs and a topology of network that gives a low training error will almost surely result in a low testing error.

Table 4. Estimation of HC levels by means of a Fuzzy Neural Systems: the 3rd column reports the inputs of the non-linear model, the related inputs have been selected by means of the fuzzy curve approach; the model's complexity is expressed in terms of the number of rules, and the estimation error on the test database is expressed in terms of RMS error (best results)

| #case | Pollutant | Inputs | Output | #Rules | Error |
|---|---|---|---|---|---|
| 1 | HC | CO, Vehicles from Messina to Villa San Giovanni, $NO_2$, NO, Vehicles from Villa San Giovanni to Messina, Wind speed (no #hour enclosed) | HC | 70 | 8.80 |
| 1' | HC | CO, Vehicles from Messina to Villa San Giovanni, $NO_2$, NO, Vehicles from Villa San Giovanni to Messina, Wind speed (no #hour enclosed) | HC | 79 | 10.76 |
| 2 | HC | All variables (no #hour enclosed) | HC | 32 | Not relevant |
| 3 | HC | All variables (#hour enclosed) | HC | 28 | Not relevant |

Some other input combinations have been investigated: several cases gave higher errors, mostly in the case of either insufficient number of inputs or inadequate "quality" of the inputs. The most relevant comment should be given on the apparent usefulness of the information related to the period number that should, in principle, be of help in including cyclicality or periodicity in the data. This kind of comment also helps understanding why we do not use any Fourier coefficient as a "feature" input of the procedure. Also, the use of a large number of inputs is impractical because of the consequent improvement of model complexity and, thus, of throughput time in real time applications. Finally, the generalization abilities of the network strongly reduce in the case of to-many inputs used. The following figures show the achieved results.

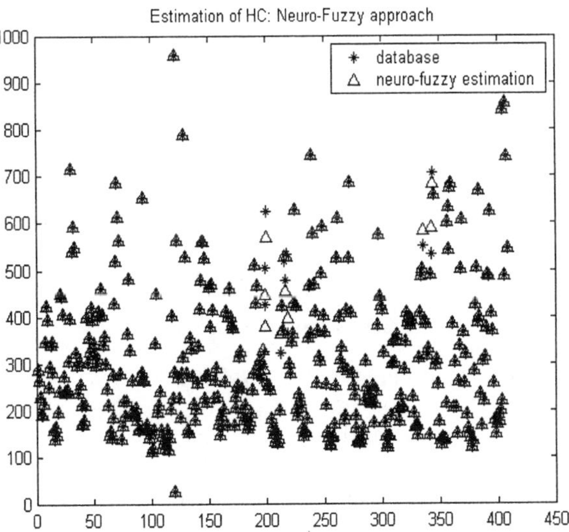


Figure 13. Target and estimated values of HC for case #1.

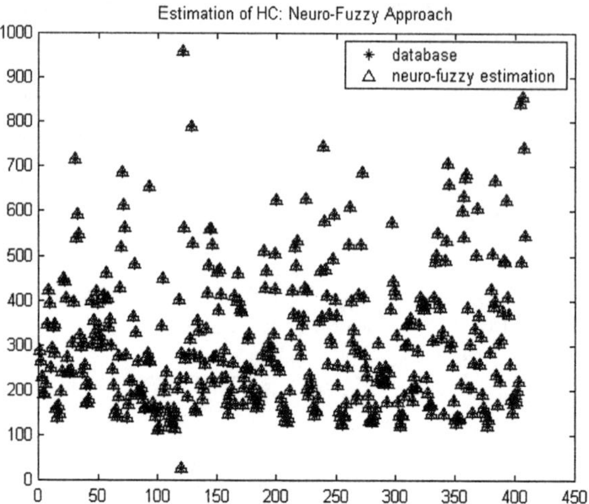


Figure 14. Target and estimated values of HC for case #2.

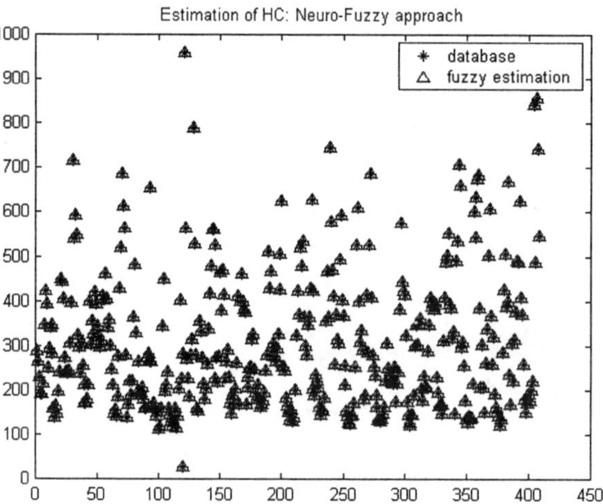


Figure 15. Target and estimated values of HC for case #3.

## 7. The Fuzzy Neural Model Predictor: Best Results

In Tables 5 and 6, we report the best results achieved by means of the fuzzy neural systems described in the previous Sections concerning the prediction of HC levels. They refer to the short-term prediction horizon (one-hour ahead) and to the (multistep) four-hour ahead longer horizon forecast. In the present application, the one-hour ahead prediction time is considered sufficient to activate the decision system. The results achieved by means of the "naï ve" approach are of interest because of the very limited number of rules needed to correctly predict the pollutant concentration evolution. Obviously, the forecasting of the pollutant levels by means of the fuzzy neural systems produces the best results because the bank of fuzzy rules is automatically extracted from the database by using the GENFIS approach and the model parameters are tuned by means of a learning procedure (ANFIS).

Table 5. Prediction of the HC Levels: "naï ve" FIS's based on the bank of rules extracted by direct inspection of the database (best results in terms of RMS error). Five previous samples of the HC variable have been used.

| #case | Pollutant | Inputs | Output | #Rules | Error |
|---|---|---|---|---|---|
| 1 | HC | $HC(t-1), ..., HC(t-5)$ ; $CO(t-1)$ ; Number of Vehicles $(t-1)$ | HC (one hour later) | 7 | 12.5 |
| 2 | HC | $HC(t-1), ..., HC(t-5)$ ; $CO(t-1)$ ; Number of Vehicles $(t-1)$ | HC (one hour later) | 9 | 12.2 |
| 3 | HC | $HC(t-1), ..., HC(t-5)$ ; $CO(t-1), ...,$ $CO(t-5)$; Temperature; $NO_2$; NO; TSP | HC (one hour later) | 13 | 35.3 |
| 4 | HC | $HC(t-1), ..., HC(t-5)$ ; $CO(t-5)$ ; $HC(t-24)$ | HC (one hour later) | 8 | 71.4 |
| 5 | HC | $HC(t-1), ..., HC(t-5)$ ; $CO(t-5)$ ; $HC(t-24)$ ; Vehicles $(t-24)$ | HC (four hours later) | 17 | 74.9 |

Table 6. Prediction of HC Levels: Fuzzy Neural Systems based on the use of both GENFIS and ANFIS approach (best results in terms of RMS error)

| #case | Pollutant | Inputs | Output | #Rules | Error |
|---|---|---|---|---|---|
| 1 | HC | $HC(t-1), ..., HC(t-5)$ ; $CO(t-1)$ ; Number of Vehicles $(t-1)$ | HC (one hour later) | 10 | 0.0133 |
| 2 | HC | $HC(t-1), ..., HC(t-5)$ ; $CO(t-1)$ ; Number of Vehicles $(t-1)$ | HC (one hour later) | 27 | Not Relevant |
| 3 | HC | $HC(t-1), ..., HC(t-5)$ ; $CO(t-1), ...,$ $CO(t-5)$; Temperature; $NO_2$; NO; TSP | HC (one hour later) | 64 | Not Relevant |
| 4 | HC | $HC(t-1), ..., HC(t-5)$ ; $CO(t-5)$ ; $HC(t-24)$ | HC (one hour later) | 57 | 0.0171 |
| 5 | HC | $HC(t-1), ..., HC(t-5)$ ; $CO(t-5)$ ; $HC(t-24)$ ; Vehicles $(t-24)$ | HC (four hours later) | 54 | Not Relevant |

## 8. Conclusions

In this paper, we have proposed the use of hybrid fuzzy neural systems for modeling and prediction of time series of pollutant concentration levels in urban air. The experimental study case is derived from the database, which was made available through the "SMAURN" EU funded project. It regards the Southern Italy small city of Villa San Giovanni. In particular, the reported results refers to the concentration of HC measured and predicted locally along the road of transit of vehicles that go to and come from Sicily. The use of basic fuzzy inference systems (FIS's) and the advantage of introducing automatic rule extraction and learning in the basic model have been assessed. The reduction of the model needed to obtain a manageable model has been carried out through the use of the fuzzy surface concept. This allowed us to rank and select the most relevant inputs for prediction. Finally, for the one-step ahead and the multi-step ahead prediction, a recursive model based on the use of a window of samples has been proposed, that supply encouraging results.

**Acknowledgments** This work has been funded by the Italian Ministry of Education, University and Scientific Research (MIUR) within the PRIN2000 project "Neural Networks for Environmental Data Processing"; part of the work has been presented by one of the Authors as a non published talk at the IJCNN2001, Washington, DC, USA.

### References

[1] D. Marino and F. C. Morabito, "A Fuzzy Neural Network for Urban Environment Monitoring System: the Villa San Giovanni Study Case", Neural Nets, *Proc. Of the WIRN Workshop*, Vietri-99, Italy, pp. 323 - 328, May 1999.

[2] M. Boznar, M. Lesjak and P. Mlakar, A neural network-based method for short-term predictions of ambient SO2 concentrations highly polluted industrial areas of complex terrain, *Atmospheric Environment*, Vol. **27B**, pp.221-230, 1993.

[3] D. Marino, F.C. Morabito and B. Ricca, "Management of Uncertainty in Environmental Problems: an Assessment of Technical Aspects and Policies"; in Handbook of Uncertainty, J. Gil Aluja, Ed., Kluwer Academic Publisher, 2001.

[4] M. Roadknight and G. R. Balls, *et al*, Modeling Complex Environmental Data, *IEEE Transactions on Neural Networks*, Vol. **8**, No.4, p. 852, July 1997.

[5] P. Werbos, Backpropagation Trough Time: What It Does and How to Do It, *Proc. IEEE*, vol. **78**, Oct. 1990.

[6] N. Moussiopoulos, H. Power and C. A. Brebbia, Eds., *Air Pollution*, Vol. **3**, Urban Pollution, Computational Mechanics Publications, 1995.

[7] W. Bach, Atmospheric pollution, Mc Graw-Hill Book Company, USA, 1972.

[8] J. Milford, D. Gao and S. Dillmann, *et al*, Total reactive nitrogen (NO) as an indicator for the sensitivity of ozone to Nox and hydrocarbons, *J. Geophys. Res.*, 1994, Vol. **99**, pp.3533-3542.

[9] A. F. Atiya and S. M. El-Shoura, *et al*, A Comparison Between Neural-Network Forecasting Techniques. Case Study: river Flow Forecasting, *IEEE Transactions on Neural Networks*, Vol. **10**, N.2, pp.402 – 409.

[10] G. E. P. Box and G. M. Jenkins, Time Series Analysis, Forecasting and Control, Holden-Day, San Francisco, 1970.

[11] J. S. R. Jang, ANFIS: Adaptive – Network-based Fuzzy Inference Systems, *IEEE Transaction on Systems, Man, and Cybernetics*, Vol. **23**, No. 3, pp.665-685, May 1993.

[12] S. l. Chiu, Fuzzy Model Identification Based on Cluster estimation, *Journal of Intelligent and Fuzzy Systems*, Vol. **2**, pp.267-278; 1994.

[13] E. Gandolfi, M. Masetti, C. Vitello and S. Zannolli, Adaptive Fuzzy Systems for Gas Microsensor Arrays to be Implemented in Air Pollution Monitoring Station: First Results, In: Advances in Fuzzy Systems and Intelligent Technologies, *Proc. of the WILF'99*, Shaker Publishing, Maastricht, The Netherlands, 2000, pp.139-148.

[14] Y. Lin *et al.*, Non Linear System Input Structure Identification: Two Stage Fuzzy Curves and Surfaces, *IEEE Transaction on Systems, Man, And Cybernetics*, Vol. **26**, N. 5, 1998, pp. 678-684.

*Systematic Organisation of Information in Fuzzy Systems*
*P. Melo-Pinto et al. (Eds.)*
*IOS Press, 2003*



# Fuzzy Evaluation Processing in Decision Support Systems

Constantin GAINDRIC
*Institute of Mathematics and Computer Science,*
*Academy of Sciences, Moldova*

**Abstract.** We propose an algorithm of forming orders portfolio when the funds are limited and multicriterion evaluations of projects by experts are fuzzy. The algorithm is used in DSS for planning of financing of scientific researches.

**Keywords:** fuzzy evaluation, decision support systems.

## 1. Introduction

The concept of decision support system (DSS) permits to combine harmoniously strict mathematical methods of optimization with the intuition and reasoning ability of the decision-maker. In most cases, when solving practical problems, the manager has to deal with incomplete and, especially, inexact data and to make decision on their base. That is why it seems to be not very natural to use, when solving problems of optimization or simulation, common methods elaborated for exact data.

The worked out by us DSS for forming portfolio of projects awaiting financing includes also an algorithm based on the approach which uses fuzzy sets.

## 2. Problem statement

Every year scientific teams and individual scientists submit their project to different foundations and state institutions for the purpose of obtaining finances. As a rule, the projects are sent to experts who evaluate them using criteria determined in advance.

Let $N$ projects be admitted to the competition. On the finite set $I = \{i_1, i_2, ..., i_N\}$ of the projects $m$ criteria (quantitative or qualitative) $c_1, c_2, ..., c_m$, each of them taking values from an ordered set. The experts evaluation can give as result some aggregated indices that show the effect $e_i'$ from the realization of the $i$-th project in case the project will be given finances at the moment under consideration. It is natural that in the orders portfolio those projects should be included which will provide the maximal efficiency and whose realization does not require the exceeding of envisaged funds.

## 3. Algorithms

The projects assume different terms of realization. It is supposed that the financing plan is worked out annually, allocating the volume $M$ of finances for the first year. The term of the realization of the $i$-th project is $T_i$. Each stage of the realization of the $i$-th projects requires the sum of $m_i^t \left( t = 1, T_i \right)$.

The total sum for the realization of the $i$-th project

$$M_i = \sum_{i=1} m_i^t.$$

The selection of projects to be included in the financing plan envisages two stages: analysis and evaluation of projects by experts, decision making.

At the first stage the set $I$ of projects is divided into subsets $I_k = \{i_1, i_2, \ldots, i_{\bar{k}}\}$, $k = 1, 2, \ldots,$ $\bar{k}$, consisting of rival projects, i. e. those ones that solve the same problem.

Let $e_{ik}^t \left( i \in I_k, k = 1, 2, \ldots, \bar{k}, t = 1, 2, \ldots, T_i \right)$ be the effect from the realization of the $i$-th project of the subset $I_k$ after $t$ years if the project will be financed. These numbers we regard as fuzzy and they are determined by experts at the first stage of the evaluation of given information.

Let $A$ and $B$ two fuzzy numbers with membership functions $\mu_A$ and $\mu_B$ respectively, then the membership functions of the numbers $C = \max (A, B)$ and $D = A+B$ are determined as follows:

$$\mu_{C(z)} = \max_{z=\max(x,y)} \min \{\mu_A(x), \mu_B(y)\}$$

and

$$\mu_{D(z)} = \max_{z=x+y} \min \{\mu_A(x), \mu_B(y)\}.$$

Let $x_i$ be the variable, which is equal to 1 if the project $i$ is included in the financing plan and is equal to 0 otherwise. Then the efficiency of the realization of the projects from the subset $I_k$ that get financing after $t$ years can be calculated as follows:

$$E_k^i = \max_{i \in I_k} e_{ik}^t \cdot x_i, t = 1, 2, \ldots, \tau_k,$$

where $\tau_k = \max_{i \in I_k} T_i$.

If $\tau_k \geq T_i$ for some $i \in I$, we will consider that $e_{ik}^t = e_{ik}^{T_i}$, for all $t \geq T_i$.

Then the total effect of the variant of the plan after $t$ years will be

$$E^t = \sum_{k=1} E_k^t, t = 1, 2, \ldots, \tau_k.$$

Taking in consideration the limited funds for the first year, we can formulate the problem [1]:

$$\begin{cases} E^t \rightarrow max, \ t = 1, 2, \ldots, \tau \\ \sum_{i \in I} m_i^1 \cdot x_i \leq M \\ x_i \in \{0,1\}, i = 1, 2, \ldots N \end{cases} \tag{1}$$

The following algorithm for solving problem (1) is proposed.

The decision make desired levels $r_t$ for each $E^t, t = 1, 2, ..., \tau$.

We obtain the following system of inequalities:

$$\begin{cases} E^t \geq r_t, \quad t = 1, 2, ..., \tau \\ \sum_{i \in I} m_i \cdot x_i^1 \leq M \\ x_i \in \{0,1\}, \quad i = 1, 2, ..., N \end{cases} \qquad (2)$$

The solutions of systems (2) may be considered admissible solutions of problem (1). As the inequalities of system (2) are fuzzy, its solutions are fuzzy sets, too. It is obvious that it is desirable to choose the solution that maximizes the membership function.

Let for $t = 1, 2, ..., \tau$, $A^t = \{y : E^t \geq y\}$. It is clear that $A^t$ is a fuzzy set with the membership function

$$\mu_{At}(y) = \max_{z \geq y} \mu_{Et}(z) \; .$$

Remark that $\mu_A^t(z_t) = \mu_D^{\,t}(x)$, where $D^t$ is the set of solutions of inequalities $E^t \quad r_t$ (1). Set of solution of problem (1) is $\quad D^t, t = 1, 2, \ldots, \tau$, and $\lambda(x) = \min_t \mu_D t(x)$ is membership function set of solution problem (1). In such a manner we are closed to the method of solving the fuzzy programming problem proposed by H. J. Zimmerman [1]. Therefor, we obtain the problem (3).

$$\begin{cases} \min_t \max_{z \leq z_1} \mu_E^t(z) \rightarrow max \\ \sum_{i \in I} m_i^1 \cdot x_1 \leq M \\ x_i \in \{0,1\}, i = 1, 2, ..., N \end{cases} \qquad (3)$$

The solution of problem (3) is the vector $x^* = (x_1^*, x_2^*, ..., x_N^*)$.

The value of function $\lambda^* = \lambda(x^*)$ may be interpreted as the degree of the membership of $x^*$ to the set of solutions of problem (1).

The decision-maker, varying the levels $r_t$, can obtain the financing plan determined by the vector $x^*$ that satisfies it.

## 4. Conclusions

We have not considered here an important stage of the expert analysis, namely determining the evaluations $e_{ik}^t$. We would like to mention only that every such evaluation $e_{ik}^t$ is the result of a set of fuzzy evaluations corresponding to a group of experts.

## References

[1] H. J. Zimmerman, Fuzzy programming with several objective functions. Fuzzy sets and systems, N1, 1978, pp. 46-55.

*Systematic Organisation of Information in Fuzzy Systems*
*P. Melo-Pinto et al. (Eds.)*
*IOS Press, 2003*



# Behavior Learning of Hierarchical Behavior-based Controller for Brachiation Robot

Toshio FUKUDA, Yasuhisa HASEGAWA
*Center for cooperative research in advance science and technology, Nagoya University,
Furo-cho, Chikusa-ku, Nagoya 464-8603 Japan*

**Abstract**. In this chapter, we introduce an adaptation and learning algorithm for a dexterous dynamical motion. It adjusts a global behavior against changes of the environment and an objective task. In this method, the relation between the coefficient of each local behavior and performance of the global behavior is supposed to be expressed by the linear equation and then the coefficients could be iteratively adjusted by using Newton Raphson method. We developed a monkey-typed locomotion robot called Brachiator III, which has 13 links and 12 joints and be able to achieve a dexterous motion like a real ape in a three-dimensional space. The primitive controller is designed in a specific case. The proposed algorithm adjusts the coefficients of each local behavior when the branch interval is changed and when the continuous locomotion is desired. We show the effectiveness of the proposed algorithm through the experiments with the Brachiator III.

## 1. Introduction

One of our desires is to realize a robot which can give us a good life, intelligently assisting us with a hard, risky, boring and time-consuming work. These useful robots should have not only a good hardware such sensors, actuators and processor and but also good software for intelligence like a human information processing. Very few points of them have been realized until now over human level. Intelligence can be observed to grow and evolve, both through growth in computational power, and through the accumulation of knowledge of how to sense, decide and act in a complex and dynamically changing world. Input to, and output from, intelligent systems are via sensors and actuators [1]. We think that there are four basic elements of intelligent system: sensory processing, world modeling, behavior generation and value judgment.

In this chapter, we focus on behavior control and on its modification algorithm: learning ability in order to perform dynamically dexterous behavior. The robot behavior is exhibited by a range of continuous actions that are performed by a robot with multiple degrees of freedom. To make such a robot perform an objective behavior, the controller is required to have the capability for nonlinear and multiple input-output relationship during its behavior. It would be very hard work to design such a controller even if some powerful learning algorithms are adopted, e.g., evolutionary algorithm, reinforcement learning algorithm, back-propagation method and so on, because of its complexity, vast searching space and the nonlinear property. In general, even if a robot has a few degrees of freedom, linearization of a robot's dynamics around the stable points or non-linear controllers such a fuzzy controller and neural network controller have been used since its dynamics is nonlinear.

On the other hand, we have to introduce another method into control architecture in order to deal with multiple input-output variables. The behavior-based approach is one of the most useful and powerful concepts that can make the designing process easier by reducing the number of degree of freedom for consideration in design process. That is we can design some controllers for decomposed simple behaviors one by one and coordinate them. Once the behavior-based controller has designed, it is not easy to adjust the controller according to a change of its environment and a desired motion.

Therefore we proposed an adaptational hierarchical behavior controller and its learning algorithm when some adjustments are required. The learning algorithm firstly identifies and linearizes the relationship between the each local and the global behavior and then iteratively determines the coefficient of each behavior controller using Newton Raphson method. We apply this adaptation method to control problem of a brachiation robot. This robot, Brachiator III, has 13 links and 12 joints and is able to take motions like a real ape in a three-dimensional space. We show the effectiveness of the proposed algorithm through the experiments using the Brachiator III.

In the following sections, Brachiator III is introduced and then we show the hierarchical behavior-based controller that can control the brachiation robot. It consists of two layers: behavior controllers on the lower layer and behavior coordinators on the upper layer. We show the online learning algorithm of the hierarchical behavior-based controller. This learning algorithm changes coordination coefficients to improve a total behavior against a change of the environment and the objective behavior.

## 2. Brachiation Robot

### *2.1 Conventional Brachiation Robots*

A brachiation is a dynamical motion for an ape or monkey to move from branch to branch swinging its body like a pendulum. Many researches about a brachiation robot have been carried out. Saito et al developed two-link brachiation mobile robot(BMR) shown in Fig. 3 [2] and proposed the heuristic learning method for generating feasible trajectory for this robot, [3]-[5]. Fukuda et al, [6] proposed the self-scaling reinforcement learning algorithm to generate feasible trajectory with robust property against some disturbances. These studies do not use a dynamics model of two-link brachiation robot directly in the learning process. On the other hand, Nakanishi et al, [8] took another approach, using target dynamics, for control an underactuated systems. The two-link brachiation robot is underactuated system with two degrees of freedom and one actuator, which is similar to the *"acrobot"* [7]. As a two-dimensional extended model, a seven-link brachiation robot is studied by Hasegawa et al. [9]. This seven-link brachiation robot is given the redundancy for the locomotion so that it is able to take a dexterous motion like a real ape but in plane. In that study, a hierarchical behavior controller is proposed in order to deal with multi-input and multi-output efficiently and in order to have learning capability. The concept of hierarchical behavior controller is based on behavior-based control, in which higher-level behaviors are generated from simpler behaviors by coordinating them. The behavior controllers and its coordinators in the hierarchical structure are generated using reinforcement learning. The learning algorithm for acquiring them is hard to apply a real robot, because of taking many trials to find and tune them. Even if a desired task or an environment including robot parameters are slightly changed, the reinforcement learning needs many trials for tuning again because it uses no structured information about the relations between the each behavior and total behavior.

Figure 1. Brachiation of the White-handed Gibbon
(Hylobates lar).

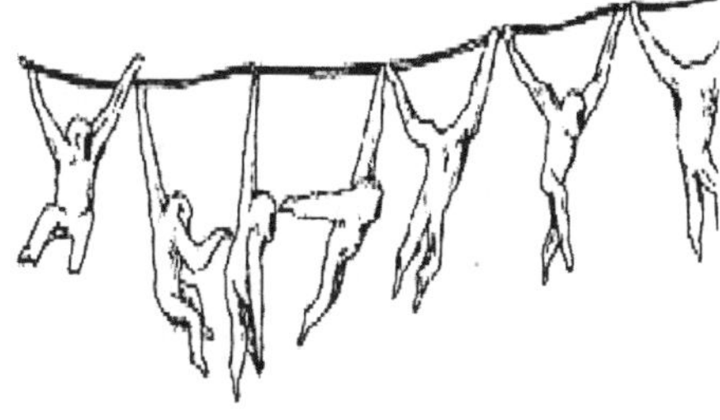

Figure 2. Locomotion of Gibbon [10].

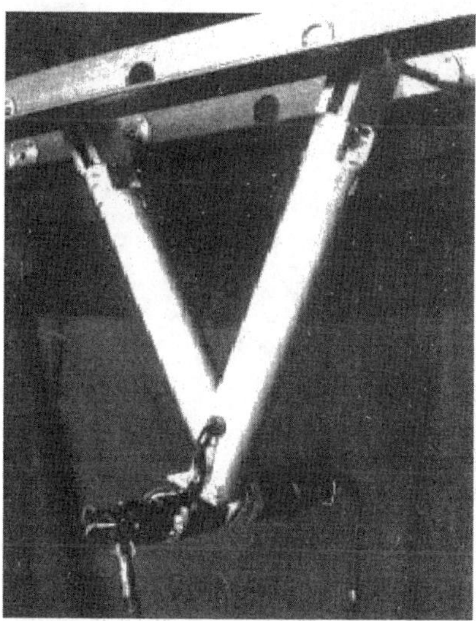

Figure 3. Brachiator II.

### *2.2 Mechanical Structure of Brachiator III*

The motion of the conventional brachiation robot is limited in two-dimensional plain. These motions are far from the real ape behaviors. Therefore we have developed the brachiation robot with 13 degrees of freedom in order to realize dynamically dexterous behaviors [11].

The objective behavior of this robot is to continuously locomote, catching the forward branch after releasing the back branch. Brachiator III has 14 motors including two motors for two grippers. Each joint is driven by DC motor through the wire so that we could conform the weight distribution to a real ape adjusting the DC motor locations. The overall look at the

"Brachiator III" is shown in Fig. 4. Figure 5 shows the brachiator III with fur, which generates much friction against the motion as a kind of diturbance. The structure and mechanism are shown in Fig. 6. The dimensions and joint angles of this robot are designed by simulating a real long-armed ape "Sherman". The weight is 9.7 [kg], which is within variance of "Sherman".

Figure 4. Brachiator III.

Figure 5. Brachiator III with Fur.

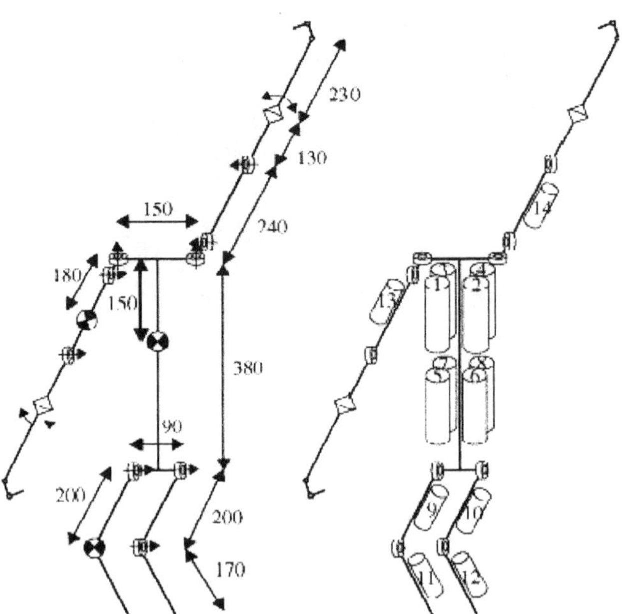

Figure 6. Mechanism of Brachiator III.

### *2.3 Motion Measurement of the Brachiation Robot Using Real-time Tracking System*

Vision sensor is very useful to measure a dynamical motion without statically constrained point, because the constrained points of this robot are switched according to the body posture. In the brachiation motion, it is almost impossible to measure the location of the tip of the free arm and the location of the center of gravity of the robot, because the slip angle at the catching grip is not directly measurable using a potentiometer or rotary encoder. We therefore use the real-time tracking system, "Quick MAG System IV", which can simultaneously measure the three-dimensional locations of the sixteen points at 60Hz sampling frequency, using two CCD cameras.

The eight measuring positions shown in Fig. 8 are chosen in order to approximately calculate the center of gravity of the robot based on the assumptions as follows,

a) The elbow of the catching arm keeps straight.

b) Both legs are controlled to behave the same motion.

c) Two joints on the shoulder are adjoining and attached on the almost same position.

Some color balls are attached for measuring of environment conditions such a target bar positions.

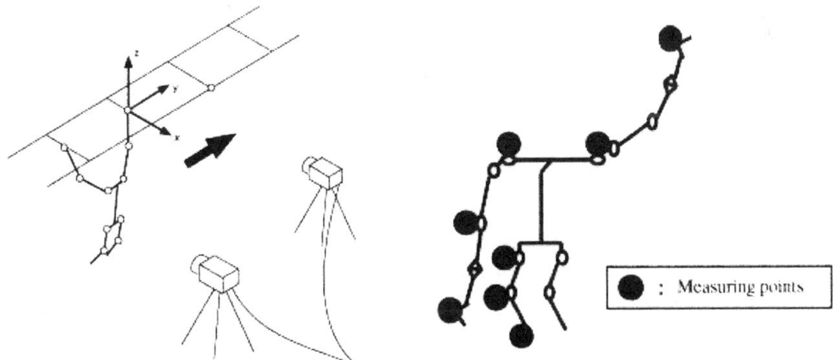


Figure 7. Coordinates System.     Figure 8. Measuring Points.

### 3. Hierarchical Behavior-based Controller

Utilization of the hierarchical behavior controller approach makes the controller designing process easier and shorter. There are some problems in this approach. One is how to adjust the behavior coordinator when the objective behavior or robot parameters are changed. We use Newton raphson method to adjust the behavior coordinator against some environmental or task changes. This method measures the effects of the local behavior controllers to the global behavior, and changes the coefficients for them in fewer trials.

The hierarchical behavior controller for brachiation robot is designed based on behavior-based approach since the robot has multiple degrees of freedom and the objective task is complex. At first, the brachiation behavior is divided into two actions: a swing action that stores the sufficient energy prior to the transfer (preliminary swing mode), and a locomotion action that is actual transfer operation (locomotion mode). After that, these actions are decomposed into the local behaviors; leg swing, body rotation I, leg stretch, body rotation II, body lifts and arm reaching. The hierarchical behavior controller is shown in Fig. 9.

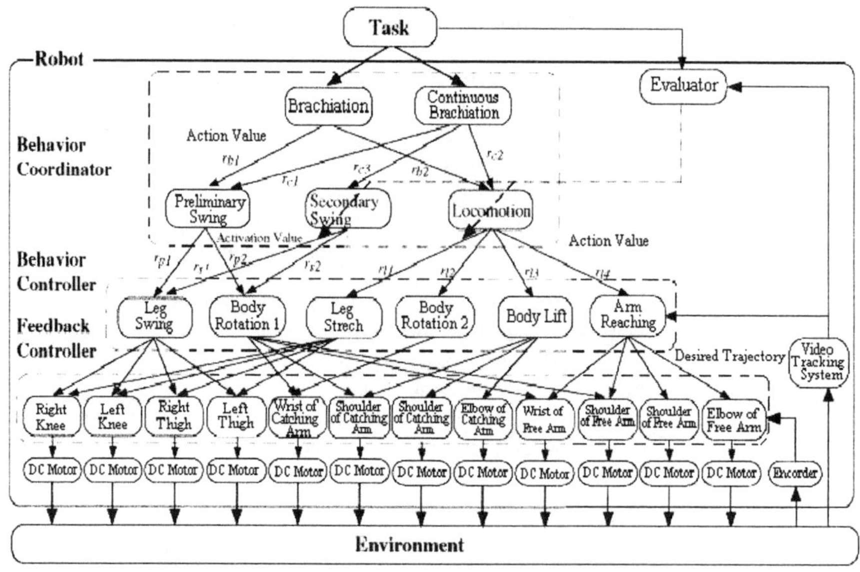


Figure 9. Hierarchical Behavior Controller for Brachiator III.

### 3.1 Rescalling the Desired Trajectories from Behavior Controller

The behavior controllers except the arm reaching behavior controller are feedforward controllers which generate the desired trajectories expressed by the cubic spline function to the feedback controllers. This desired trajectory for the actuators is rescaled by a corresponding coefficient from the behavior coordinator on the upper layer as follows,

$$yd_i(t) = r_k(y_k(t) - b_k(t)) \tag{1}$$

where $yd_i$ is desired trajectory to actuator $i$, $r_k$ is coefficient for behavior controller $k$, $y_k(t)$ is actuator trajectory from behavior controller $k$, and $b_k(t)$ is base line connecting an initial point with an end point as follows,

$$b_k(t) = b(0)(t^* - t)/t^* + b(t^*)t/t^* \tag{2}$$

where $t^*$ is finishing time of the behavior.

If multiple behavior coordinators indicate coefficients to one behavior controller, summation of these values becomes new one as follows,

$$r_k = \prod_{i \in I} r_i \tag{3}$$

where $I$ is group that indicates coefficient to the behavior controller $k$.

The feedback controller makes the corresponding actuator trace the desired trajectory from behavior controller.

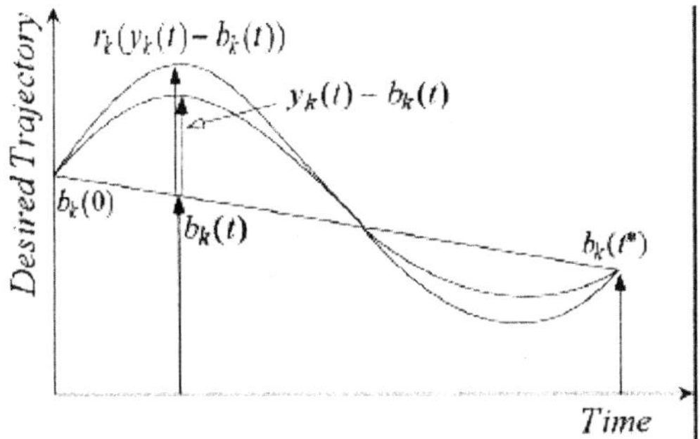


Figure 10. Trajectory expressed by the cubic spline function.

### *3.2 Adaptation for Behavior-based Controller*

The brachiation robot is controlled by the hierarchical behavior controller explained above. To adjust the total behavior against the small changes of robot parameters or change of a desired behavior, we have three parts to be updated; the fundamental behavior controller, the behavior coordinator and both of them. The easiest part is to adjust only behavior coordinator, because of the small searching space. If the structure of the robot system or the desired task is completely changed, both the behavior controller and behavior coordinator should be rearranged. The proposed method is concerning about adaptation of only the behavior coordinator. In order to adjust the behavior coordinator, we should evaluate the effect of each behavior controller to the global behavior. Therefore, in the proposed algorithm, the relations between the each behavior and the global behavior through several trials are approximately measured and then the coefficients to the behavior controller are updated based on the relations.

### *3.3 Adaptation Algorithm*

The desired trajectories for the connected actuators are calculated in the behavior controller. The behavior coordinator indicates the activation coefficients, which determine the amplitude of the desired trajectories from base line. Therefore by adjusting the activation coefficient, the global behavior could be adjustable to some extent.

The relation between change of the activation coefficient and the resultant behavior is strongly nonlinear. We however assume that the relations could be express as the multiplier of the degree of contributions and the activation values only in limited neighborhood of the current state as follows,

$$P(s) = W(s) \cdot r(s) \tag{4}$$

where *P(s)* is performance vector at step *s*, *W(s)* is a gradient matrix and *r(s)* is an activation coefficient.

The performance vector is a column of performance indices which evaluate the resultant behavior. The target activation coefficient, is derived from eq. (5).

$$e(s) = \quad p^* - p(s)$$
$$W(s) \cdot r^*(s) - W(s) \cdot r(s) \tag{5}$$
$$W(s) \cdot (r^*(s) - r(s))$$

where $e(s)$ is error vector from the desired performance, $p^*$ is a target performance vector and $p$ is a current performance vector.

This calculated activation coefficient $r^*$ is not the desired ones, because the linearized equation is adopted in the non-linear system. Therefore the target activation coefficients are searched iteratively as follows,

1. At first, evaluate its performance by making trial with activation coefficients $r$.

2. Explore the performance around neighborhoods area, $r'$, $r''$, and $r^{n-1}$. These $r$ should be linear independence each other.

3. Update gradient matrix using eq. (10), and calculate new activation coefficients using eqs. (11) and (12).

4. Check its performance with the new activation coefficients. If the behavior is not insufficient, go to step 2.

$$R(s) = (r(s), r'(s), r''(s), \cdots, r^{n-1}(s)) \tag{8}$$

$$P(s) = (p(s), p'(s), p''(s), \cdots, p^{n-1}(s)) \tag{9}$$

$$W(s) = P(s) \cdot R(s)^{-1} \tag{10}$$

$$\Delta r(s) = W(s)^{-1} \cdot e(s) \cdot \tag{11}$$

$$r(s+1) = r(s) + \Delta r(s) \tag{12}$$

## 4. Experiments

We applied the proposed method to two cases. One is the case when the objective task is changed. The other is the case when the new behavior should be generated based on the primitive behavior.

### *4.1 Adaptation to Extended Branch*

As a first case, we consider that the interval between branches is extended to 101 centimeters, while the initial hierarchical behavior controller for Brachiator III is designed to catch the branch at 90 centimeters interval. Brachiator III failed to catch the branch since a shortage of oscillation energy of the center of gravity and the body did not come up to a reachable region. We apply the proposed adaptation algorithm to adjust the locomotion behavior coordinator, which indicates four activation coefficients $r_i$ ($i=1, 2, 3, 4$) to the corresponding four local behavior controllers: Leg stretch, Body rotation 2, Body lift and Arm reaching. Four performance indices in eq. (13) are selected to represent performance of the resultant behavior; the minimum distance $g_{cog}$ between center of gravity and target branch, the body tilt angle $\theta_{body}$

from vertical direction at the catching moment, the minimum distance $d_{hand}$ in y-z plain and x-directional distance $x_{hand}$ between a free hand and a target point shown in Fig. 11.

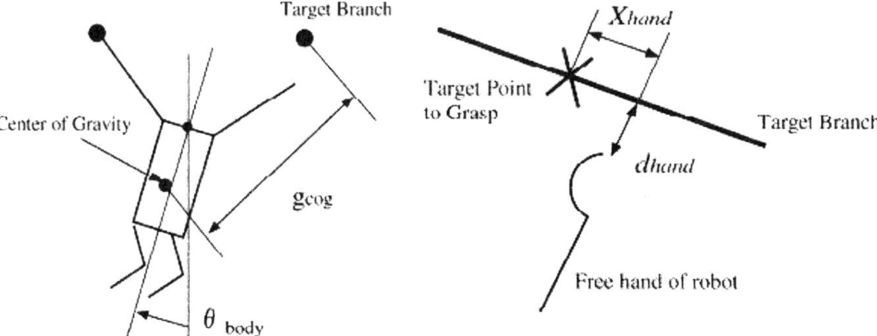


Figure 11. Evaluation Parameters.

$$P(s) = (g_{cog}(s), \theta_{body}(s), d_{hand}(s), x_{hand}(s))^T \tag{13}$$

$$R(s) = (r_{l1}(s), r_{l2}(s), r_{l3}(s), r_{l4}(s))^T \tag{14}$$

The gradient matrix is expressed by eq. (15).

$$W(s) = \begin{bmatrix} W_{l1,cog}(s) & W_{l2,cog}(s) & W_{l3,cog}(s) & W_{l4,cog}(s) \\ W_{l1,body}(s) & W_{l2,body}(s) & W_{l3,body}(s) & W_{l4,body}(s) \\ W_{l1,hand_X}(s) & W_{l2,hand_X}(s) & W_{l3,hand_X}(s) & W_{l4,hand_X}(s) \\ W_{l1,hand_d}(s) & W_{l2,hand_d}(s) & W_{l3,hand_d}(s) & W_{l4,hand_d}(s) \end{bmatrix} \tag{15}$$

The activation coefficient vectors: $r', r''$, and $r^{n-1}$ are determined in eqs. (16)-(18).

$$r' = (r_{l1}(s) + \alpha, r_{l2}(s) - \alpha, r_{l3}(s) + \alpha, r_{l4}(s))^T \tag{16}$$

$$r'' = (r_{l1}(s), r_{l2}(s) + \alpha, r_{l3}(s) + \alpha, r_{l4}(s) - \alpha)^T \tag{17}$$

$$r''' = (r_{l1}(s) - \alpha, r_{l2}(s), r_{l3}(s) + \alpha, r_{l4}(s) + \alpha)^T \tag{18}$$

The proposed algorithm is based on the linear approximation, therefore the range should be determined carefully. Three trials have been performed with different perturbations e.g. 0.05, 0.1 and 0.4.

When the perturbation is 0.1, we obtained a feasible brachiation motion against the task change. When the bigger or smaller perturbation are set, e.g. 0.05 and 0.4, the behavior did not improve. The results are shown in Fig. 12.

The transition of the activation coefficients and the performance indices through the adaptation are shown in Fig. 13, and Table 1. The adaptation algorithm obtained the feasible activation values in four steps from the extension of the branch span. Figure 14 shows the trajectories of tip of the free hand both before and after adaptations.

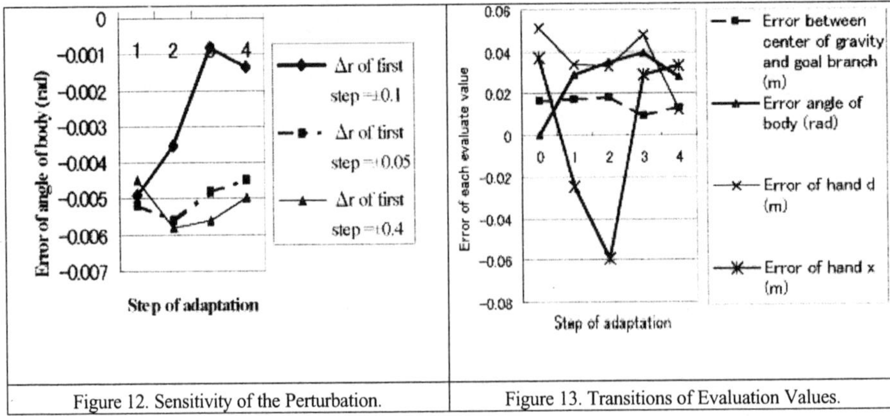

<table>
<tr><td>Figure 12. Sensitivity of the Perturbation.</td><td>Figure 13. Transitions of Evaluation Values.</td></tr>
</table>

Table 1 Transitions of Activation Values and Performance Indices for a Different Branch Interval

| Trial No. | $r_{l1}$ | $r_{l2}$ | $r_{l3}$ | $r_{l4}$ | $g_{cog}$[m] | $\theta_{body}$ [rad] | $d_{hand}$[m] | $x_{hand}$[m] |
|---|---|---|---|---|---|---|---|---|
| 0 | 1.0 | 1.0 | 1.0 | 1.0 | 0.016 | 0.0 | 0.051 | 0.037 |
| 1 | 0.855 | 0.839 | 0.996 | 1.31 | 0.017 | -0.029 | 0.034 | -0.025 |
| 2 | 0.871 | 0.779 | 0.992 | 1.32 | 0.018 | -0.035 | 0.0326 | -0.059 |
| 3 | 0.851 | 0.677 | 1.07 | 1.27 | 0.0092 | -0.039 | 0.048 | 0.029 |
| 4 | 0.897 | 0.693 | 0.993 | 1.32 | 0.013 | -0.028 | 0.012 | 0.033 |

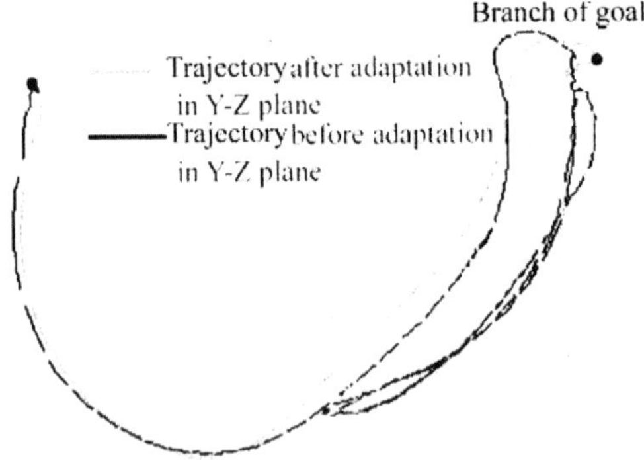


Figure 14. Free Hand's Trajectories Before and After Adaptation.

### *4.2 Learning for New Behavior*

As a learning problem, we generate the continuous locomotion which means the robot dexterously performs the first locomotion and the second locomotion without pause shown in Fig. 15. When the robot shifts from the first locomotion to the second locomotion, it keeps kinetic energy even if it catches the two branch with two arms. Therefore, the second locomotion or leg swing behavior should be adjusted in order to use the kinetic energy remaining after the first locomotion. In this case, the leg swing behavior before the second locomotion behavior is adjusted by the proposed algorithm. This behavior coordinator indicates two amplitude coefficients(eq. (19)) to the behavior controllers shown in Fig. 9. Therefore two performance indices are used for learning process shown in Fig. 16 and eq. (20).

$$r(s) = (r_{s1}(s), r_{s2}(s))^T \tag{19}$$

$$P(s) = (h_{body}(s), v_{body}(s))^T \tag{20}$$

The error transitions of two performance indices through the learning process are shown in Fig. 17. Some errors are remaining after learning process, because the number of the parameters is limited. Two amplitude coefficients are considered for an updated parameter, and the behavior controller should be updated in order to decrease these errors of performance indices. Figure 18 is a Stroboscopic photo showing the continuous motion with 33 [msec] interval.

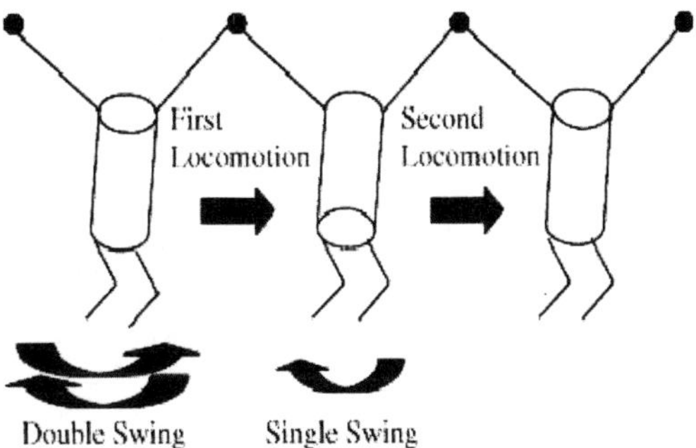


Figure 15. Sequence of Continuous Locomotion.

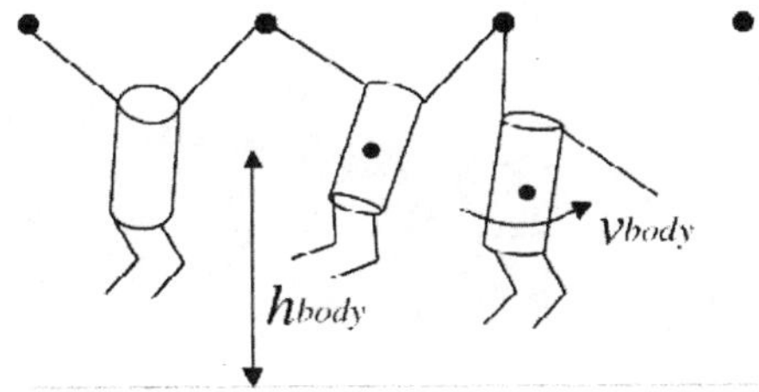


Figure 16. Evaluation Parameters for Continuous Locomotion.

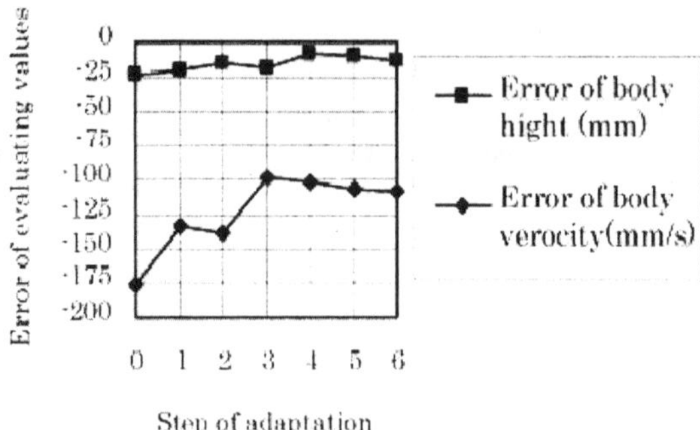


Figure 17. Transitions of Evaluation Values (Continuous Locomotion).

Table 2 Transitions of Activation Values and Performance Indices for Continuous Locomotion

| Trial No. | $r_{s1}$ | $r_{s2}$ | $\Delta h_{body}$[m] | $\Delta v_{body}$[m] |
|---|---|---|---|---|
| 0 | 1.0 | 1.0 | 24 | 176 |
| 1 | 1.13 | 1.15 | 21 | 134 |
| 2 | 1.11 | 1.25 | 16 | 139 |
| 3 | 1.14 | 1.22 | 18 | 98 |
| 4 | 1.14 | 1.23 | 9 | 102 |
| 5 | 1.18 | 1.21 | 10 | 106 |
| 6 | 1.17 | 1.22 | 13 | 108 |

Figure 18. Stroboscopic Photo of the Continuous Locomotion.

## 5. Conclusions

In this chapter, we proposed a novel adaptation method to adjust the behavior coordinator against small changes of the objective task. This method measures the effects of the each local behavior to the total behavior, and determines the activation values in the behavior coordinator for each behavior controller. We applied this adaptation method to control the monkey-like locomotion robot. We showed the effectiveness of the proposed algorithm, which could find the feasible activation values through several trials against small task changes. Experimentally it was shown that the performance of this method depends on the amplitude of perturbation, which is difficult to be adjusted because of the non-linearity of the robot behavior and measuring errors.


### References

[1] J. S. Albus, Outline for a Theory of Intelligence, *IEEE Trans. on Systems, Man and Cybernetics*, Vol. **21**, No. 3 1991.

[2] T. Fukuda, H. Hosokai and Y. Kondo, Brachiation Type of Mobile Robot, *Proc. IEEE Int. Conf. Advanced Robotics*, pp. 915-920, (1991).

[3] F. Saito, T. Fukuda and F. Arai, Swing and Locomotion Control for a Two-Link Brachiation Robot, *IEEE Control Systems*, Vol. **14**, No. 1, pp. 5-12, (1994).

[4] F. Saito, T. Fukuda and F. Arai, Swing and Locomotion Control for Two-Link Brachiation Robot, 1993, *IEEE International Conference on Robotics and Automation*, Atlanta, pp. II: 719-724, (1993).

[5] T. Fukuda, F. Saito and F. Arai, A Study on the Brachiation Type of Mobile Robot (Heuristic Creation of Driving Input and Control Using CMAC), *Proc. IEEE/RSJ Int. Workshop on Intelligent Robots and Systems*, pp. 478-483, (1991).

[6] T. Fukuda, Y. Hasegawa, K. Shimojima and F. Saito, Self Scaling Reinforcement Learning for Fuzzy Logic Controller, *IEEE International Conference on Evolutionary Computation*, pp. 247-252, Japan, (1996)

[7] M. W. Spong, Swing Up Control of the Acrobot, *Proc. 1994 IEEE International Conference on Robotics and Automation*, pp. 2356-2361, (1994)

[8] J. Nakanishi, T. Fukuda and D. E. Koditschek, Experimental implementation of a Target dynamics controller on a two-link brachiating robot, *IEEE International Conference on Robotics and Automation*, pp. 787-792, (1998)

[9] Y. Hasegawa and T. Fukuda, Learning Method for Hierarchical Behavior Controller, *IEEE International Conference on Robotics and Automation*, pp.2799-2804, (1999)

[10] E.L. Simons, Primate Evolution: An Introduction to Man's Place in Nature, Macmillan, (1972).

[11] F. Saito and T. Fukuda, A First Result of The Brachiator III -A New Brachiation Robot Modeled on a Siamang, ALIFE 5, pp.321-328, (1996).

# Filter Impulsive Noise, Fuzzy Uncertainty, and the Analog Median Filter

Paulo J. S. G. FERREIRA[1] and Manuel J. C. S. REIS[2]
[1]*Dep. Electrónica e Telecomunicações / IEETA,*
*Universidade de Aveiro, 3810-193 Aveiro, Portugal*
[2]*Dep. Engenharias / CETAV,*
*Universidade de Trás-os-Montes e Alto Douro, 5001-911 Vila Real, Portugal*

**Abstract.** Additive fuzzy systems combined with supervised or unsupervised learning have been proposed to deal with impulsive noise in discrete-time signals, but suffer from the curse of dimensionality. Faster methods, possibly based on new paradigms, would be welcome. Order statistics filters and their fuzzy counterparts provide alternative and robust ways of dealing with discrete-time data, but the idea of ranking (or fuzzy ranking) does not appear at first sight to be meaningful in the continuous-time case. This chapter investigates to which extent the concept of "sorting" is meaningful for continuous-time signals. It presents a tutorial on the basic concepts behind sorting, and applies the results to the study of the single-input single-output analog median filter. Interestingly, the concept of sorting or rearrangement, which plays a fundamental role in the development, appears naturally when trying to define the uncertainty associated with a general set membership function.

## 1. Introduction

### 1.1. Impulsive noise, additive fuzzy systems, and sorting

Impulsive noise is rather difficult to handle. To quote from [1]:
"Dealing with impulsive noise remains one of the great challenges of modern engineering. It is hard to model, predict, and filter —and yet it pervades the world."
Additive fuzzy systems have been used to deal with impulsive noise. Generally speaking, a fuzzy system $F : \mathbf{R}^n \mapsto \mathbf{R}$ stores $m$ if/then rules (if $X = A_j$ then $Y = B_j$). The if-part fuzzy sets $A_j$ and the then-part fuzzy sets $B_j$ correspond to set functions $a_j : \mathbf{R}^n \mapsto [0, 1]$ and $b_j : \mathbf{R} \mapsto [0, 1]$. An additive fuzzy system computes the output as the centroid of the summed and partially fired then-part sets:

$$B = \sum_{j=1}^{m} B'_j = \sum_{j=1}^{m} a_j(x) B_j.$$

These systems can uniformly approximate any continuous function on a compact domain. A fuzzy system approximator is equivalent to a graph cover with local averaging. Such systems can tune their rules from any source of signals or noise, using supervised or unsupervised learning. However, they suffer from the curse of dimensionality: the number of rules grows exponentially with the number of input and output variables. Other related approaches, based for example on radial basis function neural networks, face the same difficulty, and become increasingly more difficult to apply as the dimension of the problem grows.

A similar problem occurs when one seeks to apply digital median filters, or other order-statistics filters, to signals sampled at a very high frequency. These non-linear filters are known to suppress impulsive noise, while preserving sharp signal transitions, such as steps. In image processing, they are used to suppress or attenuate impulsive noise without smoothing edges, as a linear filter would do. But computing the order-statistics requires sorting, an operation whose complexity grows with the number of data. Analog implementations of these filters might lead to better solutions, but the meaning of "sorting" in the continuous-time case, and the connection between the digital and analog cases, must be clarified before proceeding. The results presented in Section 4 address this issue.

### 1.2. Fuzzy uncertainty and sorting

The importance of fuzzy set theory is a consequence of the fuzzy character of natural classes and concepts. The search for the fuzzy counterparts of certain concepts, often guided by the existing crisp models, often leads to new and fruitful directions of research. Concepts such as uncertainty, information, or entropy, are good examples of this.

Consider a set membership function $\pi(x)$, normalized so that $\sup_x \pi(x) = 1$. As usual, we may think of it as a frequency or "subjective probability", obtained by "polling experts". Independently of that, one may ask for the appropriate definition of the "uncertainty" associated with $\pi(x)$, which we will denote by $U(\pi)$. The task is simpler if $\pi(x)$ is piecewise constant (a step function), which takes only a finite number of possible values. Denote the sorted values by

$$h_1 < h_2 < \cdots h_n.$$

Note that the sorted values define a sorted (or rearranged) version of $\pi(x)$. Having sorted the step values, the definition of uncertainty can be given as [2]

$$U(\pi) = \sum_{i=1}^{n}(h_{i+1} - h_i)\log_2 M_\pi(h_{i+1}).$$

The notation $M_\pi$ denotes the distribution function of $\pi(x)$, studied in Section 2:

$$M_\pi(y) := \mathrm{meas}\{x : \pi(x) > y\}.$$

When the set membership function is not restricted to a finite number of values, but satisfies certain regularity assumptions, the correct definition becomes

$$U(\pi) = \int_0^1 \log_2 M_\pi(h)\,dh.$$

However, several questions concerning the meaning and possibility of "sorting" a general $\pi(x)$ do arise in the general case [2]. The results of Section 3 provide the answer to this question, and at the same time the natural framework for the analog median filter.

### *1.3. The median filter and sorting*

The basic ideas underlying the discrete-time median estimator and the least absolute error cost function go back to Edgeworth's note of 1887 [3], partially inspired by the previous work of Laplace. The articles by Edgeworth and Turner [3][4][5] discussed some aspects of the method, including the minimization of the sum of the absolute value of the residuals, explicitly mentioned in [5]. However, it was the work of Tukey in the seventies that lead to the extensive modern research on the digital median and median-type filters, the theory of which is now well developed.

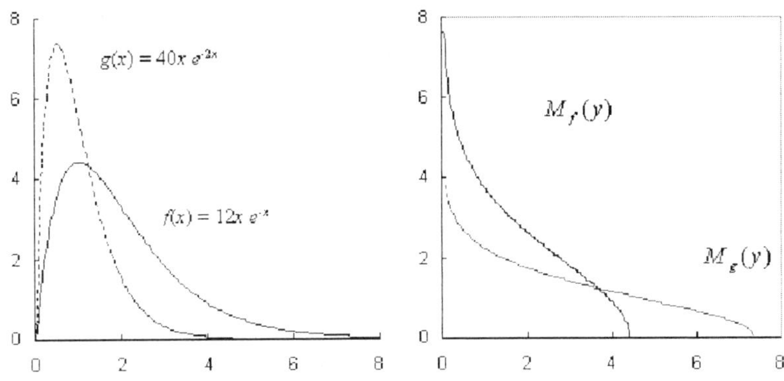


Figure 1. Left: two simple functions $f$ and $g$. Right: their distributions $M_f$ and $M_g$.

Surprisingly, the continuous-time or analog median filter remained unexplored for a long time, with the exception of [6], published in 1986, and a brief comment on it [7], of 1989. As far as we know, no attempt was made to define the meaning of "sorting" for continuous-time signals and to explore it in the context of analog nonlinear filtering until 2000.

This lack of theoretical results on the analog median filter was noted, for example, in [8]. In fact, the issues raised in [7], related to the nature and existence of root signals of the median filter as defined in [6] (without introducing the concept of sorting), were addressed only recently [9]. The properties of the analog, single-input single-output median filter, and the connection with distribution functions and sortings or rearrangements of continuous-time signals were studied even more recently, in [10].

Reference [6] gives reasons for extending the theory of median filtering from the digital to the analog domain, stressing the need for fast implementations. Analog implementations would allow simpler and faster circuits, without A/D or D/A converters, yielding large power and circuit area savings for applications such as mobile telecommunications. The need for a device to perform two-dimensional median filtering, included as a stage in the optical section at the front end of the imaging system, is mentioned in [6]. For more on the implementation issues and approaches see [8][11][12] and the notes in [10].

Bringing the theory of ranked order filters to a more symmetrical state concerning the discrete-time and the continuous-time cases seems to be desirable. The present chapter attempts to present the fundamental concepts necessary to understand the continuous-time meaning of "sorting", which plays a fundamental role in analog order-statistics filters, and is related to several other problems (including, as discussed above, the problem of defining fuzzy uncertainty).

### 1.4. Mathematical background and notation

The concept of sorting (or non-increasing rearrangement) was introduced by Hardy and Littlewood, and several results can be found scattered among books as diverse as [13][14][15][16]. However, this chapter is reasonably self-contained, and introduces the concepts necessary to understand and apply the basic theory of sorting.

The readers unacquainted with measure theory should be aware that the Lebesgue measure of an interval $[a, b]$ is its length $b - a$, and that the measure of a union of disjoint intervals is the sum of their lengths. All functions are tacitly assumed to be measurable, and the expression "almost everywhere" means, as usual, "except on a set of zero Lebesgue measure".

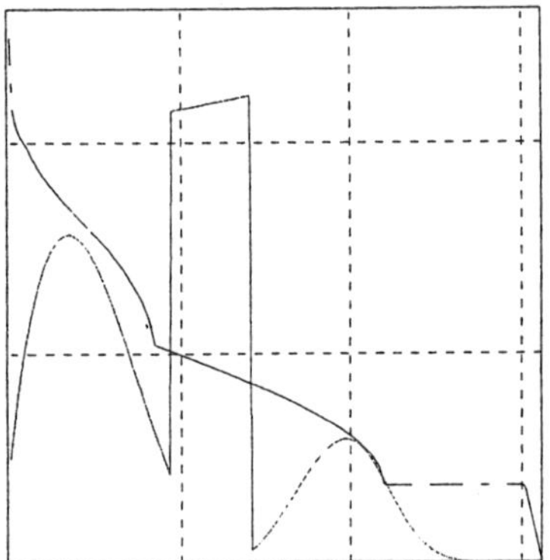

Figure 2. A more complex function (solid line) and its distribution (dashed line).

The support of a function $f$ is the closure of the set where $f \neq 0$. Note that, roughly speaking, signals with disjoint supports do not overlap. The characteristic function of $S \subset \mathbf{R}$ is denoted by $\chi_S$,

$$\chi_S(x) := \begin{cases} 1, & x \in S, \\ 0, & x \notin S, \end{cases}$$

and it is the simplest function with support $S$.

The modulus of continuity $\omega_f(\delta)$ of a continuous function $f$ is defined by

$$\omega_f(\delta) := \sup_{|a-b| \leq \delta} |f(a) - f(b)|,$$

and converges to zero when $\delta \to 0$. The condition $\omega_f(\delta) \leq \alpha\delta^\gamma$, where $\alpha$ and $\gamma$ are constants, leads to the class of functions $f$ that satisfy a Lipschitz condition of order $\gamma$ and constant $\alpha$:

$$|f(a) - f(b)| \leq \alpha|a - b|^\gamma.$$

This condition is useful only if $0 < \gamma \leq 1$. For more details, see [17], for example.

## 2. The distribution function

### 2.1. Definition and basics

Assume that somehow we sere able to associate with a given function $f : \mathbf{R} \mapsto \mathbf{R}$ a new function, denoted by $\overline{f}$, which corresponds to the idea of "$f$ sorted by non-increasing order". It is intuitively obvious that both $f$ and $\overline{f}$ should have the same "distribution of amplitudes". To render this concept precise, consider the distribution function $M_f$ associated with $f$, defined by:

$$M_f(y) := \mathrm{meas}\{x : f(x) > y\},$$

where "meas $S$" denotes the Lebesgue measure of the set $S$. It is tacitly assumed that $x$ belongs to the domain of $f$, and that $M_f$ is finite almost everywhere.

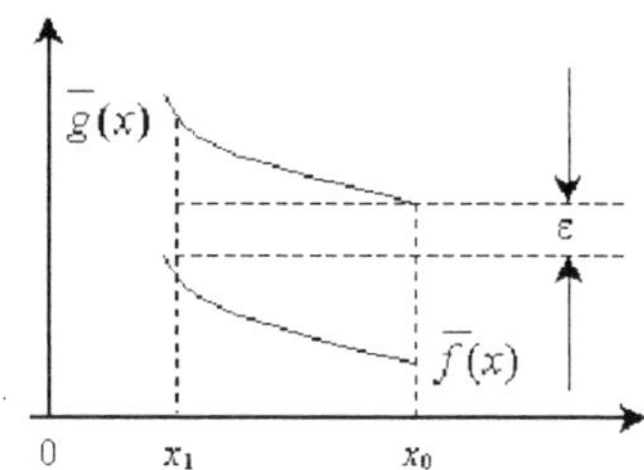


Figure 3. The sorting of a function is unique. Two distinct sortings $\overline{g}$ and $\overline{f}$ cannot be equimeasurable: if $\overline{g}(x_0) > \overline{f}(x_0)$, the left-continuity implies the existence of an interval $[x_1, x_0]$ in which $\overline{f} < \overline{g}(x_0) - \varepsilon$, and this can be used to show that $\overline{g}$ and $\overline{f}$ are not equimeasurable, a contradiction.

The distribution cannot usually be expressed in closed form. Consider for example $f(x) = ax\,e^{-bx}$. Its distribution function can be numerically computed (see Fig. 1), but there appears to be no general simple closed form expression for the roots of $xe^{-x} = y$. An even more complex case is illustrated in Fig. 2.

There are two relevant remarks at this point. First, arbitrarily modifying $f$ in a set of zero measure does not change $M_f$. In particular, the behavior of $f$ at its points of discontinuity is irrelevant. Second, the map $f \mapsto M_f$ is shift-invariant, as a consequence of the shift-invariance of the measure. Thus, the functions $f(x)$ and $g(x) = f(x - \tau)$ have the same distribution.

However, two functions $f$ and $g$ that differ on sets of positive measure may have the same distribution, even if $f(x) \neq g(x - \tau)$. For example, let $S$ and $S'$ be two sets of measure $a > 0$. Then, $\chi_S$ and $\chi_{S'}$ have the same distribution function:

$$M_{\chi_S}(y) = M_{\chi_{S'}}(y) = \chi_I(y),\tag{1}$$

where $I = (0, a)$. Two or more functions with the same distribution are called "equimeasurable", and can be considered as "rearrangements" of each other. A function $f(x)$ and its "reversed" version $f(-x)$ are equimeasurable. A more general class of equimeasurable functions can be obtained from a single function $f$ as follows: split $f$ in a number of components $f_i$ with disjoint supports, such that

$$f = \sum_i f_i,$$

and then translate, reverse, or permute the $f_i$ at will, without overlapping any two of them. The function obtained will be equimeasurable with $f$.

## 3. Sorting continuous-time signals

Now that we have defined the distribution function $M_f$ of a given $f$, we look for non-increasing (or non-decreasing) and left continuous (or right-continuous) functions, equimeasurable with $f$. To fix ideas, we consider the non-increasing left-continuous case only. It turns out that the solution is unique; we will denote it by $\overline{f}$, and call it the sorting of $f$. The second and final step is constructive: we give a formula for $\overline{f}$.

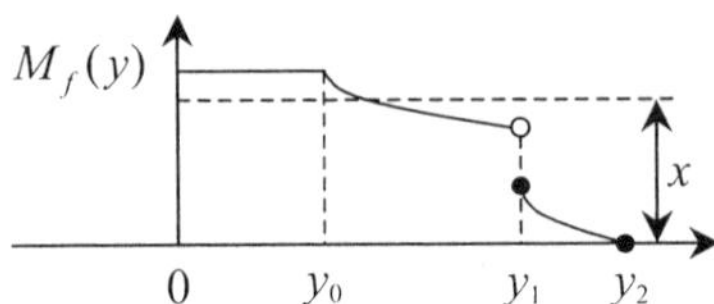



Figure 4. A (right-continuous) distribution function $M_f(y) = meas\{x : f(x) > y\}$. The sorting $\overline{f}$ of $f$ is the inverse function of $M_f$, properly defined to account for discontinuities: $\overline{f}(x) := \inf_y \{M_f(y) < x\}$.

### 3.1. Uniqueness

Assume that there exist two different equimeasurable sortings $\overline{f}$ and $\overline{g}$ of a given $f$. Without loss of generality, let $\overline{g}(x_0) > \overline{f}(x_0)$, for some $x_0$ (see Fig. 3). Pick $\varepsilon > 0$ such that $\overline{g}(x_0) - \varepsilon > \overline{f}(x_0)$. By the left-continuity of $\overline{f}(x)$ there exists an interval $[x_1, x_0]$ in which $\overline{f}(x) < \overline{g}(x_0) - \varepsilon$. Then,

$$meas\{x : \overline{g}(x) > \overline{g}(x_0) - \varepsilon\} \geq x_0,$$

while

$$\operatorname{meas}\left\{x : \overline{f}(x) > \overline{g}(x_0) - \varepsilon\right\} \leq x_1 < x_0.$$

The quantities on the left side are the values at the point $p = \overline{g}(x_0) - \varepsilon$ of the distribution functions of $\overline{g}$ and $\overline{f}$, respectively. The inequalities therefore show that $\overline{f}$ and $\overline{g}$ are not equimeasurable. This contradiction shows that (left continuous non-increasing) sorting is unique.

The uniqueness of the non-increasing right-continuous sorting can be shown using a similar method. The two sortings will be equal except at points of discontinuity (they will be equal everywhere provided they are continuous). There is an essentially unique way of sorting a given function, and all the ambiguity can be removed by imposing, for example, left- or right-continuity constraints.

### 3.2. The sorting as the inverse of $M_f$

We now proceed to the second step, the construction of $\overline{f}$. It turns out that $\overline{f}$ is, essentially, the inverse function of its distribution function $M_f$. This can be seen as follows: recall that splitting the domain of a given $f$ in a finite number of non-overlapping intervals, and permuting, translating or time-reversing the "pieces" of $f$ without overlapping them, leads to a new function which has the same distribution function $M_f$. Because $M_f$ is invariant under these operations and is non-increasing, one might be tempted to take it as the sorting of $f$. However, this is wrong: if $f(t)$ is interpreted as, says, "current as a function of time", then $M_f(y)$ would describe "time as a function of current". Swapping axes, that is, considering $M_f^{-1}$ instead of $M_f$, leads to a non-increasing function with the correct units.

The only difficulty is that $M_f$, in general, does not have an inverse in the strict sense (it may be constant over certain intervals, as in Fig. 4). Roughly speaking, there are essentially two ways of defining a generalized inverse of $M_f$: draw a horizontal line of height $x$ across the plot of $M_f$ and search for the leftmost point $y$ still satisfying $M_f(y) < x$; or search for the rightmost point $y$ still satisfying $M_f(y) > x$ (again, see Fig 4).

Both possibilities lead to the usual inverse when $M_f$ is continuous and decreasing. We adopt the first hypothesis, which leads to the definition

$$\overline{f}(x) := \inf_y \left\{ M_f(y) < x \right\} \tag{2}$$

and the left continuous sorting. Unless otherwise indicated, from now on "sorting" means "left-continuous non-increasing sorting" and therefore the function defined by (2).

It can be seen that

$$M_{\overline{f}}(y) := \operatorname{meas}\left\{x : \overline{f}(x) > y\right\} = M_f(y),$$

that is, $f$ and its sorting have the same distribution function: they are equimeasurable, as expected, and the integrals of $f$ and $\overline{f}$ are equal:

$$\int f(x)dx = \int \overline{f}(x)dx. \tag{3}$$

In fact, it is true that

$$\int F(f) = \int F(\overline{f}),$$

for any measurable function $F$ for which the integrals exist. It is helpful to compare these results with the discrete-time equivalents,

$$\sum_{i=1}^{N} f(i) = \sum_{i=1}^{N} \overline{f}(i), \qquad \sum_{i=1}^{N} F(f(i)) = \sum_{i=1}^{N} F(\overline{f}(i)),$$

which are of course true, because the sum is commutative.

### 3.3. Examples and remarks

*Example 1* The uniqueness of the sorting immediately shows that a non-increasing $f$ equals its sorting, except possibly at its points of discontinuity. The sorting of a non-decreasing $f$, on the other hand, is essentially obtained by "reversing" the function: $\overline{f}(t) = f(-t)$.

*Example 2* The distribution of the Gaussian $f(x) = e^{-x^2}$, $x \in \mathbf{R}$, is

$$M_f(y) = 2\sqrt{\log\frac{1}{y}}, \qquad y \in (0,1).$$

The sorting is $\overline{f}(x) = e^{-x^2/4}, x \geq 0,$ that is, $\overline{f}(x) = f(x/2)$ (see Fig. 5, left). This is true for any other even and continuous functions $f$ decreasing away from the origin.

*Example 3* The sorting of sinusoids can also be determined. If $f(x) = \cos x, x \in [0, 2\pi]$, then

$$M_f(y) = \operatorname{meas}\{x : \cos x \geq y\} = 2\arccos y,$$

where the function arcos $x$ is defined to be in $[0,\pi]$. The sorting is $\overline{f}(x) = \cos(x/2)$, as illustrated in Fig. 5, right. The result also applies for $f(x) = \cos(x+\theta), x \in [0, 2\pi], \theta \in \mathbf{R}$.

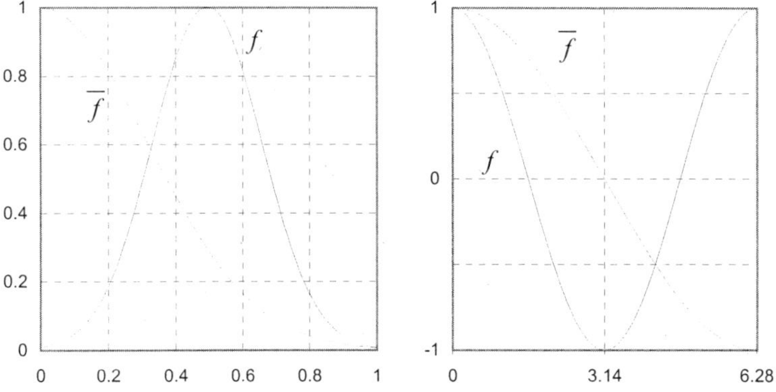

Figure 5. Left: a Gaussian-like function and its sorting. Right: one period of $f(x) = \cos x$ and its sorting $\overline{f}$. In both cases, the area under the function curve and its sorting is the same.

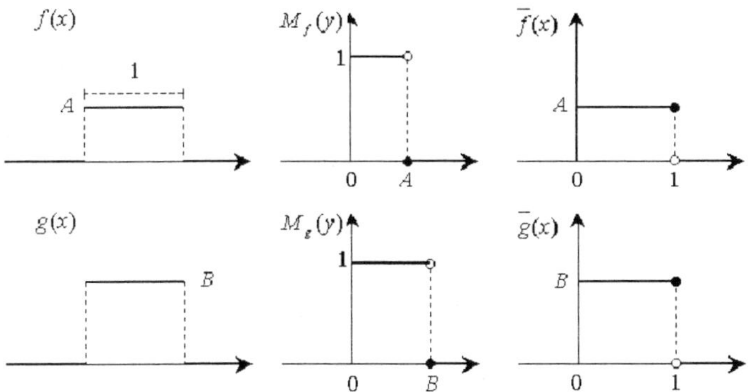

Figure 6. Two simple examples of distributions and sortings. The behavior of $f$ and $g$ at the discontinuities is not indicated because it does not change the distributions or sortings. Arbitrary translation of $f$ and $g$, or even replacement by the characteristic functions (scaled by $A$ and $B$) of any set of measure one, would also leave the distributions and sortings invariant. Note that the supports of $f$ and $g$ overlap and $M_{f+g} \neq M_f + M_g$.

*Example 4* The sorting of the characteristic function of a set of finite measure is essentially the characteristic function of an interval of the same measure (see (1) and Fig. 6).

*Example 5* Consider a step function

$$f(x) = c_n, \quad x \in I_n := (n, n+1],$$

where the $c_n$, $n = 0, 1, 2\ldots$, are real numbers. Using the characteristic functions $\chi_n$ of the intervals $I_n$,

$$f(x) = \sum_n c_n \chi_n(x).$$

It follows that

$$\overline{f}(x) = \sum_n \overline{c}_n \chi_n(x),$$

since the sorting of $f$ can be obtained from the sorted sequence $\overline{c}_n$ (see Fig. 7).

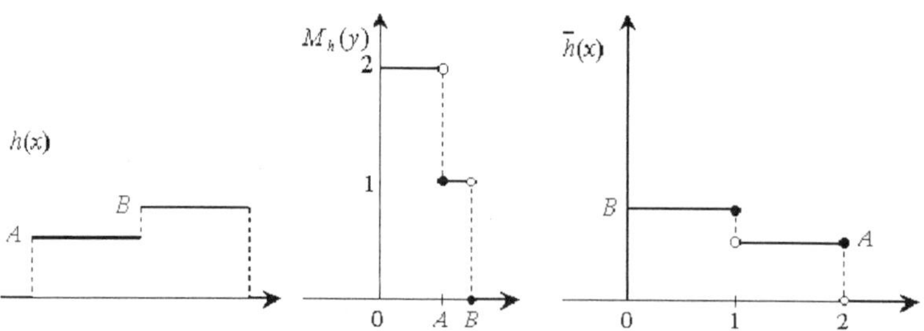

Figure 7. If $f$ and $g$ have disjoint supports then $M_{f+g} = M_f + M_g$. The addition of two properly translated copies of the functions $f$ and $g$ of Fig. 6 leads to the function $h$. The distribution $M_h$ is the sum of distribution functions, and the sorting can be obtained from the inverse (properly defined) of $M_h$, or, equivalently, by sorting. In general, the (non-increasing) sorting of a step function with values $(A, B, C \dots)$ can be obtained by sorting the sequence $(A, B, C \dots)$ by decreasing order.

The sortings often cannot be determined in closed form, as in the cases mentioned above. A simple example is $f(x) = ax\, e^{-bx}$ (the sortings are depicted in Fig. 8a, and should be compared with the distributions in Fig. 1). A more complex sorting is illustrated in Fig. 8b (compare with Fig. 2).

## 4. The continuous-time median and median-type filters

To define the continuous-time median filter, it is helpful to introduce the following notation: let $f_t(x)$ denote the result of multiplying the signal $f(t)$ by a rectangular window of duration $w$, centered at $t$:

$$f_t(x) := \begin{cases} f(x), & t - w/2 \le x \le t + w/2, \\ 0, & \text{otherwise.} \end{cases}$$

The output $g$ of the median filter is defined by

$$g(t) := \overline{f}_t(w/2).$$

Thus, the output at $t$ is obtained as a result of a two-step procedure:
1. The sorting of the windowed signal (the nonlinear step).
2. The computation of the evaluation (linear) functional at $w/2$.

Replacing the evaluation functional at $w/2$ by other functionals of $\overline{f}_t$ immediately leads to other ranked order filters:

$$g(t) := F\left[\overline{f}_t\right].$$

In the context of digital filters, this corresponds to the class of $L$ filters [18]. A family of operators depending on one parameter $a$ can be obtained using the functional

$$F_a[h] := \frac{1}{a} \int_{(w-a)/2}^{(w+a)/2} h(x)dx \, ,$$

where $h : [0,\ w] \mapsto \mathbf{R}$, and $0 < a \le w$. The corresponding filter is defined by

$$g(t) := F_a\left[\overline{f}_t\right]. \tag{4}$$

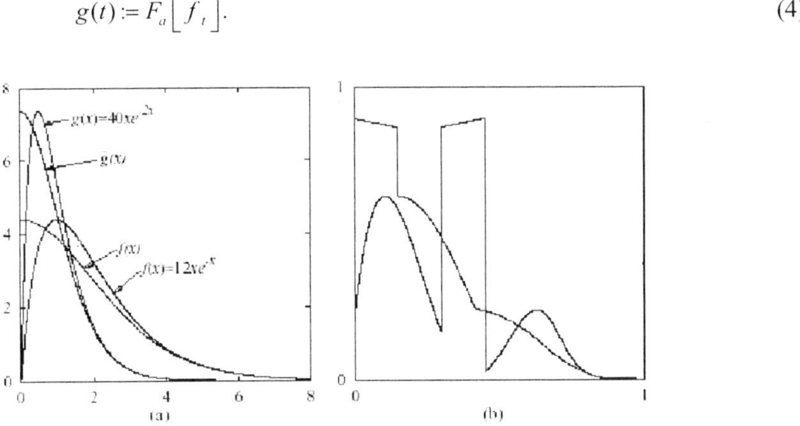

Figure 8. Left: two simple functions $f$ and $g$ and their sortings $\overline{f}$ and $\overline{g}$ (compare with Fig. 1). Right: a more complex function (solid line) and its sorting (dashed line, compare with Fig. 2).

The $\alpha$-trimmed filters described in [19] are the equivalent digital filters. When $a \to w$ this filter reduces to the (linear) moving average because of (3) (the integrals of the function and its sorting are equal).

On the other hand, when $a \to 0$, it is natural to expect that it behaves increasingly more like the median filter. In fact, if the sorting of $f_t$ is continuous at $w/2$, the absolute value of the difference between the outputs of the median filter and the filter defined by (4) satisfies

$$\left| \frac{1}{a} \int_{(w-a)/2}^{(w+a)/2} \overline{f}_t(x)dx - \overline{f}_t(w/2) \right| \le \frac{1}{a} \int_{(w-a)/2}^{(w+a)/2} \left| \overline{f}_t(x) - \overline{f}_t(w/2) \right| dx \le \omega_{\overline{f}_t}(a),$$

where $\omega_{\bar{f}_t}$ denotes the modulus of continuity of $\bar{f}_t$. But $\omega_{\bar{f}_t}(a) \to 0$ as $a \to 0$, showing that the output of the filter defined by (4) converges to the output of the analog median filter as $a \to 0$. For more related results see [10].

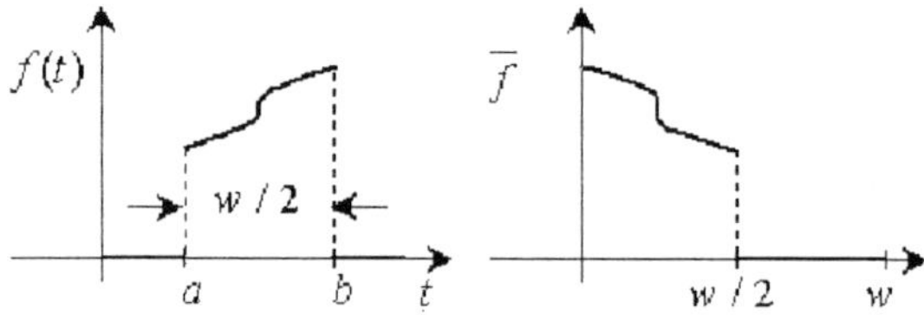


Figure 9. The sorting of the windowed version of $f$ is discontinuous at one half the window width, $w/2$, for all $t$ satisfying $a \le t \le b$.

### 4.1. The effect of the definition of sorting

Because the sorting is the basis for the definition of the median filter, different conventions regarding the sorting may in principle lead to median filters with distinct behaviors.

As an example, consider median filters based on the left- and right-continuous sortings. These sortings may differ only at countably many points of discontinuity. Hence, it might seem that the responses of the two corresponding median filters to the same input signal $f$ could only differ at countably many points (they would be equal almost everywhere). But this is false: the outputs may differ throughout intervals. For example, it can be verified that the input signal described in [7] does in fact lead to such a situation. This can be properly understood in terms of the continuity properties of the sorting [9].

Consider the example given in Fig. 9, and a running window of width $w = 2(b - a)$. Then, the sorting $\bar{f}_t$ of the windowed signal at $t$, $f_t$, will be discontinuous at $w/2$, for $a \le t \le b$. Therefore, the outputs of the median filters based (for example) on the left-continuous and right-continuous sortings will differ from each other throughout that interval.

The following question remains: is there a class of input signals for which the output of the median filter is independent of the underlying definition of sorting? Less strongly, is there a class of input signals for which the output of the median filter is almost everywhere independent of the definition of sorting? For brevity, these sets of signals will be denoted by $A$ and $B$, respectively.

The answer to both questions is affirmative. The set of all continuous functions belongs to $A$, because the sorting of a continuous function is unique and in the same Lipschitz class as the function itself [10].

However, the continuity constraint is unnecessarily strong. It is easy to construct examples of discontinuous functions with continuous sortings; the necessary and sufficient condition for the output of the median filter to be independent of the definition of the sorting is the continuity of $\bar{f}_t$ at $w/2$.

The class $B$ strictly contains $A$. To check that this is true, consider a signal similar to the one depicted in Fig. 9, but of width $b - a \ne w/2$. The signal itself is of course discontinuous, but when $b - a < w/2$ the sorting of the windowed signal will be continuous at $w/2$, for all $t$.

If $b - a > w/2$ there will be two exceptional points: $t = a$ and $t = b$. The sorting $\overline{f}_t$ will be continuous at $w/2$, for all other $t$.

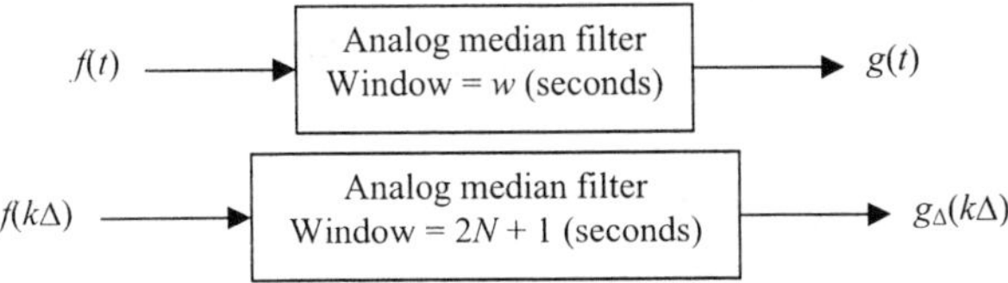



Figure 10. The connection between the digital and the analog median filters: if $g_\Delta$ is the step function obtained from the output samples $g_\Delta(k\Delta)$ of the digital median filter, and if $w = (2N + 1)\Delta$, then

$$|g - g_\Delta| \le \omega_f\left(\frac{w}{2N+1}\right).$$

### *4.2. Relations between the digital and analog median filters*

The following results on the connection between the discrete-time and continuous-time median filters build upon the sorting of step functions, examined in Section 3.3.

Let $\Delta$ be an arbitrary real number. Define

$$\chi_i(t) := \begin{cases} 1, & i\Delta < t \le (i+1)\Delta, \\ 0, & \text{otherwise.} \end{cases}$$

Any left-continuous piecewise constant function with possible discontinuities at $i\Delta, i \in \mathbf{Z}$ can now be expressed as

$$f_\Delta = \sum_i c_i \chi_i,$$

for an adequate choice of the constants $c_i$.

Consider an analog median filter with window length $w = (2N + 1)\Delta$ (the case $w = 2N\Delta$ can be handled almost similarly, see [10]). The response of the filter to the step function $f_\Delta$ at the time $t$ will be denoted by $g_\Delta(t)$. The windowed signal at any $t_n := n\Delta + \dfrac{\Delta}{2}, n \in \mathbf{Z}$, will consist of exactly $2N + 1$ steps of the step function $f_\Delta$. On the other hand, the sorting of a step function can be obtained by sorting the steps by decreasing order (see Example 5 and Fig. 7). Hence, the sample $g_\Delta(t_n)$ can be obtained by sorting the $2N + 1$ elements of the sequence $c_i$ that fall inside the analog median window. The value of $g_\Delta$ at $t_n$ is the median of these $2N + 1$ values. Thus, it is equal to the $n$th output sample of a digital median filter with a window size of $2N + 1$, which has the sequence $c_i$ as input.

Because the response $g_\Delta$ of the analog median filter to the step function $f_\Delta$ is also a step function, it is completely determined by the output samples of the digital median filter $g_\Delta(t_n)$, as explained above. In this sense, it is legitimate to mention "the step function determined by the output of the digital median filter" (see Fig. 10).

It can be shown [10] that if the sequence $f_n$ converges monotonically, then the sortings $\overline{f}_n$ will also converge. This fact and the density of the set of step functions in several function spaces establishes the convergence of the outputs of the digital and analog median filter, under suitable conditions. However, it is more useful to study the convergence with respect to the smoothness of the input signals. Therefore, we seek to obtain the order of approximation for functions in a given Lipschitz class.

Consider a continuous-time signal $f$ with modulus of continuity $\omega_f$. Let $f_\Delta$ be an approximating step function with coefficients $c_i$, obtained (for example) by sampling $f$. Then, as a consequence of the definition of the modulus of continuity,

$$\left| f - f_\Delta \right| \leq \omega_f(\Delta).$$

The bound also applies to windowed versions of $f$ and $f_\Delta$, the sortings of which, when evaluated at $w/2$, are the response of the analog median filter to $f$ and $f_\Delta$. It is shown in [10] that if $\alpha \leq f - g \leq \beta$, then $\alpha \leq \overline{f} - \overline{g} \leq \beta$. Therefore, the responses $g$ and $g_\Delta$ will also satisfy

$$\left| g - g_\Delta \right| \leq \omega_f(\Delta).$$

Because $g_\Delta$ can be obtained by applying a digital median filter of size $2N + 1$ to the sequence $c_i$, and because the window size of the analog filter is $w = (2N + 1)\Delta$, we reach the following conclusions:

*Proposition 1* Let the signal $f$ be fed to an analog median filter with window length $w$. Let the samples of $f$, taken with sampling period $\Delta$, be fed to a digital median filter of window size $2N + 1$. The output $g$ of the analog median filter and the step function $g_\Delta$ reconstructed from the output of the digital median filter satisfy

$$\left| g - g_\Delta \right| \leq \omega_f\left( \frac{w}{2N+1} \right),$$

where $\omega_f$ denotes the modulus of continuity of $f$. If $f$ satisfies a Lipschitz condition of order $\gamma$ with constant $\alpha$,

$$\left| g - g_\Delta \right| \leq \alpha \frac{w^\gamma}{(2N+1)^\gamma}.$$

Thus, the step function determined by the output of the digital median filter converges uniformly to the output of the analog median filter, provided that $\Delta \to 0$ and $N \to \infty$, with $(2N + 1)\Delta = w$ (constant).

## 5. The effect of noise

### 5.1. Noise width

We now address the problem of determining the effect of noise on the analog median filter output. In order to do that it is convenient to introduce the concept of signal width.

*Definition* We say that the signal $f$ has width $\delta$ if its sorting $\overline{f}$ vanishes for $x > \delta$.

There is in general no relation between the width of a deterministic or stochastic signal and concepts such as energy or variance. It is sufficient to observe that the width of a signal $f$ is invariant under scaling: the width of $f$ is of course equal to the width of $\alpha f$, since the supports of the corresponding sortings are equal.

To estimate the width of an impulsive signal, assume, for example, that $N$ points $\{t_i\}$ are randomly placed in an interval of length $T$. The probability of having $k$ points in a subinterval of length $\alpha$ is

$$\binom{N}{k} p^k (1-p)^{N-k},$$

where $p = \alpha / T$. If $N \gg 1$, $p \ll 1$, and $k$ is close to $Np$, this can be approximated by

$$e^{-pN} \frac{(pN)^k}{k!}.$$

When $N$ and $T$ tend to infinity, with $N/T$ constant and equal to $d$, this point process tends to a (ergodic) Poisson process with average $d$. The probability of having at most $n$ points $\{t_i\}$ inside an interval of fixed length $w$ is given by

$$P(n,d) = e^{-dw} \sum_{k=0}^{n} \frac{(dw)^k}{k!}.$$

Consider now a shot noise process, generated by the superposition of functions $h_k$, centered at the points $t_k$,

$$s(t) := \sum_k h_k(t - t_k).$$

The $h_k$ could as well be realizations of some stochastic process; we need only to assume that the measure of the support of the $h_k$ is at most $\Delta$. Consider now the windowed signal $s_t$ obtained from $s$. If the window length is $w$, the probability that $s_t$ has width less than or equal to $n\Delta$ is $P(n, d)$, as plotted in Fig. 11.

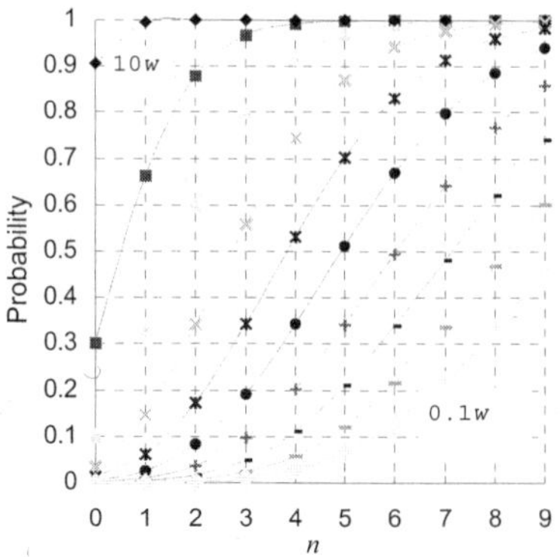


Figure 11. The probability $P(n, d)$ that an impulsive signal multiplied by a rectangular window of length $w$ has width less than or equal to $n\Delta$, plotted as a function of $n$, for several possible point densities $d$ (in the range $0.1w$ to $10w$).

In the median filter, the signals to be sorted have a maximum duration $w$ set by the window used. Their sortings will necessarily vanish for $x > w$. We will concentrate on certain types of noise signals that will interact additively with time-limited signals of width at most $w$, such as the windowed signals $f_t$ upon which the median filter is based. The relative values of $\delta$ and $w$ suggest a distinction between small width or large width noise.

- Small width signals: signals with sortings that vanish for $x > \delta$, with $\delta << w$. Such signals are allowed to have arbitrary amplitudes (not necessarily small). This model is adequate to deal with additive, impulsive noise.

- Large width signals: signals $f$ with sortings that satisfy $\overline{f}(x) = c$ ($c$ a real constant) throughout an interval of width $w - \delta$, with $\delta << w$. This model is adequate to deal with multiplicative noise (see [10] for details).

In addition to the classes of signals, we will also consider noise signals of large width, possibly up to $w$, but of instantaneous amplitude limited to $0 < \varepsilon << 1$.

Note that a windowed signal $f_t$ can be of small width for all $t$ even if $f$ itself has arbitrarily large width.

In the following sections we examine the response of the median filter to the signal-plus-noise $f + n$, evaluated at the time instant $t$, is, by definition, the value at $w/2$ of the sorting of $f_t + n_t$, the windowed signal-plus-noise. Let this response be denoted by $\overline{f_t + n_t}(w/2)$.

The absolute value of the difference between this quantity and the response of the filter to the noiseless signal, $g(t) = \overline{f}_t(w/2)$, is denoted by $E(t)$:

$$E(t) := \left| \overline{f_t + n_t}(w/2) - \overline{f}_t(w/2) \right| = \left| \overline{f_t + n_t}(w/2) - g(t) \right|.$$

We will discuss bounds for $E(t)$ in terms of the noise width and the smoothness of the input signals.

### 5.2. Additive, small width noise

Let the windowed noise $n_t$ be of small width $\delta$. Under this constraint, $f_t$ and $f_t + n_t$ differ on a set of measure at most $\delta$. The sortings of any functions $f$ and $g$ that differ on sets of measure at most $\delta$ satisfy $\overline{g}(x+\delta) \le \overline{f}(x) \le \overline{g}(x-\delta)$. Consequently,

$$\overline{f}_t(w/2+\delta) \le \overline{f_t + n_t}(w/2) \le \overline{f}_t(w/2-\delta). \tag{5}$$

If $\overline{f}_t$ is constant in a sufficiently large neighborhood of $w = 2$, the lower bound and the upper bound in (5) reduce to $g(t) = \overline{f}_t(w/2)$, and this leads to the following result.

*Proposition 2* If $\overline{f}_t$ is constant throughout an interval $I$ containing $[w/2 - \delta, w/2 + \delta]$, the output of the median filter at $t$ will be unaffected by additive noise $n_t$ of width bounded by $\delta$: $E(t) = 0$.

Let $f$ be a step function, with the steps more than $w$ apart. Assume that there is one discontinuity at $t = 0$. We wish to examine the behavior of the median filter with window size $w$ in the vicinity of this discontinuity, in reference to the previous result.

Since the steps are separated by more than $w$, the sorting $\overline{f}_t$ is continuous at $w/2$, for each nonzero $t$ satisfying $|t| < w/2$. In fact, $\overline{f}_t$ is *constant* throughout $\left( \dfrac{w}{2} - \alpha, \dfrac{w}{2} + \alpha \right)$, for all $t$ satisfying $\alpha < |t| < w/2$. By the previous result, the median filter will reject any additive noise of width less than $\alpha$, except possibly for values of $t$ closer to the discontinuities than $\alpha$.

If the instants where the discontinuities occur are known beforehand, so will the zones of uncertainty. The original step signal can then be recovered exactly, by sampling the median filtered signal at the middle point of each step.

### 5.3. Using Lipschitz conditions

Subtracting from (5) the response of the filter to the noiseless signal, $\overline{f}_t(w/2)$, leads to

$$E(t) \le \left| \overline{f}_t(w/2-\delta) - \overline{f}_t(w/2) \right|,$$

and so

$$E(t) \le \omega_{\overline{f}_t}(\delta).$$

Since the sorting of a continuous function belongs to the same Lipschitz class as the function itself, we reach the following conclusion.

*Proposition 3* If $f$ satisfies a Lipschitz condition of order $\gamma$ and constant $\alpha$, and the width of the noise $n_t$ is bounded by $\delta$, then:

$$E(t) \leq \alpha \delta^{\gamma}.$$

### 5.4. Nonnegative signal and noise

If $f$ and $g$ are nonnegative,

$$\overline{f+g}(\alpha+\beta) \leq \overline{f}(\alpha) + \overline{g}(\beta).$$

Applying this, we see that the signal and the noise are nonnegative,

$$\overline{f_t + n_t}(w/2) \leq \overline{f}_t(w/2 - \delta) + \overline{n}_t(\delta).$$

Therefore,

$$E(t) \leq \omega_{\overline{f}_t}(\delta) + \overline{n}_t(\delta),$$

and the invariance of the Lipschitz classes under sorting leads to the following conclusion.

*Proposition 4* If $f \geq 0$ and the width of the noise $n_t$ is bounded by $\delta$, then

$$E(t) \leq \omega_f(\delta) + \overline{n}_t(\delta).$$

If the width of the noise is below $\delta$, $E(t) \leq \omega_f(\delta)$, as in the previous cases.

### 5.5. Low-amplitude large width noise

Assume now that the noise is the sum of a large width, low-amplitude component $u$, $|u(t)| \leq \varepsilon$, and a small width component $v$. In this case, we have

$$\overline{f_t + v_t}(w/2) - \varepsilon \leq \overline{f_t + u_t + v_t}(w/2) \leq \overline{f_t + v_t}(w/2) + \varepsilon,$$

and so

$$\overline{f_t}(w/2 + \delta) - \varepsilon \leq \overline{f_t + u_t + v_t}(w/2) \leq \overline{f_t}(w/2 + \delta) + \varepsilon.$$

Subtracting the response to $f$ leads to

$$\overline{f_t}(w/2+\delta)-\overline{f}_t(w/2)-\varepsilon \le \overline{f_t+u_t+v_t}(w/2)-\overline{f_t}(w/2)$$

$$\le \overline{f_t}(w/2+\delta)-\overline{f}_t(w/2)+\varepsilon,$$

and we have the next proposition.

*Proposition 5* Assume that the noise can be expressed as $n = u + v$, where $|u| \le \varepsilon$, and $v$ has width at most $\delta$. Then,

$$E(t) \le \omega_f(\delta)+\varepsilon.$$

## 6. Conclusions

Additive fuzzy systems are one of the many techniques that have been used to deal with impulsive noise. They are universal approximators, equivalent to a graph cover with local averaging, but suffer from the curse of dimensionality: the number of rules grows exponentially with the number of input and output variables. Other related approaches, based for example on radial basis function neural networks, face the same difficulty. Digital median filters, and other order-statistics digital filters, are also useful to suppress impulsive noise, but the digital implementation of the sorting operation and the need for A/D and D/A conversion lead to complex implementations. Analog solutions might be the answer, but the meaning of "sorting" in the continuous-time case, and the connection between the digital and analog cases, must be clarified.

The results reviewed in this paper contribute toward this goal, but their interest goes beyond the median filter. As an example, the concept of rearrangement naturally appears when examining the meaning of "uncertainty" $U(\pi)$ associated with a set membership function $\pi(x)$. The task is simpler when $\pi(x)$ takes only a finite number of possible values, in which case the sorting poses no difficulties. But if the set membership function is unrestricted, the correct definition brings the distribution function (or the rearrangement) to scene. Thus, the same concepts that arise in the study of the rank-order concept in the continuous-time case occur in the study of certain aspects of fuzzy information theory. The concept of rearrangement is indeed a fundamental one, and it is natural that it plays a key role in several nonlinear problems.

## References

[1] H. M. Kim and B. Kosko, Fuzzy Filters for Impulsive Noise, Fuzzy Engineering, Kosko, B., (ed.), Prentice-Hall, Englewood Cliffs, New Jersey, 1997, pp. 251–275.

[2] A. Ramer and V. Kreinovich, Information Complexity and Fuzzy Control, Fuzzy Control Systems, Kandel, A., Langholz, G., (eds.), CRC Press, Boca Raton, 1994, pp. 76-97.

[3] F. Y. Edgeworth, A new method of reducing observations relating to several quantities, *Philosophical Magazine (Fifth Series)*, Vol. **24**, 1887, pp. 222–223.

[4] H. H. Turner, On Mr. Edgeworth's method of reducing observations relating to several quantities, *Philosophical Magazine (Fifth Series)*, Vol. **24**, 1887, pp. 466–470.

[5] F. Y. Edgeworth, On a new method of reducing observations relating to several quantities, *Philosophical Magazine (Fifth Series)*, Vol. **25**, 1888, pp. 184–191.

[6] J. P. Fitch, E. J. Coyle and N. C. Gallagher, The analog median filter, *IEEE Trans. Circuits Syst.*, Vol. **33**, no. 1, 1986, pp. 94-102.

[7] H. G. Longbotham and A. C. Bovik, Comments on The analog median filter, *IEEE Trans. Circuits Syst.*, Vol. **36**, no. 2, 1989, pp. 310.

[8]  S. Paul and K. Hüper, Analog rank filtering, *IEEE Trans. Circuits Syst. — I*, Vol. **40**, no. **7**, 1993, pp. 469–476.

[9]  P. J. S. G. Ferreira, Sorting continuous-time signals and the analog median filter, *IEEE Sig. Proc. Letters*, Vol. **7**, no. 10, 2000, pp. 281–283.

[10] P. J. S. G. Ferreira, Sorting continuous-time signals: Analog median and median-type filters, *IEEE Trans. Signal Processing*, Vol. **49**, no. 11, 2001, pp. 2734–2744.

[11] E. Ochoa, J. P. Allebach and D. W. Sweeney, Optical median filtering using threshold decomposition, *Appl. Opt.*, Vol. **26**, no. 2, 1987, pp. 252–260.

[12] K. Urahama and T. Nagao, Direct analog rank filtering, *IEEE Trans. Circuits Syst. — I*, Vol. **42**, no. **7**, 1995, pp. 385–388.

[13] G. H. Hardy, J. E. Littlewood and G. Pólya, Inequalities, Cambridge University Press, Cambridge, 2nd edition, 1952.

[14] A. Zygmund, Trigonometric series, Cambridge University Press, Cambridge, 2nd edition, 1968.

[15] G. G. Lorentz, Bernstein Polynomials, Chelsea Publishing Company, New York, 2nd edition, 1986.

[16] E. M. Stein and G. Weiss, Introduction to Fourier Analysis on Euclidean Spaces, Princeton University Press, Princeton, New Jersey, 1971.

[17] N. I. Achieser, Theory of Approximation, Frederick Ungar Publishing Co., New York, 1956. Reprinted by Dover Publications, 1992.

[18] H. G. Longbotham and A. C. Bovik, Theory of order statistic filters and their relationship to linear FIR filters, *IEEE Trans. Acoust., Speech, Signal Processing*, Vol. **37**, no. 2, 1989, pp. 275–287.

[19] J. B. Bednar and L. T. Watt, Alpha-trimmed means and their relationship to median filters, *IEEE Trans. Acoust., Speech, Signal Processing*, Vol. **32**, no. 1, 1984, pp. 145–153.

# Index of Terms

**P**
parallel collaborative structure 105
Petri net 161
polymorphic electronics 278, 286
possibility distribution 206, 213, 219, 226

**Q**
quantifier 212, 218, 226
quasi-ordering 63-65

**R**
re-configurable device 278
relevance 105, 157
rule clustering algorithm 123
rules clustering 118

**S**
similarity relation 71, 82-84
sociology 153
Sugeno system 138

**T**
time series prediction 194

**U**
uncertainty 21, 40, 131, 137
uncertainty-based information 21, 42-49
uncertainty principle 21, 48-50
unsharpeness 11-14
updation operator 58-59
user modeling 197

**V**
vagueness 11,14

**W**
weighted sum 143, 151, 158

# Contributors

**Krassimir T. ATANASSOV**
Centre for Biomedical Engineering, Bulgarian Academy of Sciences, Bl. 105,
Acad. G. Bonchev Str., Sofia-1113, BULGARIA
E-mail: krat@bgcict.acad.bg

**Senen BARRO AMENEIRO**
Departamento de Electrónica e Computación, Facultade de Física. Universidade de
Santiago de Compostela, Spain

**Jim F. BALDWIN**
Engineering Mathematics Department, University of Bristol, Bristol, U.K.
E-mail: jim.baldwin@bristol.ac.uk

**Armando BLANCO**
Depto. de Ciencias de la Computación e Inteligencia Aritificial, Universidad de Granada,
18071 Granada, Spain
E-mail: armando@ugr.es

**Christian BORGELT**
Dept. of Knowledge Processing and Language Engineering, Otto-von-Guericke-University
of Magdeburg, Germany
E-mail: borgelt@iws.cs.uni-magdeburg.de

**Purificacion CARIÑENA AMIGO**
Departamento de Electrónica e Computación, Facultade de Física. Universidade de
Santiago de Compostela, Spain

**Paolo FÉLIX LAMAS**
Departamento de Electrónica e Computación, Facultade de Física. Universidade de
Santiago de Compostela, Spain

**Paulo J. S. G. FERREIRA**
Dep. Electronica e Telecomunicaciones / IEETA, Universidade de Aveiro, Aveiro, Portugal

**Toshio FUKUDA**
Center for cooperative research in advance science and technology, Nagoya University
Furo-cho, Chikusa-ku, Nagoya, Japan

**Constantin GAINDRIC**
Institute of Mathematics and Computer Science, Academy of Sciences, Moldova

**Xiao-Zhi GAO**
Institute of Intelligent Power Electronics, Helsinki University of Technology, Otakaari 5 A,
FIN-02150 Espoo, Finland

**Yasuhisa HASEGAWA**
Center for cooperative research in advance science and technology, Nagoya University
Furo-cho, Chikusa-ku, Nagoya, Japan

**Marcel HRIC**
Center for Intelligent Technologies, Faculty of EE and Informatics, Technical University
Letná 9, 04001 Košice, Slovak Republic
E-mail: hric@neuron-ai.tuke.sk

**Abraham KANDEL**
Ben-Gurion University of the Negev, Department of Information Systems Engineering,
Beer-Sheva 84105, Israel. Also: Department of Computer Science and Engineering,
University of South Florida, Tampa, Florida, USA

**George J. KLIR**
Center for Intelligent Systems and Dept. of Systems Science & Industrial Eng.,
Binghamton University – SUNY, Binghamton, New York, USA
E-mail: gklir@binghamton.edu

**Kazimierz T. KOSMOWSKI**
Technical University of Gdansk, Faculty of Electrical and Control Engineering, Gdansk,
Poland
E-mail: mkwies@ely.pg.gda.pl

**Rudolf KRUSE**
Dept. of Knowledge Processing and Language Engineering, Otto-von-Guericke-University
of Magdeburg, Magdeburg, Germany
E-mail: kruseg@iws.cs.uni-magdeburg.de

**Mirosław KWIESIELEWICZ**
Technical University of Gdansk, Faculty of Electrical and Control Engineering, Gdansk,
Poland
E-mail: mkwies@ely.pg.gda.pl

**Mark LAST**
Ben-Gurion University of the Negev, Department of Information Systems Engineering,
Beer-Sheva, Israel. Also, Department of Computer Science and Engineering, University of
South Florida, Tampa, Florida, USA

**Drago MATKO**
Faculty of Electrical Engineering, University of Ljubljana, Slovenia
E-mail: drago.matko@fe.uni-lj.si

**Francesco Carlo MORABITO**
University "Mediterranea" of Reggio Calabria, Faculty of Engineering, DIMET, Reggio
Calabria, Italy

**Nikolai G. NIKOLOV**
Centre for Biomedical Engineering, Bulgarian Academy of Sciences, Sofia, Bulgaria
E-mail: shte@bgcict.acad.bg

**Walenty OSTASIEWICZ**
Wroclaw University of Economics, Wroclaw, Poland

**Abraham OTERO QUINTANA**
Departamento de Electrónica e Computación, Facultade de Física, Universidade de Santiago de Compostela, Spain

**Seppo J. OVASKA**
Institute of Intelligent Power Electronics, Helsinki University of Technology, Espoo, Finland

**David A. PELTA**
Depto. de Ciencias de la Computación e Inteligencia Aritificial, Universidad de Granada, Granada, Spain
E-mail: dpelta @ugr.es

**Manuel J. C. S. REIS**
Dep. Engenharias / CETAV, Universidade de Trás-os-Montes e Alto Douro, Vila Real, Portugal

**Luis M. ROCHA**
Los Alamos National Laboratory, Los Alamos, NM, U.S.A.
E-mail: rocha@lanl.gov

**Paulo SALGADO**
Universidade de Trás-os-Montes e Alto Douro, Vila Real, Portugal

**Peter SINČÁK**
Siemens AG, Vienna, PSE Department, ECANSE working group, Vienna, Austria
E-mail: sincak@tuke.sk

**Adrian STOICA**
Jet Propulsion Laboratory, California Institute of Technology, Pasadena, CA, USA

**Horia-Nicolai TEODORESCU**
Romanian Academy, Bucharest, Romania and Technical University of Iasi, Iasi, Romania
E-mail: hteodor@etc.tuiasi.ro

**Ioan TOFAN**
Faculty of Mathematics, University "Al.I. Cuza", Iaşi, Romania
E-mail: tofan@uaic.ro

**Ján VAŠČÁK**
Center for Intelligent Technologies, Faculty of EE and Informatics, Technical University, Letná 9, 04001 Košice, Slovak Republic. On leave from Hirota Laboratory, Tokyo Institute of Technology, Japan
E-mail: Vascak@tuke.sk, vascak@hrt.dis.titech.ac.jp

**José L. VERDEGAY**
Depto. de Ciencias de la Computación e Inteligencia Aritificial, Universidad de Granada, Granada, Spain
E-mail: verdegay@ugr.es

**Mario VERSACI**
University "Mediterranea" of Reggio Calabria, Faculty of Engineering, DIMET, Reggio Calabria, Italy

**Aurelian Claudiu VOLF**
Faculty of Mathematics, University "Al.I. Cuza", Iaşi, Romania
E-mail: volf@uaic.ro

**Ronald R. YAGER**
Machine Intelligence Institute, Iona College, New Rochelle, NY, USA
E-mail: ryager@iona.edu

**Lotfi A. ZADEH**
Berkeley initiative in Soft Computing (BISC), Computer Science Division and the Electronics Research Laboratory, Department of EECs, Univeristy of California, Berkeley, USA
E-mail: zadeh@cs.berkeley.edu

# Author Index